Springer

Berlin
Heidelberg
New York
Hong Kong
London
Milan
Paris
Tokyo

Graham Borradaile

Statistics of Earth Science Data

Their Distribution in Time, Space, and Orientation

With 177 Figures and 52 Tables

Springer

PROFESSOR GRAHAM BORRADAILE
Geology Department,
Lakehead University
Thunder Bay
Canada P7B 5E1

e-mail: borradaile@lakeheadu.ca
Fax: +1-807-3467853

ISBN 3-540-43603-0 Springer-Verlag Berlin Heidelberg New York

Library of Congress Cataloging-in-Publication Data

Borradaile, G. J.
Statistics of earth science data : their distribution in space, time, and orientation /
Graham Borradaile.
  p. cm.
Includes bibliographical references and index.
ISBN 3-540-43603-0 (acid-free paper)
  1. Earth sciences – Statistical methods. I. Title

Springer-Verlag Berlin Heidelberg New York
a member of BertelsmannSpringer Science+Business Media GmbH

http://www.springer.de

© Springer-Verlag Berlin Heidelberg 2003
Printed in Germany

Cover design: E. Kirchner, Heidelberg
Typesetting: Fotosatz-Service Köhler GmbH, Würzburg

Printed on acid-free paper   32/3141/as   – 5 4 3 2 1 0 –

To my parents, John and Margery and to my instructors

Contents

Introduction

The Goals of Data Collection and Its Statistical Treatment in the Earth Sciences

The earth sciences are characterised by loose and complex relationships between variables, and the necessity to understand the geographical distribution of observations as well as their frequency distribution. Our frequency distributions and the looseness of relationships reflect the complexity and intrinsic natural variation in nature, more than measurement error. Furthermore, earth scientists cannot design experiments according to statistical recommendation because the availability and complexity of data are beyond our control. Usually, the system we are studying cannot be isolated into discrete independent variables. These factors influence the first steps of research, how and where to collect specimens or observations. Some issues are particularly troublesome and common in earth science, but are rarely handled in an undergraduate statistics course. These include spatial-sampling methods, orientation data, regionalised variables, time series, identification of cyclicity and pattern, discrimination, multivariate systems, lurking variables and constant-sum data. It is remarkable that most earth-science students confront these issues without formal training or focused consideration.

This book presents an introductory path through data types that are rarely taught and not brought together under one cover for the benefit of students confronting their first research project. It presents the author's view of what is currently useful to earth-science graduate and undergraduate students engaged in research projects, providing a stepping stone to specialised fields. Increasingly, more branches of the earth sciences have been advanced by modern data-acquisition methods so that the average research thesis may be based on large amounts of data collected thousands of times more quickly than was possible when the instructors were graduate students. The personal computer and inexpensive commercial software provide the young thesis-writer with powerful graphic, data-processing and statistical tools that have been available only to specialists until recently. The author's laboratory mostly uses the Microsoft Excel spreadsheet and its statistical supplement Origin; SigmaPlot, SigmaStat, SigmaScan and the orientation-analysis software, Spheristat by Pangaea Scientific. Despite the availability of these powerful tools, the undergraduate and graduate student may not be able to take full advantage of them. Undergraduate curricula have become so busy that they may not include an undergraduate statistics course. Even where such courses are available they cannot normally be tailored to the special and unique needs of the earth sciences.

This is an attempt to meet that need, but it does not in any way replace a rigorous study of statistics. Rather it is a progressive introduction of methods that can be applied with caution to observations and measurements collected by a young researcher. It follows the style of excellent earlier introductions for this audience, by Davis (1973; now 2002), Cheeney (1983), Marsal (1987) and Till (1974). However, there are changes in emphasis and additional topics that have become necessary in the intervening years. Other equally excellent texts are acknowledged in the bibliography of relevant books. However, they focused either on a narrow aspect of the subject, or, where broadly based, they were more advanced treatments (e.g. Koch and Link 1970, 1971).

It is commonly said that statistics, the numbers which characterise some property of a sample, cannot remedy poor measurements. The *systematic* errors introduced by badly designed experiments or equipment, or in inappropriate procedures, are incorrigible and may be indeterminable. These are even more troublesome in the natural sciences because they are confounded with intrinsic *random* errors that express the natural variation in the system. In many cases, natural variation exceeds observation errors, quite unlike the situation in laboratory physics, chemistry or metallurgy. Greatly increasing the number of observations does not always improve decisions. We face a law of diminishing returns for such unnecessary effort, however admirable the intention.

Although our statistical applications use the same techniques as more precise physical sciences, there is a fundamental distinction. Consider the case of measurements of acoustic velocity in seawater at a certain depth, temperature and salinity. It is most likely that the scatter of replicate measurements will be small and symmetrically distributed about a mean value. Measurements closer to the mean value are recorded more often. The hump-shaped frequency distribution may be comparable to the *Normal distribution*. Thereafter follow many traditional statistical applications: regression, analysis of variance (ANOVA), and hypothesis testing, all of which use the theoretical Normal distribution as a precise mathematical model for the *random distribution of measurement error*. However, the dispersion of the distribution will largely be due to natural variation: time of measurement, water currents, proximity of a big fish, etc. This does not invalidate the applied statistical approaches using the Normal distribution, but we must not lose sight of the fact that in this, and most earth science examples, the distribution and its variance record natural variation due to heterogeneous and complex natural processes.

A still greater problem is that in many natural systems, the number of observations may be intrinsically low, not due to lack of effort but because the total number of observations available was fixed at a low value, e.g. there may only be ten volcanic vents, there may have been only four glacial advances. In these cases, the population is distressingly small. For example, determining the mean duration and variance of an interglacial period could be considered an unreasonable and not very meaningful goal.

Of course, in the natural sciences, there are many intermediate situations between the measurement of a fairly fixed value affected primarily by observation errors (e.g. acceleration due to gravity at a certain location) and the observations of a phenomenon that has only a weak theoretical relationship with its cause (e.g. sphericity of pebbles). Most geological data are closer to the latter kind. Especially in intermediate cases, we should evaluate the relative roles of measurement error and intrinsic variation in the scattered distribution.

Normally, natural observations are ideal candidates for statistical treatment. Statistical techniques address the following main issues:

1. *characterising observations with simple statistics, like mean and variance;*
2. *quantifying the confidence we have in a certain estimate, and expressing this graphically;*
3. *hypothesis testing, e.g. are the means of two samples equivalent?;*
4. *comparing sets of data: means, variances, patterns of variation, any similarities or differences of any kind;*
5. *predicting from known values (correlation, interpolation, extrapolation);*
6. *detecting an underlying relationship between variables (regression).*

It may disappoint us to learn that statistics cannot always tell us what can be. Instead, the techniques help us to understand what cannot be, perhaps qualified by a certain degree of confidence. The objectives noted above refer to two main types of data management. The frequency distributions of scalars (the number of measurements of x, plotted against x) and the plot of one variable against another (x-y scattergrams, regression lines, curve fitting, etc.) that feature prominently in traditional statistics courses. However, the earth sciences commonly deal with several issues that are not considered in the undergraduate's traditional statistics course. These concern the following:

1. *Data that cannot merely be treated by examining frequency distributions. Their geographical distribution also influences their interpretation, not only through sampling strategy but also concerning regionalised variables, a specialised subject with the somewhat misleadingly generous title of geostatistics. Simply put, location affects values.*
2. *Closed systems in which variables are not stochastic, i.e. they cannot vary randomly over an infinite range:*
   - *any compositional analysis requires variables to sum to 100%, causing their interdependence.*
   - *orientations are limited to 360° range in a plane, or a solid angle range of 4π in three dimensions.*
3. *Distributions of a stochastic variable through time. Only one measurement of the variable is available at any instant so that characterising its behaviour at any instant, or through time, is challenging. Even worse, the time coordinate is very poorly defined for a geological time series.*
4. *Distribution of orientations in space. The orientations may be of axes (undirected lines such as lineations or normals to planes), unit vectors (directions or directed lines or polarised lines) or true vectors (orientations associated with a polarity and a magnitude). The most complex aspect concerns tensors which are orthogonal sets of three axes. Each axis is associated with a magnitude defining a state or physical property at one point in space.*

The author would be grateful for readers' comments, suggestions and particularly for any examples, any of which will be gratefully acknowledged.

Acknowledgements

The author accepts responsibility for his views and any errors, but wishes to thank certain colleagues who provided invaluable input and constructive criticism. In particular Dr. Mike Jackson of the University of Minnesota and my colleague Dr. Del Li of Lakehead University read and made helpful improvements to much of the text. For reading specific parts with care, helpful criticism or suggestions, I am also grateful to Dr. David Symons (University of Windsor, Ontario), Dr. Nick Fisher, Dr. Dave Clark and Dr. Phil Schmidt (CSIRO, Australia), Dr. Isobel Clark (Geostokos Ltd., Alloa, Scotland), Dr. Bernard Henry and Dr. Max Le Goff (Institute du Physique du Globe, France), and Dr. Norihiro Nakamura (Tohoku University, Japan). I thank my friend Sam Spivak for his care and enthusiasm in the primary drafting of many of the diagrams.

Over the years, more and more geological data have been quantified and the introduction of the personal computer and powerful software has placed techniques and tools previously only available to the specialised researcher into the hands of the thesis-writer. My students have shown me the need for a simple text, drawing together introductory statistical concepts in a way that may be applied to geology and processed by modern software such as the spreadsheet Excel with its statistical extension Origin, and more specialised packages such as SigmaStat, SigmaPlot, SigmaScan and Spheristat. The students who have stimulated me at Lakehead University are too numerous to mention, but in recent years, during the gestation of this book, I would like to thank the following persons: Chrenan Borradaile, Tomasz Werner, France Lagroix, Courtney Cameron (née Goulder), Lisa Maher, David Gauthier, David Baker, Tom Hamilton, Katie Lucas, Wendy Sears, her father Dr. William Sears, and Lionel Charpentier. My research work has been possible through the continuing support of the Natural Sciences and Engineering Research Council of Canada (NSERC), and this book was gestated during their generous support over more than 25 years.

Abbreviated Glossary

Accuracy and Precision

Accurate measurements are close to the true value. Precise measurements are merely consistent about some value although their relation to the true value may be unknown, as is commonly the case in earth science. For example, if a particular physical quantity has a true value of 10 but a maladjusted or poorly designed piece of equipment reads 9, successive observations with the equipment may yield 9.01, 8.99, and 9.02 which show good precision but a systematic error near 1.0, and thus poor accuracy. A well-adjusted, well-designed instrument would, under the same circumstances, read 10.01, 9.99, and 10.02 which show the same high precision but which are now also more accurate.

The precision of a sample of measurements is inversely proportional to the spread of the measurements about their *sample mean* ($\bar{x}$). The spread or dispersion is recorded as the *sample standard deviation* (s), but for some purposes and many mathematical expressions, it is preferable to use the variance (s^2). To statisticians, precision is essentially the inverse of the variance ($1/s^2$). It is improved by increasing the sample size (n), if the sampling is unbiassed, but successive increments of precision decrease with n. Imprecision is sometimes expressed as a *coefficient of variation* ($s/\bar{x}$) but this is only useful if the values are all positive, for example in reporting abundances of elements in geochemical analysis.

Accuracy is represented by the difference between the sample mean ($\bar{x}$) and the true or population mean (μ). Precision may be improved by increasing the sample size (n) but accuracy may, at best, only indirectly be improved in this way by achieving a better definition of the sample mean ($\bar{x}$).

Central Tendency

This is the property of hump-shaped frequency distributions which causes values to group around some popular central value. Measures of central tendency include the mean, also known as the arithmetic mean; the mode, the median, the geometric mean, the harmonic mean and the weighted arithmetic mean. The arithmetic mean, or "mean" is most common and is also most efficient, statistically speaking.

Contours, Isopleths, Density Distributions

Earth scientists are intrigued with spatial distributions of observations on maps, mine-plans, and satellite images. The variation in the observations with position (e.g. UTM coordinates, i.e. Easting and Northing; or latitude and longitude) is usually the first and perhaps the most important item of interpretation. Because the observation sites are almost always irregularly spaced, it is difficult to predict (*interpolate*) a value between sites where there is no sample due to lack of exposure, unsuitable rocks, etc. This is the first and best reason to produce a set of hypothetical lines joining points on the map (not necessarily the sites themselves) with equal values of the property. Such lines, normally curving, are contours or isopleths. In advanced studies, the spatial distribution may be in three dimensions, and contour surfaces may be viewed in three-dimensional images, preferably interactively on a computer monitor.

In some cases, the predicted spatial distribution concerns the abundance of an item, such as the content of a certain mineral or oxide in a rock, so that the resulting diagram may be considered as a distribution of the concentration or density of the measured property. The isopleths may form the basis for further analysis by determining some simple mathematical description for the shape of the surfaces (e.g. *trend-surface analysis*), and therefore some more profound understanding of an underlying geological process.

Dependence and Independence

1. Observations are considered independent if the selection of one observation in no way affects the selection of another observation. The independence of observations is essential in the simple random sample that is required for most statistical work. The other main requirement of the random sample is that the observations are free from any kind of bias.
2. In bivariate systems, each observation requires two values for its specification (x, y). The value plotted on the x-axis is the independent or control variable, whereas the y-axis shows values of the corresponding dependent or response variable. The same concept applies in multivariate systems where there is more than one control (e.g. $x_1, x_2, x_3, \ldots$). Earth- and life-science systems are often so complex that we may have difficulty isolating or even identifying the dependent variables.

Digitisation, Digital and Analogue Records

Many measurements are no longer simply recorded in an analogue fashion. For example, temperature is not always recorded with a glass thermometer. This may be satisfactory in many circumstances, but where large amounts of data are collected quickly, perhaps at specific times, it is impossible for the human observer to provide a reliable record. Instead, the procedure is first modified so that the physical property is measured electrically. In the case of temperature, we would use a thermocouple. It would put out a voltage proportional to the temperature, albeit in a nonlinear fashion. The voltage would then be converted into digital form in a device that might be called a *data-logger*, whose essential component is an electronic analog-digital converter (*AD converter*). The number provided by the AD converter may then be stored and processed by a computer. This usually requires a customised computer program to convert the digital signal into a conventional temperature measurement: in the case of a thermocouple the temperature-voltage relationship requires a polynomial function for its description, depending on the metals used in the thermocouple construction.

The advantages are that data may be collected continuously, for all practical purposes, and with great consistency, avoiding all of the human problems of fatigue and systematic error. Furthermore, the data is already stored in a computer for immediate use, even for controlling an experiment by reversing the analog-digital process, sending out a voltage resulting from some computer calculation that may operate some apparatus. Such data-loggers are common in rock-mechanics experiments where analog records of loads and distortion of rock samples are acquired as analog signals, converted to numbers by the AD converter. The computer then determines a voltage designed to control the pumps which dictate the course of the remainder of the experiment .

The advantages of this technology are clear, in terms of manpower efficiency, amounts of data, and consistency. However, we must not overlook the fundamental limitation of the AD converter. The first is not too difficult and can be overcome with some electrical work. The analog signals received and that may be generated are restricted to DC signals in a quite small range, normally +5 to –5 V. In many cases, a small adjustment may be made easily to double or halve this range. The more serious issue is that the electronic circuitry fixes the range of numbers that correspond to the analog signal. For example, the simplest, common AD converter is referred to as an 8-bit converter: the range of the analogue signal is converted to a number range of 2^8, or 256. This means that the acquired or transmitted voltage can only be recognised with a precision of one part in 256, or ~0.39%. This would not suffice for the measurement of temperature via a thermocouple nor for most scientific applications. The precision of the AD converter is controlled by "n" in the expression 2^n, as shown in the following table. Of course, consideration must be given to the magnitude of a received signal. If the measured voltage exceeds the range of the device, the digital response will be saturated. For example, the eight-bit device would always yield a digital value of 256, whenever the analogue input was over scale.

Many earth science measurements are now acquired by AD converters; e.g. pH of lake water, stream velocity, ground stress in seismically active regions, strain near active faults, orientation of a drill stem underground, or borehole logging of many rock properties (porosity, permeability, electrical conductivity, etc.). We should also be aware that the digitising tablet or scanner used to input map or diagram data into the personal computer has an intrinsic precision controlled by its AD circuitry. This is of course well designed so that the precision should largely exclude human shortcomings. As an experiment in precision in general, as well as the manipulation of an AD device, you may wish to adjust your computer's mouse. Adjusting its sensitivity via the software control panel has the same effect as changing the precision of an AD converter.

Table: Analogue-Digital Precision

n	2^n	Best precision (%)
8	256	0.39
12	4096	0.024
14	16384	0.0061
16	65536	0.0015
24	16777216	0.0000059

Error of Measurement or Observation

This refers to the departure of a measurement from the true value. If a true value x_0 is measured as x, the error is $e = x - x_0$ and has the same units as x. In many circumstances, it is more convenient to use fractional error, expressed as e/x_0. This is dimensionless, and may often be expressed as percentage error $(100\ e/x_0)\%$. Care must be taken to note whether error or fractional error is used. Most statistical techniques were introduced to make the best estimates of measurements in the face of error. In this sense "error" is viewed in a negative light. However, in the natural sciences, measurement error may be negligible. The scatter in observations, values, or measurements represents natural variance and is not a bad thing at all! That is what we study.

Error: Systematic and Random

It may be already apparent that, in the statistical sense, errors are of two kinds, *systematic* and *random*. Most texts on observations focus on the laboratory sciences, and a typical explanation of the difference between the two sources of error might be as follows: If measurements of objects are made with a vernier gauge, we assume the gauge to report measurements "correctly". For example, we expect a reading of 12.01 mm to be "true" in the sense that any departure from 12.01 mm is beyond the resolution of normal eyesight. However, if the gauge had been dropped, it may show a consistently incorrect reading, a bias, of +0.2 mm. Thus we would read 12.21 mm, and every reading would be consistently increased by 0.2 mm over the true value. This is a *systematic* error. Systematic errors cannot be detected by a statistical analysis of the data, but must be determined by careful comparison of that data set with others, made with a gauge that has been more respectfully treated!

However, even with the best measuring tool, there will be variations each time one observer measures the same object (for example, due to lighting conditions, eye strain, general fatigue) so that the true value of 12.01 may be recorded as 12.02, 12.00, etc. Still greater variation may be seen if different observers attempt the same measurements. These are *random* errors, characterised by a dispersion about a central value that we may regard as the best estimate of the real value. Statistical processing may help to refine our estimate of the true value or quantify the confidence that we may place on this estimate.

In the natural sciences, especially field sciences, random errors are usually much larger than systematic ones. However, as we shall note subsequently, the random "errors" are not usually inadvertent deviations from the true value as a result of the measurement procedure, but intrinsic variation in the natural process that we are observing. Put crudely, the quality of measuring techniques and instrumentation are mostly superior to those of the data when measuring objects or processes in the Earth and Life Sciences! This caveat is not restricted to field observations.

Another important difference between systematic and random errors is that systematic errors vary little over the course of a sequence of measurements whereas random errors fluctuate rapidly depending on environmental factors.

Hypothesis

A statistical hypothesis is designed to determine whether a certain value is reasonable or not. The null hypothesis (H_0) may be that a certain value is equal to the population mean, $H_0: x = \mu$. Knowing the probability distribution of the population and depending on the type of comparison, a certain test statistic is calculated, in this case z. If z is further away from the population mean than a certain critical value $\pm z_C$ we may decide that $x \neq \mu$ and reject the null hypothesis. We would then conclude that $x \neq \mu$ at a confidence level of $100(1 - \alpha)\%$ where α is the area under the tails of the theoretical probability distribution that lies outside the range $\pm z_C$. The null hypothesis is designed deliberately so as to be rejected and is often a hypothesis of no differences. When the null hypothesis is rejected, we must accept an alternate hypothesis which is usually our desired goal.

Non-parametric Statistics

Values derived from a sample, without making any assumptions about the form of the populations are known as non-parametric, or distribution-free, statistics. Clearly, they are particularly useful when dealing with poorly understood processes, or ones for which no theoretical basis has yet been determined. Many procedures used by earth scientists are inherently comparable to either parametric or non-parametric statistical methods. For example, any contoured distribution on a map is produced without assuming a model; it is an objective simplification or characterisation of a spatial distribution. On the other hand, if we fit a surface mathematically to the mapped values, it must be constrained by some mathematical model, for example a polynomial equation. Contours are comparable to non-parametric statistics; fitted curves and surfaces are comparable to parametric statistics.

Normal Distribution and Other Theoretical Probability Distributions

The Normal distribution is a graph of probabilities of values, with probabilities plotted on the y-axis and values of measurements on the x-axis. It is symmetric and hump-shaped with tails extending to $\pm \infty$. It was devised to explain the probability distribution of random errors of measurement which are symmetrical about the mean. The standard Normal distribution is a tabled, generalised form with which any set of observations (x_i) may be compared. It is impossible to devise a table that would describe the Normal distribution for every set of Normally distributed values, since each has a different mean (μ) and standard deviation (σ). Therefore, the tabled values are of the z-variate, rather than of x. The observations, x_i, are transformed for comparison with the table using $z_i = (x_i - \mu)/\sigma$. The Normal distribution on the line just described may be extended to a plane on which a bivariate variable is plotted, requiring two values for its specification (x_1, x_2); this is the bivariate Normal distribution. The Normal distribution on the line may be wrapped around a circle to describe the distribution of orientations in a plane, although the $\pm \infty$ range is obviously sacrificed. Similarly, the bivariate Normal distribution may be wrapped over a sphere to describe an orientation distribution in three dimensions. There are many other kinds of theoretical probability distributions used for different purposes. Many are derived by sampling a Normal distribution in a special way or they are otherwise related to it mathematically. However, the Normal distribution is ubiquitous in statistical applications because it finds wide utility as a result of the Central Limit Theorem.

Parameters

Values that characterise the population are called parameters. Since the population is sometimes an unattainable data set, parameters are commonly unknown quantities that we attempt to estimate with the statistics from samples. The mean (μ) and standard deviation (σ) of the population are examples. Greek alphabetic symbols normally denote parameters.

Population

The population represents all possible examples of the phenomenon or object that we are studying. Normally, this quantity is unattainable, for example the average size of quartz grains in schists in a mountain belt. Even if we focused the study on rocks of a particular metamorphic condition, this would be unreasonable. At best, we could only deal with a sample. However, in some aspects of natural science, the population is all too easily attainable and may be unsatisfactorily small in size. For example, if we wished to study the mean diameter of all granite intrusions in Scotland, the total number of observations, representing the population, would be much smaller than the sample obtained by a physicist in some laboratory experiment. Unfortunately, in the biological aspects of natural sciences, such as taxonomy and palaeontology, *population* is an obviously misleading word. In those instances, a statistical report might use *universe* instead of population.

Probability, Probability Distribution

Probability is given by the number of times a specific event (S) can occur in (T) trials. The probability is expressed as a decimal fraction (P = S/T).

Probability can only range from 0 to 1; $0 \leq P \leq 1$. It is sometimes informally expressed as a percentage, i.e. $100(P)\%$. A probability distribution is a graph with a vertical axis representing probability and the horizontal (x)-axis representing the value (event) in whose probability we are interested. The total area under the graph is 1.0, representing all possible events. The fraction of the area under the graph between x_1 and x_2 represents the probability of randomly selecting a value x between x_1 and x_2.

Relationship

This term is loosely used to describe a correspondence between different kinds of observation. For example, on a graph, the y-values may increase linearly with x and we speak of a linear relationship. Some confusion may arise in studies of biological evolution, palaeontology, and taxonomy where the word implies a genetic or racial affinity. Apart from generally improving communication, confusion may be reduced with terms such as *similarity*, *association*, *behaviour,* or *response*. Commonly, there is sufficient technical knowledge to substitute more precise terms such as linear, power-law, exponential, random or uniform, to describe the *mathematical relationship*.

Sample

The sample represents a *number* of observations from which conclusions are to be drawn. In contrast, in everyday English usage "sample" may refer to *a single item*. In statistics, this would be considered as an individual observation or measurement. Geologists especially confuse the term "sample" with a single observation, since their "sample" usually refers to a single *specimen* of rock. It is safer to describe the field geologist's "sample" as a specimen.

Expediency usually dictates that we work with the smallest sample size that is meaningful for comparisons and for characterisation of the phenomenon we are studying. Enlarging the sample size may only improve our characterisation of the phenomenon if the sampling procedure is random. If the phenomenon that we are studying has a significant degree of uncertainty or intrinsic fuzziness, we need only enough data to define the scatter of that behaviour in a reproducible fashion. Increasing sample size unduly may not justify the effort. Where the data follow some well-known statistical model, it may be possible to select the optimum sample size after a small pilot study.

Sampling With or Without Replacement

Sampling with replacement means that every observation has the same chance of occurring in every trial. This is the case with a very large population. For example, if we sampled microfossils from every part of a completely exposed limestone mountain, there is a good possibility that we could draw a fossil with the same characteristics more than once thus repeating the observation. This is like drawing a lottery ticket from a barrel and then replacing it so that it could be drawn again.

On the other hand, if we sample macrofossils from a pail of specimens, examine each one, and then throw it aside, our small "population" becomes depleted rapidly. The remaining choices become fewer. This is sampling without replacement. Whether we sample with replacement or without is a matter of choice, depending on our goal.

Significant Figures

Whether a quantity is reported as a real number or as an integer, the number of significant figures is not arbitrary but related to the accuracy of the measurement method. For example, using a standard rule, one can easily measure to 0.5 mm. Thus, reporting a value as 25.5 mm is appropriate whereas 25.50 would be inappropriate as there is no suggestion that four figures could be significant using a millimetre scale. Note that 0.0255, 0.255, 2.55, 25.5, 255, 2550, 2.55×10^9 and 2.55×10^{-5} all have just three significant figures. Where a value is reported as an exponent (e.g. 2.55×10^9), the number of significant figures is shown in the mantissa.

The question of how many significant figures to report may require some thought. Clearly, there is no value in reporting an observation as 9.330 when there are known to be random errors of ± 0.005, since amongst many measurements a report of 9.335 or 9.325 will not affect the subsequent analysis of the data set. It is generally safe to assume that observations should not be reported with more significant figures than the measurement procedure can justify. In the natural sciences, this is especially the case since the natural variation of the data may be wide and the measuring principles reasonably precise. For example, in studying lake pollution by acid rain, individual pH readings may be reported very precisely with many significant figures. In contrast, the random variation of pH from one location to another and from one time to another within the lake may be enormous. An appreciation of significant figures is particularly relevant in geology because one deals with measurements that may be very large in everyday units. The following examples illustrate this point. For example, 25 years ago, I taught my students that the metamorphic rocks adjacent to my campus cooled off 2,750,000,000 years ago. Should I now change my lecture notes to report a cooling age of 2,750,000,025 years? It really does not matter because the best radiometric age determinations using the K-Ar series may never refine the 2,750 Ma cooling age to better than ± 5 Ma. (Some readers without a geological background may be unfamiliar with the long time intervals; we will use the convenient abbreviations ka = 1000 years, Ma = 1,000,000 years and Ga = 1,000,000,000 years.)

Having collected observations, one of the first items of interest may be the mean value. It is self-evident that the mean value cannot be reported with more significant figures than the observations. If the mean of 20 individuals, measured as 2.54, 2.56, 2.52, 2.51, etc., was calculated as 2.5302, it should be reported only as 2.53 since individual observations contain no more information than that presented by three significant figures.

How do we report derived quantities such as the estimates of dispersion like the standard deviation, or the standard error of the mean? From the example given it is clear that these will be quite small numbers in comparison with the mean 2.53. Suppose that the standard deviation is calculated as 0.04233. It is misleading to report,

mean	= 2.53
standard deviation	= ± 0.0423

even though both numbers have three significant figures. This would imply that there was greater precision in the standard deviation or dispersion estimate of the measurements than in individual measurements. Thus, we should report

mean	= 2.53
standard deviation	= ± 0.04

The reader is referred to related topics on confounding of errors in calculations in Appendix I.

Statistics are quantities that are calculated from and which characterise *samples*, e.g. mean ($\bar{x}$) and standard deviation (s). The Latin alphabet is normally used for statistics of samples.

Standard Deviation

Standard deviation is the square root of variance.

Standard Error

The standard deviation of sample means, i.e. a measure of the variability of the means of successive samples drawn from the population.

Statistical Error

A statistical (null) hypothesis is rejected when some test statistic has an unlikely value lying under the tails of its probability distribution, usually in a small area that represents 5 % probability. However, this is merely a probability. Due to the misfortunes of random sampling, it is possible that we have wrongly rejected the null hypothesis. In fact, the probability is 5 %! The 5 % probability, or area of 0.05 under the distribution is the α-error or type I error.

If we do not reject the null hypothesis, there is a probability β that we are incorrect, but this is usually not known. Loosely speaking, we "accept" our null hypothesis. That is why it is better to design tests so that their null hypothesis may be rejected decisively. The β-probability represents a type II error.

Test Statistic, Significance, Confidence Level

If our observations follow a certain theoretical probability distribution, we may calculate a value in the theoretical distribution which corresponds to some value in the sample distribution. This is the test statistic: if it exceeds a certain critical value derived in the theoretical distribution, it represents an unusual value in the probability distribution's tail. It may be rejected as an unlikely or rare value. The critical value defines one or both tails of the theoretical probability distribution under which there is an area α. This represents the probability of rejection of an extreme value (see Hypothesis,

hypothesis test) and it is called the significance value. The confidence level is $100(1-\alpha)\%$.

Variables: Scalars, Axes, Directions, Vectors and Tensors

Scalar. Most measurements concern a scalar property such as weight % of an oxide, percentage volume of a certain mineral or the mean grain size of sand. Scalars thus comprise a single unit of measurement.

Axes. In structural and sedimentological studies, the orientations of lines in two dimensions such as a fault trend, or in three dimensions such as fold hinges, may form the basis of observations. Such axes or lines are three-dimensional geometric entities, whose orientations are described with a geological shorthand convention using trend and inclination. Trend is a compass direction for which common synonyms are azimuth or declination. The inclination is a measure of steepness of a line; it may also be termed the plunge or angle of depression.

Planes. Where the orientations of planes are measured in three dimensions, the perpendicular line to the plane may be recorded as the normal to the plane. However, more complicated conventions exist that combine information on the trend of a horizontal line on the plane (= strike) and its maximum slope (= dip). Care must be taken to note which geological convention is used to record the orientation of planes, as there is commonly confusion and incomplete documentation concerning the basis for orientation systems accompanying field reports.

Directions or Unit Vectors. Orientations of linear features that possess a sense or polarity. For example, cross-beds indicate the direction of a current-direction, not just its trend. Thus, the line must be described with a trend and plunge, but there is also a sense of direction that requires a unique answer to the question, "What is the trend?". The 180° ambiguity in trend does not exist for a unit vector.

Vector. The Earth's magnetic field and paleomagnetic records in rocks are examples of vectors. Their orientation specification requires trend, plunge and sense of direction as with a unit vector. However, there is also an associated magnitude for the property.

Tensor. Some physical measurements like finite strain, electrical conductivity, seismic velocity and magnetic susceptibility do not possess a single value at a given point in a rock. Instead, the property varies with direction in which it is measured: it is said to be an *anisotropic* property. Many such properties are described by second-rank tensors that may be represented by an ellipsoid with mutually perpendicular maximum, intermediate and minimum axes. The orientations of these axes (six pieces of information) and their three magnitudes completely specify the property at one site. Although there are nine pieces of information involved in the complete specification, only six components are needed for their determination. Thus, there are only six *independent* measurements required to specify the tensor.

Presentation of Orientation Data. Orientation data in three dimensions must be plotted on some projection of a sphere. For axes and tensors, the lines in question have no sense of direction or *polarity*. They possess two possible, equally valid trends, differing by 180°. These may be recorded on one hemisphere. However, for unit vectors (directions) and vectors, the sense or polarity of the line is important. Unlike structural directions or axes, vectors cannot be arbitrarily projected onto one hemisphere for the sake of convenience. Presentations must permit the opportunity to show upward- and downward-inclined vectors and so require both upper and lower hemispheres.

Variance

Variance expresses the total variation of a group of values around their mean value. The differences of each value from the mean are combined by addition. However, in order to prevent the cancellation of positive and negative deviations from the mean, they are squared prior to summation. Standard deviation is σ for the population and s for the sample; the variances are the squares of the standard deviations (σ^2, s^2). Precision is the inverse of variance.

CHAPTER 1

Spatial Sampling

1

It is self-explanatory that whatever we sample must be representative of the object of study. However, for statistical purposes the *simple random sample* is essential and its desirability may at first seem like an obscure and unnecessary display of erudition. For example, when sampling soils, we might be able to randomly choose sites from almost any part of the study area. Unfortunately, differences in topography, land use, history of land use, microclimate, the current weather and time of day may all bias the selection of specimens drawn from the population. Since natural materials are invariably heterogeneously distributed, attempts to achieve a simple random sample are sometimes abandoned in exasperation. Let us refer to the measurements, observations or specimens collected in the simple random sample as objects. The simple random sample of objects should then possess the following qualities:

1. Selection of an object does not affect the selection of another object (*independence of sampling*).
2. Each object should have no influence on any other object (*independence of variable*).
3. Each object has the same probability of being selected (*unbiassed sampling technique*).
4. Each object is equally accessible from the population (*unbiassed natural occurrence*).

The simple random sample is more than an attractive luxury. Statistical methods are based on the frequency with which objects occur in the population of all possible objects of that type. If the objects are not independently drawn from the homogeneous population without bias, the sample statistics give a misleading impression of the population, and statistical estimates and statistical tests of hypotheses will be incorrect. Simply collecting more objects will not overcome the disadvantages of a poorly drawn sample. A small sample may be quite adequate, provided that it is a random sample in the statistical sense just outlined. This chapter will describe a few of the important problems and solutions in collecting a sample of objects from the Earth.

1.1
Sampling Versus Spatial Sampling

The statistician's *sample* is a set of objects from a larger group, or even from the *population* of all possible data of that type. By *sampling*, the statistician usually means collecting objects without bias and recording their value. Statistically, *simple random sampling* satisfies the requirements for statistical estimation and hypothesis testing. Bias and interdependence are avoided and a representative frequency distribution of the objects' values is assembled. The manipulation of the frequency distribution, comparison with theoretical probability distributions and the calculation of *statistics*, such as *mean* and *standard deviation*, and of *confidence limits* all form part of classical statistical approaches to simple random samples described in the following chapters. Unfortunately, earth scientists may use "sample" to refer to a single specimen of material. A single specimen represents one observation and should not be confused with a statistical sample that comprises many objects, in the form of observations, measurements or specimens. The value assigned to the object or observation is usually the only piece of information required for the statistical procedures in the following chapters.

Being preoccupied with their specific subject, earth science students commence their training by collecting information from different locations. The sample of observations may form the basis of some frequency distribution, but each observation corresponds to a location, so this data-acquisition process is *spatial sampling*. The complications are immediately obvious. In any natural spatial distribution, it is almost impossible that the observations at each location are independent. Consider, for example, sampling the topography of a region. The elevation at one point cannot be independent of that at neighbouring locations. They must be related by geomorphological and geological processes, the results of which are modelled by topographic contours which could be regarded as an empirical mathematical function. Thus, spatially distributed observations are not independent. Each observation is important in terms of its value, but for the earth scientist its location, which relates it to other values, may be equally important. This aspect of spatial sampling is confounded with the bias inherent in materials and the logistics of sampling in the real world.

Classical statistical training concentrates on management of frequency distributions of variables and for most undergraduates it avoids spatial distributions entirely. Analysis of spatial distributions may be found in specialised texts that first require some statistical background (e.g. Cressie 1993; Clark and Harper 2000). Whereas most geologists never need to analyse spatial distributions, they do need to know how best to acquire spatially distributed data. The first goal of this chapter is to understand how to choose a sampling strategy that best meets our needs in the face of natural obstacles. The second goal is to illustrate that earth-science variables usually have spatial dependence on some scale. The third goal is to show that, with suitable sampling, a clear pattern of spatial variation may be recognised and characterised in various ways, for example by contouring.

1.2
Introductory Concepts

In classical statistics, the most elementary appreciation of data requires its characterisation with a commonly occurring value, usually the *arithmetic mean* and the dispersion of observations around the mean with the *variance*, which is the square of the *standard deviation*. We hope that these sample *statistics* are reliable estimates of the true values, the *parameters* of the population from which the sample was drawn. To this end, it is essential that the sample is constructed by randomly selecting observations and that the observations are independent of one another. This is not necessarily a barrier to processing our values, collected over some area or volume, as if they were merely a frequency distribution of numbers. However, in connection with their attribute of location, we cannot regard them as independent variables. Usually, the degree of dependence between values decreases as the separation of their locations increases. In geology, the scale upon which the association is predictable may vary from the microscopic scale to that of continents, as examples will show.

For simplicity, this chapter and the following four chapters use *univariate* data, variables described by a single number. However, the procedures can be generalised to cope with *bivariate* data (x, y paired values),

multivariate data (e.g. *x, y, z*) and more complex variables (orientations, directions, vectors, tensors). Paradoxically, earth scientists usually become familiar with such measurements before they consider how to sample them. Regardless of the observations' complexity, the determination of statistics from the sample is complicated if their location attribute must be retained in the analysis. More usually that is not required, but the recognition of any meaningful pattern or spatial association of the variables usually is almost always of interest. Its documentation may depend critically on the way in which the observation sites are chosen.

Spatial-sampling strategies must therefore be chosen so that they do not bias or restrict the choice of data in unfavourable ways. In most cases, idealised strategies are unrealistic because natural scientists take data where they can find it. Still, we may choose a sampling pattern to suit our purpose and minimise natural complications that may thwart our goals.

Spatial sampling accumulates data from an area represented by a microscope thin section, a slab of rock or a map, or over a volume from combined drill-core or mine-plan information. This chapter presents spatial sampling more formally than is in the experience of most earth scientists who acquire the knowledge intuitively and gradually through their career. Spatial sampling requires documenting the variable as well as its location using *x-y* coordinates for a plane projection or using (*x, y, z* = elevation) for a volume. Resampling may provide the simplest assurance of successful characterisation. If the pattern of data can be reproduced in subsequent sampling campaigns, the pattern is meaningful inasmuch as it is adequate to overpower distracting spurious information, sometimes termed noise and errors of observation or measurement. That is not a guarantee, as natural data distributions are sometimes biassed in favour of certain values of the variable. For example, certain rock types may crop out more frequently simply because they are more resistant to weathering; the high probability of recording their presence does not reflect their true abundance. Even measurement errors may be reproducible in the sense that certain regional conditions (e.g. swamps, mosquitoes, precipitous cliffs) may constrain the observer's effectiveness.

Even where the observations are free from intrinsic bias, many factors complicate the implementation of a spatial-sampling plan. Vegetation, water, ice or rocks may conceal the appropriate materials. Moreover, there may be logistical restrictions to making observations at locations that would be most satisfying from the statistical point of view; Antarctic expeditions and extraterrestrial exploration would be extreme examples. Finally, re-sampling in a subsequent campaign may be impossible if the limited material was destroyed or removed in the first sampling campaign.

1.3
Examples of Spatial Distributions

Spatial variations upon all scales are found throughout earth-science studies but we must not automatically expect a clear spatial pattern. Suppose we record one value of an index of soil-fertility at many sites, plotting the values on a map. The map of fertility values may be examined and some pattern may be detected and contoured. However, if we were to measure soil fertility not once but ten times at every site, we could determine the variance at each site. If the within-site variance is similar in magnitude to the regional variations or larger than them, we can be confident that we did not record any significant regional variations and therefore soil fertility may be reasonably uniform. In the analysis of variance (ANOVA), this notion is described as the contrast between *within-sample variance* and *between-sample variance* (Chap. 4). It is surprising how often geologists compare outcrops without considering the variation within the outcrops. We will proceed on the assumption that we have verified that within-site variance is smaller than spatial variation of site means. Any spatial variation we recognise may therefore be significant.

Imagine that we measure the mean diameters of sand grains by sieving at various localities on a wide beach. One could establish a sampling grid and collect data as frequently as required, at regular intervals. The general aim is to space sampling sites at closer distances than the wavelength of the variation one wishes to study. If sedimentary structures are observed with a scale of metres on this beach, the sampling should be done at intervals of no more than tens of centimetres. Where we sample on a grander scale, for example, that represented by a geological map, we may also need to pay attention to the vertical dimension. Equally spaced stations in *x-y* coordinates may place the observation at different elevations (*z*) that may affect the outcome of the study in unpredictable ways. For example, geochemical sampling of a granite pluton in the Andes may permit specimens to be taken at any predetermined map location due to almost unlimited exposure of fresh rock. However, if the granite at high mountaintops is contaminated by the former roof rocks to the magma chamber, the topographic *z*-variation of the granite may spoil the areal pattern of any geochemical variation in the map projection (*x-y* plane). Apart from soils, the availability of most earth materials is at the whim of nature so that grid sampling is infeasible. For example, in a poorly exposed area of the forested Canadian Shield, geochemical sampling of granites could not be performed systematically on a grid. The probability of the target rock-type granite cropping out at a grid intersection would be vanishingly small.

Utilising naturally available sites, such as rock outcrops, provides a spatial distribution that is usually far from random. For example, where rocks crop out along lakeshores and streams, their distribution may be quite unsuited to the needs of science. The scientist may want an unbiassed distribution of samples, in which case a *random spatial distribution* is preferred. For other purposes it may suffice that there are equal numbers of samples per unit area as in a *uniform spatial distribution*. For ease of interpretation or contouring, the researcher may prefer the regular pattern of sites located at intersections of a hypothetical grid to yield a *systematic spatial distribution*. Natural occurrences are distributed non-randomly and non-uniformly. For example, softer rocks are under-represented due to differential erosion. Where spatial distribution is not random, any attempts to characterise the data set are thwarted to some degree. Consider the construction of contour lines to show where we expect to find equal concentration values of an economic mineral observation. Unfortunately, the contours pass mostly over terrain where there are no sites, contour values being interpolated between sites where measurements are available. If the occurrence of sites was biassed by differential erosion, the positions of the contours would obviously be affected. Contours are most easily and reliably constructed where the sites are uniformly and densely distributed, but a random distribution is better than a biassed one. We will consider some specific geological situations that affect the design of spatial sampling. It is probably unrealistic to achieve an ideal scheme very often in nature, but being aware of the pitfalls and advantages of different approaches alerts us to the magnitude of potential problems in interpreting the data. Sampling is like religion: many of us believe in it but few of us achieve its ideals.

1.4
Sampling Scales and the Periodicity of Geological Features

Knowing where to collect data is not a trivial matter. Unfortunately, with hindsight and some statistical treatment, we may see that we should have sampled differently or with differently spaced sites. Idealistically, one might propose *systematic sampling* at each intersection point of a hypothetical grid, such as certain UTM map coordinates. This would certainly sim-

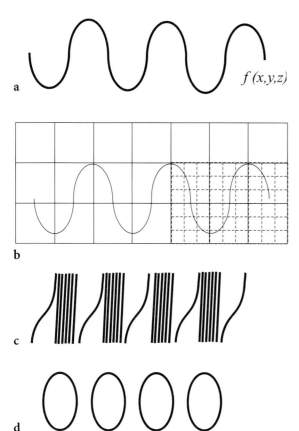

$f(x,y,z)$

Fig. 1.1 a–d. Periodic structures and their effects on geological sampling schemes. **a** Idealised sine-wave variation. **b** Sampling on the larger grid would fail to recognise the variation: sampling on the smaller grid scale would suffice. Generally, the sample grid spacing should be 10 to 20% of the variation's wavelength for successful identification. **c** Periodic variations in the schematic form of ductile shear zones. **d** Periodic variations in the form of spaced intrusions, ore deposits or pebbles, etc.

considers a sampling traverse along which observations are approximately regularly spaced. Consider Fig. 1.1, which shows a periodic variation of some geological feature, in the form of a sine wave. This could represent a series of folds, heterogeneous displacements on shear zones or the spacing of diapirs or plutons (Fig. 1.1 b–d). Sampling at the same spacing as the wavelength shown in Fig. 1.1a would produce a false estimate of homogeneity as one would record the same orientation of a fold limb, shear zone or the centre of an intrusion in every site. Thus, if systematic sampling is preferred, observations should be spaced at less than 20% of the wavelength of the structure (Fig. 1.1b). However, the actual periodicity of the geological structure may vary at a higher frequency than the observation spacing. The wavelength, λ, is inversely proportional to frequency, f. Ideally, observation spacing should be much less than the wavelength of the structure we hope to study (Fig. 1.2a). The highest detectable frequency that we may detect, the Nyquist frequency (f_{NY}), is half the observation frequency, i.e. $f_S = 2 f_{NY}$ (Fig. 1.2b). Where structures vary more rapidly than the observation frequency, false low-frequency variations may appear as an artefact of the method (Fig. 1.2c). Inadequate observation frequency is often evident from measurements along a traverse or down a well log. In preparation for a gravity study of a mineralised basic sill, density measurements were made at intervals of approximately 50 m over the 860 m log (Fig. 1.2d). The sill terminates at a depth of ~ 620 m and, since the errors of individual measurements are small, we may infer that there is magmatic differentiation, concentrating dense minerals (4000 kg/m³) near the base. However, the apparently periodic density variations from 0 to 400 m are probably spurious. Since each data point corresponds to a turning point, we should probably say no more than that density varies in the range of 2500–3000 kg/m³ over that interval.

Non-systematic sampling could avoid the above-mentioned bias. Instead of sampling every grid intersection, one could randomly select the intersections at which specimens are taken. This would be achieved by numbering the intersections and selecting them on the basis of random numbers. In principle, one might select a site simply by throwing dice to select the grid-intersection. In practice, random locations are selected by using a suitably scaled random number obtained from a table of random numbers or from a computer's random-number generator.

Neither random nor systematic spatial sampling is practicable in most field studies. The selection of rock specimens is dictated by paucity of exposure, and mountainous regions add the complication of the

plify contouring of the observations and make interpolation between stations equally sensitive over the study area. Logistics rarely permit such sampling, and the biassed spacing of the sites could systematically overlook some features. Even in the case of 100% rock exposure, systematic geological sampling may be impossible due to rough terrain, variable weathering, unavailability of the target material, etc.

A further consideration is the spatial frequency with which the studied rocks vary. For example in a reconnaissance study it may be appropriate to use a helicopter and minimise landings for financial and mechanical reasons. Landing to take granite specimens every 2 km may produce an agreeable looking distribution map. However, if the geology varies at a similar frequency, the observations could all represent the same feature. For simplicity, the following example

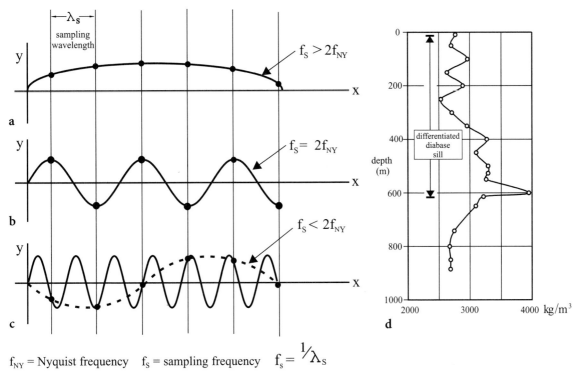

f_{NY} = Nyquist frequency f_S = sampling frequency $f_s = \dfrac{1}{\lambda_S}$

Fig. 1.2 a–d. The spatial frequency of sampling sites may affect the recognition of the true periodicity of some actual geological variation. **a** The frequency of the natural variation is documented accurately where the sampling frequency (f_s) is much higher. **b** The minimum sampling frequency that could identify the natural periodicity is twice the Nyquist frequency (f_{NY}). **c** Spurious frequencies may be inferred where sampling frequency is less than twice the Nyquist frequency. **d** Density variation from drill core through a gravity-differentiated sill. High-density material associated with mineralisation is fairly obvious from the high densities ~580 m depth. However, the oscillatory variation in density from the surface to ~300 m is probably an artefact of sampling

sample sites' altitude. The geologist must collect where nature permits access. We should not be too cynical here; it is too easy to deplore the statistical limitations imposed by natural site distributions. However, the earth scientist usually has a wealth of supplementary techniques and knowledge that help to "fill in the gaps". For example, geologists make very accurate maps, as verified subsequently by drilling or geophysical surveys, even though there may be less than 1% of the bedrock exposed. However, we must concede that nature usually biasses the rock exposure so that observations of any specialised quantitative geological data may be less reliable than the geological map. We have noted that hard rocks are preferentially exposed whereas softer rocks erode and become concealed by superficial cover. Thus, a study of the regional variation of the density of sedimentary rocks or their deposition depths will be biassed towards the values corresponding to weathering-resistant rocks. The geologist's intuition may be more important in planning sampling than any idealistic procedure based on a mathematical model. Nevertheless, an understanding

of theoretical plans enables us to understand the limitations of interpretation.

1.5
Geological Examples of Spatial Variation

Some major geological processes occur on such a grand scale that they were successfully identified without any considerations of sampling strategy. The spatial variations in the following examples were successful for the two reasons that we have already mentioned. First, the spatial variations were on a larger scale than any individual field study. Thus, normal field sampling was at a much higher frequency than the natural variation. Secondly, due to the immense scale of the variations, they were recognised in numerous studies, in adjacent and overlapping areas by different research groups, thus providing verification by re-sampling. The spatial variations could not be explained when most of the data was collected. Subsequently, Plate Tectonic

Fig. 1.3a, b. Regional variations in scalar observations controlled by Plate Tectonic processes, here principally influenced by the depth to the subducted Pacific Plate. **a** Ishihara's line separates regions of ilmenite-bearing granites from magnetite-bearing granites. **b** Boundaries between sodic-alkalic, calc-alkaline and tholeiitic volcanic rocks

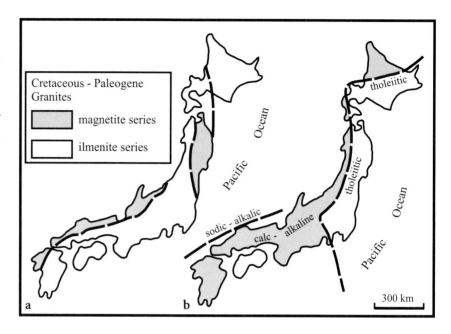

theory and the discovery of associated processes have explained the spatial patterns.

The first example, from Japan, shows a clear spatial distribution of variation in accessory mineral content of granites. In some areas, ilmenite content exceeds magnetite and in others magnetite is the dominant accessory iron oxide (Fig. 1.3a). A line parallel to the subsequently discovered Pacific plate boundary separates the two subtly different types of granite. This phenomenon is now explained by magmatic processes that depend upon distance from the offshore trench where subduction commences and the influences of depths of partial melting on oxygen fugacity and the oxidation state of the iron oxide minerals. Because the depth to the underlying subduction zone is proportional to distance from the ocean margin, the boundaries between contrasting volcanic compositions are also parallel to the plate boundary (Fig. 1.3b). The ilmenite-magnetite boundary is geophysically important because magnetite's susceptibility is much larger than that of ilmenite, so that aeromagnetic anomalies are much stronger over magnetite granites. Also, magnetite is the prominent carrier of paleomagnetic signals whereas ilmenite does not carry paleomagnetic signals.

Another large-scale example of quantitative data that support but partly predate the concept of plate tectonics is shown in Fig. 1.4, for the western seaboard of North America. In (a) the distribution of subtle variations of granitic rocks is shown; quartz-diorite varieties occur closer to the ocean than quartz monzonites. Petrologists subsequently explained this by a phenomenon related to the depth of partial melting of the underlying subducted plate. Thus, objective, quantified observations of the rock composition revealed a phenomenon related to plate-tectonic processes, long before plate tectonics was fully understood. Subsequently, still more subtle aspects of granite petrology also reveal distributions that are parallel to the Pacific plate margin, and relate to the depth of the underlying subduction. For example, strontium isotope ratios and the distribution of S-type granites in California (Fig. 1.4b) lie parallel the plate boundary.

The previous examples of spatial distributions throw light onto major processes in the earth and are based on sampling huge areas. Data sites in these cases, like in most geological situations, are dictated by the degree of exposure of the rock and the suitability of the outcrop for sampling. For example, weathered outcrops might be excluded if the study required geochemical work. If the target is granite and its regional variation, then sampling is clearly limited to the location of granites, which in turn is dictated by factors beyond the control of the geologist. For example, the emplacement mechanism and the ascent path of the magma control where granites are exposed. Moreover, expense or difficulty of access may prevent the study of some areas. Studies of very large regions such as the two previous examples are less affected by sampling heterogeneity simply because so many workers in so many different

Fig. 1.4a, b. Regional variations of granitic rocks, whose spatial distribution is related to distance from the Pacific margin. They are due to processes associated with subduction of the Pacific Plate. **a** Separation of regions dominated by quartz monzonites and quartz diorites, respectively. **b** A critical strontium-isotope ratio isopleth for granitic rocks and the region dominated by granites of the S-type

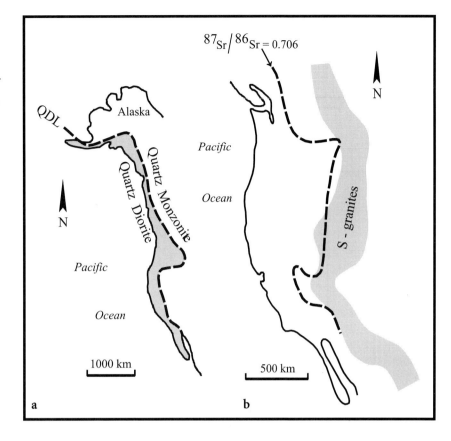

terrains have merged data on an unplanned but *much smaller sampling spacing* than the spatial variation of the phenomenon. Thus, important, independently corroborated concepts can be discovered from unplanned sampling. Indeed, the previous examples could not have been planned, as the scale of the variations and the Plate Tectonic theory were not understood at the time much of the data was collected.

This does not recommend unplanned sampling. It is more effective and more scientifically sound to sample according to a design that reduces any bias and therefore increases the chances of discovering meaningful variation. Some notion of the scale of the natural variation and its periodicity is enormously valuable in designing a spatial-sampling scheme. This depends on the relative scales of the variation we are investigating and the scale of heterogeneity of the area. Matters usually become progressively more sensitive as the size of the sampled area decreases. For example, in a field study of an area of several square kilometres, lithological variation and topographic elevation may play havoc with the establishment of any rigorously designed sampling pattern. A simple example of a quan-

tifiable project, where sampling defies planning is shown in Fig. 1.5. The study examined the distribution of pebble diameters in tributaries of the Susquehanna River. Clearly, the pebbles must be located in stream channels so that the geographical distribution of observations is intrinsically biassed. Nevertheless, with even limited geological intuition, one may recognise groupings along the lengths of channels that show similar diameters, and that the tributaries east of the Susquehanna river have larger clasts than those from the main channel leading to the NE corner of the mapped area. Clearly, to reject this data because the observations were not randomly distributed would be throwing out the baby with the bath water. Without local and subject-specific knowledge, it would be difficult to interpret the spatial distribution of clast sizes. Statistical approaches must always be applied with common sense, and this is especially so in geology where the data sets may be very complex, the number of variables very large, and where there may be no simple underlying mathematical relationship between the observations and the processes or principles which account for them.

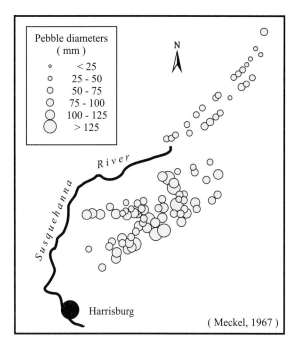

Fig. 1.5. Spatial distribution of mean pebble diameters in the Susquehanna River basin, Pennsylvania. Sites are distributed along river channels associated with different clast sizes

1.6
Site Distributions

Systematically distributed data sites on a *regularly* spaced grid would facilitate simple analysis (Fig. 1.6a). For example, interpolation would be equally sound in all parts of the region and contours could be drawn everywhere with similar confidence and ease. Some allowance might be necessary if there was strong topographical relief; otherwise observations would not represent sampling on one plane. Instead, the observations might be influenced by geological processes or differing rocks above and below a certain topographic level. For a region with 100% exposure and the presence of the target lithology at every outcrop, dense random sampling would be ideal (Fig. 1.6b). In this case the *x* and *y* coordinates (Easting and Northing) might be selected with the help of a random number from a calculator, a computer or from a set of statistical tables. Random numbers (*U*) are usually available in the range $0 \leq U \leq 1$ so that they must be multiplied by some value to scale them to the user's convenience, e.g., the factor may simply be a map scale. This procedure is performed for both the Easting and the Northing to determine the data-collection site on the grid. An example of a randomly distributed set of sites is shown in

Fig. 1.6b. The possibility of satisfying this sampling strategy in most field studies is negligible. In reality, nature might restrict available materials to stream courses, yielding a distribution such as that shown in Fig. 1.6c, reminiscent of the distribution of pebble sizes along river channels shown in Fig. 1.5.

Clearly, geological and logistical considerations are usually unfavourable for systematic or regular sampling (Fig. 1.6a). However, as long as there is *uniformity* to the distribution of data sites, a similar level of interpretation should be possible across the area. A uniform distribution is one in which there is a constant site density, i.e. a constant number of sites per unit area. It is different from a random distribution, but it goes some way to reducing sampling bias. A truly *random* pattern is more demanding; it requires that the location of any site has no influence on the position of another site. A uniform site distribution may suffice for most purposes.

We may evaluate the uniformity of a distribution of *N* sites fairly simply. The map should be divided into a grid so that each of the *k* grid segments is expected to contain five or more sites. The expected number (E_i) of sites in each segment is given by (N/k). For each grid segment (designated *i*) we determine the following expression:

$$\frac{(O_i - E_i)^2}{E_i} \tag{1.1}$$

This expression represents the disagreement between the expected (*E*) and observed (*O*) numbers of sites, normalised as a fraction of the expected number. The difference term is squared, so that positive and negative discrepancies will not cancel one another when we sum (Σ) the discrepancies for all cells (*i* = 1 to *k*) as given by the following definition of the χ^2-statistic.

$$\chi^2 = \sum_{i=1}^{k} \frac{(O_i - E_i)^2}{E_i} \tag{1.2}$$

We shall see in Chapter 5 that this commonly used statistic is independent of the actual nature of the data distribution. It is called a *non-parametric* or *distribution-independent statistic* since it involves no assumptions about the distribution of data. A small χ^2-value indicates that the sites are fairly uniformly spaced and it is possible to assess the significance of the uniformity, as explained Chapter 4. The test requires at least 50 observations with no less than five *expected* (E_i) observations in any cell. Adjacent cells may be combined to give $E_i \geq 5$, or even arbitrarily shaped cells may be used as long as they are contiguous and provide coverage of the entire area (e.g. see triangular cells,

regular sample pattern   random pattern $\chi^2 = 8.2$ fluvial pattern $\chi^2 = 37.3$
(uniform at 95% level)   (not-uniform at 95% level)

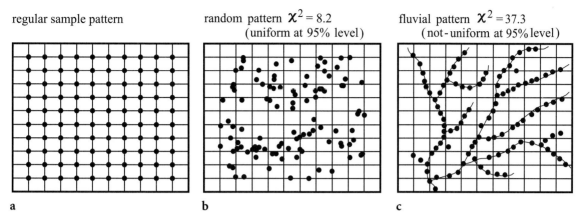

a   b   c

Fig. 1.6a–c. Sampling procedures using a map grid. **a** Idealized systematic sampling on a rectangular grid. **b** A randomly chosen number, scaled to the map, is used to predict the x and the y (east and north) coordinates of sampling sites. **c** Typical geomorphologically controlled site distribution with outcrops located along stream channels. The χ^2-test is a robust statistical procedure that can be used to compare the site distribution with a uniform spatial distribution (further, see Chap. 5). Uniformity of sample coverage is useful even in the absence of an ideal, unbiassed, random site distribution

Fig. 10.19). O_i and E_i are then recorded for each of k cells; k is the number of cells, not sites (N).

For simplicity, Fig. 1.6 used a square grid, with dimensions to provide a suitable number of cells, and minimum numbers of expected observations per cell. As in all applied statistics, some judgement is involved, and knowledge of the application is always important. For example, on what scale is any departure from randomness expected? Figure 1.6b shows a random distribution of sites whose χ^2 value is 8.20; this is not too large to rule out a uniform spatial distribution. In Chapters 4 and 5, we shall learn that if χ^2 exceeds a certain critical value, one must reject the hypothesis that the sample is drawn from a uniform distribution. The actual critical value of χ^2 must be considered in conjunction with the degrees of freedom v of the sample, defined as $(k-1)$ where k is the number of cells used in counting observations. The critical values are available from tables or graphed values (e.g. Table 5.6, Fig. 5.8d).

In practice, field sampling is usually strongly influenced by accessibility or topography, for example along mine adits, road cuts, or in river valley exposures. The fluvial pattern of sites shown in Fig. 1.6c has $\chi^2 = 37.3$ which fails the test for uniformity at a 95% confidence level, if that were not obvious by inspection. Although it may be impossible to overcome the natural limitations, we are at least aware of the extent of the shortcomings of the data set, and this may curb overenthusiastic interpretation.

A real example is provided in Fig. 1.7a. Over an area of 319 km² attempts are being made to determine the flow directions of magma within an ophiolite dike se-

quence in the Troodos complex of Cyprus (D. Gauthier and Borradaile, in progress). Despite strong topographic relief and good exposure, economical access limits sampling sites to ravines and mountain roads. Consequently, biassed sampling could interfere with the detection of spatial patterns in magma flow directions, from which one may infer the presence of underlying fossil magma chambers near an old ocean-floor spreading axis (see Fig. 1.20). It is preferable that the observation sites were not biassed in any way; so a uniform spatial distribution would be agreeable. A *uniform spatial distribution* expects the number of sites to be equal in equal areas. Therefore, the map was divided into 53 equal squares of 2.5×2.5 km and the $N = 420$ sites were tallied in each of $k = 53$ squares, that we shall call *classes*. If the spatial distribution is uniform, we would expect an average of $E_i = 7.92$ sites/class. (This is good because the χ^2-test needs $E_i \geq 5$). The calculation gives a moderately small value, $\chi^2 = 6.88$; one realises that a small value supports the view that the observed and expected number of sites per class are similar. In other words, it favours the hypothesis that the spatial distribution is uniform. In Chapters 4 and 5, we will learn that this can be turned into a rigorous decision. At a 95% confidence level ($\alpha = 0.05$), Table 5.6 shows that for the spatial distribution to be significantly non-uniform, $\chi^2 > 68$ with 52 degrees of freedom of the test ($v = 52 = k - 1$). Clearly, the calculated value is much smaller and we reject the hypothesis that the spatial distribution is non-uniform.

Where the number of sites (N) is smaller, it may be necessary to group the map squares to arrange the

Fig. 1.7 a, b. Site distributions. **a** UTM grid at 2.5 km spacing for the SE part of the Troodos ophiolite complex of Cyprus. 420 sites were investigated to determine magma flow directions within dikes of the sheeted dike complex. The χ^2 statistic is a measure of all the differences between an observed distribution and some expected distribution. Here, χ^2 is too large for the distribution to be considered uniform, at a 95% confidence level. **b** Distribution of kimberlite and alkaline intrusions in southern Africa. The χ^2-test requires ≥ 5 observations in each of k cells, but cells may be of different size and shape. Therefore, the original 200×200 km cells were grouped as shown in *heavy outline*. The value of the χ^2 statistic indicates that kimberlite-distribution is not different from that expected of a uniform spatial distribution at the 95% level

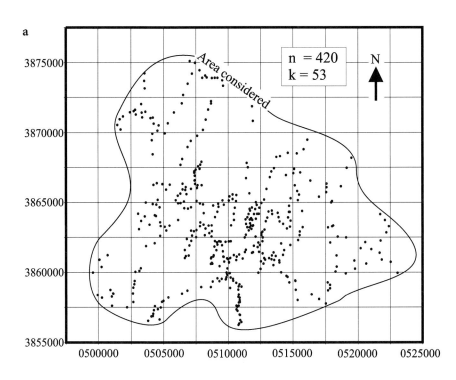

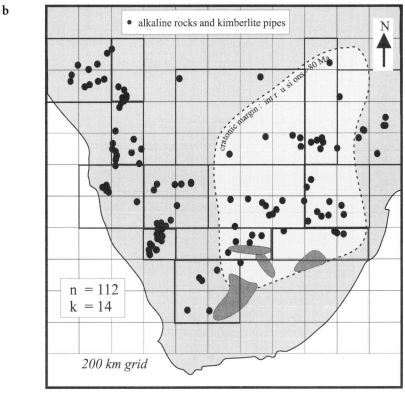

classes (k) so that $E_i = (N/k) \geq 5$. This was necessary in parts of the example in Fig. 1.7b which shows the distribution of diamond-bearing and related igneous intrusions in southern Africa (Mitchell 1986). It is reasonable to wonder whether the distribution is uniform or not. Uniformity might imply an absence of crustal heterogeneity and anisotropy of emplacement distribution. That would prompt us to consider mantle controls on the distribution. The sites are counted initially in sixty, 200×200 km quadrants, but with only $N = 112$ sites, the number of classes (k) must be reduced to ensure that at least 5 counts are expected in each. Many possible groupings are possible; one arrangement of $k = 14$ cells is shown in Fig. 1.7b for which $\chi^2 = 15.8$; this is small enough that we can reject the hypothesis of non-uniformity, at a 95% confidence level. In fact, χ^2 must be >22.4 to reject the hypothesis of non-uniformity with 14 classes ($\upsilon = 13 = k - 1$), using Table 5.6.

A different approach to assess uniformity or randomness of site distribution is to compare the observed spacing with the theoretical model of a uniform or random spatial distribution (King 1969; Davis 1973; Bartlett 1975; Cressie 1993). A uniform distribution of N sites over a region of area A would yield an expected site density, $\lambda = (N/A)$. A uniform distribution is unlikely in nature and it may be undesirable as a planned scheme for sampling. Nevertheless, it is a useful unbiassed introductory model that serves as a reference for some more realistic spatial distributions. We are aware, of course, that a non-uniform distribution is not necessarily random (e.g. river-bed sampling, Fig. 1.6c). Our aim is to understand the distribution of local site densities (n/a), where n is the number of sites in a small subarea of area a.

When independent, random values (x) are recorded as "counts per unit interval", such as site density; they may follow the Poisson probability distribution (Chap. 3). It was applied by King (1969) to predict the frequency distribution of *sites per subarea*. Its y-axis shows the probability, ($0 \leq P \leq 1$), with which x occurs. In this application, $x = n/a$ is the local reported site-density per unit area. The Poisson distribution shows that the probability that x sites lie in a small unit area is given by:

$$P(\lambda) = \frac{\lambda^x e^{-\lambda}}{x!} \qquad (1.3)$$

where λ is the mean site density for the whole region (N/A). A characteristic of this theoretical distribution is that its mean and variance share the same value (Chap. 3 and Table 3.7). If one accepts that the site density is random, the formula may be used to determine the probability of any given site density ($x = n/a$), i.e. of 1 per sub-

area, of 2 per subarea, etc. This may then be compared with the observed distribution of x using the χ^2-test.

1.7 Site Spacing

The randomness of site distribution may also be approached by considering the spacing of closest neighbours. Intuitively, one may realise that the average separation of randomly distributed, neighbouring sites would tend toward some common value. Depending on the application, problems may arise with very small separations. For example, the finite size of objects may prohibit very close spacing, leading to the concept of an *inhibition radius* within which no occurrences are possible (Bartlett 1975). However, a general expression is available to describe the probability distribution of nearest neighbour separations (d) on a plane where, as before, λ represents the regional site density (N/A) or mean site-density probability:

$$P(d) = (2\pi\lambda d)e^{-\pi\delta d^2} \qquad (1.4)$$

It may be shown that the expected shortest distances (Δ) between randomly distributed neighbours have the following mean ($\bar{\Delta}$) and standard deviation (σ_Δ), where m is the number of *distances* recorded between closest neighbour pairs.

$$\bar{\Delta} = \left(\frac{1}{4\lambda}\right)^{\frac{1}{2}} \qquad \sigma_\Delta = \left(\frac{0.0683}{m\lambda}\right)^{\frac{1}{2}} \qquad (1.5)$$

If the observed average distance between neighbour sites, $\bar{d}$, is similar to $\bar{\Delta}$ predicted by the above equation, the observed spatial distribution of sites may be random. The standard deviation may be used to estimate our confidence in the predicted mean separation of nearest neighbours.

Simpler evaluations of uniformity of the sites' spatial distribution are obtained by comparing observed and expected mean site separations using either of the following ratios:

$$\left.\frac{\bar{d}}{\bar{\Delta}}\right\} \begin{matrix} 1 = \text{uniform} \\ 0 = \text{coincidence} \end{matrix} \qquad \left.\frac{m\sum d^2}{\left(\sum d\right)^2}\right\} \begin{matrix} 1.27 = \text{random} \\ 0 = \text{coincidence} \end{matrix}$$

$$(1.6)$$

Zero values correspond to no separation and therefore coincidence of the sites. The finite values of 1.0 and 1.27 correspond respectively to uniform and random distributions of nearest neighbour separations in the two expressions. However, these statistics can have higher values where the sites are systematically distributed on some lattice pattern. Such statistics may be valuable in themselves for preliminary interpretation. However, they may be evaluated to assess the suitability of the data for further processing, for example, by contouring or more sophisticated treatment, such as trend-surface analysis.

A uniform or random spatial distribution of sites may be beneficial where some simple trend is known to exist or for reconnaissance work where there is no prior knowledge. However, the consequences of different spatial-sampling strategies are discussed more fully in the next section.

1.8
Locally Planned Sampling Schemes

The previous section discussed random and uniformly distributed sampling sites. Their strict requirement is rare and is often very difficult to satisfy. In contrast, many studies actually require sites restricted to certain locations due to restricted research interests, accessibility or intrinsic geological control of the distribution. Under these circumstances the notion of planned sampling may be applied to parts of the region, for example in discrete traverses.

Systematic sampling (Fig. 1.8a) is the most elementary and ideal plan that one might conceive. It has an advantage where one wishes to document gradational changes, but it is rarely possible to achieve in practice because observations may not be possible at arbitrary locations. This problem affects almost all spatial-sampling schemes, including those mentioned subsequently. However, one of the important issues we must also address is that our sites should be selected independently of each another, in a random fashion so as to minimise any bias in sampling. This is considered below.

Consider observations recorded along a fully exposed stream bed. These could be selected *at random*

sites along the section, shown conveniently as a straightened line in Fig. 1.8b, in the hope of reducing any bias in their spatial distribution. This would be a fair strategy where the phenomenon being investigated is homogeneous on that scale. Suppose, however, there is some geological heterogeneity to the section, such as a preponderance of sandstones at one end of the section. Of course, clusters occur in any random distribution. If some observations clustered in the sandstone section, the resulting sample would be biassed towards sandstone data. Thus, *hierarchical* or *stratified* sampling may be appropriate (Fig. 1.8c). Hierarchical sampling breaks the sampling area into homogeneous parts that are then randomly sampled as separate subunits (e.g. part P, part Q, etc., Fig. 1.8c).

It is assumed usually that any sensible spatial variation will be attributable to a single cause, giving rise to a simple spatial pattern. However, earth science commonly presents more complex situations in which the measured variable at each site is also influenced by some secondary control. For example, Au concentrations at the surface should focus on the underlying ore body, the primary control being distance from the deposit (Fig. 1.8e). A secondary control such as fault-induced permeability may disperse the Au concentrations as shown. These compound the influences of distance from the underlying deposit with those due to fault-related permeability. The spatial variation cannot be isolated to a single variable as in an ideal case (Fig. 1.8d). There are several ways to overcome this. First, one could normalise the Au concentration by the permeability of the specimens. Second, one could restrict consideration to Au values from sites of similar permeability (*paired sampling* of Chap. 4). In either case, it may be possible to compute "corrected" Au concentrations whose contours would better locate the underlying gold deposit (Fig. 1.8f).

Initially, most of us propose traverse sampling, either along naturally controlled paths such as stream courses, or along artificial lines cut through a forest. The latter permits systematic sampling but is expensive and normally only used in the final stages of ex-

Fig. 1.8. Sampling strategies, shown for convenience along lines in **a** to **d**; of course, the concept could apply to sampling along a nonlinear traverse such as a stream course, or to a mapped distribution (**e**). **a** Systematic sampling; sites chosen at regular intervals. **b** Simple random site spacing; each location is independent of the others. **c** Stratified or hierarchical random site locations, in which sites are randomly located in each of several segments (P, Q, R...). This has the advantage that random clusters of sites need not fall in an unfavourable part of the entire traverse, e.g. where suitable material is unavailable. **d** For the preceding strategies, it is implicitly assumed that at each location x_n the observation is of a single-valued scalar variable that is independent of any other control. **e** Observations may be influenced by two controls; here Au abundances are controlled primarily by some focused underlying economic target, but the spatial distribution shows subordinate influence due to fault-related permeability. **f** If concentrations are paired with samples of similar permeability, the underlying source may be located more precisely (for more about paired sampling see Chap. 4). **g** Cluster sampling may be dictated by economy of time or resources; sampling regions, preferably chosen randomly (A–E), are used for detailed local sampling (*grey dots*)

a Systematic sample traverse

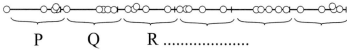

b Simple random-sample traverse

c Stratified (hierarchical) random-sample traverse

P   Q   R ....

d location x_n ; observation y_n scalar stochastic variable

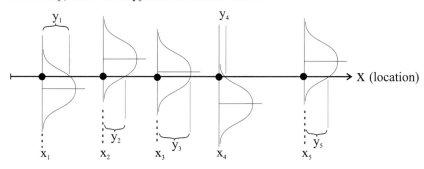

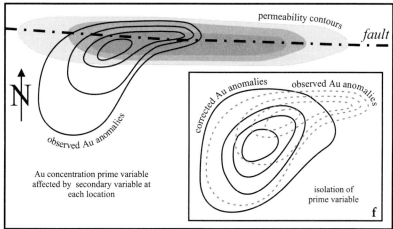

permeability contours

fault

corrected Au anomalies

observed Au anomalies

observed Au anomalies

Au concentration prime variable
affected by secondary variable at
each location

isolation of
prime variable

f

e

cluster sampling

B

E

A

D

C

g

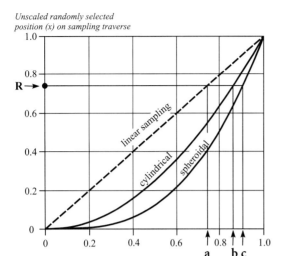

Unscaled randomly selected position (x) on sampling traverse

x scaled so that sample is representative of
a) linear position b) annular area c) volume of spherical shell.

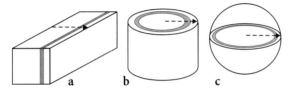

Fig. 1.9. Selection of randomly distributed sample sites. Suppose the relative location of a site along a traverse is fixed by x, which could be selected randomly as R between 0 and 1. This could then be scaled to an actual distance on the ground by multiplying by the traverse length in say, kilometres. **a** In most sampling schemes the sampling frequency is not biassed by the volumes that the sample is supposed to represent (*shaded* in sketch). In some cases, sampling frequency may be adjusted so that each potential sample site represents a volume comparable to other sites (Cheeney 1983). **b** For a vertical cylinder, successive sample sites should represent annular segments of similar volume. This requires increased sampling frequency towards the rim of the cylinder, predicted by the quadratic curve shown in the graph. **c** For spherical bodies, sampling frequency must increase still more rapidly towards the margin, following a cubic relation in the graph. Such schemes might be applied successively to sampling granites with different structural styles (see Figs. 1.10, 1.11).

ploration or in the evaluation of a known resource. Traverse-sampling strategies are one-dimensional. This may be appropriate where we wish to sample across the strike of some unit, perpendicular to some geologically significant feature such as the inclusion trails in the Donegal granite (Fig. 1.10). Then the observations within the sample represent some infinite slab perpendicular to the traverse (Fig. 1.9a). However, if we wish to sample along the radius of a vertical cylindrical body, it may be appropriate to telescope observation spacing towards the perimeter. The observations would then represent similar areas of the circular outcrop (Cheeney 1983). In that case observation fre-

quency should increase radially from the centre of the cylinder outwards, according to a quadratic relationship (Fig. 1.9b). Along a radius traverse of unit distance, the position of a site could be randomly chosen by assigning it a random number between 0 and 1, e.g. 0.75. This would then be stretched appropriately to compress sites toward the perimeter of the circular outcrop (≈ 0.85; Fig. 1.9b). The distance of the observation site from the centre of the intrusion is then fixed by multiplying 0.85 by the length of the traverse, for example in kilometres. Similar considerations may be applied when specimens must be taken from an approximately spherical body, and we wish the spatial-sampling plan to produce observations that represent equal volumes of the geological body. In that case, sites are even more closely compressed toward the perimeter of the body (Fig. 1.9c).

Unfortunately, logistics commonly dictate that observation sites force *cluster sampling* upon us, contrary to any desired theoretical spatial-sampling plan (Fig. 1.8g). For example, the number of helicopter landings per day is normally restricted in reconnaissance surveying. Thus, observations are clustered on the scale of the survey although it may be possible to choose the location of each cluster. Of course, each cluster of observation locations may be planned on a very local level to give the best information to satisfy the needs of the survey. For example, within each cluster, local sampling might be planned according to any scheme. Cluster sampling is disguised in many forms in the earth sciences. For example, information retrieved from drill core represents cluster sampling. Each geographical (x, y) location may provide a large sample of observations. This is not necessarily a bad thing at all. For example, if the rocks are horizontally stratified, this yields information about all formations from the "cluster" of observations represented by the well log. Within the well log, a locally imposed scheme, e.g. random and hierarchical schemes, might be most useful. Unfortunately, drill core is often logged systematically, at regular intervals, which may yield misleading information (Figs. 1.1, 1.2).

The selection of an appropriate, planned spatial-sampling scheme requires some prior geological knowledge and is not normally relevant in reconnaissance work. Consider the variety of different sampling strategies that could be applied in geochemical or petrological studies of the Donegal granites of Ireland (Pitcher 1993; Fig. 1.10). Possible sampling strategies include ideas embodied in Figs. 1.8 and 1.9 and relate to the differing emplacement modes, evident by inspection of the regional map (Fig. 1.10). The main Donegal granite is a tabular body influenced by a regional strike. A simple random sampling traverse is

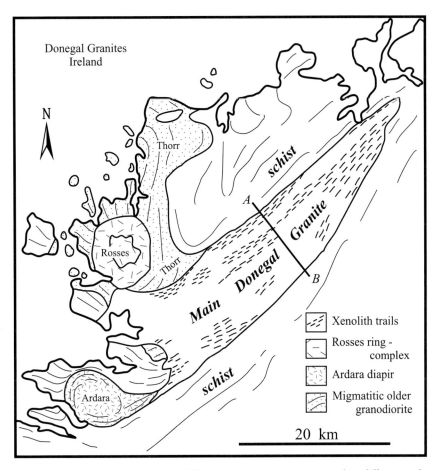

Fig. 1.10. From inspection of the map it is clear that the different Donegal granite intrusions show different modes of emplacement (Pitcher 1993) that affect sampling strategies. The main Donegal pluton shows regional structural control, following the strike of the country rock with aligned and partly assimilated xenolith trails. A traverse across the short dimension of the body could be representative and sample spacing could be random with each sample site representative of a similar strike-parallel volume (cf. Fig. 1.9a). The Thorr, Ardara and Rosses plutons are subcircular in plan. Radial traverses are recommended, but if the samples are to be representative of a cylindrical volume as in the stoped Rosses pluton, site spacing frequencies might be increased quadratically towards the margin so that they represent comparable volumes (see Fig. 1.9). The Ardara pluton may be a subspherical body so that each sample site would be representative of a spherical shell; sampling frequency should therefore increase even more rapidly towards the margin (see Fig. 1.9c). Further complications arise because the Ardara and Rosses plutons show discrete internal contacts that require hierarchical sampling in separate parts of the traverse

reasonable (Figs. 1.8b, 1.10). A little feedback from preliminary geological knowledge could refine the sampling strategy. It is known that the main Donegal granite conforms to the regional trends and the xenoliths preserve a ghost stratigraphy. Some tabular heterogeneity may therefore be expected in a cross-strike traverse. The initial observations may be collected along streams, using a hierarchical-linear type of traverse (Fig. 1.8b), which prevents unfortunate concentrations of sites where a particular stratigraphic formation was assimilated. Since the trend of the country rocks is NE-SW, the traverse would be best made in a general NW-SE direction, perpendicular to the strike of the country

rocks (Fig. 1.10, A–B). Depending on the success of that pilot study, subsequent sampling could be pursued using a more refined geographical distribution.

The Rosses pluton is cylindrical in form and shows discrete petrological phases perhaps due to stoping (Fig. 1.11a). In any case, it is both radial and discretely heterogeneous. To overcome any bias due to the heterogeneity, hierarchical sampling is recommended, in other words, selecting specimens randomly in segments across each phase (Fig. 1.8b). If the observations are to represent equal annular segments equally, sites should increase in frequency toward the margin, according to a quadratic relationship (Fig. 1.9b).

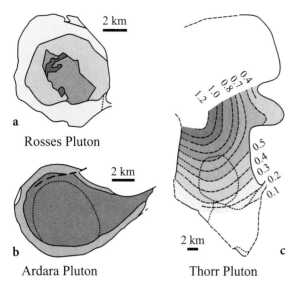

Rosses Pluton

Ardara Pluton   Thorr Pluton

Fig. 1.11a. Rosses pluton shows discrete internal lithological contacts due to separate magmatic pulses. A continuous traverse across this pluton would mix samples from quite different magma pulses and it would probably be more appropriate to use hierarchical sampling within each zone of the pluton. **b** The Ardara pluton shows characteristics of a forceful intrusion, with concentric foliation in the outer phase. The two phases, shown with different shading, would require separation for hierarchical sampling. However, the inner phase is compositionally zoned and could be sampled randomly or linearly along a traverse. **c** The Thorr pluton shows concentric gradational compositional variation as shown here by a value recording feldspar content. Systematic (regularly spaced) sampling would record the gradational variation more faithfully than randomly chosen sites

The Ardara pluton shows evidence of forceful intrusion, with concentric foliation and concentrically flattened xenoliths. This "balloon" inflation pattern suggests a spherical shape for the body. It also possesses some discrete concentric phases (Fig. 1.11b). A radial traverse would be sensible but sampled hierarchically in segments across each discrete phase. However, if we need samples that are unbiased with respect to volumes, we may wish to take account of the approximate spherical shape. In that case, towards the edge of the pluton, observation frequency should be greatly increased, according to a cubic equation (Fig. 1.9c).

The Thorr pluton changes progressively in composition as shown, for example, in feldspar content (Fig. 1.11c) and in colour (Fig. 1.12b). In this case, a systematic spatial-sampling pattern is beneficial, but if the concentration gradients are unknown, this should not be in one traverse direction. Traverses in several directions or an attempt at areal coverage would be better. Random sampling may fail to document the smooth gradational pattern. Clearly, the different

structural and petrological characteristics of the Donegal plutons require different sampling schemes to produce meaningful and representative patterns from the different bodies. It is clear that none of these schemes is suited for reconnaissance purposes. In each case, prior field knowledge is needed to choose the best sampling scheme.

1.9
Monte-Carlo "Sampling"

Where we wish to study the behaviour of a system for which a well-known mathematical model is known, we can generate *pseudo-data* to perform simulations and provide a model against which actual observations may be compared. This is explained in terms of a single variable here, but the process can be applied to synthesise an x-value and then a y-value, thereby simulating a map distribution.

To give a trivial example, in one dimension, if bed thicknesses are believed to be normally distributed, we can generate a pseudo-data set of bed thicknesses for comparison, quite simply, using the theoretical Normal distribution. One selects, at random, a number between zero and unity to represent a probability under the Normal distribution. One then notes the corresponding standardised variable z for the Normal distribution (Chap. 3, Table 3.5), which will proxy for bed thickness. Multiplying z by some appropriate factor provides model bed thickness (x). This procedure may be repeated as often as required, generating a normally distributed, hypothetical set of bed thicknesses. However, this statistical experiment does not produce a "sample" as understood by natural scientists. The numbers generated are not real data, although the terminology of some authors may be slightly misleading in that respect. *Pseudo-data* is a more appropriate term, alerting the reader to the true nature of the experiment. Pseudo-data generated randomly by such a Monte-Carlo simulation may be designed to follow any theoretical distribution. The procedure will be more easily grasped after reading Chapters 3 and 4. If we used the procedure to produce normally distributed (x, y)-pairs, they could be mapped and contoured to produce a mound-shaped contour pattern following the bivariate Normal distribution (Chap. 7).

Monte-Carlo simulations are also used in certain specialised fields of earth science where samples are restricted to a few observations and where no theoretical model can be proposed for the population. The approach has staunch proponents and opponents on

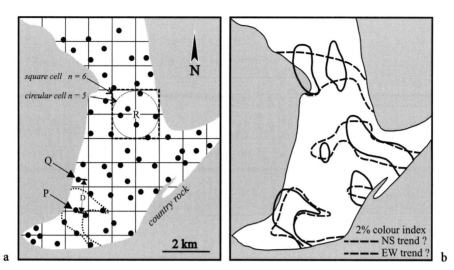

Fig. 1.12 a, b. Colour variation in granites is a field guide to their composition. This classic early study of the Thorr granite (Donegal) provides a useful introduction to the concept of contouring (Whitten 1963). **a** Distribution of sample sites, at which a colour index was recorded (%). The 1-km grid was established as a basis for contouring. Contours may simply be constructed by interpolating linearly between sites with lower and higher values. For example, in the SW corner, the 2% contour can pass smoothly through several sites where the index = 2%, but its position between Q (1.7%) and P (3%) is determined by proportions to be located at a distance $D(2-Q)/(P-Q)$ from Q towards P, where D is the separation of P and Q. Generally, superior and more representative contours are obtained by considering artificially determined positions, e.g. R, and calculating the mean value of adjacent observations. These may lie in a square cell or a circular cell, centred on R. **b** Conflicting interpretations of the compositional variation due to the paucity of data and subjectivity of contouring procedure. These are the results of the simplest manual contouring using visual linear interpolation. The two possibilities arise from one contoured without preconceived notions and one contoured with a knowledge of the north-south structural grain of this granitic body

subject-specific grounds but is statistically valid. The technique is known as *bootstrap sampling*. The application is considered to be most valuable where less than 25 original measurements are available. This original, real, small sample of n observations is *re-sampled with replacement*. This means that n observations are selected from it randomly to produce a new pseudo-sample but are replaced, so that the same values may be drawn in subsequent trials. In the pseudo-sample, or *re-sample*, some of the original measurements may be omitted, and some may be repeated. A computer program generates at least 200 re-samples in this way. The means of each of the many re-samples produce a synthetic distribution from which we may estimate the confidence in the true mean of the population. Of course, a cynic might note that if the original sample was representative of the population, a bootstrap technique might be unnecessary. Bootstrap techniques require specialised computer programs and are generally only used in specialised applications, for example, to determine confidence limits for mean orientations (Chaps. 9 to 11). Some earth scientists are very concerned about its application because of uncertainty about how well imperfectly sampled natural observations represent the population. The basic idea of boot-

strapping is described in Chapter 4, and applications are shown in Chapters 9, 10 and 11.

1.10
Geometrical Limitations to Sampling Imposed by Geological Structure

Previously, we recognised that sampling must take into account the variations we wish to detect, as well as their position and elevation. The reader had perhaps realised that until now, the discussion naively assumed that the bedrock was reasonably *homogeneous*, which in the geologists' jargon might refer to a massive granite outcrop that is lithologically similar throughout. In contrast, a sequence of alternating beds would form a *heterogeneous* body because different locations within the body would have different compositions (e.g. limestone, shale, sandstone). Some geologists refer to the physical constitution of the components of a rock as its *fabric*. Thus, a sedimentary sequence would show a location fabric whereas a granite of uniform composition would not.

A homogenous body shows no variation between data sites but it may possess an internal "grain" that af-

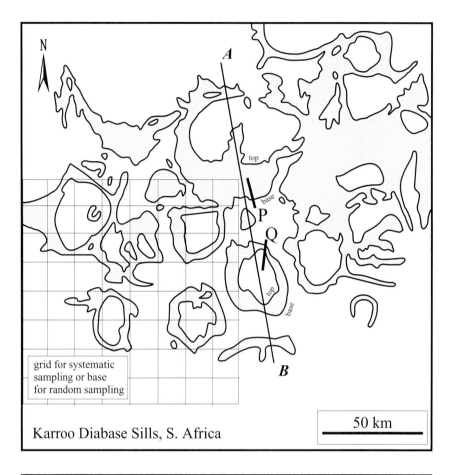

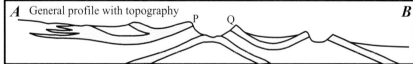

Fig. 1.13. Map and cross section of the Karroo diabase sills, South Africa. This simple and obvious example shows how structural control on outcrop distribution affects the sampling strategy. In reconnaissance work the general distribution may be detected from a random or systematic sampling pattern based on an (x, y) grid, which is shown for part of the region. For more detailed studies another scheme might be chosen, depending on the nature of the investigation. For example, vertical sections through sills (P, Q) may be targeted for petrological sampling, or bases of sills may be selectively sampled to study cumulate textures and potential concentrations of heavy economic minerals. For paleomagnetic work, either tops or bases of sills may be targeted for rapidly chilled specimens that carry a reliable thermoremanent magnetization

fects certain observations according to the orientation of a sampling traverse. For example, xenoliths may be aligned in a pluton so that recognition may be favoured in traverses perpendicular to the pluton. Any geological body that shows such directional control on its properties is said to be *anisotropic*. Those who have studied mineralogy know that most minerals are anisotropic because their crystal lattices have systematic arrangements of atoms that are differently spaced in different directions. For an everyday analogy, Ched-

dar cheese is fairly isotropic. Wood is anisotropic because it possesses fibres that give it different physical properties in different directions. However, both cheese and wood are reasonably homogeneous.

When we sample rocks in the field, we must consider fabric and structure in these broad terms. Heterogeneity determines where we make measurements; anisotropy affects the orientation in which the observation is made. The elements of these ideas are illustrated by an example that may appear too obvious at first sight. Consider the Karroo diabase sills of South Africa (Fig. 1.13). Thick, approximately horizontal sills crop out over a large area, producing a map pattern that geologists immediately associate with layered rocks forming mesa-like topography (Fig. 1.13, map and section). Let us suppose that such geological features were covered with vegetation or superficial deposits so that the bedrock must be inferred from soil specimens, isolated outcrops or drill holes. The following steps might be followed.

Stage (1) involves a reconnaissance study for which some sampling grid may be established with a grid mesh determined by balancing time, money and resources against the required detail. Part of such a grid is shown. Without the benefit of a pilot study, we might choose a systematic sampling pattern, perhaps landing a helicopter at the coordinates of each grid intersection; this would certainly simplify analysis. A random sampling pattern based on randomly selected x-coordinates and y-map coordinates might also be satisfactory in this case. If the sampling were done by surface transport, or if helicopter landing sites were restricted, multiple traverses would normally be used to acquire data. From the observations, it might then be possible to construct a map of the distribution of the sills as shown in the map of Fig. 1.13.

Stage (2) sampling would satisfy some specific need for further study. We have established the distribution of the lithologies of interest; now what further information do we want from them? If we wish to understand the petrology or geochemistry of the sills, a logical sampling pattern would focus on their vertical differentiation. The outcrop map would be compared with the topography (see profile, Fig. 1.13) to locate vertical sections (e.g. P, Q). More specifically, we may be interested in the concentration of metals of economic interest that sometimes concentrate near the base of sills. Thus, sites should be located along the lower contact of the sills. For paleomagnetic work, rapidly chilled igneous rocks are sometimes preferred, so either the upper or lower sill contacts could be traced as sampling traverses. Such focused spatial sampling is far from being systematic or random on the original reconnaissance grid.

Stage (3) may normally be taken for granted. It concerns the mundane process of collecting the material or information at a site. For example, a rock specimen may be taken from loose regolith, solid bedrock or from material retrieved by drilling, such as percussion fragments or intact cores. Whatever material is removed, the mechanism for removal should be unbiassed and free from arbitrary technical or logistic constraints. Nothing should get in the way of science. It is remarkable how many of us select bedrock fragments, armed only with a hammer. The use of a chisel greatly increases the range of available specimens.

The sill example leads us to the more general case of the *cut-effect*. How does the orientation and shape of a geological body influence our sampling? In the case of a bed, the most common item of interest is variation perpendicular to the layer since this relates to the passage of time as the bed accumulates. In other cases, lateral variation may be important, for example the change in sedimentary facies may reveal different paleo-environments at the same stratigraphic level. However, for the most common case, it is important to recognise a section perpendicular to bedding. Sections oblique to bedding will reveal neither the true dip nor the true thickness. Apparent dips and apparent thicknesses bias any study that needs to relate observations to stratigraphic position and thus to time.

A more complex example of this occurs in the case of deformed strata where beds have changed shape, thickness and orientation due to tectonic events. Such effects may also be due to syn-sedimentary or soft-sediment deformation. A study of the de-straining and unfolding of a major fold in the Scottish highlands is shown in Fig. 1.14. (The view in the upper figure is not a map but a plunge profile, or down-plunge view inclined at about 40° to the horizontal.) The spatial distribution of sites is very unsatisfactory because unusual, deformed sedimentary structures must be found from which the strain can be determined. The effects of strain are then mathematically reversed to reconstruct the stratigraphy, as shown in the lower part of the figure. Clearly, sites must be chosen very carefully, and it is probably fruitless to consider planning a spatial-sampling scheme using the post-strain distribution of rocks. As noted earlier, one must take suitable data wherever it can be found. There are many structures that interfere with spatial-sampling plans, including heterogeneously strained structures such as folds, shear zones and strain shadows around plutons. Heterogeneous structures fail to preserve the straightness or parallelism of any original lines or planes. In contrast, the preservation of such features constitutes homogeneous strain. (The loose term *deformation* encompasses more aspects of tectonic change than just strain. It includes also volume change, rigid rotation and translation, all of which may be of importance in sampling deformed terrains.) A further issue is the *scale* of the heterogeneity in a geological structure. A heterogeneous structure at one scale may be homogeneous at another. Where smaller folds occur, the distances between data sites must be reduced to produce a similar information base to that from larger scale folds (Fig. 1.14; cf. Fig. 1.15).

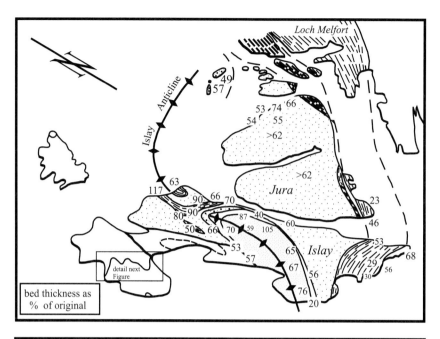

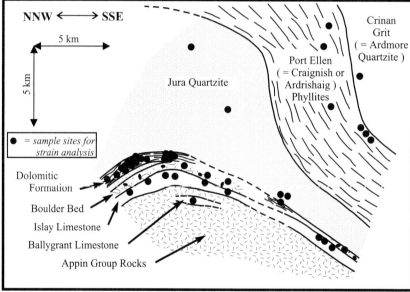

Fig. 1.14. *Above* The Islay Anticlinal nappe, a major fold in Scotland. It is an oblique areal view, which shows sites at which the strain of the rock was determined from strained sedimentation structures. From the strain ellipsoid at each site the bed thickness was determined as a percentage of the original value (Borradaile 1979a). The site distribution is extremely biassed, by lithology and the scarcity of suitable strain markers. Moreover, to use the strain estimates to reconstruct the pre-strain shape of the fold shown *below*, we confront a much more significant problem: the sites crop out on a topographic surface, not on the profile plane of the fold, which was used for de-straining the fold and determining the relative changes in bed thicknesses. (Fig. 1.17c illustrates the oblique projection of a fold profile onto a horizontal surface)

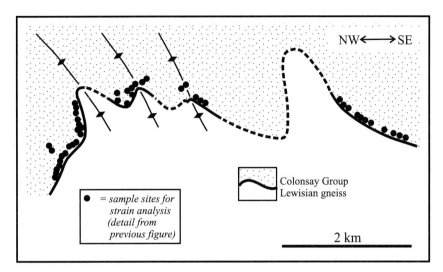

Fig. 1.15. A plunge profile shows the spatial distribution of strain estimate localities, from smaller folds shown in the boxed area of the previous figure. The uneven distribution of sites is due to the serendipitous occurrence of suitable deformed conglomerate at the unconformity with the underlying basement. Here the use of sites from an outcrop plane is less troublesome, because the scale of the folds is smaller than in the previous figure, and the corresponding distances between the outcrops and their projected positions on the hypothetical profile plane are smaller

1.11
Fabric Influences on Observations: The Cut-Effect

Petrofabric refers to the geometrical constitution of a rock, especially anisotropic rocks and usually but not exclusively confined to strained rocks. Some of the more important aspects include the shapes of grains and grain boundaries, the crystallographic orientation or dimensional orientation of grains, and the nature of grain contacts. These are especially important in understanding processes of solid-state change that accomplish the preferred orientation of minerals during the flow of rocks during metamorphism and tectonic strain (Nicolas and Poirier 1976; Poirier 1985). However, they are also considered in studies of magmatic flow and other processes in igneous rocks (Vernon 1976) and of depositional processes in sedimentary rocks.

The rocks' petrofabric and anisotropy may strongly influence the way in which we sample a three-dimensional body from a two-dimensional surface. These problems are collectively described as *cut-effects*. Consider the trivial example of an assemblage of identically sized, spherical grains all in contact to form a rigid framework (clast-supported as opposed to matrix-supported). A section through this assemblage would give the false impression that the grains varied in size and that they did not touch, and it would exaggerate the contribution of intergranular white space, representing an

intergranular matrix (Fig. 1.16a). All this is an artefact of a two-dimensional view of a three-dimensional assemblage. Clearly, this considerably affects any sampling procedure. In reality, even for a specimen of sediment, the addition of complicating factors such as variation in grain size and mineral composition of the grains compounds these problems significantly (Fig. 1.16b). How do we estimate the likelihood of certain minerals being in contact? What is the relation of pore volume to empty area in a two-dimensional thin section? Since a section rarely passes through a grain centre, how can we estimate the distribution of grain sizes? Even the question of estimating the abundance of grains is influenced by the cut-effect and is generally overestimated by visual inspection of low concentrations, especially if the objects are darker than their background (Fig. 1.16c). Some of these questions are too difficult to answer satisfactorily from observations on a plane, and quite different approaches are pursued in practice. Furthermore, as discussed next, these issues affect sampling across planes from the scales of microscopic thin sections up to maps of any scale.

In field geology, the best-known cut-effect is of the true versus apparent dips of planar structures (Fig. 1.17a). Any surface not perpendicular to a plane reveals an apparent dip, which is less than the true dip. The dip angles are related by a simple trigonometric relationship presented in the nomogram of Fig. 1.17b. In the study of thin sections of textures, this principle makes interfacial angles deceptive. For example, a com-

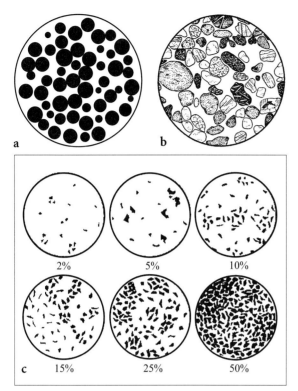

Fig. 1.16a–c. The cut effect refers to various unrepresentative views of three-dimensional features on a plane. **a** Plane section through identical, close-packed spherical grains gives the illusion of grain-size variation and the false impression that grains do not touch. **b** Microscope thin section confounds the cut effect with real grain-size variation; interpreting grain-size variation, the nature of grain contacts and porosity are therefore non-trivial. **c** Visual estimates of grain abundances are affected by the cut effect as well as grain-size variation and spatial distribution. Generally, low abundances of more easily observed phases may be overestimated

mon feature of statically annealed, thermally metamorphosed or slow-cooled plutonic rocks is that three grains of the same mineral commonly have grain boundaries meeting in a "triple junction" with the interfacial angles of the grains being 120°. The cut effect ensures that the angular relationships will not be preserved unless the plane of observation was by chance perpendicular to all three of the grain boundaries (Fig. 1.17d). Usually, the section is oblique and distorts interfacial angles (Fig. 1.17e). A corollary to this is that the curvature of grain boundaries may be overestimated or underestimated. Planes of observation necessarily underestimate any interfacial angle that is <90° and overestimate any larger dihedral angle.

Quantifying observations from thin sections, polished rock surfaces and digital images now uses digital procedures embedded in very sophisticated software

packages. This is a quantum leap from the 1970s when such data was collected manually, the only concession to automation being the "point-counter" which mechanically advanced thin sections across the microscope stage to sample them systematically in equal x or y increments! Nevertheless, the sophistication of some commercial software packages is so great that it is difficult to assess the procedures and assumptions used in the algorithms. However, there are systems dedicated to petrographic equipment (e.g. Zeiss) and for microscope-independent scanner software (e.g. Jandel, Sigmascan-Sigmastat).

Greater problems are met where observations are required from surfaces through assemblages of objects that show a preferred orientation, such as aligned crystals in metamorphic or igneous rocks, or strained pebbles in a tectonically deformed conglomerate. There are two aspects to this. First, the objects may all be of different shapes, and these shapes may be related to the orientation. For example, in a strained polymictic conglomerate, limestone pebbles will be more elongated than quartzite pebbles, reflecting the ductility of the respective lithologies. However, the strain process rotates pebbles so that the most elongated examples may be most closely aligned with the preferred direction, usually a schistosity or extension lineation. Thus, there is a complicated orientation distribution with different shapes dictated by pebble type and by degree of rotation. Even if the strained objects are identical in size and properties, such as feldspar crystals in a tuff, they will show a variety of orientations as some phenocrysts will rotate more than others, depending on their original orientation and on the matrix consistency. The resulting petrofabric can be described by a fabric ellipsoid, but its determination and any quantification or characterisation of the fabric require a careful choice of the sampling surface (Fig. 1.18). At the simplest level, determining the relative proportions of limestone versus quartzite pebbles depends not merely on their size, but also on their orientation. Thus, the direction of the sampling traverse or orientation of the sampling grid is critical. On a structural note, the appearance of a fabric depends on the orientation of the two-dimensional observation surface. One side of a block may appear to show a lineation and another side may appear to show no alignment (Fig. 1.18). At least two, carefully chosen orthogonal surfaces are needed to estimate the relative strengths of planar and linear fabric components (schistosity and lineation; S and L; Flinn 1965). Particular care is needed when making thin sections of rocks with preferred grain or mineral orientations; an unfortunately oriented cut may exaggerate or conceal an alignment.

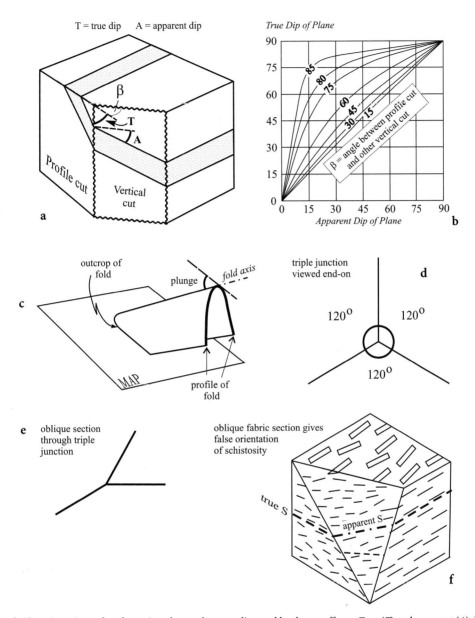

Fig. 1.17a–f. The orientations of surfaces viewed on a plane are distorted by the cut effect. **a** True (*T*) and apparent (*A*) dip of an in-clined layer. The true dip of a plane is measured on a vertical cut parallel to its maximum inclination. On any other vertical cut, the apparent dip is less than the true dip (a nomogram for use in this situation is found in Chap. 10, Fig. 10.6). **b** The relation is given graphically, each curve corresponding to the angle between the vertical cuts that show the true and apparent dips. **c** Outcrop of a plunging fold does not provide sample sites distributed on the more useful profile plane (see also Fig. 1.14). **d** Cut effect on interfacial angles: a triple junction between three grains only shows true 120° interfacial angles when viewed along their mutual line of contact. **e** General oblique views of a triple junction distort appearance of interfacial angles. **f** Planar fabric defined by a preferred orientation of objects produces a spurious orientation of the true foliation when viewed on any arbitrary plane not perpendicular to the foliation (see also stereographic example, Fig. 10.20b)

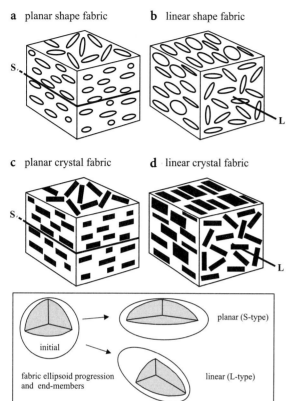

a planar shape fabric   **b** linear shape fabric

c planar crystal fabric   **d** linear crystal fabric

planar (S-type)

initial

fabric ellipsoid progression
and end-members

linear (L-type)

Fig. 1.18a–d. Three-dimensional orientation distributions dictate the appearance of a rock's fabric on differently oriented planes. Simple orientation distributions usually show orthorhombic symmetry varying anywhere from linear (*L*) fabrics to planar ones that show a foliation or schistosity (*S*). The L–S fabric scheme was designed to describe the degree of preferred orientation and the inferred shape of the strain ellipsoid (Flinn 1965). Depending on the orientation of the planar surface used for observation, the same specimen may appear to show a strong alignment of grains or none at all. This reminds us to be particularly careful in choosing the orientation in which to make thin sections or prepare cut surfaces

1.12
Interpreting Spatial Variation of Scalar Values from Samples

1.12.1
Contouring

A very common type of data in earth science yields some scalar measurement such as a concentration (wt%) or grain size, at various locations. The interpretation of spatial variation of data on or projected onto a plane surface such as a map, or for that matter on a graph, is simplified if discrete values are generalised in

some way to produce contours that represent hypothetical lines of equal value. Topographic maps show the most contours, known as topographic contours. The process of generalising on the basis of discrete, dispersed measurements should subdue spurious details due to measurement errors, location errors and even real physical differences that are too minor to be considered. The simplest contouring procedure involves interpolating the required contour value between a higher and a lower spot measurement. For example, consider the Thorr granite example where it is desired to produce a 2% colour-index contour. The sites are shown in Fig. 1.12a, and part of one manually constructed contour is shown in the SW corner of the granite. For the most part its construction was relatively simple as there were observations equal to the desired value (2%), through which sites the contour passed smoothly. However, in the space between two sites whose values were P and Q, one could linearly interpolate the position of the 2% contour. Q was slightly less than 2% and P was larger, so the 2% contour was proportionately closer to Q. If the distance between the two measurement sites was D, the 2% contour is located at a distance $D(2-Q)/(P-Q)$ from Q in the direction of P.

That simple approach is sometimes inadequate or it may be invalid, as in the case of a distribution of points over a sphere (Appendix II, Fig. AII-5). Instead, it is generally beneficial to blend information from adjacent sites, recording their average value at a location, such as R. For simplicity, we may select measurements within squares of a sampling grid, or within a mobile counting circle (both centred on R, Fig. 1.12a). The average of all the *n* observations in that cell is then considered more representative of that location. Contour lines may then be constructed as described in the previous paragraph using the average values in cells, rather than the original measurements. To refine the position of contours, we may add extra contour values by moving the grid or counting cell to any location. The pattern of contours becomes simpler and more diffuse as the grid sizes increase. This is comparable to increasing the coarseness of a filter, and it has the effect of emphasising the overall trend at the expense of detail. In some cases, the size of the counting cell is varied from site to site so as to include a constant fraction of the total sample (*n*). Thus, the sampling fraction, e.g. 0.1*n*, is constant for each cell, and the cell is defined as an area which includes the appropriate number of sites. This may be useful if site distribution is uneven, as it somewhat overcomes the problem of having overpopulated subareas contributing to certain counting positions. Some uranium concentrations in surface deposits from France (from Soyer, in De Vivo 1984) have

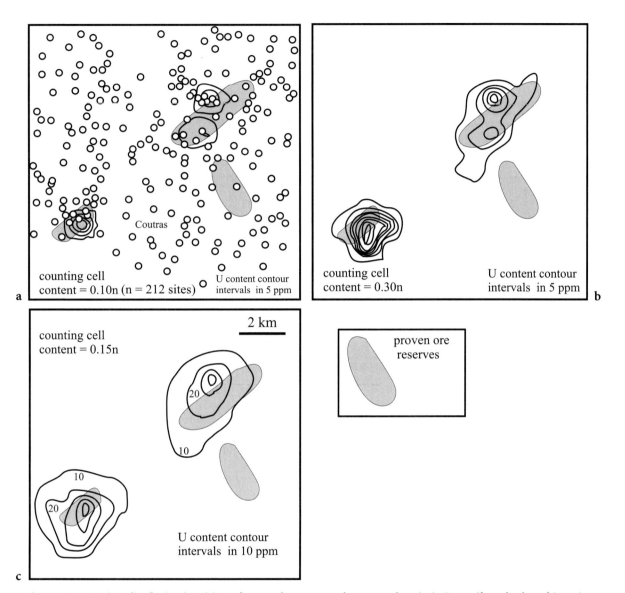

Fig. 1.19a–c. Uranium distribution (ppm) in surface samples near some known ore deposits in France (from the data of Soyer in De Vivo et al. 1984). Observations were contoured with counting cells that included 10, 15 and 30% of the adjacent sites, respectively (**a, b, c**). Contours in **b** may most closely match the subcrop ore reserves, but some offset is understandable due to topographic relief affecting the correspondence between surface samples and bedrock. (Contoured in SigmaPlot-SigmaStat)

been contoured in this way with counting cells containing adjacent data from 10, 15 and 30% of the total number of sites ($n = 212$). The sites are reasonably well distributed and the three different filtering techniques all identify contour peaks close to the subsequently proven underlying reserves (Fig. 1.19a–c). Counting cells using the adjacent 15% of sites seem to correspond closely with the ore subcrop. Some offset between subcrop and high surface concentrations may be attributable to hill creep.

Common sense tells us there must be a balance between filtering out too much detail and overgeneralising trends. The great strength of a contouring approach is that nothing is assumed about the actual distribution; it does not have to conform to any model or preconceived pattern. Contours are merely simplified graphic representations of the data that add objectivity to description, interpretation and qualitative analysis. With a computer program the spatial density of observations may be detected more sensitively. Instead of just counting observations falling within an arbitrary cell, each observation may be *weighted* according to its distance from the centre of the counting cell, at which the observation's frequency will be recorded. A common practice is to weight data inversely proportional to their distance from the centre of the cell, so that more distant sites have proportionately less influence. With all forms of contouring, care has to be taken with *contour edge effects* at the perimeter of the contoured region. Contours near the margins of the data-collection area become unreliable or difficult to interpret if they are constructed with some weighting proportional to neighbours. Having no neighbours artificially reduces the weighting of marginal sites. For that reason, contours should not be constructed too close to the edges of the data distribution, or the adjacent unsampled area could be filled with dummy values close to the local mean. Whereas contours are objective in the sense that they demand no preconceived mathematical form to which the data is fitted, in practice they may carry inherent weaknesses attributable to the process by which they are constructed. Principal limitations on their accuracy are controlled by the contouring mechanism, degree and method of filtering, weighting of data at a site, spacing of data sites and edge effects of the sampled area.

Without any statistical formality, intuition tells us that there must be uncertainty in the position of any contour. Consider a contour on a map representing 5%wt concentration of iron (Fe). Counting mean concentrations in certain cells across the map and then interpolating the 5% value, e.g. halfway between a cell with 7.5% and another with 2.5% fixes the position of the contour. Even in the absence of errors of measurement or errors in locating the position, we appreciate that we are dealing with a naturally variable phenomenon; the addition of one more site could change the mean value in the cells and thus change the position of the 5% contour. This uncertainty is formulated more easily in the following way. We have constructed a 5% contour, but uncertainties in the data mean that the contour at that location may have a deviation $\Delta = +0.3$, meaning that the actual value of the position of the contour is 5.3%. For reasons that will become apparent later (Chap. 4), the value assigned to any contour should have an associated uncertainty that is non-systematic and simply related to the number of observations in the counting cell (n) and their standard deviation (s). It can be shown that there is a 95% probability that the true value of the variable may differ from the contour estimate by as much as $\Delta = \pm (1.96\, s/\sqrt{(n)})$. Inspection of the formula shows that the best way to reduce the uncertainty is to increase n. Although this could be our motto in all aspects of processing observations, we are aware that this can only be achieved by finding more data sites within the area of the counting cell or by increasing the area of the counting cell. The former may be logistically impossible and the latter produces very generalised coarse contours that may remove too much interesting detail. Contouring is all about compromise. Cell counts are more representative and contours are somewhat easier to construct where the data sites cover the area with a uniform density. The contours should not claim to provide more detail than the observations. It is usually helpful to show observation sites as well as contours so that the reader can evaluate the contours for himself. For example, we should suspect over-interpretation if the contours are more closely spaced than the observations. Edge effects may be suspected where contours continue to the edge of the sampled area. Such contours may follow aimless paths particularly for low values. Edge effects may be minimised by disregarding contours near the boundaries of the sampled area. Similar unstable contours appear anywhere within the map where the contoured value is a near-zero value or where sites are particularly sparse. For example, suppose we wish to contour a map of the anisotropy parameter Tj (e.g. Chap. 11) which may range from -1 to $+1$. Obviously an area with few data will produce Tj contours ~ 0 that would imply a special kind of magnetic fabric for which $Tj = 0$. However, those near-zero contours would be meaningless artefacts due to paucity of data.

Computers have introduced considerable consistency and objectivity in contouring, and the procedure of contouring is the essence of objectivity; it does not assume any theoretical model to which the data should conform. Attempts at manual contouring may inadvertently introduce bias. Whitten (1963) showed that prior knowledge of the north-south tectonic grain of the Thorr granite predisposed the interpretation to yield slightly more N-S trending contours than when the contouring was performed without that knowledge (Fig. 1.12b).

Understanding spatial distributions is complicated further where the measurements are not scalars but require more than one value for their specification. This subject is rarely discussed in statistics texts, but is a very common situation in geology, geophysics, geography, oceanography and meteorology. Consider the variables in increasing steps of complexity for a geologist (Table 1.1).

First, the observation may be an *axis* or orientation, for example a fold axis or some other lineation in the rock. This requires two items for its specification; trend ($0° - 360°$ in map view) and plunge ($0° - 90°$), the latter

Table 1.1. Mathematical forms of measured physical properties

Name	Description	Examples
Scalar	Any simple value, with or without dimensions	Elemental abundance, grain-size, mass, length
Axis	Any non-polar line orientation	Fold hinges, normal to bedding
Unit vector or "direction"	Directed line (polarised) with no associated magnitude	Cross-beds giving paleo-wind direction
Vector	Direction associated with a magnitude	Velocity of glacier Wind velocity Paleomagnetic vectors (= remanent magnetism)
Second-rank tensor	Variation of a three-dimensional property at a point, summarised by three, orthogonal, principal axes, each associated with a separate magnitude. Described by an ellipsoid shape and orientation if magnitudes all of same sign	Strain Stress Electrical conductivity anisotropy Anisotropy of magnetic susceptibility (AMS) Anisotropy of anhysteretic remanence (AARM)

always being arbitrarily recorded as downwards (+ve). Either value could be contoured or one value could be contoured and "weighted" according to the value of the other. For example, if the plunge is very steep, the trend is very inaccurate; in the extreme case a vertical plunge may be assigned any trend from 0° to 360° with equal validity. Therefore, one might wish to assign a low weighting to the trends of axes with plunges >85°, reflecting their lower importance.

An example of a spatial distribution of structural orientations is shown in Fig. 1.20 (Gauthier and Borradaile, work in progress). In the eastern part of the Troodos ophiolite of Cyprus, a dike swarm that formerly fed the ocean floor is exposed. The shape and distribution of the magma chambers that fed the dikes are unknown but it may be inferred from the flow direction of magma in the dikes, which is inferred precisely, quickly but indirectly from the anisotropy of magnetic susceptibility (AMS). This procedure, in essence, determines the preferred orientation distribution (Chap. 10) of the flow-aligned minerals. Above a magma chamber the flow should be vertical (up); away from the chamber, the flow in the dikes should be lateral. Thus, the steepness of the flow alignment in the dikes should indicate the boundaries of the underlying fossil magma chambers. The mean orientation at a site is not simply calculated as an arithmetic mean (Chap. 2) but requires a more involved procedure that sums unit vectors (Chap. 9, Fig. 9.12). The same is true for the flow lineations determined from AMS (Fig. 1.20d, e). Furthermore, the mean orientation at any site is influenced by those at adjacent sites, weighted inversely with distance. Such procedures with orientation data are now routine (e.g. Spheristat software of Pangaea Scientific). As we increase the diameter of the counting cell, minor details of heterogeneous orientation of the dike swarm subside, almost as though we are looking at the distribu-

tion through fogged spectacles (Fig. 1.20a–c). However, the greater filtering also produces considerable smearing around the boundaries, essentially filling in the map where there were no observations (Fig. 1.20c). Some compromise must be reached on the basis of the researcher's judgement, the 5 km-diameter counting circle seems reasonable, so the corresponding flow axes are considered to be the most reasonable (Fig. 1.20b, d). The short lineation arrows in Fig. 1.20d represent steep flow axes, and the longer arrows represent less steeply inclined flow axes. One might conclude that the dikes were fed by small magma chambers, somewhat elongated north to south and ~4 km in the east-west dimension. The conventional contour map of the inclination values loses the visual impact of symbolic flow lineation arrows although it conveys the information more precisely (Fig. 1.20e).

Orientation data may be more complicated than the axial orientations discussed in the previous example. Complexity is incremented where the orientations possess polarity, such as paleocurrent direction. This would be recorded by a declination between 0°–360° and some inclination on the bed. However, an E-W trending scour mark must be assigned a direction of either west or east because currents have a *sense of direction*. This type of orientation-measurement is referred to as a *direction* or *unit vector*.

The next complexity brings us to the *vector* whose direction has an associated magnitude. Now three items must be considered, declination, inclination with polarity, and magnitude. We could contour any of the three values on the map and weigh them with either or both of the two remaining items, according to our needs. Examples include velocity of glacial ice, wind and paleomagnetic vectors.

The *second-rank tensor* is the most complicated variable that we shall encounter (Chap. 11). For example,

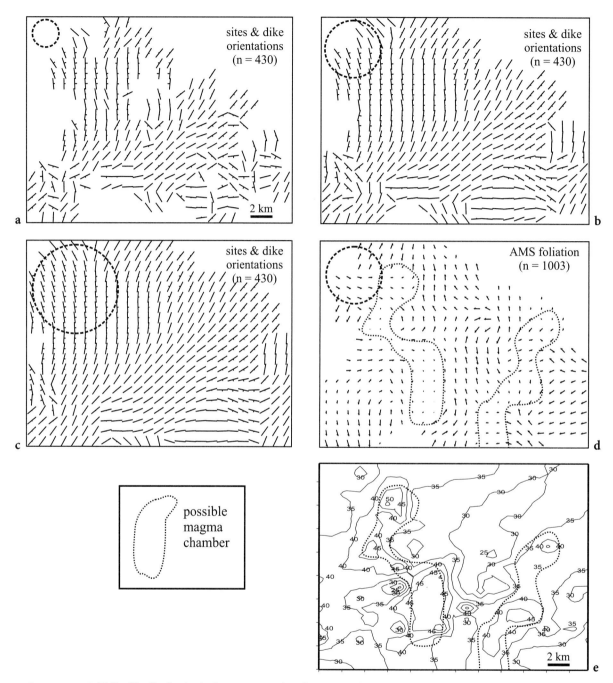

Fig. 1.20a–e. Ophiolite dike distribution in the eastern Troodos of Cyprus in which magmatic flow axes are inferred from anisotropy of magnetic susceptibility (AMS, Gauthier and Borradaile, in prep.). **a, b, c** Mean dike orientations and **d** mean flow lineations at any site are not calculated from arithmetic means. Since they are orientations, they must be averaged by the methods discussed in Chapter 10. **a–c** As the counting circle for averaging dike orientations is increased, local heterogeneities are progressively subdued. **c** Appears to overgeneralise the distribution and expand inferred orientations well beyond the mapped limits shown in **a**. Selecting **b** as the best compromise, the corresponding flow lineations are contoured in **d**. Steep lineations (*short arrows*) indicate vertical flow and may overlie and define the locations of the magma chambers that fed the dikes. *Long arrows* indicate gentle flow directions within dikes, peripheral to magma chambers. **e** Conventional contours of inclination of flow lineation. **a–d** were contoured in Spheristat software (Pangea Scientific), **e** was contoured in SigmaStat-SigmaPlot

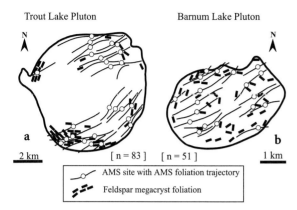

Trout Lake Pluton

Barnum Lake Pluton

2 km   [n = 83]   [n = 51]   1 km

AMS site with AMS foliation trajectory

Feldspar megacryst foliation

Fig. 1.21 a, b. A study of anisotropy of magnetic susceptibility (*AMS*) in two adjacent granite plutons reveals cryptic preferred grain orientations (Borradaile and Kehlenbeck 1994). However, spatial interpretation is complicated since the distribution of sites is uneven. Still worse, the information at each site is described by a tensor (see Chap. 11), represented by an ellipsoid whose three axes define the magnitudes and orientations of the three principal susceptibilities. Unfortunately, little of the AMS information is actually used on maps. Typically, as here, authors only plot the orientation of the maximum-intermediate plane of the AMS ellipsoid, known as the magnetic foliation. Here, it is compared with the alignment of feldspar megacrysts.

stress, strain, petrofabric and petrophysical measurements may be described by second-rank tensors, commonly conceptualised by the shape and orientation of their magnitude ellipsoid. (However, not all second-rank tensors may be visualised in that way.) The orientations of the axes and their dimensions represent three principal values and principal orientations of some physical property, such as strain. Therefore, at each site we have six possible numbers to consider. Which should be contoured, which others should be used to weigh the representative number that will be counted in a cell, and then averaged with other values to construct a contour? In mapped distributions of tensors, usually most of the original information from each measurement is ignored. For example, the cryptic orientations of minerals in rocks may be determined very easily from their anisotropy of magnetic susceptibility (AMS). This is described by a tensor, but most maps present only limited information, such as the mean intensity (arithmetic mean of the three principal values), the orientation of the maximum axis or the orientation of the plane containing the maximum and intermediate axes (Fig. 1.21).

1.12.2
Trend-Surface Analysis

Although it is not used as much presently, *trend-surface analysis* is a useful way to characterise spatial dis-

tributions. Specialised computer programs are marketed, but general statistical packages and some spreadsheets permit basic trend-surface analysis under the guise of surface fitting in three dimensions. Contouring goes out from a completely objective approach, making no assumptions about the spatial distribution of data; it is comparable to a non-parametric approach in formal statistical analysis. On the other hand, trend-surface analysis assumes some model for the generalized distribution of measurements (z) in (x, y) coordinates. For simplicity, we will only consider measurements that are scalar values (z). From an initial inspection of the data or some reasonable theoretical inference, the simplest mathematical model is assumed for a surface that would approximately fit the values of z at each point. Usually a polynomial surface is chosen and the most reasonable model is considered to be the simplest polynomial with the best fit. Trend-surface analysis benefits from some understanding of the controls on the variable z, based in subject-specific knowledge. For example, if a geomorphologist has elevation measurements at various points, and has good reason to suspect that a tilted peneplain is present, he could characterise that by fitting the simplest possible trend surface, a plane. A plane is considered a first-order trend surface because the observed value (z) is dependent only on constants and on first-order terms in x and y, e.g.

$$z = a + b_1 x + b_2 y \qquad (1.7)$$

The next most complex surface is a quadratic surface in which the observation depends on first- and second-order terms of the map coordinates (x, y):

$$z = a + b_1 x + b_2 y + c_1 x^2 + c_2 y^2 \qquad (1.8)$$

Commonly, useful trend surfaces in the earth sciences range from third ($m = 3$) to sixth ($m = 6$) order:

$$z = a + b_1 x + b_2 y + c_1 x^2 + c_2 y^2 \dots$$
$$\dots + m_1 x^m + m_2 y^m \quad \text{or} \quad z = T^m(x, y) \qquad (1.9)$$

Fitting such an equation to the data set uses a regression technique and is necessarily a computer-based technique that selects the best surface, for which the observed magnitudes (z_0) show the smallest absolute differences with z at that site. This is a regression technique that minimises the differences ($z_0 - z$) for the whole data set (Chap. 6). Because the deviations may be positive or negative, we minimise the squared sum of their differences, ($z_0 - z$)². The polynomial equation is the most popular equation to fit, and computer software readily permits the degree n to be varied, although other functions may be used. Polynomial trend surfaces of higher than sixth-order polynomials are rarely used as they rarely introduce the desired generalisation and simplification.

It is rarely one's intention to exactly match the data set with a surface, nor is it implied that the polynomial equation actually relates the observations to some causative process. The trend surface is merely a descriptive tool that will clarify trends and isolate outliers from the trend that may identify important localities. Thus, trend surfaces should appear simpler than raw data that is contoured empirically. However, this depends somewhat on the contouring technique (cell size, weighting, contour interval) versus the assumed trend-surface complexity given by the highest index m of the polynomial function.

Uniformly or randomly spaced sites make it easier to construct trend surfaces; clusters of information in small areas give undue emphasis to some local detail. For example, in economic geology, ore assays may be sparsely scattered over a region but may be very common near a potentially extractable ore deposit. This non-random spatial distribution of observations may be more amenable to trend-surface analysis if data clusters are replaced with averaged values.

As we increase the order of the trend-surface polynomial (m) towards the number of data points, the trend-surface becomes less of a generalisation and more like a mathematical specification of the original data set without any simplification. To make the necessary mathematical regression procedure work, the number of data points (n) must exceed the degree of the polynomial (m). Moreover, to produce any sensible trends we must have $n \gg m$ as we shall see below. Usually $m \leq 6$ for the surface can be simply interpreted.

Clearly, judgement must be used to decide what order of polynomial surface to fit in the reasonable range $m = 1$ to 6. This should be guided by some geological intuition of the complexity of the underlying process. Normally, most fundamental geological processes produce trend-surface distributions that require low-order polynomials for their sensible description. In geomorphology, a tilted peneplain, plateau or wave-cut bench would be accurately represented by the simplest trend surface which is a plane ($m = 1$). Most geological phenomena requiring representation in the form of a surface can be successfully generalised without undue loss of detail by a polynomial of sixth order or less. For example, petrological variations in magmatic rocks and the permeability of aquifers may have satisfactory trend surfaces of second to fourth order. Although coefficients for higher order terms (c, d, e) become smaller as the index (m) of the variable increases, the higher order terms still greatly affect the trend surfaces due to their larger power (m).

A simple example of trend-surface analysis uses the Thorr granite colour-index variation that was contoured in Fig. 1.12. Colour variation is a useful field guide to petrological variations in homogeneous igneous rocks, but sparse outcrops make it difficult to determine meaningful trends from contours. The same data are modelled by trend surfaces in Fig. 1.22 (Whitten 1963). The simplest surface in (a) is a plane, indicating values decreasing steadily in a direction slightly north of east. This is referred to as a *linear fit* because the form of the equation is linear, or first order, meaning that there are no indices $m > 1$. A higher order surface provides a closer fit to the data, for example a quadratic surface

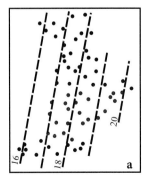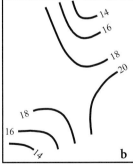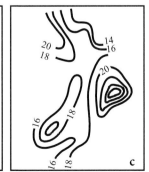

Fig. 1.22. Colour index is a field guide to subtle petrological features. This example is from the Thorr granite, Donegal (Whitten 1963). Simple map contours of the colour index were shown in Fig. 1.12. Trend-surface analysis assumes that the variation follows some mathematically predictable form. The mapped values may approximately fit some polynomial surface that may be determined by trial-and-error. Here are shown: **a** planar surface, or first-order polynomial, fitted to data; **b** quadratic surface; **c** sixth-order polynomial surface. The higher the polynomial order, the better the hypothetical surface fits the observations, but the less successful it is in generalising the data set for the purposes of description and prediction. In earth sciences, polynomial surfaces of order higher than five are rarely useful

($m = 2$, Fig. 1.22b) or a cubic surface ($m = 3$). Such functions provide simple relief surfaces with culminations and depressions that conform more closely to the actual data. In this example, the quadratic surface yields a trough-shaped distribution. Successively more complicated surfaces fit the data better at the loss of simplification of the entire data set. Usually, a sixth-order surface is the highest that one would consider on practical grounds, balancing detail against generalisation. This is shown in Fig. 1.22c. In the case of the Thorr granite, it preserves the depressed values in the colour-index trough but also acknowledges the peak values in the eastern part of the area that were obscured by the simpler, lower order surfaces. Choosing the appropriate order for the polynomial surface is a matter of judgement and greatly affects the appearance of the trend surface and any subsequent interpretation or description. Computer programs make it too easy to produce a high-order surface that obscures the trend. An example of an over-fitted trend surface is found in the spatial distribution of undesirable sulphur and ash concentrations in a coal seam (Gomez and Hazen 1970). The authors' trend surface involved 26 coefficients with cosine, sine, exponential, quadratic and cubic functions of the variables. Such detail hardly isolates any trend. As one worker noted, *"insisting on having uncorrelated errors means that the function has to twist and turn a lot"* (Armstrong 1998). In other words, although a surface may be found to fit observations remarkably well, it will fail to provide broad generalisations or simplifications of the spatial variation from which some useful conclusions might more easily be drawn.

When an appropriate trend surface is defined, its polynomial equation predicts z at any location, but that differs from the observed value (z_0) by some small local component, ε. That may be considered as a deviation or "error" from the trend. The observed value (z_0) is related to the trend-surface function of its location (x, y) and the local deviation (ε) by:

$$z_0 = T^m(x, y) + \varepsilon \qquad (1.10)$$

Here, m is the highest order involved in the trend-surface polynomial. If our attempt was to produce a trend surface that provided a faithful mathematical description of the surface, the terms ε should be randomly distributed with respect to the trend surface like any measurement error. They should follow the Normal distribution (Chap. 3). Such a surface is useful, if it is our goal to provide a surface that may be used to predict a value (z_0) at some unsampled location (x, y). However, commonly we require a simpler trend surface to isolate a background, larger scale variation. Its subtraction from the raw data

may then yield local anomalies, termed residuals, due to interesting local processes. This is an exploration philosophy, in which local deviations (ε) from the trend surface contain useful information about a meaningful departure from the trend. For example, one may isolate an ore deposit, hydrocarbon reservoir or, as in the next example, a sedimentary basin. The residuals may therefore be regarded as valuable information, not merely spurious normally distributed errors from a perfect-fitting surface. A common procedure is to determine the trend surface using the first term of this expression. This isolates a general polynomial trend. That surface is then subtracted from the actual observations to yield a separate map of residuals (ε), which correspond to anomalies that may have some natural significance.

This is illustrated with a study of the relief of the basement of a sedimentary basin just east of Bratislava, CZ (Merriam and Robinson 1970) in Fig. 1.23. Progressively higher order surfaces are more complex and imitate the original data set more closely. They still identify what may be broadly regarded as trends, yielded by the first term of the previous expression. In such a study, the polynomial is neither intended to provide a mathematical description of the basement surface nor to have any physical significance. The trend surface's use is apparent when it is subtracted from the original observations to yield residuals (ε). For example, if we believe that an overall regional tilt modified the subcrop topography, the planar surface (linear fit) may be a useful trend surface. Subtracting that from the actual observations produces the map of residuals, which locates the potentially important sedimentary basins more readily (Fig. 1.24).

Selecting the appropriate degree (m) for the trend surface requires subject-specific knowledge and some common sense. It is possible to refine the decision by calculating whether a certain order of polynomial fits better than another. Unreasonably increasing m merely produces an intimidating but useless trend surface that replicates the observations. For a meaningful trend surface $n \gg m$ and the geographical distribution of the n observations should be reasonably uniform if the surface is to be representative. To compare the significance of a selected polynomial trend surface with another, it is possible to apply a simple statistical test (F-test, Chap. 4). Basically, this compares the deviation of the observations from the value predicted by the polynomial, using a ratio of variances. The F-statistic can help us to decide, at an appropriate confidence level, whether the trend surface should be rejected or not rejected as a reasonable model for the observations. For future reference, the interpretation of F depends on its degrees of freedom ($v_1 = 1$ and $v_2 = (n - 1) - m$). To achieve any F-value likely to produce acceptable agreement between the trend surface and the

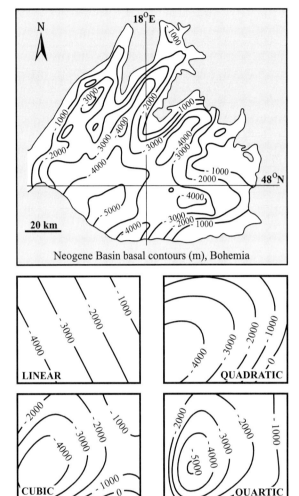

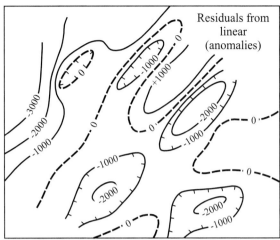

Fig. 1.24. The difference between the actual depths of the basement and the depths predicted by the planar trend surface yields the residuals shown here. These define anomalies that isolate local sedimentary basins. It was not assumed that the linear-trend surface of Fig. 1.23 was a valid model of any theoretical process, but rather was a contouring tool that helped isolate the anomalies

Fig. 1.23. *Top* Depths in metres to the basement of a Neogene sedimentary basin in Bohemia (Merriam and Robinson 1970). *Below* are shown their best-fitting polynomial trend surfaces of first, second, third and fourth order. The first-order polynomial describes a plane surface. As it is described by a linear equation, it is sometimes known as a linear-trend surface which is somewhat confusing. Low-order polynomials emphasise the overall trend, whereas high-order polynomial surfaces characterise detail. The choice of polynomial surface depends on the goals of the investigation, as shown by the example in Fig. 1.24

data, we require $v_2 \geqslant 30$. Given that the highest degree polynomial we should apply is $m = 6$, we see that the minimum $n \geq 40$. Given the further constraints of the edge effect, problems of clustered observations, and possible errors of position and measurement, all of which distort ε, we should really only attempt trend-surface analysis with much larger n.

We have seen that there are really two common applications of trend-surface analysis that may not always be explicitly distinguished. The most practical use is fitting some surface to natural observations (abundance of a mineral, depth to a stratigraphic horizon, etc.). In this case, the equation for the trend surface is not intended to describe the effects of any geological process that accounts for the observations. However, the differences between the trend-surface prediction and the actual observations produce *residuals*. These anomalies may correspond to the location of geochemical anomalies that may reveal ore deposits, or depth anomalies may locate sedimentary basins.

The alternative use of trend-surface analysis is to identify some underlying physical control of a natural process. For example, the application to the distribution of colour values in a granite pluton is due to compositional variation (Fig. 1.22). In that case, the trend surfaces identify some spatial variation that has some overall physical significance, perhaps varying degrees of assimilation and solid-state diffusion or even primary magmatic processes, such as fractional crystallisation, magma mixing and flow. The mathematical description of the surface is in itself useful but might have some interpretive and predictive value.

Polynomial surfaces are smooth approximations to the data points fixed in three dimensions. They necessarily shorten the arcs connecting data points. In certain situations this may be unreasonable. For example, if the trend surface is fitted to a basement relief that is faulted, the trend surface will subdue or conceal the expression of the fault. It is difficult for polynomial trend surfaces to model the features that require a step in the surface. That may be accommodated in an entirely different approach using the *Fast Fourier Transform (FFT)*. Complex surfaces may be described by summing many trigonometric waveforms of different wavelength, amplitude and phase, and will be discussed in connection with time series in Chapter 8. Fourier summation of regular waveforms can provide a precise description of any complex variation, even steps like fault scarps but it will only be valid as long as the form of the investigated data may be decomposed into regular waveforms. The FFT model can fit the observations as closely as desired. It could also be used to designate a trend from which actual observations may be subtracted to estimate the local anomalies as residuals. In the next Chapter, it will be shown that spatial variations may be studied in a quite different manner called *kriging*. Kriging does not assume a form for the surface that describes the variation; instead it only uses the actual mapped observations directly. It requires an understanding of variance, so discussion is postponed until the end of the next chapter.

CHAPTER 2

Central Tendency and Dispersion

The most commonly required characterisation of any sample size of measurements is a statistic representing a value that usually occurs commonly, and in the centre part of the range of observations. The *arithmetic mean* is most important, but other important central values will be described. Conversely, the second most useful statistic describes the degree to which observations are scattered. *Variance*, the square of standard deviation, is the most powerful descriptor of dispersion. The following summary explains these concepts in earth-science contexts. Descriptive measures like mean and variance may be calculated for any sample of measurements, without any theoretical knowledge of the population from which the sample is drawn. However, the nature of the populations may affect their usefulness. In this chapter, through Chapter 4, it is important to note that we deal with observations that are *scalars*, measurements encapsulated in a single quantity. Quantities like 2.1 m, 4 kg, and dimensionless values like 0.03 and 7% are scalars. More sophisticated treatment is needed in order to describe central tendency or dispersion of orientations, or of quantities that include orientation in their specification, like vectors (e.g., velocity) or second-rank tensors (e.g., stress, strain). Those issues are addressed in Chapters 9, 10 and 11. William Borough introduced the concept of an average value in 1581, although it was applied to compass readings which, ironically, strictly require a different treatment as they are orientations, not scalars (Chap. 9).

2.1
Central Tendency: Mean, Median and Mode

Measures of *central tendency* are the most simple statistics; they are single values that usefully characterise a distribution of measurements. They may be calculated for any distribution but are representative and most useful if the frequency distribution of observations is hump-shaped. In other words, the observations most commonly cluster around a certain unique value, which is most popular, although exceptions will be discussed (e.g., Figs. 2.1f, 2.2, 2.3). This humped distribution is graphed with the measured variable as the horizontal axis (x) and the number of observations with that value as the vertical axis (y). The vertical axis is usually labelled as frequency (n), indicating the number of measurements at x, or in a small range of x. The frequency may also be plotted as a normalized value in percent or as a fraction of unity, facilitating comparisons between data sets and also with theoretical frequency distributions (e.g., Chap. 5). Theoretical distributions do not describe a finite number of measurements but rather the abstract mathematical form that would be taken by all possible measurements, without practical limitation. In that case, the vertical axis of a theoretical distribution represents the probability of x, $P(x)$, occurring in a small interval of x. $P(x)$ ranges from 0 to 1.0. The unit area under the theoretical distribution curve represents 100% of the observations, the total density of the probability function, $P(x)$. This point may be better appreciated after inspecting Figs. 2.4 and 2.5.

Frequency is the number of observations within a small interval of x, giving the height of the distribution at that x-value. If the x-interval is finite, it may be called a bin or a class. Values of frequency, plotted on the y-axis against the corresponding x-values, constitute a frequency-distribution graph. If the frequency distribution has one prominent peak, indicating the most commonly occurring x-values, it is said to be unimodal or hump-shaped. Many natural phenomena are recorded as measurements that have humped frequency distributions. For example, multiple measurements of a single object will show variation on either side of the true value, caused by small, random measurement errors. The measurements tend toward a single value, located centrally in the range of x-values. However, sometimes two processes interfere causing the measurements to cluster around more than one popular value. The multiple peaks of an X-ray diffraction experiment provide a common example (Fig. 2.1b). Such multimodal distributions

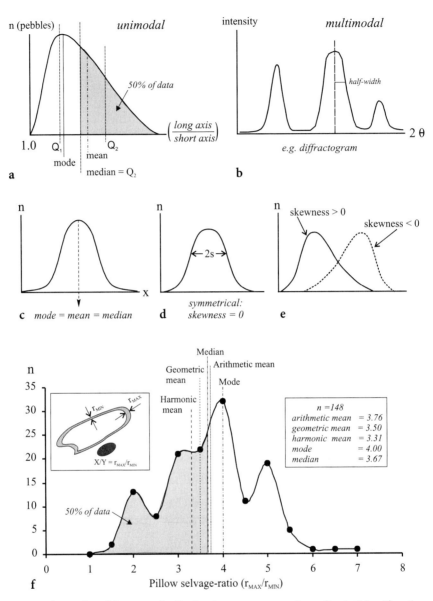

Fig. 2.1. a An asymmetric, hump-shaped frequency distribution. Asymmetry causes the median (= Q_2), arithmetic mean and mode to differ. Q_1, Q_2 and Q_3 are the first, second and third quartiles. 25% of the observations have values < Q_1; 50% have values < Q_2 and 75% have values < Q_3. **b** Multimodal distribution (several peaks), $x_{1/2}$ is the half-width of a peak at half its height, a simple measure of its dispersion. **c** Symmetrical hump-shaped distribution showing equivalence of mean, mode and median, as for example, the Normal distribution. **d** Dispersion of a sample following the Normal distribution is represented by sample standard deviation (s); σ would represent the standard deviation of the population, i.e. the parent distribution from which the sample is drawn. **e** Asymmetric or skewed distributions. **f** Ratios of tectonically deformed pillow selvages; in the undeformed state the ratio = 1.0, but after tectonic strain the selvage ratio (r_{MAX}/r_{MIN}) may be closely related to the finite strain ellipse ratio (X/Y) which is of value to structural geologists. The arithmetic, geometric and harmonic means are given; the latter may be preferred as a measure of central tendency in this situation

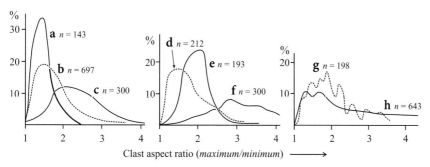

Fig. 2.2. Examples of frequency distributions of pebble aspect ratio from Borradaile (1987). **a** Pebbles in slumped conglomerate, Texas. **b** Pebbles in monomict Triassic conglomerate, Spain. **c** Modern beach pebbles, Oamaru, NZ. **d** Re-sedimented pebbles, Italian Flysch. **e** Accretionary lapilli in Proterozoic volcanic tuff, northern Ontario. Multi-modal shape distributions are shown by pebbles whose shapes are influenced by sedimentary abrasion as well as another competing process: **f** modern beach pebbles of jointed and schistose material, Waianakarua, NZ; **g** tectonically deformed quartz sandstone pebbles, Loch Awe, Scotland; **h** tectonically deformed polymict pebbles, Seine River, northern Ontario

are difficult to treat unless the data are deconvoluted to isolate the response of different effects, each of which may be uniquely associated with one of the peaks. A graphic example of deconvolution is given in Chapter 8 (Fig. 8.9). In the case of the diffractogram, the vertical axis would be labelled "intensity" as there is no simple record of counts directly. Thus, a subject-specific substitution for "frequency" is permissible and does not prevent the analysis of frequency distributions. However, for concepts involving probabilities, frequency must be standardised so that the area under the curve is unity. This is achieved by plotting, as y, the fraction of the values that have a certain value x. This ensures that the area under the curve represents 100% of the sample. Any fraction of the area is then proportional to a percentage of observations (Fig. 2.4c).

Although the frequency-distribution of measurements is humped, it may not be symmetrical about the most common value (Fig. 2.1). For humped distributions with minor asymmetry, the three common descriptors of central tendency, mean ($\bar{x}$), mode and median (Q_2) are approximately related by the following simple, empirical approximation, for which there is some theoretical support.

$$\frac{\bar{x} - \text{mode}}{\bar{x} - Q_2} \approx 3 \qquad (2.1)$$

The influence of skewness on the relationships among mean, median and mode for unimodal distributions is shown in Fig. 2.1a, and a multimodal distribution is shown in Fig. 2.1f. Where the tendency towards the central value is unbalanced for larger and smaller measurements, the distribution is said to be skewed. Skewness is positive if the tail is longer to the right, negative if it is to the left (Fig. 2.1e). Some basic definitions follow.

2.1.1
Arithmetic Mean

Central tendency is mostly recorded by the arithmetic mean, obtained by adding the values of all observations (x_i from i = 1 to n) and dividing the sum by the number of observations (n). When people speak of a mean value, they invariably imply an arithmetic mean. This mean is very sensitive to even one or two extreme values, the so-called *outliers*, which pull the arithmetic mean towards them. For this reason, other measures of central tendency may be preferred for specific purposes and they are described below.

The formula for the arithmetic mean introduces us to the *summation notation*. We denote each observa-

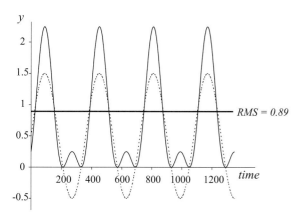

Fig. 2.3. The Root Mean Square (*RMS*) is a useful representative value for the magnitude of a variable that varies in sign. The original variable (*dashed curve*) is squared (*smooth curve*) yielding alternating large and small peaks. The *horizontal line* shows the root of the variable's mean squared value (RMS value = 0.89)

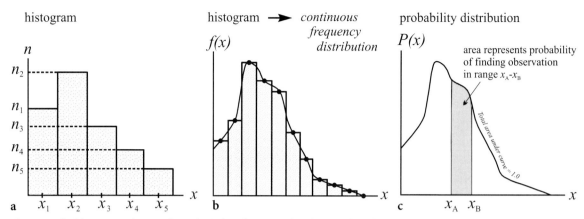

Fig. 2.4. a The histogram or bar graph as a basis for a frequency distribution of x-values. Each column's height represents the number of observations in the corresponding x-interval. **b** A continuous frequency distribution formed by joining the midpoints of histogram columns. **c** A probability distribution with the y-axis scaled so that it gives the probability of finding a corresponding x-value. The fractional area under the curve between two values x_A and x_B gives the probability of finding a value in that range

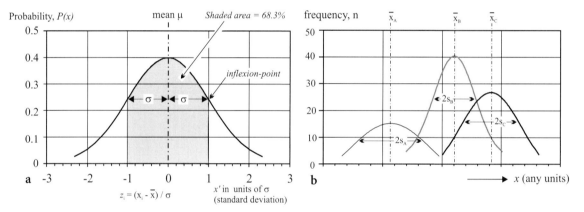

Fig. 2.5. a The Standardised Normal Distribution with a mean of zero and standard deviation (σ) of unity. Note the inflexion points where the curve changes from convex-up to convex-down are located at a distance σ from the mean, thus permitting an estimate of σ from the frequency distribution of any normally distributed variable. A sample of x-measurements may be standardised into a universally recognisable Normal distribution, by replacing x_i with a dummy variable $z_i = (x - \mu)/\sigma$. If the sample was randomly drawn from a population of x that is normally distributed, the frequency distribution of z will perfectly match the shown Normal distribution. **b** Samples (A, B, C) with different means ($\bar{x}$) and standard deviations (s). Standardisation can transform them to the form shown in **a**, for direct comparison with the theoretical Normal distribution

tion as x_i where i is an integer index that identifies the value in the sequence measured. The shorthand Σ symbol (Greek uppercase S to abbreviate Sum) indicates the sum of several values. In the following definition of the arithmetic mean, the sum is of n values of x_i. That is divided by n to give the arithmetic mean:

$$\bar{x} = \frac{\sum_{i=1}^{n} x_i}{n} \qquad (2.2)$$

Summations may become more complicated, and the following three simple examples illustrate *summation*

theorems that may be useful, especially when performing the calculations by computers in program loops. k is a constant and x, y and z are variables counted from $i = 1$ to $i = n$.

$$(1) \quad \sum_{i=1}^{n} k = nk$$

$$(2) \quad \sum_{i=1}^{n} kx_i = k \sum_{i=1}^{n} x_i \qquad (2.3)$$

$$(3) \quad \sum_{i=1}^{n} (x_i + y_i + z_i) = \sum_{i=1}^{n} x_i + \sum_{i=1}^{n} y_i + \sum_{i=1}^{n} z_i$$

The examples may appear trivial but their application is useful to simplify the summation of expressions, e.g.,

$$\sum_{i=1}^{n}(3x_i^2 + 2x_i + 7) = 3\sum_{i-1}^{n}x_i^2 + 2\sum_{i-1}^{n}x_i + 7n \qquad (2.4)$$

2.1.2
Median and Mode

The *median* is the middle x-value when all values are listed in ascending order (Fig. 2.1a). Thus, there are equal numbers of observations above and below that value. It is also known as the second or middle quartile, Q_2. Half of the measurements exceed Q_2 in value, and half are less than Q_2. The *mode* is the x-value of the peak of the unimodal distribution, corresponding to the peak frequency (n). It is the value of x that occurs most frequently. For a perfectly symmetrical unimodal distribution, the mode = mean ($\bar{x}$) = median (Q_2). The Normal distribution is one example of a distribution with this property (Fig. 2.5). In a positively skewed distribution the mode will be less than the mean and the median, in a negatively skewed distribution the mode is the largest of the three statistics.

The arithmetic mean ($\bar{x}$) of the following sequence of nine observations is compared with its median (Q_2) and mode. The value (27) may be considered an unlikely value, in a statistically unbiassed sample. Such an *outlier* might be omitted, leaving a more symmetric hump-shaped distribution of values in the range (5…11).

(5, 6, 7, 7, 8, 9,	Mean	Median	Mode	($n = 9$)
10, 11, 27)	= 10	$Q_2 = 8$	= 7	

On omitting the outlying value, (27), the measures of central tendency change to:

(5, 6, 7, 7, 8, 9,	Mean	Median	Mode	($n = 8$)
10, 11)	= 7.88	$Q_2 = 7.5$	= 7	

Note that the mean is sensitive to outliers, being strongly attracted toward an outlier and away from the mode.

The median is one of four *quartiles* that divide the distribution into four equal numbers of observations. 25% of the observations have values less than Q_1, 50% have values less than Q_2 (= *median*) and 75% have values less than Q_3 (Fig. 2.1a). Quartiles are just specific, popular *percentiles*. The pth percentile is a number that exceeds the values of p% of the observations; (100-p)% of the observations have values larger than this number.

2.1.3
Weighted Mean

For specific purposes in a study, one may wish to change the influence that certain observations have on a mean value. Giving them more or less *weight* will change their importance. The importance of an observation is changed by multiplying its value x_i by a certain weight-factor w_i. Then, ($w_i x_i$) replaces x_i in the equation used to calculate the mean. Note that w_i may be different for every observation and may be determined by a complicated formula related to the value or some other attribute of x. A value is decreased in importance if $w_i < 1.0$, and increased in weight if $w_i > 1.0$. Weighting may be applied in calculating the arithmetic mean as well as other kinds of average value for scalars and also for more complex variables like vectors and tensors. Examples of reasons for weighting observations could include the following:

1. reducing the contribution made by values with larger errors of measurement ($w_i < 1.0$)
2. increasing the importance of values within a certain range ($w_i > 1.0$)
3. increasing the role played by observations from rocks of a given formation or age ($w_i > 1.0$)

It may be realized that multiplying observations by weight factors has the same effect as changing the numbers of observations. This is not sensible, so we design the weights to sum to unity. Alternatively, when working with an arithmetic mean, we may simply divide the sum of weighted observations by the sum of their weights:

$$\text{weighted arithmetic mean} \quad \bar{x}_W = \frac{\sum_{i=1}^{n}w_i x_i}{\sum_{i=1}^{n}w_i} \qquad (2.5)$$

2.2
Other Averages

The arithmetic mean described above is intuitive and most familiar. It is the simplest characteristic of *central tendency* for any frequency distribution of measurements, but it is most meaningful if the distribution is humped. Subsequently, "mean" refers to arithmetic mean unless otherwise stated. It is so familiar that we sometimes overlook other equally valid ways of characterising a sample with one representative central value. Although the arithmetic mean is the most universally applicable representation, there are several branches of science, engineering and technology where there are also different single-value descriptors of a distribution's central tendency. In some cases, the arithmetic mean may be misleading or even incorrect.

For example, the *geometric mean* is an expression of central tendency given by

$$\text{for}\quad x_i > 0 \qquad G = \sqrt[n]{x_1 \cdot x_2 \cdot x_3 \ldots x_n}$$

$$\text{or}\quad G = \sqrt[n]{\left[\sum_{i=1}^{n}(x_i)\right]} \tag{2.6}$$

$$\text{or}\quad \log_{10} G = \frac{1}{n}\sum_{i=1}^{n}\log_{10}(x_i)$$

It is not quite as intuitive as the arithmetic mean but may be very helpful in certain situations dealing with positive, non-zero values such as shape, aspect or concentration *ratios*. For example, the relative lengths of axes, or aspect ratios, describe grain shapes. Where the maximum, intermediate and minimum diameters of an object are *a*, *b* and *c* respectively, the most representative diameter is given by their geometric mean, $(a.b.c)^{1/3}$. More complicated examples occur in strain analysis where structural geologists study the change of shape of certain "strain markers", usually described by an aspect ratio. For example, Robin (1977) addressed the problem of determining the strain of such objects, using aspect ratios of their outlines, even where these are of irregular shape. He showed that their orthogonal dimensions, parallel to arbitrary coordinate axes, define a ratio that is close to unity for weakly strained objects. The objects are strained, their shape ratios are >1 and their geometric mean approximates the actual strain ratio. This simple method may be remarkably accurate for studying the strain of objects that defy a solution by other methods. A geometric mean is usually meaningless if any measurement is zero or negative, and it is sensitive to small variations of small values; this explains its effectiveness with shape ratios that always exceed unity but which are small numbers.

A still less intuitive expression of a central tendency is given by the *harmonic mean*. It is suitable where the values are positive and non-zero, and it finds useful applications where the quantities are expressed as reciprocals. It is defined as:

$$H = \frac{n}{\displaystyle\sum_{i=1}^{n}\frac{1}{x_i}} = \frac{n}{\dfrac{1}{x_1}+\dfrac{1}{x_2}+\dfrac{1}{x_3}+\ldots+\dfrac{1}{x_n}}$$
$$\text{for}\quad x_i > 0 \tag{2.7}$$

A geological application is found in structural geology, again in the field of strain analysis. Lisle (1977) showed that the harmonic mean of the shape of passively strained ellipsoidal objects is a reasonable approximation to their strain-ellipse ratio. It may be very accurate where natural structures satisfy the passive-behaviour criterion, where there is little contrast in physical properties and the shape of the marker represents the shape change of the rock, e.g., ooids in limestone, reduction spots in shale, or a sandstone pebble in a sandstone matrix. This is an especially useful, simple device in structural geology because a hand calculator may provide an estimate of strain in the field from as few as 30 measurements of deformed objects.

A common application of the harmonic mean is to determine an average magnitude for velocity (i.e. average speed) where the speed varies over different distances. We must consider the average speed only in terms of total distance travelled and the total time required. For example, if a glacier moves 1 km at 6 km/a and over another kilometre at 3 km/a, the average speed is given by:

$$\frac{\text{total distance (km)}}{\text{total time (years)}} = \frac{2}{\dfrac{1}{6}+\dfrac{1}{3}} = 4 \ \text{km/a} \tag{2.8}$$

This is the harmonic mean. Because different distances are travelled in the same reference time unit, the familiar arithmetic mean (4.5 km/a) would incorrectly represent the average speed and the harmonic mean must be used. The relationship among the arithmetic, geometric and harmonic means is:

$$\bar{x} \geq G \geq H \tag{2.9}$$

Both G and H are sensitive to small values and meaningless if a zero value occurs. An example of the three types of mean is shown for a study of tectonically deformed pillows (Fig. 2.1f). Undeformed pillows have selvages of constant thickness. Tectonic deformation produces maximum and minimum rim thicknesses whose ratio may define the ratio of the finite strain ellipse (Borradaile and Poulsen 1981; Borradaile 1982). For those data, the three common types of mean are shown in Fig. 2.1f, but which is most representative? We are dealing with a ratio, so that one may suspect that the geometric or harmonic mean could be useful. Moreover, subject-specific knowledge shows that strain markers are unreliable for large strain-ellipse ratios, as heterogeneous strain occurs (the selvages vary irregularly in thickness). Moreover, interesting subtle differences occur between markers at low strain (i.e. $X/Y \rightarrow 1$). These are further good reasons to favour the geometric or harmonic mean, as they are sensitive to small values and to small variations.

The *Root Mean Square (RMS)* or quadratic mean is an average that is useful where observations vary in sign but for which an estimate of the average magnitude is desired. Its most familiar use is in the study of waveforms such as alternating electrical currents, where the magnitude (y) varies with time (t). It is defined as:

$$RMS = \sqrt{\overline{y^2}} = \sqrt{\frac{\sum_{i=1}^{n} y_i^2}{n}} \qquad (2.10)$$

This is shown in Fig. 2.3 where the variation with time is based on the sine function. However, the "average magnitude" of any natural, regular waveform that produces variation on either side of zero may be represented readily in this manner. This is very useful for estimating the goodness-of-fit of a line or plane in regression or trend-surface analysis where the deviations from the fitted function, called residuals, may take on either positive or negative values (e. g., Chap. 6). A specific direct use of this average could be applied to the cyclical deviations of magnetic declination due to the periodic secular variation of the geomagnetic field. The RMS value then informs us of the average *magnitude* of its deviation regardless of its sign with respect to a "zero" reference, easterly declinations being positive and westerly declinations being negative.

2.3
Histograms and Frequency Distributions

A histogram, also known as a bar graph, records the number (frequency) of observations (f). These are plotted on the vertical (y) axis for some variable whose values are plotted on the x-axis (Fig. 2.4a). The height of the column represents the number of observations falling into a small range ($x_2 - x_1$). The interval represents the width of the column and is sometimes referred to as a "bin", or class, into which a certain number of discrete observations fall. Clearly, for most purposes, including simplicity of visualisation, the bins should be of equal width. One is at liberty to choose the bin boundaries, but the relative advantages of wide bins that contain a worthwhile number of observations must be balanced against the accompanying loss of detail. For further evaluation, certain statistical tests require a minimum number of observations, e. g., the χ^2-test expects at least five counts per class (Chap. 5).

To complete the bar graph or histogram, the number of measurements in that bin, f, is plotted on the y-axis at the top centre of the bin, located at (y; [$x_2 - x_1$]/2). It is a readily visualised representation of the frequency distribution of x-values in discrete columns. For the histogram shown (Fig. 2.4a), the mean value of the sample is given as follows:

$$\bar{x} = \frac{(f_1 x_1 + f_2 x_2 + f_3 x_3 + \ldots + f_m x_m)}{n} \quad \text{or} \quad \bar{x} = \frac{\sum_{i=1}^{m} f_i x_i}{\sum_{i=1}^{m} f_i} \qquad (2.11)$$

Σ represents the summation for each bin, from $i = 1$ to m, the number of observations in each bin being f_i. Histograms are firmly entrenched as the simplest presentation of a distribution of frequencies. Nevertheless, we should note that the finite width of each column means that an *area* represents the number of observations in that interval or bin. The notation here was chosen for the explanation of Fig. 2.4a and the relation of the mean value of the values (x) to the number of their counts (n). However, for continuous frequency distributions, explained below, the vertical axis representing the frequency of observations is generally more useful as a frequency, $f(x)$ given as a percentage or fraction of unity (Fig. 2.4b). It is easier to compare histograms if they are standardised to show relative frequencies. Each column then represents a fraction of the total number of counts.

It may appear self-evident that there is a 100% chance of finding an observation somewhere in the distribution, so the unit area under the curve also represents the certain probability (= 1.0) of finding an observation in the distribution. The vertical axis represents the probability, $P(x)$, of observing a certain value x. It therefore follows also that the area under the curve between any two points, x_A and x_B, some fraction of unity, represents the probability of finding an observation in that range by random selection. This very important concept is essential to our understanding of statistical methods (Fig. 2.4c), distributions and tests (Chaps. 3 and 4).

2.4
Histograms and Their Relation to Continuous Frequency Distributions

Commonly, data are collected in discrete intervals or *bins*, the limits of which give the width of the columns of a histogram. The simply constructed histogram may be convenient and sufficient for small amounts of data. It is, however, useful to have a smoothly varying graphical representation of the data. This is not merely for aesthetic purposes, and it would be misleading to produce such a diagram from a small amount of data. However, this is appropriate for large samples of measurements or observations ($n > 30$). Certain experiments using automatic methods of data collection such as data loggers, or formerly chart recorders, produce a continuous curve directly. For example, earthquake activity is monitored continuously on seismograms and energy spectra are monitored as a continuous curve with many kinds of analytical equipment. We should not lose sight of the fact that such presentations still comprise discrete measurements at some level, but the intervals between the observations are normally too small to concern us.

With sufficient sampling, and column widths that become vanishingly narrow, the histogram approaches a *continuous frequency distribution.* In practice, the continuous frequency-distribution graphs may be produced from histograms by smoothly joining the midpoints of their columns as shown in Fig. 2.4b. That example is somewhat irregular in form. This could be an intrinsic property of the phenomenon that has been measured, but it may also be the result of inadequate sampling. Fortunately, many geological processes produce distributions that are much simpler in shape; humped and nearly symmetrical, similar to Fig. 2.5. Their central tendency is uniquely recognised and the dispersion of measurements is symmetrical, leading to a simple understanding of variance. The most important and familiar symmetrical, hump-shaped distribution is the Normal distribution, sometimes abbreviated ND and also known as the Gaussian distribution (Fig. 2.5). However, the mean and variance discussed in this chapter are useful for any hump-shaped frequency distribution. The Normal distribution describes the distribution of measurements subject to random errors from the true value. It may be equally valid in the natural sciences where measurements scatter due to the imprecision of some natural controlling process or to the variability of some natural phenomenon.

In the case of measurements subject to random error, observations tend to lie close to a central value, the "true" value, resulting in a single peak. Departures from the central value are equally likely to overestimate or underestimate the mean so that the curve is symmetrical. Uncommon observations that greatly underestimate or overestimate the true value are far from the peak and few in number, resulting in low tails. The bell shape of the symmetrical hump gives rise to the collo-

quial term *Bell curve.* The theoretical form which random errors of measurement follow is more properly called the *Normal distribution,* which is perfectly symmetrical and bell-shaped about its mean. In contrast, *systematic errors of measurement* would displace the mean from the true value by an unknown and perhaps unknowable amount. This would result in shifting the whole curve to the left or right of the true mean. The theoretical Normal distribution is very similar to many distributions of natural measurements, which also show a strong tendency to a central, mean value, and progressively fewer occurrences at values far above, or far below the mean. Although the Normal distribution explains errors of observation nicely, measurement errors are not always the reason for the distribution of natural observations. The measurements on a natural system may be both precise and accurate but their distribution may still follow the Normal distribution. This is an intrinsic property of many natural processes to favour a central value corresponding to the peak of the distribution, with unusual values being progressively less common as they deviate further from the mode. It does not indicate that the extreme x-values are in error, they are just uncommon in the natural scheme of things due to the convergence of natural processes towards a common goal. As we shall see in a later example, the porosities of sandstones may fluctuate around a certain mean value. A few sandstones may have much higher, or much lower porosities but those observations are equally valid. They merely represent unusual cases due to special geological reasons, such as variations in deposition, cementation and compaction. The simplicity of the Normal distribution arises from its symmetry, giving an equivalence of mode, mean and median as in Fig. 2.1c. We may safely state that the Normal distribution is the most important frequency distribution. This is not just because it explains random errors of observation and measurement, nor that it models the intrinsic variation of many natural processes. It finds wider application because many natural distributions may be mathematically or graphically transformed to be similar to the Normal distribution for ease of analysis (Chap. 5). Moreover, several other important frequency distributions are derived by sampling the Normal distribution while others share a fundamental theoretical relationship with it (Chap. 3, Table 3.7; Chap. 5).

With many data sets, a pattern similar to the Normal distribution may be clear even with small sample sizes ($n < 30$). The most common departure from the Normal distribution that still produces a curve of simple appearance is a skewed hump (Fig. 2.1e). The quantification of skewness finds relatively little use in geology, but the sense of skewness may be relevant. The *lognormal distribution* is a positively skewed theoretical distribution that is followed by some earth science observations. Many observations that possess a finite lower limit are skewed. For example, grain size and permeability have a distinct lower boundary of zero and are therefore usually positively skewed. Skewness may arise in other distributions with a fixed, non-zero lower boundary such as the aspect ratio of a sand grain, or certain physical properties such as rock density or magnetic susceptibility. The detection limit of the observation technique may also impose arbitrary lower limits on the observation's value (e.g., earthquake magnitudes, geochemical analysis). Common distributions with fixed upper and lower limits include porosity, or any quantity that is expressed as a concentration, such as the proportion of feldspar in sandstone. These may be expressed in a zero to 100% range. Some concentrations, such as trace elements, are so diluted that modern analytical techniques routinely permit their measurement in a few parts per million (ppm) or parts per billion (ppb). Obviously, the maximum concentration of a measurement in ppm can range up to 10^6 whereas typical trace elements may have abundances < 100 ppm. Thus, the reported values are sufficiently small that the potential range can be considered practically infinite. This is fortunate because many statistical techniques require theoretical models in which the variable has a potentially infinite range.

Some samples that illustrate natural frequency distributions are given in Fig. 2.2. These represent the distribution of shape ratios of pebbles and volcanic lapilli, expressed as the ratio of the lengths of *long/short* axes. All three axes may be important to describe clasts, but for simplicity only the *maximum/minimum* ratio is shown here. Moreover, the relevance of the ratio depends on the closeness of the clast to an ellipsoid shape. Another source of error is that for nearly spherical objects, the error in the ratio is relatively large. However, the errors of measurement are small in comparison to the range of natural variation, a common occurrence with field data. All the samples show a skewed distribution because it is not possible to have a shape ratio < 1. The skewness is positive because the constraints on the upper limit of shape ratio are rather vague; clast length is reduced by attrition during sedimentary processes. Pebbles that are of one rock type and of homogeneous texture give well-defined, smooth frequency distributions (Fig. 2.2a–e). The width of the hump increases with the heterogeneity of the rock type comprising the pebbles. We will see that the width of the hump is an expression of variance and can be

quantified with the *standard deviation* (*s*). The mean or mode would correspond to pebbles with the more typical properties that influence shape. The peakedness of the curve, which could be quantified by kurtosis, is influenced by the anisotropy of the source rock for the pebbles. For example, schists give more elongated pebbles than limestones. However, even with quite large samples, such a simple measurement as a clast-shape ratio may not produce a unimodal distribution (Fig. 2.2f–h). In the case of the modern beach pebbles of Fig. 2.2f, this is because the source rocks are jointed and schistose. Thus, two factors compete to control pebble shape: sedimentary abrasion and inherent anisotropy. The result is a multimodal distribution. Another source of multimodal distributions in conglomerate pebble shapes occurs where tectonic deformation has modified the pebbles' shapes, long after sedimentary deposition (Fig. 2.2g, h). Here, the pebble shapes' multimodal distributions are due to three competing factors: intrinsic anisotropy, depositional processes and reshaping by tectonic deformation. The original shape depends on the abrasive qualities of the source rock including its anisotropy and also on the sedimentary processes. The tectonic reshaping process depends on the metamorphic conditions, the pebbles' ductility and the pebbles' matrix. Here, multiple processes produce multimodal frequency distributions that are difficult to characterise by a single, simple statistic. Further work would require measuring different properties, sampling differently to simplify the variables under consideration, or deconvoluting the frequency spectrum so that each mode could be considered separately (e.g., see example, Fig. 8.9f, g). However, it is fortunate and remarkable that most natural processes produce simpler, hump-shaped frequency distributions.

We occasionally overlook the fact that the most commonly applied statistical model, the Normal distribution, is inapplicable to many geological data sets for the reasons given above. In some instances it may be possibly to mathematically transform the original data set's *x*-values so that the frequency distribution of transformed values may be more similar to the Normal distribution (Chap. 5).

Where geological data appear to conform to the Normal distribution, we require more than a cursory comparison for rigorous work. Intuitively, adjusting the scale of the natural data axes by a standardisation procedure may facilitate comparison of the observed distribution with the theoretical Normal distribution. Unfortunately, elsewhere in science, standardisation procedures, used to reduce dissimilar variables to a common scale for comparison, may be referred to as "normalization". That would cause unfortunate confusion with the statistician's *Normal distribution*, so that *standardisation* in our context refers to transforming a sample distribution to facilitate comparison with the Normal distribution.

A standardisation procedure will be described in detail later, but for now it suffices to show that the shape of any distribution may be standardised by scaling the x-axis in terms of the fundamental unit of dispersion of the x-variable. The basic unit of dispersion is the standard deviation, (s for sample, σ for population), described below (e.g., Fig. 2.5). Furthermore, the distribution may be centred on zero so we can translate the observed frequency distribution until its peak is also at $x = 0$. If our observed frequency distribution closely follows the Normal distribution, standardisation will achieve several things. The standardised mean will be located at zero, the curve will be squeezed or stretched so that its standard deviation will be one unit of the horizontal scale, and it will have the bell shape and symmetry of the theoretical Normal distribution (Fig. 2.5 a). Note that the frequencies are also standardised as a fraction of unity so that they represent probabilities (1.0 represents 100% of observations). Thus, the standardised y-axis represents the probability of observing the associated x-measurement. Samples that are Normally distributed may possess different means and different standard deviations and may even be in different units of measurement (Fig. 2.5 b). Nevertheless, their frequency distributions are readily compared after standardisation. If the sample is drawn from a population that follows the Normal distribution, its standardised form will exactly match Fig. 2.5 a.

2.5
Dispersion: Interquartile Range, Half-Width and Standard Deviation

We previously noted that in any set of natural observations (x_i), values scatter about the mean due to the intrinsic variation of the natural processes, usually with minor contributions due to errors of observation. Where we are concerned with the natural variation due to some physical process, this may be best termed *dispersion*. However, where the variation is due primarily to errors of observation or measurement, this aspect of departure of data from the central value is inversely proportional to *precision*. In fact, precision is sometimes defined as the inverse of the variance.

At the most elementary level, the dispersion of some observations might be indicated by the difference between the largest and smallest values, i.e. the *range*.

This is not a very reliable indicator as it depends on only two values. It is a logical improvement to consider more of the data, for example by using the *semi-interquartile range*. One determines the lower (Q_1) and upper quartiles (Q_3); 25% of the data have x-values lower than Q_1 and 25% of the data have x-values greater than Q_3 (Fig. 2.1 a). The semi-interquartile range (Q) is a simple measure of the dispersion of the data:

$$Q = \frac{Q_3 - Q_1}{2} \tag{2.12}$$

This measure of dispersion focuses on the range required to encompass the central 50% of data. The standard deviation may be estimated approximately from $Q = 2/3\sigma$. The semi-interquartile range (Q), median (Q_2) and upper and lower quartiles (Q_1, Q_3) may be very useful in sedimentology as they can define the degree of sorting of grain sizes from size intervals obtained by sieving. In effect, $\sqrt{(Q_3/Q_1)}$ is a simple sorting coefficient which, when plotted against the median grain size (Q_2), reveals the minimum possible degree of sorting of grain sizes (Marsal 1987). The *box-and-whiskers plot* shows Q as the length of a rectangular box, with a line across it to mark the median (Q_2), and error bars protruding from the ends of the box to show the range (e.g., Fig. 2.6 d). This conveys the range, central tendency, skewness and dispersion without fuss. Another attribute may be considered in companion with the main variable by assigning different widths to the boxes. Thus, in Fig. 2.6 e, a box-and-whiskers plot presents neatly the variation in modern crustal heat flow, while the width of the box indicates the age range of the crust (Vitorello and Pollack 1980). These observations are considered *bivariate*: two quantities are associated with each observation.

Another simple statistic records the height-width ratio of a humped distribution. It finds specific use in certain aspects of earth sciences and is simply defined as the *breadth of a symmetrical, hump-shaped curve at half the height of its peak*. This *half-width* ($x_{1/2}$) value is useful for comparison of curves, assuming that they are reasonably symmetrical (Fig. 2.1 b). One must determine the height of the peak relative to the background level of the values and then measure the breadth of the curve at half this peak value. It is a simple quantifying measure used, for example, to compare the crystallinity of clays and mica crystals from their X-ray diffractograms. The more crystalline the minerals, the narrower are the X-ray diffraction peaks that correspond to reflections from crystallographic planes. Crystallinity increases with the degree of meta-

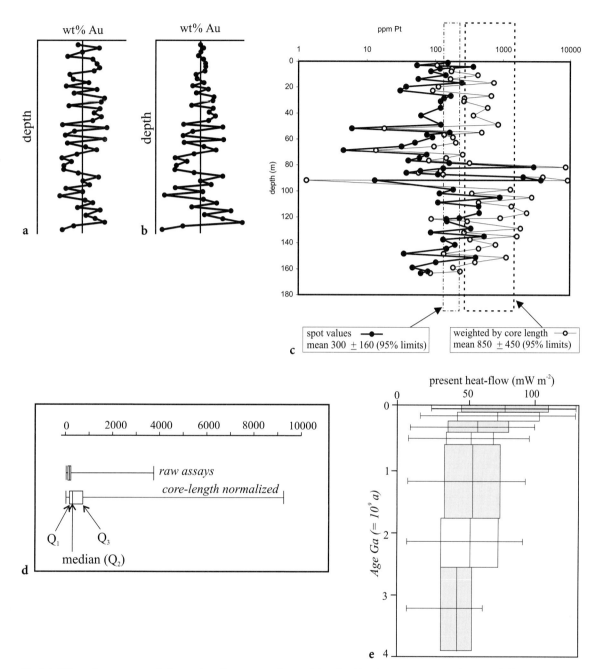

Fig. 2.6. Where the sequence of observations forms part of a spatial series (that in some ways may be treated like a time series, Chap. 8), treating it only as a frequency distribution may suppress useful information. For example, in a drill core, the position or depth control on assayed values is important, not merely their frequency distribution. **a** Where the mean and variance do not vary with depth, there is little problem. **b** Where variance changes with depth, a single value for variance suppresses useful information. **c** Real drill-core assays may be more complex, showing changes in variance and mean with depth, and the results may differ if the assays are weighted according to the core lengths that they characterise. **d** Box-and-whiskers plot; a linear-scale diagram of the range, median and semi-interquartile range. Those simple quantities may be determined easily and give a clear impression of the central tendency and skewness of ore concentrations from the core assays of **c**. **e** Box-and-whiskers plot of present crustal heat-flow value. Here, the width of the box indicates the value of another attribute, in this case the age range of the crust. Older crust has lower heat flow for tectonic and radiogenic reasons

morphism. This is a useful tool in assessing the potential of sedimentary basins to preserve petroleum; the optimum maturation of hydrocarbons may have been exceeded if overheating has occurred, as indicated by the degree of crystallinity of micas. In a totally different context, the half-width value is used also in exploration geophysics to interpret the shape of magnetic or gravitational field anomalies. With surprising simplicity, the anomaly's half-width has a direct mathematical relationship to the depth of the anomalous geological body that causes the local deflection of the magnetic or of the gravitational field. The half-width may be taken for a peak or trough value.

Of course, the range, semi-interquartile range and half-width use relatively little of the data present in the sample. For reconnaissance work they are useful, but it makes sense to use all of the data wherever possible. A more sophisticated measure of dispersion is given by the variance. $(x_i - \mu)$ is the deviation of one observation x_i from the mean value (μ). Summing $\Sigma(x_i - \mu)$ would not be useful because the deviations may be both negative and positive, cancelling each other out to some degree. Thus, the *variance* is obtained by squaring the deviations to preserve their collective influence:

$$\sigma^2 = \sum_{i=1}^{n} \frac{(x_i - \mu)^2}{n} \tag{2.13}$$

The *standard deviation* is then σ.

For a finite sample of n measurements, rather than the entire population, the *sample standard deviation, s,* is given by:

$$s^2 = \sum_{i=1}^{n} \frac{(x_i - \bar{x})^2}{n-1} \tag{2.14}$$

In the denominator, the appearance of $(n-1)$ introduces the important concept of degrees of freedom. Simply put, this is the number of *independent* observations. A simple explanation is that we knew the mean in order to calculate the standard deviation, so that only $(n-1)$ of the original observations are needed in order to write out a complete list of the x_i.

For observations that follow the Normal distribution, the points of inflexion of the frequency distribution provide a simple graphic interpretation of standard deviation. The inflexion points, at which the curve changes from convex-upward to convex-downward, are located at $\bar{x} \pm s$. Thus, if one suspects the distribution is of the Normal type, the x-distance between an inflexion point and the mean permits a visual estimate of the standard deviation.

In some circumstances the dispersion about a mean value must be interpreted carefully. Its usual application is intended for frequency distributions. However, these statistics are sometimes used in situations where they may suppress useful information or even produce misleading conclusions. This is apparent in the case where the observations also possess some attribute, such as position in a sequence. We shall consider assays down a drill core and note that time series may also illustrate a similar feature (Chap. 8). In the simple drill-core assay of Fig. 2.6a, the variation in value with depth is fairly consistent: departures from the mean value are equally common, similar in magnitude and equally likely to scatter above or below the mean with progression along the sequence of measurements. For simplicity, the mean is also constant, implying an absence of any progressive trend. More commonly, there will be changes in the scatter, or variance with depth, as in Fig. 2.6b. Although the variance could be the same for both drill-core assays when treated as frequency distributions, it is clearly misleading to quote a single value for variance in the case of Fig. 2.6b. The parts of the log with higher variances are important economically because they also have richer ore. A further consideration in the treatment of such data is that the interpretation can change somewhat if discrete assays are weighted according to the lengths of drill core that they represent. Richer parts of the core are sometimes assayed more frequently and the frequency with which rich assays are recorded is as valuable as the high assay values. Weighting of observations according to their importance in a conventional frequency distribution is automatic, e.g., if there are ten values in an interval, it is twice as important as an interval with five values. However, when the observations form part of some geologically meaningful series, e.g. measurements along a traverse, assays along a drill core, or observations in a time sequence (Chap. 8), it may be useful to weight the observations according to the interval that they represent. After all, if a high concentration of an element is representative of a long core section, it is more important than if it is representative of only a short core section. This is apparent for Fig. 2.6c, where spot values of platinum at discrete sites suppress the true variability and richness of the deposit. When normalized according to the core lengths that they represent, the Pt-concentrations give a more optimistic view of the richly mineralized section.

Chebyshev's rule estimates the range containing a certain percentage of observations, related to the standard deviation. Although it provides limited information, it is very general, being valid for any frequency distribution. It may be applied to either a sample as here, with mean ($\bar{x}$) and standard deviation (s), or an entire population. Where k is the number of standard deviations (s) and $k > 0$, we expect the following fraction of observations to lie in the interval ($\bar{x} \pm ks$):

$$\geq \left(1 - \frac{1}{k^2}\right) \qquad (2.15)$$

The result is not particularly meaningful for $k = 1$ which informs us only that more than zero observations lie within one standard deviation of the mean. However, we see that $\geq 75\%$ lie within two standard deviations of the mean and $\geq 89\%$ lie within three standard deviations of the mean.

Finally, a simple expression defines the *coefficient of variation* (C). The advantage of this statistic is that it combines information on both mean ($\bar{x}$) and variance (s) as a ratio:

$$C = s / \bar{x} \qquad (2.16)$$

This is meaningful if all the measurements (x_i) are either positive or negative, and it is most useful where variances are large. It is commonly used to characterise abundances in analytical chemistry, where it may be expressed as a percentage, 100C%.

2.6
Review of Some Statistics for Frequency Distributions

The concepts of variance, skewness and kurtosis are systematically related as increasing powers of the summed differences from the mean. Together with the mean, these are successive statistical moments:

First moment
arithmetic mean $= \bar{x} = \Sigma(x_i)/n$
(For the population use μ instead of $\bar{x}$, and perhaps N instead of n. However, care is needed in earth science; sometimes N refers to the number of sites, and n refers to the number of observations.)

Second moment
sample variance $= s^2 = (\Sigma(x_i - \bar{x})^2)/(n - 1)$
population variance $= \sigma^2 = (\Sigma(x_i - \mu)^2)/(n)$

Although they are not used further in any great detail, the other moments occasionally used are:

Third moment
skewness $= [\{\Sigma(x_i - \bar{x})^3\}/n]/s^3$
(Asymmetry of the hump: mode $\neq$ mean, + ve skewness if long tail to right)
Alternatively $=$ (mean – mode)/s

Fourth moment
kurtosis $= [\{\Sigma(x_i - \bar{x})^4\}/n]/s^4$
(peakedness of hump)

2.7
Detection Limits and Precision with Analytical Data

Major elements (O, Si, Al, Fe, Mg, Mn, Ca, K, Na, Ti) are present in concentrations exceeding 1%, minor elements form 0.1–1% and trace element concentrations are usually $\ll 0.1\%$. Some of the most interesting chemical analyses return very small concentrations of economically or academically interesting trace elements in *parts per million* (ppm) or even *parts per billion* (ppb = mg kg^{-1} = ng g^{-1}). Such small concentrations obviously draw attention to the level of confidence that we may place in the result. Analyses of major elements in rocks usually fail to sum precisely to 100%. Major element analyses commonly total 100.7 or 99.6%, for example, reflecting the analytical error. Each of the components is usually present in quantities >1% by weight, so that we should not place too much importance on individual major-element concentrations reported as <1%. This may give an initially discouraging and unfair view of geochemical analyses because many interesting aspects of the science concern low-abundance trace elements present in ppm. However, almost all of that discrepancy is in the major-element concentrations and the methods of analysis ensure that trace elements are detected with similar or better relative sensitivities. Most of the errors are analytical, but where analysis is by energy-dispersive analysis for example, errors may be introduced by assuming an oxidation state for Fe, and by assuming a concentration for H$_2$O although based on reasonable experience and knowledge of mineral structure. Certainly, the imprecision of some major element analyses may approach a fraction of 1% by weight. However, that is satisfactory for most applications of major element analyses. Gill's (1997) review text nicely describes the issues of analytical uncertainty and the reader is recommended to review it for details of sources of error in geochemical analysis.

At a very simple and condensed level, analytical errors vary with the element under consideration, its concentration, heterogeneity of its spatial distribution, the method of analysis and the matrix (e.g., rock, soil, water or vegetation). Mineral analyses using energy-dispersive techniques are further subject to interference from neighbouring grains: if spectra are similar for the grain and its surroundings, it will be difficult to isolate the contribution from the analysed grain. Collectively, these sources of uncertainty share the concept of a *detection limit*. This is the smallest possible indication (e.g., weight, electronic signal, spectral amplitude) that can be recorded *above background* by the analytical procedure. The *detection limit* is usually set at $\bar{x}_B + 3s_B$ where $\bar{x}_B$ and s_B are mean and standard deviation of sufficient background readings. Clearly, the detection limit is an overly optimistic estimate of the smallest possible concentration that may be considered reliable. In fact, determinations are usually only considered significant if they are at least double the detection limit. Thus, reliable concentrations must exceed the *limit of determination*, conventionally taken as $\bar{x}_B + 6s_B$. There is a great variation in detection limit, depending on the choice of analytical method and the nature of the matrix (Table 2.1). Clearly these should be of major concern in the evaluation of trace concentrations of different elements. It is not always possible to make direct, straightforward comparisons of low concentrations obtained by different methods or in different materials.

The spatial distribution of the analyte is another major concern in geochemical analysis, especially for low concentrations. For example, if the analyte is heterogeneously distributed through the specimen, we can only be sure of determining a representative concentration by analysing a large quantity of material. That may be logistically or financially unreasonable. This *nugget-effect* arises later in this chapter; rare materials inevitably occur sporadically. It is intrinsically difficult for nature to spread them evenly though the specimen. We must also consider the size of the particles in which the analyte occurs. Sample size increases dramatically as analyte concentration decreases and target particle size increases. If we assume 30 target particles constitute a satisfactory sample, the specimen size required depends on the concentration and mass of the target particles, as shown in Fig. 2.7. This is an optimistic nomogram requiring particles of target particles of uniform mass, homogeneous spatial distribution, and all equally representative of the analyte. Clifton et al. (1969) derived a similar relationship in an application to native gold concentration; they assumed 20 target particles were sufficient, which was reasonable in their case, as the target particle completely represented the analyte.

2.8
Sample Size

How many measurements are enough? Unbridled enthusiasm might suggest as many as possible. However, there is a point at which well-intentioned effort meets the law of diminishing returns. Moreover, if the sample does not possess the statistical properties of an ideal simple random sample (as defined in Chap. 1), the extra work of collecting a very large sample will be counterproductive. In general, fewer, more careful measurements serve us better, although that strategy does not help much with complex distributions or those comprising intrinsically noisy observations. For a variable that follows the Normal distribution, or even a fairly symmetrical hump shape, $n = 30$ is usually the optimum number of measurements required to define a statistic such as the standard deviation. Under those conditions, the sample's standard deviation (s) is a satisfactory estimate of the population standard deviation (σ). In planning a study, one often needs to estimate a minimum sample size to obtain a certain precision. For example, suppose we wish to estimate the mean μ, within a certain precision unit, Δ. This may be estimated from the sample mean's standard error ($\sigma/\sqrt{n}$). With 95% confidence, for a large sample ($n \geq 30$), the precision unit is defined in terms of the standard error

Table 2.1. Detection limits ($\bar{x} + 3s_B$) for selected elements according to media, and to analytical technique

	NAA method, different media					Silicate rocks, different methods			
	Air	Water	Vegetation	Sediment	Rock	AAS-flame	AAS-graphite furnace	ICP-AES	ICP-MS
	(ng m⁻³)	(µg l⁻¹)	(ppb)	(ppb)	(ppb)	(ppb)	(ppb)	(ppb)	(ppb)
Fe	20	0.6	3.00	105	50	7.5	0.0150	6.2	3.5
Mg	600	800	19.00	800	10	0.3	0.0003	30.0	0.5
Ca	200	2000	17.00	3000	200	1.5	0.0150	10.0	8.0
Na	40	50	0.16	360	10	0.5	0.0005	26.0	1.2
Cd		0.5	0.50			1.5	0.0003	3.5	0.025
Zn	1	2	0.20	220	10	1.5	0.0030	4.0	1.0

Fig. 2.7. This nomogram provides optimistic estimates of the minimum specimen mass required to provide sufficient (~30) mineral grains (= target particles) for chemical analysis assuming the concentrations shown in ppm for the specimen. Important assumptions include the homogeneous distribution of target particles, their uniform mass, and their uniform content of the target analyte

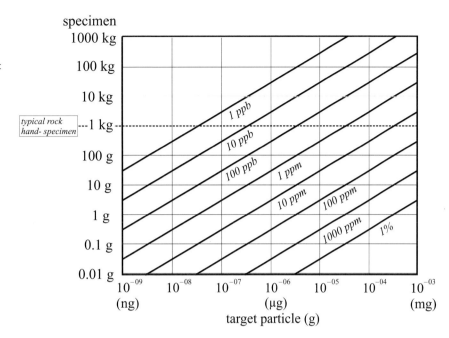

by the following equation:

$$\Delta = \frac{1.96\sigma}{\sqrt{n}} \qquad (2.17)$$

Thus, to achieve that precision, the minimum sample size is:

$$n = \left(\frac{z\sigma}{\Delta}\right)^2 \qquad (2.18)$$

The z-value is a critical value of the standard Normal distribution that defines a certain confidence limit. If the precision must be defined with a 95% confidence limit, $z = 1.96$. If 99% confidence is required, $z = 2.58$. This will be understood after reading Chapters 3 and 4. If the population does not follow the Normal distribution, in particular if it is not even hump-shaped, a larger sample may be required.

We shall see that such estimates are in themselves dependent upon sample size because the previous equations were established assuming that the population standard deviation (σ) was suitably estimated from some pilot sample. Where the pilot sample has $n < 30$, the estimate of σ is given by ts rather than s. The t-statistic is given by the t-distribution, whose shape depends on n and it will be discussed in Chapter 4. For our purpose, s obtained from a small pilot sample ($n \leq 30$) may be used in conjunction with a t-value appropriate for the confidence level, and n is substituted

in the previous equation to indicate a minimum sample size in a further study to achieve the desired precision unit, Δ:

$$n = \left(\frac{ts}{\Delta}\right)^2 \qquad (2.19)$$

A further consideration of sample size comes into play when making decisions based on the value of a certain statistic. This is discussed under hypothesis testing, and the power of the test, in Chapter 4. The effectiveness of the decision-making process is determined by sample size, amongst other things, and it may be necessary to use large samples ($n > 30$) for sensitive comparisons. In carefully controlled laboratory experiments, it may be possible to predetermine a sample size that maximises the efficiency of the decision-making process, as described in Chapter 4. That is rarely possible in the natural sciences.

2.9 Mean and Variance of Spatially Distributed Samples: Kriging

In the earth sciences, we commonly must know how to determine the mean and variance of some value which is not distributed along the line from $-\infty$ to $+\infty$ as in previous frequency distributions. Instead, like many earth science variables, the values of the observations are influenced by their spatial location. The mean and variance for a small subarea are mainly influenced by local values of the observation. More distant observations have decreasing influence on our estimate of the mean and variance for the small area in question. This has enormous practical implications in economic and environmental geology. For example, ore concentrations or pollution estimates may only be available from discrete sites. From that information, how do we estimate the mean value and variance at some location that have not, or cannot, be sampled? In mining, this comes down to the basic question: which blocks should be mined and which should not? Two approaches spring to mind. We have some theoretical model that may fit the spatial distribution of values and which may be used to interpolate the values we require at unsampled locations. Statisticians call this a *parametric* approach, as it requires knowledge of statistical parameters of the population. In broadest terms, *trend-surface analysis* falls into this category and may be used to predict spatial behaviour of a variable (Chap. 1). Alternatively, we make no assumptions about the population and work only with the data in hand. Statisticians call this a *non-parametric* approach. For the purpose of predicting the behaviour of spatially distributed values, the *kriging* technique is a good example. However, simply contouring values on a map also provide a simple *non-parametric* summary (Chap. 1).

Earth scientists deal intuitively with very complex mapped distributions due to their subject-specific training; indeed this book commenced with that topic. However, in the science of statistics, *regionalised variables* form an advanced and specialised field, applications of which are beyond the needs of most workers. Applications include characterising map distributions of elemental concentrations, ore grade, coal quality, water-table elevation, soil quality and even topographic elevation. The observations are represented by single values (scalars) that vary with location. At each location the values may fall anywhere in a range that may not be simply related to that at adjacent locations. Therefore, even if a subjectively chosen function could be found to describe the spatial variation, comparable to a trend surface (Chap. 1), it would be inappropriate. It would disguise or subdue sudden variations that may be of importance, especially in economic applications where local concentrations and their variability are controlled by localised processes (e.g., ore formation or porosity in hydrocarbon reservoirs). Moreover, geological trend-surface analysis mostly describes the variation in two dimensions (x, y), usually a map projection. In contrast, geochemical,

petroleum reservoir and ore studies often require estimations of concentrations or physical properties in three dimensions from samples taken at different elevations, e.g., in drill core. Such applications are inappropriate even for three-dimensional trend surfaces that overgeneralise the spatial variation. We shall focus on examples using mapped distributions of ore grade (= concentration) since this is the most common application, and a very important one economically. Thus, the value must be known at locations (x, y, z) corresponding to N, S and depth coordinates.

Kriging, named after D. G. Krige, is the name given to a complex spatial-averaging technique that can determine the mean and variance of grade at any point P, from unequally spaced surrounding sites where grade has been determined. In mining geology, kriging the mean and variance of a spatially distributed concentration at various locations is referred to as *geostatistics*. Unfortunately, to some people, this term means anything to do with statistics and geology.

We will consider the use of kriging applied to mining, but the principles apply to spatial distributions in environmental work, groundwater studies, petroleum-reservoir studies and biology. The geological examples may use some strange units. Ounces per ton, wt % per foot of drill core, pounds per tonne, or carats per tonne are retained where the physical quantity is in familiar units of economic interest. However, these are all disguises for simple abundances that are dimensionless, e.g., percentage of a major element, parts per million (ppm) of gold, or parts per billion (ppb) of palladium. Of course, all concentrations have a finite, zero to 100 % range, and an increase in one component causes a decrease in another. This is the intractable *constant-sum* issue; it is preferable to deal with components that are completely independent variables, each with a very large potential range. The constant-sum issue is considered further in Chapter 7. However, kriging is mostly used in applications that deal with low abundances (say, in ppm), and the effective range is so large in comparison (1 million ppm) that constant-sum issues do not trouble us.

Let us return to the problem of estimating mean and variance of a value at an unsampled location from known values nearby. Students may at first wonder why so much attention is needed in this subject. One can be forgiven asking, "why not just determine mean and variance for values from adjacent sites?" Unfortunately, a typical drill core through mineralised terrain reveals that rare commodities vary erratically with position (Fig. 2.6a). Moreover, as one approaches richer ore, variance typically increases (Fig. 2.6b). It is important to weight assayed values by the size of the sample that

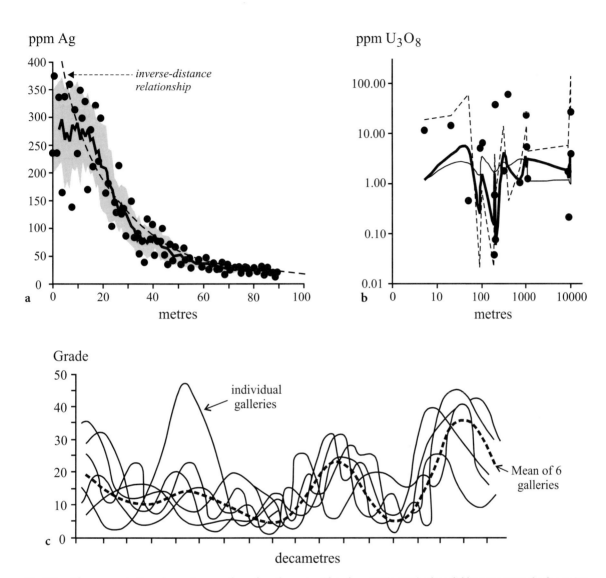

Fig. 2.8. **a** Silver concentration along a traverse through a silver mine (data from Davis 1973). The *solid line* represents the five-point moving average (Chap. 8, Fig. 8.5a for definition). This determines the arithmetic average from the value at a point and the adjacent two values on either side. The *shaded zone* represents the $\pm\,\sigma$ limits. Note that standard deviation increases and becomes more variable as the ore grade increases. **b** Grade variation in a traverse through a uranium deposit (data from Sandefur and Grant 1980). The curves show the different methods of estimating variation (moving averages, trend-surface analysis, kriging): kriging proved to be superior, shown by the *heavy line*. **c** Ore-grade variation along six parallel mine galleries and their mean. (Bukrinski 1965)

they represent (Fig. 2.6c). Ideally, a background value is clearly identifiable, the mean is approximately constant, and at any site the scatter can be potentially as great as at any other site. Unfortunately, a characteristic of ore deposits is that the richer the concentration, the greater is its variability. Clearly, the location of the adjacent specimens must somehow be factored into a sensible estimate for the mean ore grade and concentration variation at the unsampled site.

Detailed observations along a typical traverse approaching rich ores are instructive. Consider the many assays of ore along a traverse in Fig. 2.8a. It is clear that there is some geographical control on the mean value and on the variance. One notes that the standard deviation increases with ore grade. This is a common feature for the abundances of many rare materials. Gold is the classic example where high concentrations (nuggets) occur close to low concentrations and the richer ore is more variable in grade and its location is more difficult to predict from adjacent concentrations. Clearly, global averages and variances are not very helpful for the traverses shown in Fig. 2.8. With distance from the high-grade ore, mean grade and its variance decreases, inspection of Fig. 2.8a suggests something close to a simple inverse relationship with distance (d) such that grade is proportional to d^{-1}. However, a variety of inverse-distance functions have been proposed with grade proportional to d^{-m} where m may be 1/2, 2, or 3 depending on the empirically recognised sensitivity to the separation distance (Clark and Harper 2000).

An example of grade variation in a traverse through a Mexican silver deposit illustrates this in Fig. 2.8a (after data in Davis 1973). Actual data points scatter considerably. Straightforward local averaging reveals the increase in grade towards the main ore zone, and the shaded area illustrates a one-standard-deviation (σ) scatter on either side of the mean. As is evident in most mining situations, the concentration's variance increases dramatically with the mean value. The practical consequences are serious because estimating the grade in the richest part of the deposit dictates where pillars should be left and where ore should be excavated. It is essential that the lower grade ore be left behind, but it is difficult to predict grade precisely when its variance increases so dramatically in the richer ore. This aspect of heterogeneity is sometimes referred to as the *nugget effect* because the location of very rich ore is difficult to predict and loosely related to adjacent concentrations. Figure 2.8a hints at some other fundamental problems. First, if we need to know the ore concentration and variance at one point, how far afield do we consider data? And what distance relationship controls the variation in grade? Clearly, there will be some distance beyond which the variation in grade has little effect on estimating the statistics at the point in question. This is sometimes referred to as a *sill effect*. Secondly, this leads to our questioning whether an overall background value may be assumed for the deposit at all. Various methods of interpolating the variation in grade produce differing results, as shown with data from traverse through a uranium deposit in Wyoming (Fig. 2.8b). One may wonder whether variations in ore grade can show any reproducible or reliable pattern if there is such heterogeneity in their concentration. However, during exploration and extraction, empirical relationships are established between ore grade, calorific value of coal, or hydrocarbon concentration, and the separation of the sites at which they are determined. Numerous post-extraction studies have confirmed that the kriging technique is most successful for estimating mean and variance of concentration from observations made at different distant sites. An example of six traverses through an ore deposit shows that each gallery has idiosyncratic variations in concentration. Nevertheless, the mean of the six traverses retains the essential features of the six individual traverses (Fig. 2.8c). It is the cryptic pattern concealed under chaotic local detail that makes kriging so successful.

Where the mean value does not vary greatly with distance (e.g., Fig. 2.8b, c), *Ordinary Kriging* permits us to estimate the mean and variance at any point in a fairly easily comprehended manner, as will be explained below. However, there are more complex versions, including *Universal Kriging* that is used where there is a marked trend in mean concentration across the mined area. A further complication in nature is that the grade statistics may be anisotropic. In other words, estimates of mean concentration and variance may vary with the direction along which data is collected. There is consequently no universally valid kriging procedure. Spatial heterogeneity (nugget and sill effects), anisotropy and trend vary from situation to situation, and each economic problem must be treated as a separate case study. The kriging computer program may even be customised for each mine or part of a mine, using locally obtained data to refine its predictive power based heavily on sensitivity to separation distance of assay sites. Some elementary aspects of ordinary or simple kriging are considered, but the reader is directed to Clark and Harper (2000) for a definitive account accompanied by useful computer programs. The subject becomes rapidly quite complex, requiring specialised computer programs of which a few limited or generalised examples are available commercially (e.g., Clark and Harper 2000). However, the principles can be grasped at an elementary level, graphically and with a spreadsheet calculation. Here, the discussion is limited to ordinary kriging, applicable to grade variations without a significant large-scale trend over the area.

The simplest and best known estimation method for local values of a regionalised variable is due to Krige (1951), and thus known as kriging. It is the subject of various texts (Matheron 1971; Journel and Huijbregts 1978; Clark 1979; Isaaks and Srivastava 1989; Cressie 1993; Armstrong 1998) and compilations of articles (e.g., Merriam 1970; Royle et al. 1980), but the account by Clark and Harper (2000) is the best starting place for earth or life scientists. Let us designate the regionalised variable as q, measured at various locations indicated by the subscript "i".

The first and simplest statistic we wish to estimate is the mean value, at some point, determined from adjacent values q_i that are irregularly dispersed in an (x, y)-plane. Kriging in mining exploration studies is extended to estimate statistics from adjacent locations scattered in three dimensions (x, y, z). The kriged mean value is a sophisticated improvement of the simple *moving average* described in Chapter 8. The moving average does not consider all the observations, only those from local sites. It may be recomputed at different locations using overlapping sites. Unfortunately, the simplest moving average at each position considers adjacent sites equally and completely ignores those beyond. This is clearly not appropriate for ore-grade variations (Fig. 2.8) or any spatially distributed, low-abundance material. In fact, Krige first pursued moving averages to determine the relative contributions of adjacent localities with a graduated template that he placed over a map to give greater emphasis to sites according to their proximity. However, kriging is a much more powerful group of techniques that account for the continuous variation in influence of surrounding sites and their complex dependence on distance. Figure 2.9a shows simplistically the way in which difference in ore grade at a pair of sites depends on their separation.

For example, if we have N assays in total, we may wish to estimate the mean grade $\bar{q}_i$ characteristic of a 1000-m³ block of ore that we intend to recover. From our knowledge of the variability of ore deposits, the grade at a site located at 2 km distance is unlikely to be relevant. Thus, we must make some decision to consider only observations of grade that are likely to influence $\bar{q}$. For example, we may decide to consider any assays made within 20 m, or 100 m. The number of observations we use, n, may be less than the total available. We may feel obliged to extend the consideration until n is suitably large. More commonly we may seek q_i for quite considerable distances (d_i), perhaps ten times larger than the dimension of the extracted mining block, however, we would weight the significance of the assays by some factor w. The *weighting factor* will decrease with the distance of the adjacent point (i) and the simplest version is an inverse-distance weighting with $w_i = (d_i)^{-1}$. (Such a relationship might apply to the data shown in Fig. 2.8a.) Since our estimate is somewhat like a mean value of q, or *kriged mean* $(\bar{q}_K)$, the weighted values of n adjacent known values (q_i) must sum to a value similar to a conventional arithmetic mean. It follows that the sum of the weighting factors must be unity. The best estimate T is then given by a simple sum of the product of weighted known values $(w_i \, q_i)$. For a distant site, $(w_i \, q_i)$ is very small; for a nearby site, it is larger. Thus, the estimated value T from n adjacent assays would be given by:

$$\bar{q}_K = T = \sum_{i=1}^{m}(w_i \cdot q_i) \text{ where } \sum_{i=1}^{m}(w_i) = 1 \qquad (2.20)$$

For this reason, the T-estimator which we informally call a *kriged mean*, omits a factor $(1/n)$ from its equation. The summation is not of values but of fractions of values, cleverly modified by the weighting function so that their sum directly gives an "average" without dividing by the sample size.

Clark (1979) provides a very clear account of kriging and one of her early examples is reproduced here in Fig. 2.9c. For a group of sites, it shows the difference between a simple arithmetic mean and the kriged mean which includes the effects of spatial distribution. Kriged values are normally desired for points such as P and considered representative for some block, not necessarily rectangular, and usually referred to as a polygon. The polygon may be one of interest for extraction, or as a potential pillar to be left behind as supporting waste, depending on the estimated, kriged-mean grade $\bar{q}$. As noted, when inspecting the mine traverse of Fig. 2.8a, some decision must be made concerning the number of adjacent points (n) under consideration. This requires specific local knowledge and judgement, and cannot be generalised.

Whereas the mean grade is important in the decision to extract or leave the block of ore, we cannot be confident in that value without a knowledge of its variance. Clearly, this is a more complex matter than the variance of a group of numbers. Our values are of *regionalised variables* whose locations in (x, y) or even (x, y, z) space dictate their influence on the variance at the point of interest (P in Fig. 2.9c). This becomes a very sensitive issue in connection with ore deposits because experience shows that variance increases with grade and decreases with the size of the block sampled.

Variance is concerned with differences between observations and in this case that depends on their separation in space. We have an estimate T, the kriged mean $(\bar{q})$, at the unsampled site P. To estimate its associated variance, we must see how the value differences vary with separation of the sites. Consider just two sites A and B at which the values observed are q_A and q_B. Their difference $(\Delta q = q_A - q_B)$ is some unknown function of their separation Δh. We have no precise mathematical model for the distance-grade relation so we must investigate all the data that we have from all the relevant sites in the study area. This non-parametric approach permits us to establish an empirical relationship for the degree of mutual influence between ore grades at different locations.

Consider all the available values (q_i) and their locations (Fig. 2.9a). There is a difference $(q_i - q_{i+j})$ between a pair of assays separated by a distance Δh. If we select all pairs of values with separations similar to Δh,

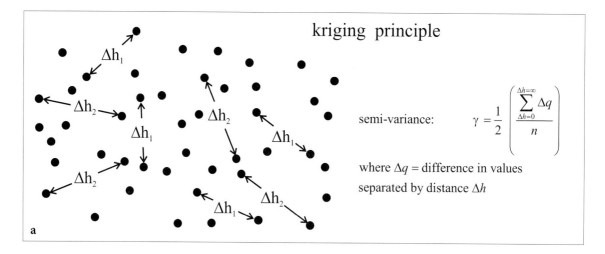

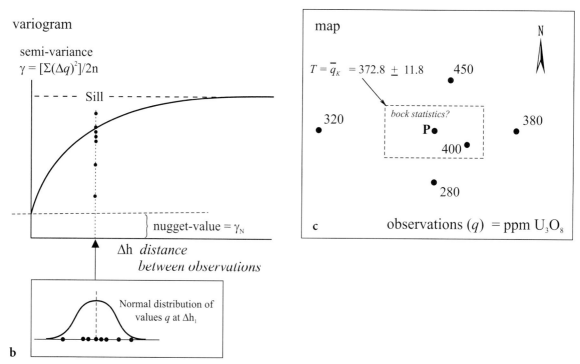

Fig. 2.9. Kriging is a special application of the concepts of mean and variance to regionalised variables. Regionalised variables are those in which the location of the measurement is extremely important. **a** Kriging requires the determination of the average difference in observations (Δq, e.g. ore-grades) at points separated by similar distances (Δh). This is repeated for different separations to determine the influence of separation on the variation in values for n pairs of observations. **b** The variogram plots semi-variance (γ) against site separations (Δh): usually at large distances, there is a plateau (= sill) indicating the separations beyond which there is negligible influence of one site's value on the other. At very short site separations ($\Delta h \sim 0$) there is usually some minimum value of variance, an intrinsic variability or nugget-effect. The variances at a given separation distance are most meaningful if the distribution of value differences (Δq) is symmetrical, e.g. approximately normally distributed. **c** An example from Clark (1979) estimates the mean and variance of grade at P in a mining block from the surrounding known values

their variance should be given by the following expression:

$$\frac{\sum_{i=1}^{n}(q_i - q_{i+j})^2}{n} \qquad (2.21)$$

Traditionally, and perhaps for mathematical convenience, we manipulate half this value to define the *semi-variance* as:

$$\gamma = \frac{\sum_{i=1}^{n}(q_i - q_{i+j})^2}{2n} \qquad (2.22)$$

Of course, this variance is only meaningful if the scatter in values (q) associated with each separation Δh follows a symmetrical hump-shaped distribution like the Normal distribution (Fig. 2.9b). (Transformation techniques exist that overcome this problem, e.g., Chap. 4, Fig. 4.11; Chap. 6, Table 6.1.)

For the introductory example of U_3O_8 concentrations in Fig. 2.9c, kriging for point P yields an estimate T which for simplicity we could loosely call a kriged mean $\bar{q}_K = 372.8 \pm 11.8$ ppm (Clark 1979). The kriged values differ from the simple arithmetic mean and conventional standard deviation of adjacent sites that give 366 ± 66.9 ppm. The semi-variance helps us to make specific local recommendations about the number of adjacent assays that should be considered and how to weight them according to their distance from the site at which a mean and variance are required. This is usually achieved visually with the *variogram*, a graph of semi-variance γ against separations Δh (Fig. 2.9b). The separation distances are also termed the *range* (Δh); for a point at distance (x, y) from that under consideration, $\Delta h = \sqrt{(x^2 + y^2)}$. As the range increases, the semi-variance increases towards a plateau (= *the sill*). The values at sites separated by such distances have no influence on each other. Beyond the

threshold of the sill, variance γ remains constant with increasing site separations Δh. There may be some intrinsic variance, even for very close sites (small Δh). This is easily remembered as the *nugget effect*. It expresses the concept that rare commodities usually show larger changes in variance in grade at short ranges (Fig. 2.8). Consequently, for a real ore deposit the variogram may not start at the origin but at some finite intercept γ_N (Fig. 2.9b).

A thought experiment demonstrates kriging, using a table of values that represent minable blocks with different grades. Assume that the actual assay values for 100 blocks are as shown in Table 2.2. Could we have estimated this spatial distribution of values from one assay, shown in bold type, in each of just 25 blocks? A three-dimensional representation of the 100 assays is shown in Fig. 2.10a. If we only had access to 25 polygons, shown by the data encased in boxes in Table 2.2, the polygon method would have taken the one assay (bold font values in Table 2.2) as representative of that block. The polygon-interpreted distribution would not be a good approximation of the actual distribution (Fig. 2.10b). In contrast, kriging would estimate the other three values in each block, basing the estimate on the adjacent known assays, closely reproducing the distribution of the entire 100 observations (Fig. 2.10c). This experiment has been performed in real life, comparing predicted values with those actually determined after the blocks were mined (e.g. Sandefur and Grant 1980; Krige and Magri 1982; Blackwell and Johnston 1986; Raymond and Armstrong 1986). The value of kriging may also be shown on a smaller scale by comparing it with other methods in detailed sampling traverses (Fig. 2.8b). The kriging method provided the most realistic estimate of true spatial variations in grade in that case study. From the same study, a comparison of simply contoured map values of uranium concentration shows unrealistic spikes and gradients on which one would not wish to plan extraction. It is clear that the contouring procedure is affected ad-

Table 2.2. Example of abundances (ppm) grouped into blocks of four (see Fig. 2.10)

	x or E-W									
y or N-S	1000	1000	1000	1000	**1005**	1005	**1005**	1000	1000	1000
	1000	**1000**	1000	1015	1015	1010	1015	1010	**1000**	1000
	1000	1000	1020	**1022**	1025	1030	1035	1033	1034	1010
	1000	1005	1019	1028	1026	1035	**1038**	1033	1025	1015
	1000	1000	1015	1025	**1025**	1033	1030	1025	1020	**1010**
	1000	1000	1005	1020	1020	1025	**1020**	1010	1005	1000
	1000	1000	1000	1015	**1015**	1020	**1015**	1005	1000	1000
	1000	1000	**1000**	1010	1010	1015	1010	1000	1000	1000
	1000	1000	1000	1000	**1005**	1005	**1005**	1000	**1000**	1000
	1000	**1000**	1000	1000	1000	1000	1000	1000	1000	1000

Fig. 2.10. A hypothetical example of spatial variation in concentration of 100 mining blocks (data in Table 2.2). **a** Values for 100 individual blocks. **b** Polygonal method: based on one value for each of 25 blocks. **c** Kriged values from 25 observations

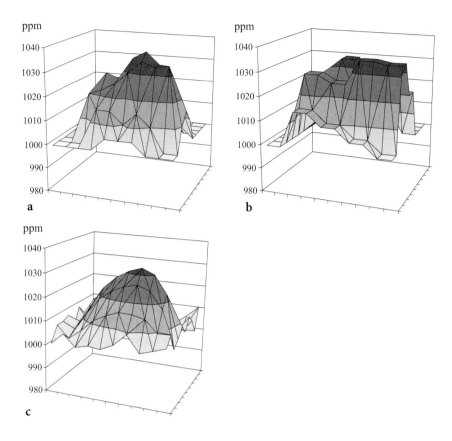

versely by heterogeneity, and a kriged map is much more meaningful and was subsequently verified by mining (cf. Fig. 2.11 a, b).

The complex regional behaviour of a variable such as ore grade, with severe heterogeneity (nugget effect) and a spatially limited dependence on adjacent values (sill effect), can be estimated successfully from discrete values by kriging. A corollary of the nugget effect is that the variance of grade will increase as the size of assayed blocks decreases. Thus, there is considerable danger in estimating reserves of blocks that may easily attain a volume of 10,000 m³ from exploration based on drill core with a total volume of a few m³. Kriging provides the safest means of interpolating and contouring grade through a deposit from values known at specific sites. However, the spatial distribution of known values may greatly influence the success of the kriging procedure. Mining will be much more efficient where the krige-estimated grades for mining blocks are close to their true grades.

In conclusion, we should note that kriging has some important differences from trend-surface analysis which was discussed at the end of Chapter 1, and they have been extensively compared (e.g. Watson 1972).

Trend-surface analysis makes an assumption about large-scale variation and therefore suppresses local variation that may be important. Kriging makes no assumptions about the geometrical form of the entire data set. It is purely empirical, estimating statistics at a site that are influenced strongly by surrounding known values. It is true that one must decide how to weight the relative importance of the heterogeneity of local and distant values, balancing the nugget and sill effects. Conversely, trend-surface analysis assumes a complex mathematical equation to describe the *entire* data set, usually without the need for any theoretical justification. Some judgement is required as to the complexity of the assumed mathematical description and the trend surface considers all data equally, simultaneously suppressing the influence of local anomalies. In ore mining, the latter are especially important. Trend surfaces are therefore less suitable to the evaluation of ores and other economic deposits where concentration variations are very irregular in their spatial pattern and must not be smoothed out. Kriging is superior in ore-concentration evaluation and in predicting the spatial variation of similar regionalised variables. However, economic deposits may vary vastly in scale

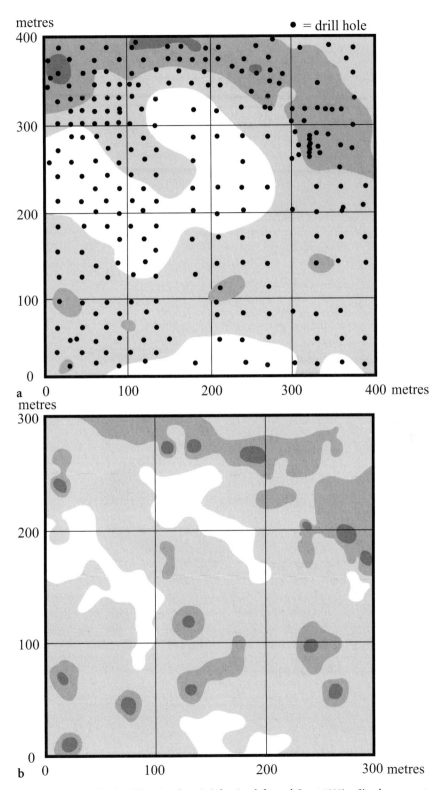

Fig. 2.11 a, b. Maps of U_3O_8 concentration in a Wyoming deposit (after Sandefur and Grant 1980). **a** Simple map contours of drill-hole values; drill-hole locations indicated by *dots*. **b** Map of kriged values at 4-m intervals

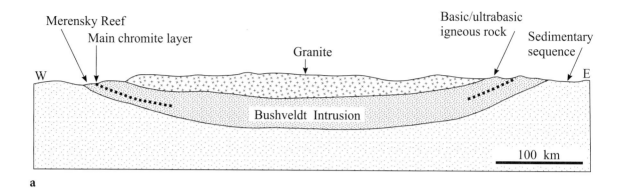

a

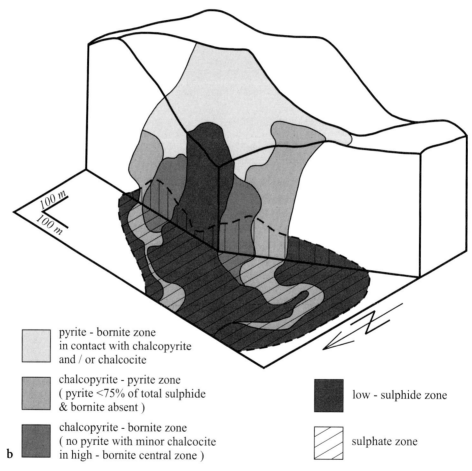

pyrite - bornite zone
in contact with chalcopyrite
and / or chalcocite

chalcopyrite - pyrite zone
(pyrite <75% of total sulphide
& bornite absent)

chalcopyrite - bornite zone
(no pyrite with minor chalcocite
in high - bornite central zone)

b

low - sulphide zone

sulphate zone

Fig. 2.12. The scale and heterogeneity of economic mineral deposits are very variable. Contrast **a** the simple stratiform Bushveld complex, similar over hundreds of km² with **b** the complexity of a typical porphyry-copper deposit that varies heterogeneously on a scale of hundreds of metres. The sampling strategy should be carefully planned to achieve the maximum benefit for cost from any assessment procedure (see Chap. 1)

and degree of complexity, which strongly influences sampling logistics. For example, the mineral deposits of the vast and structurally simple Bushveld Complex present few problems in sampling and determining representative mean values and variances of economic commodities. The body is essentially a sub-horizontal sill, amenable to designed schemes in spatial and bore-hole sampling (Fig. 2.12a). In contrast, some economic deposits, like porphyry-copper deposits, are strongly heterogeneous. Mineralisation values show gradational or disjunctive changes, controlled by heterogeneity of rock type, heterogeneity of mineralisation process, three-dimensional structural complexity, and aniso-tropy (Fig. 2.12b). Whether kriging or a less sophisticated evaluation process is used, it will be quite misleading if the field-sampling scheme is designed without due regard for scale, heterogeneity and also perhaps anisotropy. Although spatial-sampling strategies were introduced in general terms in Chapter 1, it is important to realise that they may be more critical if the statistical-estimation technique is complex, such as in kriging.

CHAPTER 3

Theoretical Distributions: Binomial, Poisson and Normal Distributions

3

Although observations of natural processes and phenomena in the earth sciences may combine many complex and poorly understood factors, it is remarkable that their frequency distribution may closely follow one of a few theoretical models. Generally, a theoretical distribution may be useful as an idealisation or approximation for interpolation and for comparisons. More specifically a theoretical model provides equations from which useful statistics such as mean, variance and confidence estimates can be calculated. The theoretical probability distribution also permits statistical hypotheses to be tested.

The theoretical frequency distribution provides a probability-density distribution that we can use to predict the probability of occurrence of certain values. We showed previously that the area under the curve between x_A and x_B represents the probability of occurrence of a value in that range (Fig. 2.4c). Since all x-values must lie under the curve, the total area under the curve represents a 100% probability or the absolute certainty represented by a probability of 1.0. The forms of probability distributions vary widely, as this chapter will show. They are hump-shaped and may be symmetrical or skewed either positively or negatively. However, some simple general theorems are applicable to all probability distributions. Chebyshev's rule, described in Chapter 2, is universally applicable, and the empirical rule (Chap. 5) is valid if the distribution is mound-shaped and reasonably symmetrical. These estimate the probability that an observation occurs in a certain range without any assumptions or knowledge of the actual probability distribution (Table 3.1).

Table 3.1. Probabilities of certain ranges for x in an unspecified probability distribution

Probability that x lies within this distance of mean where σ = standard deviation	Chebyshev's rule (Chap. 2)	Empirical rule (symmetric mound-shaped distributions; Chap. 5)
$\pm\sigma$	≥ 0	$\approx 68\%$
$\pm 2\sigma$	$\geq 75\%$	$\approx 95\%$
$\pm 3\sigma$	$\geq 89\%$	$\approx 100\%$

Many commonly recognised natural distributions are similar to the Normal distribution. This is not usually because the observations vary due to measurement error, for which the Normal distribution was devised. Instead, natural processes often distribute values non-systematically around some central value to form a hump-shaped frequency distribution that is similar to the Normal distribution. Even where the data do not follow the Normal distribution, their *sample means* approximately follow the Normal distribution (*Central Limit Theorem*, Chap. 4). This justifies the use of Normal-distribution statistics to estimate statistics and confidence limits from many data sets, even where the population's distribution is unknown.

This chapter describes, in order of increasing popularity, the binomial, Poisson and Normal distributions which may be considered as fundamental theoretical models of greatest value in earth science. Other distributions are then described, such as the t, F and Chi-squared (χ^2), which may be derived by sampling a Normal distribution. They are very useful in testing the significance of statistical decisions (hypothesis testing, Chap. 4). The χ^2-distribution also finds applications in goodness-of-fit tests between frequency distributions and any type of curve (Chap. 5). The chapter closes noting the interrelationship of the fundamental distributions.

3.1
The Binomial Distribution

This theoretical distribution describes the frequency with which a discrete observation has a certain attribute, as opposed to it not having that attribute. For example, n fossils may be counted of which some are mature and some have immature forms. The *attributes* would be maturity and immaturity. Where the attribute occurs in the population with a probability p, it is shown by $100p\%$ members of the population, and it is present in pn members of a random sample. Alternatively, we may say that the intrinsic probability of the attribute occurring in any single observation is always

p. The observations are normally referred to as *trials*, as the observations may be compared to repeated attempts to draw a playing card of a certain value in a game of chance. In the example of mature/immature fossils, we may be seeking mature forms. That would be the attribute of interest and we would consider it a *success* if found. The number of *successes* is x and the number of trials is n. If we examined 1000 fossils ($= n$) randomly and found 750 to be mature forms ($= x$), the individual success probability would be $p = 0.75$. The probability of not finding the attribute of interest is $(1 - p)$. The probability of x successes in n trials is given by the probability density of the binomial distribution:

$$\frac{n!}{x!(n-x)!} \cdot p^x (1 - p)^{n-x} \qquad (3.1)$$

Probabilities always lie in the range $0 \le p \le 1$. This probability distribution concerns *dichotomous* variables, those that possess only two possible values such as presence/absence, yes/no, on/off or true/false. Common geological examples of dichotomous properties are the presence or absence of a fossil, whether shale succeeds or does not succeed sandstone, and whether the geomagnetic field polarity is Normal or Reversed. Interesting worked examples are found in Cheeney (1983) and Marsal (1987). Generally, the individual

success probability (p) is unknown but it may be determined by counting the number of successes (x) in a large number of random trials (n). If we regard success as the geochronologist finding a zircon in a crushed rock, $p \ll 0.5$ and the binomial distribution is skewed positively. If a sedimentologist defines a success by finding a quartz grain in a crushed quartzite, $p \gg 0.5$ and the binomial distribution will be skewed negatively. The simple and special case where $p = 0.5$ may be modeled by flipping a fair coin (Figs. 3.1, 3.2a, b). The forms of some binomial distributions with different p-values are shown in Fig. 3.2.

For simplicity, we will first consider the situation where the individual-event probability is $p = 0.5$. Thus, the attribute will occur in half of the population or in half of any suitably large sample. This occurs where the individual event success is completely random. It may not be a useful model for a process in which we are interested, but it provides a useful basis for comparisons. The $p = 0.5$ situation is simulated by tossing an unbiassed or *fair* coin (Fig. 3.1). Heads and Tails must occur with equal probabilities of 0.5 in any single trial because the ideal process is completely random. The distribution is produced by setting $p = 0.5$ in the above formula; for this special case the binomial distribution is symmetrical about the mean (Fig. 3.2c). Some simple statements may clarify the use of the binomial distribution and its terminology:

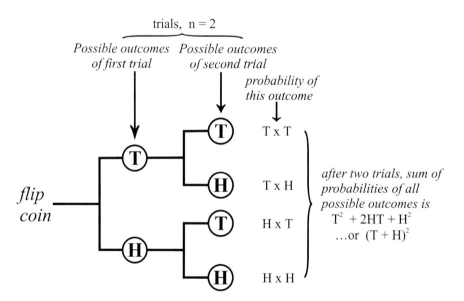

Fig. 3.1. Successive outcomes of an event that has only two possibilities, True/False, illustrated by the analogy of tossing a fair (unbiassed) coin, to yield heads or tails (H/T). The first trial has two possibilities, the second has four, etc. The equation gives the possible outcomes after $n = 2$ trials. This illustration may be used to show the behaviour of a binomial distribution, which finds application in examples such as the presence/absence of some attribute such as a certain fossil, rock type or event at different points in a sequence

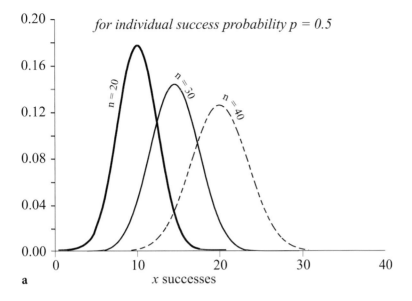

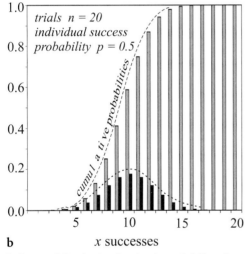

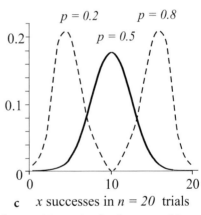

Fig. 3.2. a The binomial distribution showing the probability of successes, where $p = 0.5$, meaning that the two possible outcomes are equally likely in each trial. Shown for $n = 20$, 30 and 40 trials. **b** Also for $p = 0.5$, the cumulative probability of x successes in $n = 20$ trials, compared with its continuous probability distribution. **c** The number of successes predicted in $n = 20$ trials by the binomial distribution where the likelihood of success in any single trial is $p = 0.2$, 0.5 and 0.8

1. The observations (n) are termed trials (= attempts to determine the presence of the attribute, think of flipping a coin to obtain heads).
2. p is the intrinsic probability of success in every single trial, e.g. $p = 0.5$ for obtaining heads when flipping a fair coin. In the population, or after many trials, the proportion of successes is $100p\%$; i.e. the attribute was present in np trials.
3. The outcome of each trial must be independent of any other trial (again, think of successive coin flips)

4. The trials must represent sampling *with* replacement. For example, suppose we make replicate counts of fossils from a spoil heap in a quarry. If we wish to determine the number of mature forms from a group of fossils, the removal of each specimen would bias the remaining proportion of mature individuals. Each sample should be returned to a random location in the pile after its examination.
5. In most natural science applications $p \neq 0.5$, and in many of interest, p may be very small.

The binomial distribution is understood easily where the two possible outcomes of each trial (observation) are equally likely ($p = 0.5$), as in any perfectly random process. Flipping fair coins provides a good example; the outcome may only be heads (H) or tails (T). After one flip, there are two possibilities (H, T); the probability of either H *or* T is the sum of their individual probabilities (0.5), i.e. 1.0. The probability of the first flip being H *and* the second flip being H is the product of their probabilities; H^2, 0.5×0.5 or 0.25. In Fig. 3.1, the possible outcomes of the first flip are either H or T, i.e. $(H+T)^1$. After the second flip the total number of possible outcomes is a sequence of two tails being thrown $(T \times T)$ *or* tails following heads $(T \times H)$ *or* heads following tails $(H \times T)$ *or* heads following heads $(H \times H)$. In other words, the sum of the probabilities of all possible outcomes after two coin-flips is $T^2 + 2TH + H^2$ which may be written as $(T+H)^2$. After n trials the total possible number of outcomes is $(T+H)^n$ which may be written out explicitly using the binomial expansion, hence the name of the distribution. As shown in Table 3.2, each term of the expansion has a value that gives the probability of achieving a certain number of successes after n trials where p and $q = (1-p)$ are the probabilities of success and of failure respectively, in any single trial.

Inspecting the previous equation reveals the meaning of the terms p^x and $(1-p)^{n-x}$; they influence the probabilities that the attribute will be present in x observations out of n trials with a probability p, and in the remaining $(n-x)$ observations with probability $(1-p)$. It is useful to know some simple statistics, where x is the number of successes in n *trials* and the individual event probability of success is p. The mean and variance of the number of successes (x) are given as follows:

$$\bar{x} = np \quad \sigma^2 = np(1-p) \qquad (3.2)$$

It is also convenient to work with the number of successes x_p ($= np$). The mean and variance of the number

of successes follows, and for future reference a simple expression is given for the 95% confidence limits for x_p. Confidence limits are explained in Chapter 4.

$$\bar{x}_P = p \quad \sigma_p^2 = \frac{p(1-p)}{n}$$

and 95% confidence limits are $\quad \bar{x}_P \pm 1.96\sqrt{\dfrac{p(1-p)}{n}}$

$$(3.3)$$

Statistical tables for consulting binomial probabilities are quite cumbersome because the form of the distribution depends on sample size (usually $n < 30$), the number of successes (x) and on p. However, the probabilities are now readily available in spreadsheets, the calculation from the formula is not too forbidding for special cases anyway, and the expressions for mean and standard deviation all make the binomial distribution relatively easy to use. For most reliable uses, we prefer our sample to have $np \geq 5$. For $np < 5$, the binomial distribution is almost indistinguishable from the Poisson distribution which differs in having large numbers of events (n) in each of which the probability (p) of success is very small. The Poisson distribution is discussed next in this chapter.

Examples of the probability density of the binomial distribution are shown for $p = 0.5$ and for trials of $n = 20$, 30 and 40 in Fig. 3.2a. The probability distribution is shown as a continuous curve giving the probability of x successes in n trials. Most often, the probabilities are tabulated in cumulative form, i.e. the probability of achieving $\leq x$ successes in n trials. This is illustrated graphically in Fig. 3.2b. In order to determine the probability of achieving exactly x successes, one must subtract adjacent cumulative probabilities. This is illustrated by an example in Table 3.3 which gives the cumulative probabilities for anywhere up to 20 successes in anywhere up to 25 trials for $p = 0.5$. The actual integer number of successes is obtained by multiplying the cumulative probability ($\leq x$ successes) or individual probability ($= x$ suc-

Table 3.2. Probability of successes after n trials from terms of the binomial expansion

n Trials (observations)	Binomial expression for n trials	Binomial expansion where $q = 1 - p$, where p is probability of success in any one trial
1	$(p + q)^1$	$p + q$
2	$(p + q)^2$	$p^2 + 2pq + q^2$
3 [a]	$(p + q)^3$	$p^3 + 3p^2q + 3pq^2 + q^3$
4 [b]	$(p + q)^4$	$p^4 + 4p^3q + 6p^2q^2 + 4pq^3 + q^4$

[a] The value of $3p^2q + 3pq^2 + q^3$ represents the probability of achieving two or less successes (i.e. 2, 1 or 0 successes, being the powers of p) after three trials.

[b] The value of $4p^3q$ represents the probability of achieving three successes after four trials.

Table 3.3. Cumulative binomial probabilities of x successes in n trials for $p = 0.5$, e.g. (a) the probability of ≤ 10 successes in 20 trials is 0.59; (b) the probability of 10 successes in 20 trials is (0.59–0.41) or 0.18

n	Successes x																			
	0	1	2	3	4	5	6	7	8	9	10	11	12	13	14	15	16	17	18	19
2	0.25	0.75	1.00																	
3	0.13	0.50	0.88	1.00																
4	0.06	0.31	0.69	0.94	1.00															
5	0.03	0.19	0.50	0.81	0.97	1.00														
6	0.02	0.11	0.34	0.66	0.89	0.98	1.00													
7	0.01	0.06	0.23	0.50	0.77	0.94	0.99													
8	0.00	0.04	0.14	0.36	0.64	0.86	0.96	1.00												
9		0.02	0.09	0.25	0.50	0.75	0.91	0.98	1.00											
10		0.01	0.05	0.17	0.38	0.62	0.83	0.95	0.99	1.00										
11		0.01	0.03	0.11	0.27	0.50	0.73	0.89	0.97	0.99										
12		0.00	0.02	0.07	0.19	0.39	0.61	0.81	0.93	0.98	1.00									
13			0.01	0.05	0.13	0.29	0.50	0.71	0.87	0.95	0.99	1.00								
14			0.01	0.03	0.09	0.21	0.40	0.60	0.79	0.91	0.97	0.99								
15			0.00	0.02	0.06	0.15	0.30	0.50	0.70	0.85	0.94	0.98	1.00							
16				0.01	0.04	0.11	0.23	0.40	0.60	0.77	0.89	0.96	0.99	1.00						
17				0.01	0.02	0.07	0.17	0.31	0.50	0.69	0.83	0.93	0.98	0.99						
18				0.00	0.02	0.05	0.12	0.24	0.41	0.59	0.76	0.88	0.95	0.98	1.00					
19					0.01	0.03	0.08	0.18	0.32	0.50	0.68	0.82	0.92	0.97	0.99	1.00				
20					0.01	0.02	0.06	0.13	0.25	0.41	0.59	0.75	0.87	0.94	0.98	0.99				
21					0.00	0.01	0.04	0.09	0.19	0.33	0.50	0.67	0.81	0.91	0.96	0.99	1.00			
22						0.01	0.03	0.07	0.14	0.26	0.42	0.58	0.74	0.86	0.93	0.97	0.99	1.00		
23						0.01	0.02	0.05	0.11	0.20	0.34	0.50	0.66	0.80	0.89	0.95	0.98	0.99		
24						0.00	0.01	0.03	0.08	0.15	0.27	0.42	0.58	0.73	0.85	0.92	0.97	0.99	1.00	
25							0.01	0.02	0.05	0.11	0.21	0.35	0.50	0.65	0.79	0.89	0.95	0.98	0.99	1.00

Small numbers of successes that occur with $\leq 2.5\%$ probability

Large numbers of successes that occur with $\leq 2.5\%$ probability

cesses) by n and then rounded down to the next lower integer.

The binomial distribution is commonly used in counting exercises where the number of expected tallies must be estimated. For example, in modal analysis of rocks, either in thin section or from mineral separations or sieved sediment samples, we may need to know the probability of finding a certain number of grains of a particular mineral (x successes) from a collection of grains (n trials). We must know the probability p of a success in any single trial. We have the advantage of accessing any probabilities for any (n, x) combination from a spreadsheet, whereas former generations of students were limited to published tables of selected probabilities.

Suppose that we wish to study single crystals of feldspar from a leucogranite, comprising quartz and feldspar. Those minerals may be difficult to distinguish in a mineral separation so it may be of interest to know with what probability the x feldspar grains will occur in a given sample of n grains. Suppose 50% of the grains in the granite are feldspar, i.e. $p = 0.5$ for the separated grains, if no bias is introduced by the crushing

process that liberates the minerals. If we have a sample of $n = 25$ grains, what is the probability of it containing ten or fewer feldspars ($x \leq 10$)? Table 3.3 under the appropriate (x, n) values shows a probability of 0.21. In other words, there is a 21% probability that samples with $n = 25$ contain ≤ 10 feldspars. It is more likely that we are interested in having ten *or more* feldspars, from the Table the probability of ≤ 9 successes is 0.11. Thus, the probability of ≥ 10 feldspar grains is (1–0.11) or 89%.

This simple introduction to binomial probabilities also provides an opportunity to understand the use of the areas under any probability distribution. In Fig. 3.2b, a binomial distribution is shown cumulatively, for $n = 20$. For simplicity $p = 0.5$, each trial has equal likelihood for its two outcomes, and therefore the distribution is symmetrical. Inspection of the cumulative plot makes it clear that the probability of 20 *or fewer* successes in 20 trials is of course 100%, we may expect 0, or 1, or 2, ... or any number of successes up to 20. Zero successes are as unlikely as 20 successes and the corresponding areas under the discrete-probability curve are vanishingly small. However, either 5 or 15 successes would occur with a probability of ~2%. This may be read as the height of the columns on the discrete plot, or by subtracting the difference between adjacent columns on the cumulative plot.

Further inspection of Fig. 3.2 reminds us of the concept that will be used extensively and is applicable to all other probability distributions. We understand that the area under the graph between x_1 and x_2 represents the probability that a value may occur in that range. Now, we can predict the probability of observing x in the range x_1 to x_2 from the fraction of the area under the graph between those values. This will introduce us to the standard practice of consulting probability tables that give probabilities (as a fraction of unity) for certain x-ranges. Therefore, let us reconsider Table 3.3 in connection with another example. For simplicity, consider a range of successes, x, starting at the lower limit $x = 0$ upwards, for samples of $n = 25$ trials. Interestingly, we note that the distribution encloses approximately 95% of its area (probability = 0.95) for ≤ 16 observations. This means that if we were to repeat our experiment many times, randomly drawing 25 grains at a time (and replacing them after the count), in 95% of the experiments we would find $x \leq 16$ feldspar grains. The choice of the probability 0.95 for ($n = 25, x = 16$) in the Table was convenient and fortunate because it introduces the idea of the popular *95% confidence limit*. Alternatively, since we are seeking large numbers of feldspar grains (= successes), we may note the complementary 5% probability of our finding >16 feldspars in a sample. We shall meet the 95% confidence limit repeatedly through earth science statistics in many kinds of problems with many different probability distributions. It is a reasonable compromise satisfying practical, theoretical and subject-specific needs for most natural systems, more of which will be discussed in Chapter 4.

Our example of estimating the success of finding desired minerals in aggregates used $p = 0.5$ for simplicity and by analogy with our introduction to the binomial distribution from flipping a coin (Fig. 3.1), for which the two possible outcomes must be equal (heads, tails = 50% probability). More commonly, in geological counting problems we may be searching for rare items of low abundance, e.g. accessory minerals such as zircon used in U-Pb geochronology, or iron oxide minerals of interest in rock-magnetism and paleomagnetic studies. Thus, let us inspect binomial probabilities for $p = 0.01$, and $p = 0.05$ where the intrinsic probability of success in any single trial is only 1 or 5% respectively (Table 3.4). The tabulation is cumulative, as in the previous example. Suppose our geochronologist, or paleomagnetist crushed many kilograms of rock in the hope

of extracting at least one grain of his target accessory mineral. There may be good discipline-specific reasons to infer the abundance in the population to be 1%, i.e. the binomial probability $p = 0.01$. In each and every trial (observation) there is only a 1% chance of successfully finding the target mineral. From the bottom row of Table 3.4, we read that if the sample size $n = 30$, the probability of our colleague finding at least one grain is 0.96 (96%). However, this includes the possibility of recording zero grains, which is not very useful. The probability of finding just one grain of interest is found by subtracting the cumulative probabilities for one and for zero grains, or $(0.96 - 0.74) = 0.22$ or 22%.

An application of the binomial model is appropriate when we wish to estimate the reliability of a counting procedure based on a small sample. Suppose that the number of counts or observations is n and the number of successful counts that show the desired attribute is m. From the sample, the proportion of successes is then $x_P = (m/n)$. The proportion (x_P) is distributed according to the binomial model, and from the preceding equation we may estimate the range around x_P, within which the true proportion should lie with 95% confidence. The preceding equations permit the construction of a nomogram from which the 95% confidence limits on x_P may be estimated (Fig. 3.3). For example, suppose that we are interested in the abundance of quartzite pebbles as a clue to the maturity of a depositional environment. We sample 100 pebbles from a conglomerate. We observe that 20 are quartzite pebbles and 80 are not, giving the estimated proportion of successes, $x_P = 0.20$. How confident are we that this is close to the true proportion in the population? Using Fig. 3.3, follow a vertical line up from $x_P = 0.20$ to meet the two curves for $n = 100$. The lower curve gives the lower confidence limit as ~ (-0.07), the upper curve gives the upper limit as ~ $(+0.06)$. These bounds define the *95% confidence limits* on the observed proportion x_P. Thus, our estimate of the fraction of quartzite pebbles for the population is $0.20^{+0.06}_{-0.07}$ with 95% confidence. Note the following:

1. upper and lower confidence limits are unequal for $p \neq 0.5$.
2. Fig. 3.3 is valid for the 95% confidence limits ($\alpha = 0.05$ significance level, see Chap. 4)

Applications of the binomial distribution can be found in other situations, not literally related to counting procedures. For example, suppose 20 outcrops of a potentially cyclothymic sequence show shale-sandstone and sandstone-shale contacts. We must assume outcrop bias is absent, i.e. the outcrops must represent independent observations of contacts. Is there a tendency

Table 3.4. Cumulative binomial probabilities for x successes in n trials with $p = 0.01$, $p = 0.05$

Trials (n)	x Successes								
	For $p = 0.01$			For $p = 0.05$					
	0	1	2	0	1	2	3	4	5
2	0.98	1.00		0.90	1.00				
3	0.97	1.00	1.00	0.86	0.99	1.00			
4	0.96	1.00	1.00	0.81	0.99	1.00	1.00		
5	0.95	1.00	1.00	0.77	0.98	1.00	1.00	1.00	
6	0.94	1.00	1.00	0.74	0.97	1.00	1.00	1.00	1.00
7	0.93	1.00	1.00	0.70	0.96	1.00	1.00	1.00	1.00
8	0.92	1.00	1.00	0.66	0.94	0.99	1.00	1.00	1.00
9	0.91	1.00	1.00	0.63	0.93	0.99	1.00	1.00	1.00
10	0.90	1.00	1.00	0.60	0.91	0.99	1.00	1.00	1.00
11	0.90	0.99	1.00	0.57	0.90	0.98	1.00	1.00	1.00
12	0.89	0.99	1.00	0.54	0.88	0.98	1.00	1.00	1.00
13	0.88	0.99	1.00	0.51	0.86	0.98	1.00	1.00	1.00
14	0.87	0.99	1.00	0.49	0.85	0.97	1.00	1.00	1.00
15	0.86	0.99	1.00	0.46	0.83	0.96	0.99	1.00	1.00
16	0.85	0.99	1.00	0.44	0.81	0.96	0.99	1.00	1.00
17	0.84	0.99	1.00	0.42	0.79	0.95	0.99	1.00	1.00
18	0.83	0.99	1.00	0.40	0.77	0.94	0.99	1.00	1.00
19	0.83	0.98	1.00	0.38	0.75	0.93	0.99	1.00	1.00
20	0.82	0.98	1.00	0.36	0.74	0.92	0.98	1.00	1.00
21	0.81	0.98	1.00	0.34	0.72	0.92	0.98	1.00	1.00
22	0.80	0.98	1.00	0.32	0.70	0.91	0.98	1.00	1.00
23	0.79	0.98	1.00	0.31	0.68	0.89	0.97	1.00	1.00
24	0.79	0.98	1.00	0.29	0.66	0.88	0.97	0.99	1.00
25	0.78	0.97	1.00	0.28	0.64	0.87	0.97	0.99	1.00
26	0.77	0.97	1.00	0.26	0.62	0.86	0.96	0.99	1.00
27	0.76	0.97	1.00	0.25	0.61	0.85	0.96	0.99	1.00
28	0.75	0.97	1.00	0.24	0.59	0.84	0.95	0.99	1.00
29	0.75	0.97	1.00	0.23	0.57	0.82	0.95	0.99	1.00
30	0.74	0.96	1.00	0.21	0.55	0.81	0.94	0.98	1.00

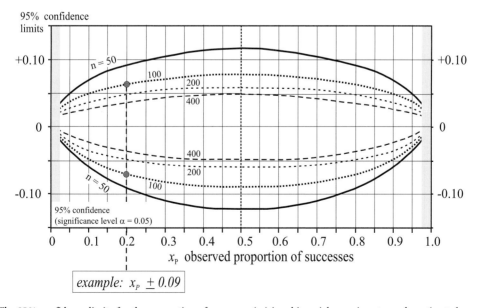

Fig. 3.3. The 95% confidence limits for the proportion of successes (x_P) in a binomial experiment may be estimated approximately from this nomogram. For example, if 100 pebbles are counted and 20 are quartzite, how confident are we that the population actually contains 20% quartzite pebbles? Follow the vertical line from the observed proportion of successes ($x_P = 0.20$) and note the intersections with $n = 100$. The lower one gives the lower confidence limit $\sim (-0.07)$, the upper curve is intersected at the upper confidence limit of $\sim (+0.06)$. Thus, the true proportion of quartzite pebbles in the population should lie in the range $0.20^{+0.06}_{-0.07}$.

for shales to succeed sandstone, or is the order entirely random? If the order is random, the probabilities of shale following sandstone or sandstone following shale are both 0.5, comparable to the heads/tails outcome of a coin-flip. Let us hypothesise that shale follows sandstone, and for each individual contact there is an intrinsic probability of $p = 0.5$ that it will meet that criterion successfully.

What is the probability that we would record five or fewer shale-over-sandstone contacts? Consulting Table 3.3 we may read along the row $n = 20$, under the $x = 6$ column we see the probability of ≤ 5 successes is 0.06, or 6%. Therefore we may infer that if we only found five shale-over-sandstone contacts, our initial hypothesis is unlikely to be correct. At the other extreme, if we repeatedly took samples of 20 outcrops in a region and we found that out of each sample of 20 trials there were on average ≤ 13 such contacts, the associated cumulative probability is 0.94 (Table 3.3: $n = 20$; $x = 13$). Thus, *up to* 13 shale-over-sandstone contacts could be expected with a 94% probability. Conversely, there is only a 6% probability that one could expect > 13 shale-over-sandstone contacts. Thus, if there is indeed a long-term tendency for shale to succeed sandstone, there is only a 12% probability that a sample of 20 outcrops would show either < 6 or > 13 such transitions. This again touches on the notion of hypothesis testing about which more is said in Chapter 4. Cheeney (1983, p. 40) discusses a similar example ($p = 0.5$) and tabulates directly the critical numbers of successes (x) for various numbers of trials (n). However, a simple ladder boundary in Table 3.3 separates "unusually small" x on the lower left of the Table that occur with a probability ≤ 2.5%, for the associated number of trials (n). Similarly, a ladder boundary isolates "unusually large" numbers of successes that should occur with a probability ≤ 2.5%.

3.2
The Poisson Distribution

This probability distribution may provide a useful description of rare events such as the mutation rates of fossils, the occurrence of a storm event in a stratigraphic sequence, radioactive decay rates, excursions of the geomagnetic field and the frequency of major earthquakes. *The events must be independent and may occur only one at a time, and their probability of occurrence increases with the time elapsed since the last event.* The events cited are characterised by a low probability of success and appear mostly as catastrophic versions of a frequent phenomenon. For example,

storms are very frequent but rarely of sufficient intensity to be preserved in a stratigraphic record. This encapsulates the essence of the Poisson distribution, which might apply to many earth science phenomena, given the vastness of geological time. The drawback with recognizing geological Poisson processes is that rare events may easily escape detection.

The probability distribution of values following the Poisson model depends on the number of events x recorded in some interval and λ, the mean number of events that occurs per unit of measurement. For example, λ may be the long-term average of some event per unit time, e.g. number of particles emitted per second during radioactive decay, the number of turbidity flows per Ma off the continental shelves, or the number of major seismic events per Ka. Less common, the observation may be spatially distributed. For example, the following spatial distributions have been compared with the Poisson model: number of oil wells per km^2, the number of meteorite falls per km^2 and the count of worm burrows per m^2 on a bedding surface.

The following expression gives the probability density, $P(x)$, of the Poisson distribution in which mean and variance are both given by λ ($\mu = \lambda$ and $\sigma^2 = \lambda$):

$$P(x) = \frac{\lambda^x e^{-\lambda}}{x!} \tag{3.4}$$

The Poisson distribution may be considered a limiting case of the binomial distribution, where the number of trials (n) is large but the probability of success in any individual trial (p) is small (np is then $\approx \lambda$).

Examples of Poisson distributions may be found in geochronology (Faure 1986). First, we are familiar with the fact that radioactive decay constitutes a sequence of spontaneous rare events controlled by the atomic nucleus, unaffected by any environmental physical or chemical factors, and completely independent of one another. The decay of an atom in any time interval is a rare event and the counts of such events in a unit of time are highly variable. Mostly we find very few decay events per unit time, and the distribution has a long low tail representing the rare time intervals in which there are unusually large numbers of decay events. This example has all the features required of a Poisson process, resulting in a positively skewed frequency distribution of decay counts per time unit. Other distributions may approximate the Poisson model. For example, paleomagnetism recognises that the geomagnetic field has switched polarity, approximately every million years, over long periods of geological history. The intervals over which the polarity remains constant are of considerable geophysical interest. It appears that the interval lengths approximately follow the Poisson distribution

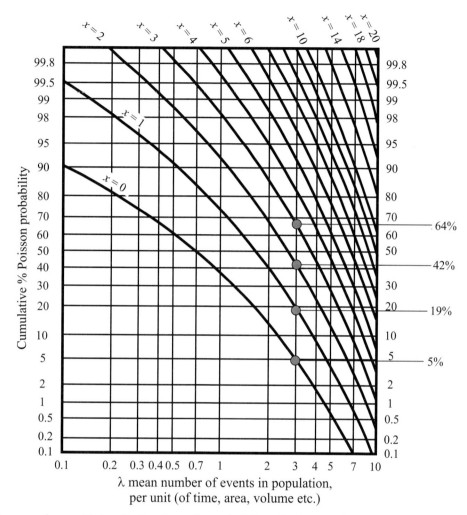

Fig. 3.4. Nomogram for use with data distributed according to the Poisson distribution. Where μ is both mean and standard deviation of the data, the percentage probability of x successful trials is given by the curves. For example, for data with $\mu = 3$, 5% of data have zero successes, and 20% of data have one or less success (interpolate along the vertical line $\mu = 3$)

where the mean interval length is $1/\lambda$ (McElhinny and McFadden 2000). Weaker approximations to Poisson distributions may be found in stratigraphy. For example, one might test whether the frequency of fossils per unit time was approximately Poissonian, using sediments with constant deposition rates as a time marker.

We may use the Poisson distribution to determine the probability of occurrence of a rare event with the aid of a nomogram (Fig. 3.4, after Marsal 1987). Suppose a large section of the continental shelf has a uniform stratigraphic succession and no strong structural anisotropy. One might expect the occurrences of commercially viable oil reservoirs to follow a Poisson distribution. Ideally, we would like to exploit several oil reservoirs in the same sector for logistical reasons. We

must know the mean concentration in the population, for example from the whole continental shelf, and seismic exploration may suggest there are 3 viable reservoirs per 100-km² sector. Thus, knowing $\lambda = 3$, what is the probability that *at least two* useful oil reservoirs would be found in any given sector? Consulting Fig. 3.4, follow $\lambda = 3$ upwards; we see there is a 5% probability of finding zero reservoirs. Further, there are probabilities of 19, 42 and 64% of finding ≤ 1, ≤ 2 and ≤ 3 reservoirs respectively. The probability of discovering exactly three reservoirs is given by subtracting the probabilities for ≤ 3 and ≤ 2 successes, i.e. 64% – 42% = 22%. As with many probability tables, the cumulative aspect demands attention. For example, with $\lambda = 3$ reservoirs per 100 km², the probability of ≤ 9 reservoirs being found in one sector is close to 100%; however, this statement also includes the possibilities of anywhere between eight and zero successes! It may be more useful to know that the probability of eight discoveries per sector is very small, approximately 0.3% (99.9–99.6%).

3.3
The Normal Distribution

The results of many physical processes and geological observations follow a symmetrical hump-shaped frequency distribution that is very similar to the Normal distributions, sometimes named after its originator Karl Gauss (1777–1855). The theoretical form of the distribution was introduced to explain the pattern of non-systematic errors of observation and measurement, mainly in the laboratory sciences. Such a distribution requires a physical property (x) whose value can range from $-\infty$ to $+\infty$, with x-values showing a strong

tendency to cluster around their mean $(\bar{x})$. Actual observations are equally likely to undershoot or overshoot the mean, so that the frequency distribution is symmetrical with long tails corresponding to rare values, far from the mean. The dispersion about the mean is defined by the variance (σ^2) or by the standard deviation (σ). The Gaussian curve has a symmetric hump shape with exponentially decaying flanks given by an equation of the form:

$$y = ce^{-ax^2} \tag{3.5}$$

Here, $c > 0$, $a > 0$ are constants, and the hump has a peak value, $y = c$, located at $x = 0$. Inflection points mark the change in curvature, from upward convex to downward concave, and occur at $x = \pm (2a)^{-1/2}$. These points are important visual clues in any natural distribution because they correspond to the x-values that pinpoint the standard deviation (σ) graphically (Fig. 3.5). This equation may be compared with the formal definition of the Normal distribution in Eq. (3.6).

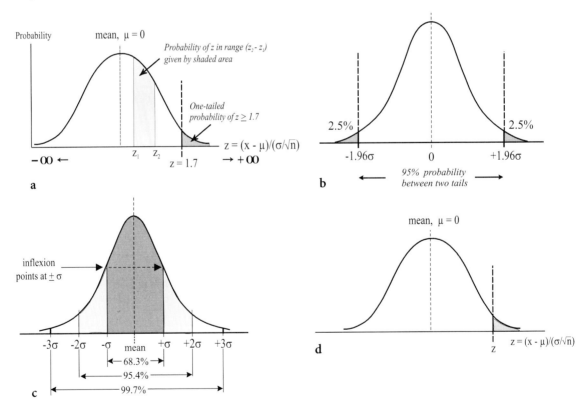

Fig. 3.5. **a** The area under the Normal distribution from $z = 1.7$ to $+\infty$, *shaded*. The z-variate is a standardized version of the measured value x, by the formula shown as the x-axis label. **b** Sometimes two extreme cases are considered: the two tails are *shaded* and each represents 2.5% of the observations that lie further than 1.96σ from the mean. **c** Specific values of z, in units of standard deviations (σ) that encompass certain percentages of observations symmetrically about the mean (see Table 3.6). **d** The *shaded area* under the Normal distribution from z to ∞ represents the probability or percentage of observations lying in the range z to ∞. Such values are directly given in Table 3.5

For distributions that strictly adhere to the theoretical model, the probability of occurrence for any given *x*-value is precisely predictable. Therefore, for visual comparison of different sets of data, even in different units of measurement or for hypothesis testing, it is convenient to report ranges of *x*-values as multiples of the standard deviation σ. We shall see below, and again in Chapter 5 that statistical comparisons between the frequency distributions of different data sets may be made completely universal by *standardisation*.

For any observed distribution of values, the validity of statistics based on the theoretical Normal distribution depends entirely on how closely the natural distribution follows the theoretical model. This may be overlooked. Statistics and the results of statistical tests based on the Normal distribution are almost universally quoted, although there may not be enough observations to determine whether their distribution is Gaussian. Clearly, there should be enough observations in the sample to establish that it follows the Normal distribution before using any statistical decisions requiring a perfect Normal distribution, such as hypothesis tests (Chap. 4). Nevertheless, it is common to read geological reports based on small samples ($n < 30$) in which confidences of values are expressed in multiples of σ with corresponding percentage probabilities extracted from the tables for the Normal distribution. There are good reasons why this is acceptable. First, the Normal distribution explains the dispersion of measurements subject to any random fluctuations, whether they are intrinsic features due to process variations or due to errors of observation. Second, the means of successive samples are approximately Normally distributed, regardless of the form of the population's distribution (*Central Limit Theorem*, Chap. 4). This greatly increases the applications for Normal-distribution statistics, permitting us to estimate mean values and their confidence without knowing the distribution of the population. Third, the applications are still greater if we can take advantage of the *additive property of the Normal distribution*; the sum of normally distributed independent random variables also follows a Normal distribution.

Sometimes we must use some procedure derived from the Normal distribution despite concern that the sample was not drawn from a normally distributed population (e.g. when comparing means of multiple samples, analysis of variance or ANOVA, see Chap. 4). Concerns may be alleviated if the original values may be transformed numerically so that they more closely follow the Normal distribution. Several types of transformation are described in Tables 5.8 and 6.1, but the most useful in this context are described next. The principle is to transform the data so that they are more like a Normal distribution in appearance, permitting evaluation by Normal-distribution techniques. Adding a constant to the data, to shift the origin and ease comparisons, may also enhance comparisons. When statistics are determined from the transformed distribution, they must be back-transformed into original *x*-units. The following two examples involve a common earth science situation where the data range is finite.

First, although the upper limit may be large enough not to cause concern, data may cluster near a lower boundary, as with trace-element concentrations. The *lognormal distribution* has such a form so it is reasonable to use it as a basis for *logarithmic transformation*. Thus, where *x* is the original data, we set a dummy variable $x' = \log_{10}(x)$, and its frequency distribution may then be sufficiently similar to a Normal distribution to permit analysis with Normal-distribution techniques.

Second, concentration values may spread over most of the finite range, for example mineral contents in a rock, or major element concentrations from a whole-rock analysis. At best, such data can only give a very compressed imitation of a Normal distribution. We may re-express the data to produce a distribution amenable to Normal-distribution statistics using the *arcsine transformation*. One divides the percentages by 100 to express them as a fraction of 1.0 yielding (x), then form a dummy variable $x' = \sin^{-1}(x)$. The distribution of x' will be more symmetrical, like a Normal distribution and occupy only a small part of the transformed range.

Assuming that the observed distribution is approximately Normal or, at the least, symmetrically hump shaped, it will combine both natural variation in the system and random errors of observation. Thus, mean and variance blend measurement errors with the inherent variation in the controlling geological process. In contrast to laboratory sciences, geology is fortunate that most observations have small measurement errors compared with the natural variation in the system. Each study must be considered on its own merits to ascertain the degree to which measurement error and natural variation are confounded in the calculated statistics.

Following the form of the Gaussian equation given previously, the probability density, $P(x)$, of the Normal distribution is described by:

$$P(x) = \frac{1}{\sigma\sqrt{2\pi}} \cdot \exp\left[-\frac{1}{2}\left(\frac{x-\mu}{\sigma}\right)^2\right] \qquad (3.6)$$

The *y*-axis represents the probability of recognising an observation of value *x*, with mean value μ and variance

σ^2. These are the population parameters, and we may estimate them from a sample by the corresponding statistics: $\bar{x}$ and s^2. Probability is graphed and tabulated in the range zero to unity, and the total area under the curve is consequently unity. The tabulated variable is usually given as z, rather than x, for reasons that will become apparent. We are normally interested in the probability of recognising observations in a given range, perhaps between z and $+\infty$, represented by the shaded area in Fig. 3.5a. Areas (= probabilities) between any two limits z_1 and z_2 are obtained by integrating the above expression between those limits. (The principle was shown for a non-Normal sample in Fig. 2.4c.) Areas under the Normal distribution are available in published tables and from statistical programs and spreadsheets. The areas represent probabilities and are tabulated cumulatively, usually from z to $+\infty$. Thus, the probability of finding an observation in the range z_1 to z_2 is determined by subtracting the tabled areas between them and $+\infty$; $(z_1$ to $z_2)$, see Fig. 3.5a. A simplified version of the table of probabilities for the Normal distribution is given in Table 3.5. It

is a perfectly general *standardised* distribution that may be applied to any normally distributed x-variable, regardless of its units of measurement, or of the values of its mean and standard deviation. The original measurements (x) are adjusted by arithmetic to give the z-variate which has a mean = 0 and standard deviation = 1. Thus, z conforms exactly to the units of the theoretical Normal distribution and the fractional area between z and $+\infty$ represents the probability of an observation lying in that range (Fig. 3.5d). Shorthand exists for describing probability distributions, for the *standard Normal distribution* it would be *ND(0,1)* where the values in parentheses describe the mean and standard deviation. Other probability distributions may be conveyed in this format. Since the Normal distribution is symmetrical about a mean, $z = 0$, we only need to tabulate probabilities for half the distribution, usually from zero to $+\infty$. In practice, the range tabulated is from $z = 0$ to $z = +3$ because 99.7% of observations following the Normal distribution lie in that range. Spreadsheets usually return the areas under probability distributions cumulatively. In Fig. 3.5c, the

Table 3.5. Areas under the standardised Normal distribution representing probabilities (P) from $+z$ to $+\infty$ z is in units of standard deviations (σ) (Fig. 3.5d)

$z = (x_i - \mu)/\sigma$	P	$z = (x - \mu)/\sigma$	P	$z = (x_i - \mu)/\sigma$	P
0.00	0.5000	1.00	0.1587	2.00	0.0228
0.05	0.4801	1.05	0.1469	2.05	0.0202
0.10	0.4602	1.10	0.1357	2.10	0.0179
0.15	0.4404	1.15	0.1251	2.15	0.0158
0.20	0.4207	1.20	0.1151	2.20	0.0139
0.25	0.4013	1.25	0.1056	2.25	0.0122
0.30	0.3821	1.30	0.0968	2.30	0.0107
0.35	0.3632	1.35	0.0885	2.35	0.0094
0.40	0.3446	1.40	0.0808	2.40	0.0079
0.45	0.3264	1.45	0.0735	2.45	0.0071
0.50	0.3085	1.50	0.0668	2.50	0.0062
0.55	0.2912	1.55	0.0606	2.55	0.0054
0.60	0.2743	1.60	0.0548	2.60	0.0047
0.65	0.2578	1.65	0.0495	2.65	0.0040
0.70	0.2420	1.70	0.0446	2.70	0.0035
0.75	0.2266	1.75	0.0401	2.75	0.0030
0.80	0.2119	1.80	0.0359	2.80	0.0026
0.85	0.1977	1.85	0.0322	2.85	0.0022
0.90	0.1841	1.90	0.0287	2.90	0.0019
0.95	0.1711	1.95	0.0256	2.95	0.0015
		1.96	0.250	3.00	0.0013

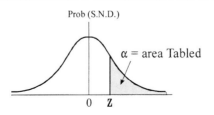

Prob (S.N.D.)

α = area Tabled

0   Z

areas lying under the curves within $\pm\sigma$, $\pm2\sigma$ and $\pm3\sigma$ of the mean represent the probabilities of recording observations in those x-intervals.

Later, we shall see that this may be applied to answer questions such as, "What is the probability of finding an observation that exceeds a certain value, x?" When we have learned to transpose the real unit of measurement (x) into the standardised units (z) of Fig. 3.5a, the answer would be represented by the *shaded* area under that curve (the right-hand tail). Table 3.5 gives such areas, as fractions of unity, which when multiplied by 100 yield the percentage probabilities of values exceeding z. The use of just one-half of a probability distribution leads eventually to a *one-tailed test*. We could also use the *lower tail* to determine what percentage of measurements occurs with x-values *below* a certain level.

For example, suppose the gold content in drill core has a mean concentration $\mu = 714$ ppm, and that the standard deviation, $\sigma = 121$ ppm. For commercial reasons, the prospector is interested in samples with concentrations >920 ppm; this is regarded as the critical value. What is the probability of such samples occurring? This requires transforming the critical value into a z-value so that we may compare it directly with the Normal distribution:

$$z = (x - \mu)/\sigma$$

or $z = (920 - 714)/121 = 1.7$

The area under the *standardised* Normal distribution, between $z = + 1.7$ and $+\infty$, represents the probability of observations being present in that range. From Table 3.5 we see that the area above $z = + 1.7$ is 0.0446 so that there is a 4.46% probability of measurements >920 ppm. This probability is represented by the shaded area under the right-hand tail of the curve in Fig. 3.5a. From the practical standpoint we should realise that this is not necessarily the same as *the probability of finding such samples* which suffer from bias in non-random specimen selection in the field (see Chap. 1).

Another common application of this principle uses both tails of the distribution. For example, in a potash mine, the optimum exploitable bed thickness to accommodate the machinery may be 2 m with a margin of ±0.20 m permitted by the extraction equipment. Thus, we need to determine what percentage of suitable seams exist with thicknesses between 1.80 and 2.20 m ($=2.00\pm0.20$ m). This excludes the extreme thicknesses corresponding to the *two tails* of the distribution. We know the mean, $\mu = 2.00$ m. From a large sample, we estimate the standard deviation, $\sigma = 0.102$

from a sample of measurements. We may standardise, determining the z-variate for the edge of the upper tail:

$$z = (x - \mu)/\sigma \quad \text{or} \quad z = (2.20 - 2.00)/0.102$$

Therefore, $z = 1.96$ for the lower limit of the upper tail. Thus, by interpolation from Table 3.5 we see that $\sim2.5\%$ of bed thicknesses should exceed 2.20 m. By symmetry, $\sim2.5\%$ of potash layers should be thinner than 1.80 m. This is indicated diagrammatically for the normalised scale (z) of the measured variable (thickness) in Fig. 3.5b, showing the two shaded tails of the distribution, under which the area represents the percentage of unsuitable, extreme bed thicknesses. This convenient example uses z-values $= \pm 1.96$ which correspond to the 95% limits about the mean (Table 3.6).

Table 3.5 gives the probability of finding a sample in the shaded area under the curve of Fig. 3.5d, between z and $+\infty$. However, for ready use of such a table, our data must always be *standardised* to facilitate a universal comparison as introduced informally above. The standardisation procedure of the observations (x_i) involves three steps:

1. subtract the mean, μ from each value: ($x_i - \mu$)
2. divide each by the standard deviation, σ: ($x_i - \mu$)/σ
3. now any observation x_i has been expressed in standardised form: $z_i = (x_i - \mu)/\sigma$

This dummy variable is traditionally called the *z-variate*. Subtracting the mean from each value centres the new distribution about a zero value. Dividing by the standard deviation makes the dummy variable z universal, in the sense that it is in units of standard deviations. A sample distribution transformed in this way may be readily compared with any other standardised distribution, or with the standard Normal distribution. Different mean values, different variances, or different

Table 3.6. Normal distribution, proportion of observations lying within certain ranges, symmetric about the mean value (σ = standard deviation). The upper three rows may be applied to many hump-shaped, symmetric, empirical distributions

Range about mean e.g. from $-\sigma$ to $+\sigma$	Observations included range (%)	Fraction of observations included in range (Normal distribution probability)	Fraction excluded from range
$\pm\sigma$	68.3	0.683	0.317
$\pm2\sigma$	95.4	0.954	0.046
-1.96σ to $+1.96\sigma$	**95.0**	**0.95**	**0.05**
$\pm3\sigma$	99.7	0.997	0.0027
$\pm4\sigma$	~100	0.999937	0.000063
$\pm5\sigma$	~100	0.99999943	0.00000057

units of measurement of the original variables do not hinder such comparisons. More commonly, single values of a variable are transformed in this way to facilitate comparison with the Normal distribution and to take advantage of the tabled probabilities (Table 3.5), as in the previous examples.

The area under the standardised distribution is 1, representing 100% of all possible observations. Where $x = -\infty$, the probability of finding an observation is zero. Between $-\infty < x < \infty$ all possible data are found so that the *cumulative* probability is 1 (100% of observations occur in this range). Using standard deviations as a universal reference unit permits scientists to refer to certain ranges of values about the mean, which can be readily compared with other studies. Certain ranges are chosen that define popular percentages of observations, equivalent to the probability of their occurrence. For example, the *one-sigma limits* about the mean (thus, $\mu \pm \sigma$) enclose 68.4% of data; the most popular ranges are shown in Table 3.6 and Fig. 3.5c. Note that the $\pm 1.96\sigma$ limits define the range that includes 95% of observations and will be used commonly in tests of significance (Chap. 4).

Table 3.6 and Fig. 3.5b–c illustrate the use of the sigma limits for symmetrical or *two-tail* tests. These consider values that lie beyond a certain multiple of σ in both senses, both positive and negative. The fractions are derived by integrating areas under the Normal-distribution curve *between* the boundary x-values giving the range in multiples of σ (Table 3.6). The 2σ *(two-sigma)* limit is most commonly used with laboratory-based measurements whereas the σ-limit (*one sigma*) is more realistic for the treatment of some field observations that have greater intrinsic variability and which may be measured less precisely. Figure 3.5b shows the popular 95% limits and a one-tailed test; 95% of observations fall within $\pm 1.96\sigma$ of the mean value. We may wish to note that 2.5% of observations exceed $+1.96\sigma$, and 2.5% are less than -1.96σ; these correspond to the two tails of the distribution. An example using one tail is shown in Fig. 3.5a and was discussed earlier. Since the Normal distribution is symmetrical, simple arithmetic relates the percentage probabilities for one-tailed and two-tailed tests when using Table 3.5.

Commonly, we like to know the x-range that would include 95% of our data. Table 3.6 and Fig. 3.5b indicate the corresponding z-range is -1.96σ to $+1.96\sigma$ for a two-tailed situation. It is sobering to realise that this means there is a 5% chance that any observation lies outside this range. Nevertheless, this value is particularly common in assessing significance of results as it is a pragmatic balance between the rigour bestowed by high confidence and the appreciation of the impreci-

sion of geological data. (There are also theoretical reasons for preferring confidence levels near 95%; Chap. 4.) The 95% confidence-level range is sometimes used synonymously with the term 2σ-level (two-sigma) range in geological literature, although the 95% limits are actually $\pm 1.96\sigma$. More seriously, we should state explicitly whether a one-sigma, two sigma or 95% confidence limit is used in reports and diagrams; this is sometimes omitted. Whereas a one-sigma limit may be appropriate for many field measurements, geochronology merits a two- or three-sigma limit. Some parts of the geological community adopt a certain confidence level by convention and quote their results without specifying at what level the imprecision was determined.

The Normal distribution is a mathematical model of dispersion commonly applied in geology, although its value may be reduced if the observed distribution is non-Normal. For example, it is difficult to estimate the population mean and variance if the sample distribution is notably skewed. Even if the distribution is at least a roughly symmetrical hump shape, many geological variables do not range from $-\infty$ to $+\infty$. For example, porosity or the weight% CaO must lie between 0 and 100%. These present part of an irresolvable issue known as the *constant-sum* problem to which we return in Chapter 7. In these limited or closed systems, the use of the Normal distribution may be warranted only where the observed data range is small, e.g. 20 to 30% within the maximum possible range of 0 to 100%. Otherwise, the finite range truncates the tails of the distribution. For example, the distribution of major oxides' abundances may fall only within the 100% range, and the frequency distribution of abundances cannot be Normal. In contrast, trace elements occur in low abundances, usually in a few parts per million (ppm). Thus, the range of observed values (e.g. 50 to 300 ppm) is small enough that the potential range of 1,000,000 ppm may be considered a fair approximation to infinite. Thus, the closed nature of the system does not cause a significant practical problem. Constant-sum (= closed-system) problems also arise with observations restrained to a semi-infinite range, such as pebble-aspect ratio varying from 1 to $+\infty$, and grain size varying from zero to $+\infty$.

3.4
The Links Between Certain Theoretical Probability Distributions

It will be of interest to note the fundamental continuous probability distributions are interrelated so that

Table 3.7. Relationships between some fundamental probability distributions

Probability distribution	Density	Limits	Mean	Variance
Uniform (U)	$1/(\beta-\alpha)$	$\alpha \le x \le \beta$	$(\alpha+\beta)/2$	$(\beta-\alpha)(\beta-\alpha+2)/12$
Gamma (Γ)	$\dfrac{1}{\Gamma(\alpha)\beta^\alpha} x^{\alpha-1} e^{-\frac{x}{\beta}}$	$\alpha > 0;$ $\beta > 0;$ $x > 0$	$\alpha\beta$	$\alpha\beta^2$
χ^2	$\dfrac{1}{\Gamma\left(\frac{\mu}{2}\right)2^{\mu/2}} x^{\left(\frac{\mu}{2}-1\right)} e^{-x/2}$	μ integer $> 0;$ $x > 0$	μ	2μ
Normal (ND)	$\dfrac{1}{\sigma\sqrt{2\pi}} \cdot \exp\left[-\dfrac{1}{2}\left(\dfrac{x-\mu}{\sigma}\right)^2\right]$	All $x;\ \sigma > 0$	μ	σ^2
ND *Standard Form*	$\dfrac{1}{\sqrt{2\pi}} \cdot \exp\left[-\dfrac{1}{2}(z)^2\right]$	all z	0	1
Poisson	$\dfrac{e^{-\lambda}\lambda^x}{x!}$	$x = 0, 1, 2\ldots;$ $\lambda > 0$	λ	λ
Exponential	$\dfrac{1}{\beta} e^{-\frac{x}{\beta}}$	$x > 0;\ \beta > 0$	β	β^2
Binomial	$\dfrac{n!}{x!(n-x)!} p^x (1-p)^{(n-x)}$	$x = 0, 1, 2, \ldots, n;$ $0 < p < 1$	np	$np(1-p)$

for extreme values of certain parameters, two distributions may appear quite similar. For example, a binomial distribution with a large number of trials (n) and small individual success rates (p) for each event, is indistinguishable from a Poisson distribution whose mean $\mu = np$. As the mean (μ) of the Poisson distribution increases, it approximates a Normal distribution.

The Gamma distribution (Γ), although used rarely directly in geology, may be compared to an umbrella encompassing the Normal and Poisson probability distributions. Its probability density is defined in Table 3.7 together with formal definitions of the probability densities of the commonly used continuous probability distributions. Whereas these theoretical details are unimportant to most geologists, they may help us to appreciate what sample properties dictate when one distribution is preferable to another. This may help in the choice of the simplest appropriate starting model in a study (Johnson 1994). Chatfield (1978) shows some details of the theoretical links between χ^2, t, F, Normal, Poisson and Γ-distributions, and a good overview of most distributions is found in the appendix to Lee's book (1989). The similarity between the χ^2-, t- and F-distributions and the Normal distribution can be better appreciated when we realise that they may be generated by repeatedly sampling the parent Normal distribution (Till 1974). As a consequence, for certain sample sizes and degrees of freedom, the χ^2-, t- and F-distributions may appear indistinguishable from the parent Normal distribution.

CHAPTER 4

Statistical Inference: Estimation and Hypothesis Tests   4

Confidence intervals around the mean are of great practical importance as they convey the importance of our results and usually require few assumptions about the nature of the population. Moreover, these simple statistics are readily visualised on diagrams where they may appear as error-bars or confidence ellipses around the mean value (e.g. Figs. 4.2, 7.3, 9.9, 10.8a,b, 10.12). We are reminded that "in general, though not always, estimation is more important than significance testing" (Chatfield 1978, p. 165). Thus, we first address the concept of *confidence limits*. This is followed by the subject of *hypothesis testing* which permits us to qualify a decision at some level of confidence. For that we must know the form of the probability distribution from which the sample was drawn. Hereafter, it is important that we deal only with a *simple random sample*. This means that each observation is selected without bias and has no influence on the selection of any other observation. Those precautions require specific knowledge of the subject being studied.

4.1
The Central Limit Theorem

Commonly, we record measurements of a new phenomenon with no knowledge of the distribution of its population. How then may we evaluate our sample? Fortunately, the *Central Limit Theorem* rescues us. From the unknown population, we draw repeated random samples, each with n observations, and determine their means. The symbol $\bar{x}$ is traditionally used for the *sample-mean* but here, for clarity, the symbol $\bar{x}_S$ is used here to emphasise designation of a sub-sample's mean. Sampling is performed with replacement, i.e., the sample would be returned to the population after determining its mean, so that all the original observations are always available for subsequent sampling experiments. Each experimental sample mean, $(\bar{x}_S)$, represents an "observation", somewhat like any individual unbiased measurement subject to random fluctuation. Thus, just like individual measurements subject to random error, the sample means $(\bar{x}_S)$ tend to follow the

Normal distribution. Furthermore, it can be proven that the individual sample means $(\bar{x}_S)$ approximately follow the Normal distribution, *regardless of the form of the population from which they are drawn.* Moreover, the average (μ_x) of the sample means $(\bar{x}_S)$ closely approximates the population mean (μ). We shall see that the approximations improve with the size of individual samples (n_s), requiring only that the population has a finite variance (σ^2).

An experiment with hypothetical data corroborates this theorem in Fig. 4.1a. The population shown is clearly not Gaussian; it is not even hump shaped. Nevertheless, repeated samples comprising only two observations $(n_s = 2)$ produce a hump-shaped distribution. As the size of the samples increases, the distribution of sample means becomes closer to the Normal distribution, especially for $n_s > 20$.

The Central Limit Theorem can be stated more formally. Consider a population with mean μ and standard deviation σ, but which need not follow the Normal distribution. Replicate samples from the population have sample means $(\bar{x}_S)$ that show a central tendency to μ. The *standard deviation of the sample means* $(\bar{x}_S)$ is given by $(\sigma/\sqrt{n_s})$ and is known as the *standard error*, where σ is the standard deviation of the population and n_s is the size of each sample. *Thus, it is possible to determine confidence limits and error bars for a wide variety of data without any assumptions about the distribution of the population from which they are drawn, provided the population has a finite variance.* This is achieved simply by calculating the means of multiple samples. Without any knowledge of the form of the population, one obtains quantitative estimates of certainty that make graphical reports of earth-science data more objective and easier to evaluate. The Central Limit Theorem is illustrated with an example of 959 gold assays that clearly do not follow a Normal distribution (Fig. 4.1b, data from Koch and Link 1970). Like many low-concentration abundances, there are high frequencies of low concentrations and a long positively skewed tail of high values. Rich ore is scarce, uneconomic concentrations are more common. As a caution to the evaluation of this example, we

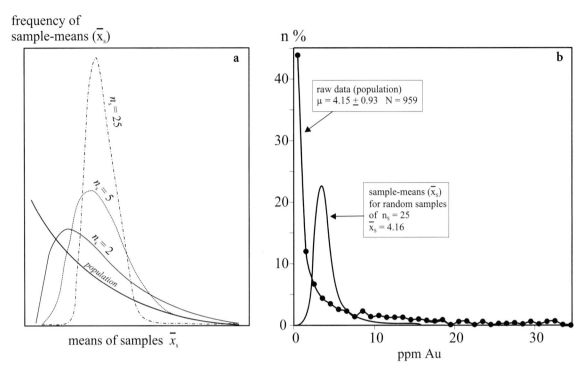

Fig. 4.1. The Central Limit Theorem reveals that the means of samples drawn randomly from any population approximately follow the Normal distribution. **a** In this theoretical example, the population clearly does not follow the Normal distribution but the sample means do form hump-shaped distributions. As the size of samples (n_s) is increased, the distribution of sample means becomes more symmetrical and more closely follows a Normal distribution with a mode closer to the population mean. **b** A population of N = 959 gold assays from Homestake mine are not normally distributed (data of Koch and Link 1970). However, when random samples of n_s = 25 assays are selected, their means ($\bar{x}_s$) follow a Normal distribution more closely, centred on the mean (μ) of the population

should recall from Chapter 2 that low abundances commonly approach the detection limit of the analytical method. There could even be a peak just above zero concentration. However, that would not affect this demonstration. For our purposes, it is sufficient that the population of 959 values is clearly not Normal. Nevertheless, when 500 random samples of 25 assays are taken, the frequency distribution of the samples' means ($\bar{x}_s$) is hump shaped and almost symmetrical (Fig. 4.1b). The sample means are approximately normally distributed and they are also approximately centred on the population's mean. This experiment was repeated using different sample sizes (n_s). As individual samples increase in size, their means draw closer to the population mean. Moreover, the standard deviation of the sample's means (σ_x) decreases as ($1/\sqrt{n_s}$).

The most important results may be summarised and generalised as follows (Hoel 1971; De Vore and Peck 2001). Consider a sample of n_s observations drawn randomly from a population of N x-values. The population may follow any kind of unspecified distribution that has a mean μ and standard deviation σ. The

means of many such samples (*sample means*, $\bar{x}_s$) have an average of $\mu_{\bar{x}}$ and standard deviation of $\sigma_{\bar{x}}$. The Central Limit Theorem then provides us with the important general results;

1. $\mu_{\bar{x}} = \mu$
2. if $n_s \geq 30$, the distribution of *sample means* ($\bar{x}_s$) follows the Normal distribution regardless of the form of the population of x. If the population of x is reasonably symmetrical and hump shaped, $\mu_{\bar{x}} \approx \mu$ for $n_s \geq 15$.
3. $\sigma_{\bar{x}} = \sigma/\sqrt{(n_s)}$ (Note if (n_s/N) > 0.05 this may be invalid; a rare case where smaller samples are appropriate, see also *bootstrapping* later).

One cannot underestimate the generality, significance and the optimism that the Central Limit Theorem brings to the statistical estimation of natural observations. We cannot always assume a form for a distribution of natural observations and yet we may proceed with basic statistical estimation from a small number sample means. It is even more encouraging that the Central Limit Theorem is successful with populations that are asymmetric or lacking a peak (McClave et al. 1997). The applications are still greater when we take advantage of the *additive property of the Normal distribution* which explains that the sum of normally distributed independent random variables must also follow the Normal distribution. As a consequence, the combined variance of independent variables (A, B) is given by the sum of the individual variables variances;

$$\sigma^2_{A+B} = \sigma^2_A + \sigma^2_B.$$

4.2
Estimation

In general, estimation refers to point estimates and interval estimates. Here we are concerned with estimating *intervals of confidence* about a mean value. For simplicity, we adopt the common earth-sciences convention of a 95% confidence level. For the Normal distribution, 95% of observations fall in the range $\pm 1.96\sigma$ (Tables 3.5, 3.6), which will explain the presence of the factor 1.96 in the following expressions. Recall that if multiple samples were taken, 95% of the intervals should contain the population (true) mean represented by μ. The expressions below may be generalised for other confidence levels by substituting the appropriate z-value that bounds the rejection region of area α, under the Normal distribution (Table 3.5), shaded in Fig. 3.5b, d.

4.2.1
Standard Error of the Sample Mean

The concept that sample variance represents the dispersion of values about the sample mean is now familiar. In a similar fashion, we may calculate the variance of the sample means calculated from each of several samples. For any distribution of x_i, we may calculate the variance σ^2, and the Central Limit Theorem shows that the sample means $\bar{x}$ follow the Normal distribution with a variance σ^2/n. Thus, the *standard error of the mean* is $\sigma/\sqrt{n}$. In effect, this is the *standard deviation of sample means*. Of many samples, 68% should have means with a confidence interval defined by $\bar{x} \pm (\sigma/\sqrt{n})$. In 68% of trials this interval will contain (μ), the mean of the population from which the observations (x_i) were sampled. Usually, we require a closer estimate of the range that contains the population mean, for example, the range that contains the true mean with 95% confidence is a common choice.

4.2.2
Confidence Interval that Contains the Population Mean

A simple manipulation of the sample's mean and standard deviation can provide an estimate of the "true" or population mean within a 95% confidence interval. A similar procedure is adopted for estimates of other population parameters such as the slope of a regression line or the value of a correlation coefficient (Chap. 6.)

If the distribution is Normal (ND), as predicted for sample means according to the Central Limit Theorem, we know that values corresponding to the extreme tails of the ND should be rejected. At the 95% confidence level, we would reject the 2.5% of values at the upper end of the range of values, as well as the smallest 2.5% of values. Under the standardised Normal distribution, 2.5% of the area lies above $+1.96$ and 2.5% lie below -1.96. Therefore, to express 95% confidence in a result, we wish to exclude values for which the corresponding z-values exceed $|1.96|$. This leaves us with the z-range that encloses 95% of the probability distribution, $-1.96 < z_{0.95} < 1.96$. Therefore, the sample mean $(\bar{x})$ must be standardised to permit a valid comparison with z-values, and we write the 95% confidence interval as:

$$-1.96 < \left(\frac{\bar{x} - \mu}{\sigma/\sqrt{n}} \right) < +1.96 \tag{4.1}$$

rewriting the two halves of the inequality:

$$\left(\bar{x} - \frac{1.96\sigma}{\sqrt{n}} \right) < \mu < \left(\bar{x} + \frac{1.96\sigma}{\sqrt{n}} \right) \text{ thus the} \tag{4.2}$$

95% confidence interval for μ is given by

$$\bar{x} \pm 1.96 \frac{\sigma}{\sqrt{n}}$$

These are the confidence estimates, defining the interval that contains the population mean (μ), with a 95% probability. Alternatively, we may say that for 95% of samples, this interval will contain the true mean, the population mean.

This approach may be generalised and extended in two ways. First, we may express a confidence interval at any level, not just 95 or 68%. We recall the complementary relation between the confidence level and the significance level, α, which is defined by the rejection area under the tails of the Normal distribution. The factor 1.96 (for $\alpha = 0.05$) may be replaced by the appropriate percentage point of the Normal distribution for other confidence levels. Thus, in general, for a confidence level $100(1-\alpha)\%$, the confidence interval for μ is given by

$$\bar{x} \pm z_{(\alpha/2)} \frac{\sigma}{\sqrt{n}} \quad (n \geq 30) \tag{4.3}$$

$$\bar{x} \pm t_{(\alpha/2)} \frac{s}{\sqrt{n}} \quad (n < 30) \quad \text{select t with } (n-1)$$

The second generalisation of this procedure extends its application to other distributions. Confidence intervals are mostly estimated for the Normal distribution since they are applied most often to sample means that must follow the ND. Consequently, they require the z-variate and its familiar critical values, $z = \pm 1.96$, to bracket the true mean with 95% probability. However, sometimes confidence intervals are required for samples drawn from non-Normal parent distributions. For example, if $n < 30$, the sample standard deviation is a poor estimate of σ and the t-probability distribution (Table 4.5) is more appropriate than the Normal distribution. We then substitute the critical value of t for z in the previous equation and replace σ with its estimate from the sample, s. In consulting the t-Table, we use $(n-1)$ degrees of freedom. The dependence on sample size then becomes clear for the following values (for the 95% confidence level, $\alpha = 0.05$);

$t_{(\alpha/2)} = 1.96$ for very large n
$t_{(\alpha/2)} = 2.04$ for $n = 30$
$t_{(\alpha/2)} = 2.09$ for $n = 20$
$t_{(\alpha/2)} = 2.23$ for $n = 10$
$t_{(\alpha/2)} = 2.57$ for $n = 5$

4.2.3
Confidence Intervals by the Bootstrap Method

A computer procedure may be used to estimate the confidence interval for the mean of an unknown population from a small, random sample ($n \leq 25$) that is representative of the population. The *bootstrap* approach is non-parametric, assuming that all the necessary information is carried in the small sample. Knowledge of the population is not required. It is logically discussed here but mainly used with orientation data (Chaps. 9 to 11), where theoretical models for the population are usually unknown but nevertheless necessary for most statistical estimates and tests.

Using the computer's random number generator, the original sample is repeatedly *resampled* to simulate other potential trials. The pseudo-samples are always drawn from the original, undepleted sample of obser-

vations. For example, consider a sample of six original observations (x_i) with mean $\bar{x}_0$:

A	B	C	D	E	F	$(n = 6, \bar{x}_0)$

We need to select one of six observations, as if by throwing a die, so we must choose a random number in the range from one to six. A computer may be used to give a random number U which will lie in the range $0 \leq U \leq 1$, for example $U = 0.673$. The integer part of nU will give a value that may be used to select an observation from the original sample, with 0.5 added to prevent rounding down, thus $m = INT (nU + 0.5)$.

Thus, the mth value in the original sample will be the first value for the bootstrap sample. The observation is returned to the sample and the random selection procedure is repeated for the next pseudo-observation for the bootstrap sample. This means that the same observation could be drawn again, a procedure known as *sampling with replacement*. It is like drawing one of six lottery tickets from a bin but then replacing it so that the same ticket could be chosen again in the next draw. The procedure is repeated five times and the first bootstrap sample is complete when we have six pseudo-observations such as:

A	B	B	D	F	F	$(n = 6, \text{mean } \bar{x}_1)$

The bootstrap pseudo-sample or *resample* may omit or duplicate some of the original observations due to the randomness of their selection. The mean of the first bootstrap sample is calculated ($\bar{x}_1$).

This preceding experiment is repeated between 200 and 1000 times. In each experiment, a bootstrap sample of $n = 6$ pseudo-data are selected with replacement, and the mean of that bootstrap sample is recorded. Suppose that 1000 experiments were performed. The 1000 bootstrap means are now sorted in ascending order of magnitude, $\bar{x}_{MIN}, \bar{x}_{MIN+1}, \dots \bar{x}_{MAX}$.

The basis of the confidence range is that we wish to identify extreme values and exclude them from consideration. The 95% confidence range excludes the extreme values that constitute 5% of the distribution ($\alpha = 0.05$), the 2.5% smallest values and the 2.5% largest values. Therefore, we want to exclude the 25 smallest values and the 25 largest values of the 1000 bootstrap means. The 95% confidence range of bootstrap means would therefore be bounded by $\bar{x}_{MIN+26}$ and $\bar{x}_{MIN+974}$ at the lower and upper limits, respectively. In general, for N bootstrap resamples, whose means are listed in *ascending order of magnitude,* then the confidence limits for the population mean μ are given by:

$$\overline{X}_{\left(\frac{\alpha}{2}\right)N} < \mu < \overline{X}_{\left(1-\frac{\alpha}{2}\right)N} \qquad (4.4)$$

Effectively, the bootstrap technique assumes that the sample is the population. It is recycled in a complex manner to attempt to generate a pseudo-population, which should give reliable confidence estimates, if the original sample was representative and random. Pulling yourself up by your own bootstraps is not meant to be easy. The following much older approach places great importance on a small sample but requires a little knowledge of the population, however poor that knowledge might be.

4.2.4
Estimation and Bayesian Statistics

Classical statistics used above for estimation are based on the use of moments and maximum likelihood estimates using sampling theory. An alternative and old approach, initiated by Reverend Thomas Bayes (1702–1761), published posthumously in 1763, introduced the notion of conditional probabilities in prediction. Applied to modern statistical estimation, this provides a kind of feedback in which information from the sample is used to revise the population parameters. In other words, the population parameters are regarded somewhat like random observations, themselves subject to improved estimation. Bayesian techniques have not intruded far into earth science, although updating population parameters as new random samples are acquired seems a logical advance (Morgan 1968; Smith 1988; Lee 1989). This would especially be the case in earth science where the population is commonly poorly constrained in the first place and samples are sometimes small. In the case of time series that we discuss in Chapter 8, data are incremented one item at a time so that the instantaneous sample has $n = 1$ and the Bayesian approach seems a welcome advance (Spall 1988).

The approach involves a *prior distribution* which provides some previous knowledge of the population, e.g. its mean and variance (μ_0, σ_0^2). We then use the sample's mean and standard deviation ($\overline{x}$, s) and sample size (n) to revise the population parameters. Our goal is to determine an improved estimate of mean and variance, as *posterior statistics*, which combines information from the prior distribution and the sample.

Lee (1989) explains lucidly how the feedback mechanism is fundamentally related to the increase in precision as knowledge is gained from the sample. This assumes that our variable is at least approximately normally distributed, permitting use of the additivity theorem; combined Normal distributions produce another Normal distribution. If we recall that the precision is the inverse of variance ($1/\sigma^2$), and proportional to (n) for the sample, it is almost intuitive that the posterior precision combines the precision of the prior distribution and of the sample. This is shown in the first of the following pair of equations and permits us to calculate the posterior standard deviation (σ_p). In the second equation, we see that the posterior mean (μ_p) combines the prior and sample means, weighted according to their precision and sample size.

$$\frac{1}{\sigma_P^2} = \frac{1}{\sigma_0^2} + \frac{n}{s^2} \qquad \frac{\mu_P}{\sigma_P^2} = \frac{\mu_0}{\sigma_0^2} + \frac{n\overline{x}}{s^2} \qquad (4.5)$$

These equations may then be used to derive the *posterior population parameters* (μ_p, σ_p) which can be used to estimate confidence limits in the usual way. This usually produces tighter estimates with a posterior mean located between the prior and sample mean. For example, suppose experience shows we believe that the concentration of Pt in assays in a mineralised zone has:

1. *population mean and standard deviation (Bayesian prior distribution)* ($\mu_0 \pm \sigma_0$) = 950 ± 25 ppb.

A random sample of ten more sites gives:

2. *sample mean and standard deviation* ($n = 10$) ($\overline{x} \pm s$) = 850 ± 50 ppb.

Combining that information according to the previous equations gives:

3. *Bayesian posterior distribution, mean and standard deviation* ($\mu_P \pm \sigma_P$) = 879 ± 13 ppb.

The effects of relative precision among the sample and prior distributions are clear when we consider the following list. Line (1) shows that even a relatively imprecise sample effectively controls the posterior mean more than the prior distribution. As the standard deviation of the sample decreases, i.e. as its mean becomes more precise, the posterior mean rapidly converges toward the sample mean, in lines (1) through (3).

	Prior mean ± SD	sample ($n = 10$) mean ± SD	Posterior mean ± SD
1	950 ± 50	850 ± 100	862 ± 17
2	950 ± 50	850 ± 50	853 ± 9
3	950 ± 50	850 ± 25	851 ± 5

Bayesian statistics are increasingly popular in fields where sample sizes are necessarily small, such as medical clinical studies. By the year 2000, one-third of all clinical studies used Bayesian approaches. Of course, great care is needed in interpretation. The very concept of accumulating extra observations and integrating them into the prior distribution implies that conditions do not change as information is accumulated. Lee (1989) quoted the example of early geochronological work on the Ennerdale granite in the English Lake District, and for illustration in an introductory example, effectively used a sample size of one. First, this example shows the use with the smallest possible sample. Second, it shows the importance of understanding the subject matter in order to determine whether the sample may in fact be merged directly into the prior distribution. The earliest determinations, using the K/Ar method, gave an age and standard deviation of:

K/Ar   370 ± 20 Ma

Subsequently, the Rb/Sr method, regarded as an improved technique, yielded an older age with higher precision:

Rb/Sr   421 ± 8 Ma

Combining the information using the Bayesian approach yields a supposed "best estimate" or posterior estimate of

K/Ar and Rb/Sr   414 ± 7 Ma.

We may safely conclude that the sample has a strong influence on the prior statistics. However, how do we interpret the result geologically? It has since become known that the K/Ar method is a clock set at rather low temperatures whereas the Rb/Sr method has a higher blocking temperature. Therefore, the younger K/Ar date fixes the age of the granite at a later stage in the cooling history than the Rb/Sr age. The new method is not "better". It dates a different phenomenon. Was it meaningful to blend the information? That really depends on our goal. Neither technique was appropriate if the intention was to fix the crystallisation age of the igneous rock. That requires a radioactive product trapped in the rock from the earliest stages of crystallisation, such as provided by the U/Pb series in zircon. The preceding example really just blended two different cooling ages, each equally valid for the ages at which the rock cooled through different blocking temperatures.

Bayesian statistics do not use the concept of a true frequency distribution with which to compare a sample. Instead, small-sample statistics improve our imperfect knowledge represented by prior distribution parameters. The resulting posterior statistics may be used in the classical fashion to estimate confidence limits, but it may be misleading to use terminology like "95% confidence limits". After all, we are only improving our ideas and have no sound frequency distribution on which to base our estimate. For this reason, Bayesian statisticians use terms such as *high-density region (HDR)*, *credible interval* or *Bayesian confidence interval*. Lee contrasts the difference in emphasis between the classical 95% confidence interval and the HDR, in the following manner, where the superscript tilde (~) identifies quantities that are considered "most" variable.

Classical Approach, most faith in population

$$-1.96 < \frac{\mu - \tilde{x}}{\sigma} < +1.96$$

Bayesian Approach, more faith in sample

$$-1.96 < \frac{\tilde{\mu}_{POST} - x}{\tilde{\sigma}_{POST}} < +1.96$$

For simple situations as previously used in this chapter, the differences may be considered of academic importance. In more complex situations, Bayesian statistics may lead to quite different conclusions from the classical approach. The remainder of this Chapter, and indeed this book, is concerned with classical statistics, which is most useful in the majority of earth science projects.

4.2.5
Error-bars and Confidence Regions on Graphs

In these early chapters, we considered observations represented by a single value and recorded in frequency distributions. However, many useful data are bivariate; they may be presented graphically in the form of two values (x, y). Estimates of one variable may give many pairs of (x, y) values in which x or y or both scatter about a central value. Often, the scatter is normally distributed, leading to simple useful estimates of variance and confidence about the mean value, just as with frequency distributions of a single variable in preceding Chapters.

On (x, y) graphs the mean value of a pair of variables (x, y) could be plotted with *error-bars* passing through the point (x, y). The lengths of the error-bars are proportional to some estimate of scatter, for example, standard error or standard deviation. Where the uncertainty in x is independent of that in y, and vice versa, the error bars are parallel to x and y respectively, and the confidence region is an ellipse symmetrical with the graph axes. For example, let us inspect the strains of a slate, determined from the distortion of sedimentary structures. The strain or shape change of the rocks was determined on the scale of outcrops using many observations of certain sheared sedimentary dikes. From this, one can determine what shortening is expected (y) in any given direction in the rock from the rotation of sand dikes and compare it with independent estimates (x) of their shortening from their small-scale folding (Fig. 4.2a). The errors associated with the two types of strain estimate are independent; thus, the error ellipses are parallel to the (x, y)-axes. Chapter 7 shows a similar example where the thickness of the lower crust and the intensity of aeromagnetic anomalies are graphed (Fig. 7.3). In these cases, the *covariance* is said to be zero; uncertainty of the x-variable is quite independent of the uncertainty in the y-variable. Covariance is a property that will be discussed in connection with linear regression (Chap. 7). It should be noted that the ellipticity of confidence regions can be an artefact of scaling of the axes, which is mentioned further below. Scales may even be chosen that could make some confidence regions appear circular around the data point, even if the measurements of x and of y have different precision. Figure 4.2e shows error ellipses that vary from elongate parallel to the x-axis through circular shaped, to elongate in y. This results from large variations in the uncertainties associated with the dependent and independent variables. This archeomagnetic data set also illustrates the dangers of over-interpretation. The large error-bars and small number of data points make it difficult to justify many of the turning points in the archeomagnetic intensity curve. The fact that some turning points are based on a single datum may cause concern.

In situations where there is non-zero covariance of x with y, the error-bars and the axes of the elliptical confidence region are not parallel to the graph axes. This is a typical feature of the concordia diagram, used in geochronological work with Uranium-Lead isotopic ratios. Due to the interdependence of the two isotopic ratios ($^{206}Pb/^{238}U$ and $^{207}Pb/^{235}U$) in the decay series, the error-bars are inclined to the coordinate axes and the confidence ellipses are thus not parallel to graph axes (Fig. 4.2b, c). The inclination of the confidence ellipse with the x-axis is related to the variances of x and y correlation coefficient and will be discussed under the *bivariate Normal distribution* in Chapter 7. In the case of U-Pb geochronology, the long axis of the confidence ellipse is controlled by U/Pb uncertainty whereas the short axis relates to uncertainty in the ratio of the lead isotopes (York 1969; Davis 1982).

Earth sciences very often involve complex physical and chemical systems with poorly understood interactions between variables. Sometimes, appropriate diagrams reveal confidence regions informally, in a semi-quantitative manner. An example is provided by the distribution of (P–T) conditions of diamondiferous kimberlite inclusions versus those that are barren (Fig. 4.2d). The data points have not been subject to statistical treatment, and the dispersion shown is by visually approximated ellipses. Nevertheless, from our understanding and appreciation of statistics, it is clear where the mean conditions lie for these two petrological cases and what are the valid ranges of pressure and temperature. The interdependence of the variation in P and T is also obvious from the inclination of the confidence regions with the graph axes. Uncertainty in the distribution is defined informally, simply by plotting the observations directly (Finnerty and Boyd 1984); an explicit understanding of confidence ellipses and covariance was not required. Clearly, we should not underestimate intuition founded in a knowledge of the subject, even in the absence of formally determined confidence regions. Moreover, it shows a use of confidence regions qualitatively defined by natural variation, not precision in the statistical sense.

The geochronology literature provides some very instructive examples on the careful analysis of errors. Work determining the U-Pb ages of metamorphosed rocks of the Canadian Shield of Northern Ontario shows the careful analysis of errors in three categories (Davis 1982; Davis et al. 1982; Davis and Edwards 1986; Faure 1986; Thurston et al. 1991). The first is experimental error, usually rated at 0.1 to 0.5% for the determination of the isotope ratios from mass spectrometry. This is a standard error for the mean. The data yield a line (isochron) whose slope defines the critical isotopic ratio. This is the second and greatest source of uncertainty in the final result, because the natural scatter in the data produces some variation in possible isochrons. Thus, combining the *slope $\pm$ (Δslope)* of the isochron with the decay constant λ (and its inherent error) yields the age and its errors. The combination of sophisticated laboratory techniques is so refined that results are remarkably precise. The 95% confidence limits are quoted about the mean so that typical results for the age of volcanism in Northern Ontario yield values such as $2734.8^{+4.0}_{-3.1}$ Ma (Davis et al. 1982; Corfu and Andrews 1987). Note that the errors may be asymmetric, as in the example just quoted. The reason for this is that the best age estimate is given by the intersection of the concordia curve with the line through the data points (Fig. 4.2c). There is a small uncertainty in the slope of the line. Consequently, its possible intersections with the concordia curve are asymmetrically distributed on either side of the best intersection. Techni-

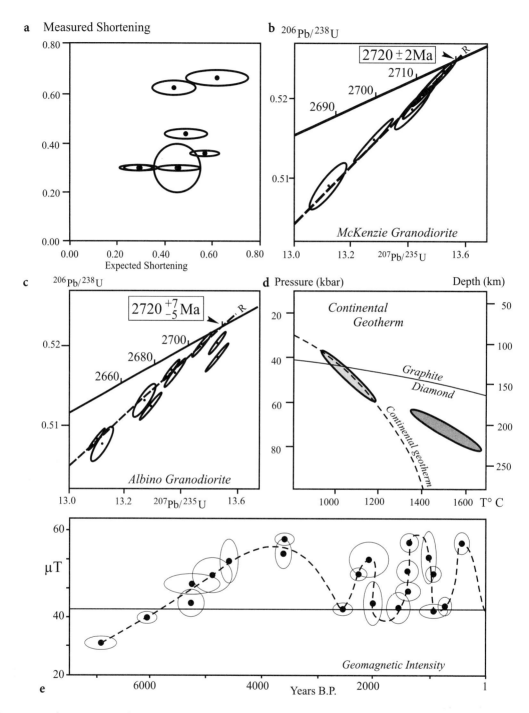

Fig. 4.2. Error bars proportional to standard error of mean, or multiples of standard deviations ($\sigma, 2\sigma$, etc.) define a confidence ellipse that is parallel to the graph axes if the errors in x and y are free to vary independently (zero covariance). **a** Estimates of tectonic shortening determined from two independent methods (x, y axes) (Borradaile 1974, 1979a,b). **b** U/Pb isotopic ratios used in geochronology are interdependent, thus the variation in one ratio influences the other; therefore error ellipses are not parallel to the x-y axes. The confidence ellipses define regions of radius 2σ. **c** Another geochronological concordia diagram; regression line intercepts the isotopic-ratio line obliquely giving *asymmetric limits* for determined age (Davis et al. 1982, Davis and Edwards 1986). **d** Petrological data *not subject to rigorous statistical treatment* but whose distribution illustrates the principles of a confidence region and the non-zero covariance of the variables. **e** Geomagnetic intensity variations over archeological time, southern Australia (Barbetti 1983). The sizes of the confidence ellipses may not justify the curve interpreted, especially where turning points are based on one measurement

cal improvements have now halved these errors in many cases so that current research articles may be justifiably titled "high precision U-Pb dates". There are few field-derived geological data that can be reproduced with a precision ≈ 0.1 % at the 95 % confidence level, and certainly no other geological age-determination techniques rival this level. We should recall that such work always assumes that the original sources of error are normally distributed, and that sufficient experiments have been performed to establish a certain precision, e.g. 750 ± 2 Ma. Multiple sources of random error are manageable because the *additive rule* ensures that when normally distributed errors from several sources are confounded, the combined error contributions also follow the Normal distribution. A more pressing concern is that in some branches of earth science, including geochronology, it may not be possible to acquire sufficient original observations to characterise their distribution. For example, we may wish to establish the age of a volcanic episode but, if there are only five different lava flows, our best estimate will be based essentially on a small sample, which is also the population for the purposes of that study.

We should not lose sight of the fact that confidence limits express uncertainty due to the variation of natural processes as well as random measurement errors. Statistical methods cannot detect *systematic* errors due to poor sampling or measurements, or from consistently inaccurate equipment. For example, determining the age of the minerals within a pebble may be as precise as the examples discussed above. However, this provides only the age of the pebble's source, not the age of the conglomerate in which the pebble is found. Obviously, pebble ages are systematically older than the hosting conglomerate bed. Even the most careful statistical treatment could not uncover the true age of the conglomerate. As is commonly the case in earth science, specific subject knowledge is needed to resolve this type of problem.

For many geological reports, graphical presentation of error-bars and confidence regions is satisfactory and sometimes the best way to indicate the uncertainty or confidence of the conclusions. Even when we inspect published reports that omit formal statistical analysis, our understanding of dispersion and confidence gained from statistics allows us to appreciate the information appropriately (e.g. Fig. 4.2 d). Hypothesis testing provides a better grasp of the validity conclusions, as we will see later. Unfortunately, earth scientists do not evaluate the significance of their results as rigorously or as commonly as other scientists.

Finally, and also in reference to age determination, we note that evaluation of error is critical where the

method produces a calibration curve that is non-monotonic and thus has the potential for ambiguity. The archeomagnetic calibration curve and the C^{14} calibration curve do not associate a value of the observed variable with a unique age. Instead, the measured values oscillate and each one may be assigned to a different age. Normally, other criteria eliminate the spurious possibilities. The radiocarbon calibration curve is shown in Fig. 4.3a; ΔC^{14} varies rapidly with time, and oscillations are more noticeable than the trend. ΔC^{14} may be translated into a *radiocarbon age* by a direct formula, but this must be adjusted by a calibration curve that accounts for atmospheric variations in carbon-isotope proportions through time. The resulting radiocarbon ages thus also oscillate, and a detailed part of the master curve is shown in Fig. 4.3b. This shows ages determined from a site in Eastern England: the radiocarbon age is 2078 B.P., which intersects the calibration curve at a *calendar age* of B.C. 65. However, the 2σ limit which is traditionally used in C^{14} age determinations allows any age between the dashed lines that define the 2σ confidence band. The limits intersect the curve at points *a, b, c* and *e*. "*d*" is the most probable intersection, but the range permitted would be from the "*a*" intersection at B.C. 165 to the "*e*" intersection at B.C. 30. Thus, the quoted age would have *asymmetric* 2σ limits and would be given as B.C. 65^{+100}_{-35}.

Consider error-bars ($\pm \Delta x, \pm \Delta y$) on an (x, y) observation. In some previous examples, we noted that it might be instructive to encompass the confidence region with an ellipse that has Δx and Δy as its semi-axes. Where the units of measurement of x and y are equal, and where the scales of the axes are equal, the eccentricity of the ellipse directly indicates the relative precision of x and of y. This may be too obvious, but the principle can be deceptive when we deal with maps, especially with global phenomena where the map projection is an important concern and for certain kinds of observation. For a scalar observation, represented by a single number, as with mineral abundance for example, contours of those values will reflect the intrinsic variation of the measured property but also the distortion due to the map projection. For example, on a scale of metres or kilometres (Figs. 1.20–1.24) the variations are unaffected by projection problems. On continental scales (e.g. Figs. 1.4, 1.7), regional variations are distorted by whatever map projection is chosen. Some map projections require momentary pause for their appreciation, especially at high latitudes. For example, using a continuous plane projection like Gall's, Antarctica is exaggerated to appear as one of the largest continents. Nevertheless, one may argue that the problems are not serious with careful inspections because every-

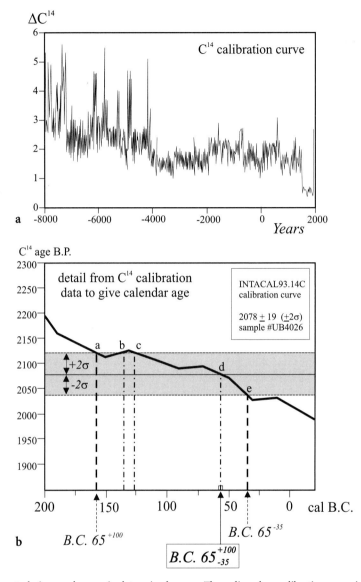

Fig. 4.3. The radiocarbon technique and errors in determined ages. **a** The radiocarbon calibration curve is far more complex than the monotonical decay shown by other decay series used in geology. A given ΔC^{14} measurement usually yields a non-unique solution for the material's age. Ambiguity is resolved using specific knowledge of the problem or cross-referencing with another independent method. Dendrochronology may not strictly be considered as an independent technique as it is also used in the construction of this calibration curve. **b** Typical application of the radiocarbon age-determination technique shows how a symmetrical $\pm 2\sigma$ confidence band usually gives rise to asymmetric corresponding age errors: these are B.C. 65^{+100}_{-35} in the example shown (Borradaile et al. 1999b)

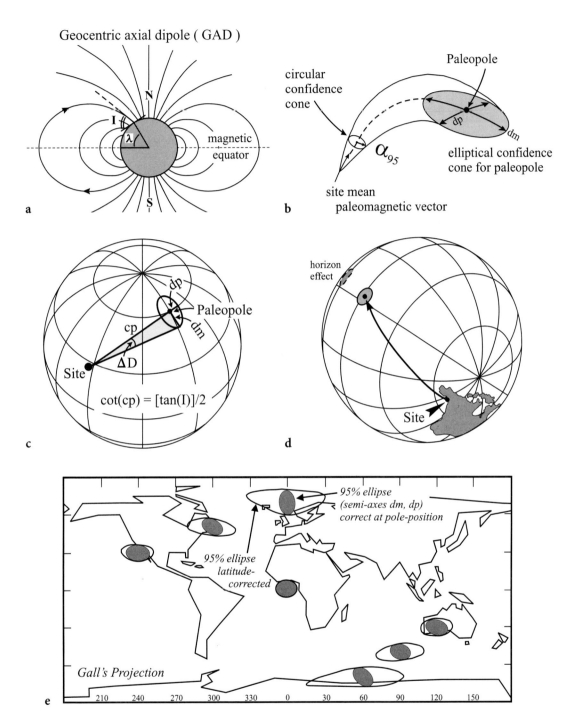

Fig. 4.4. Special problems arise in the plotting of errors (confidence cones) on spherical projections. This is particularly obvious for paleopole positions on maps. Even where a circular cone defines the angular uncertainty of the magnetic vector, the range of corresponding pole position will generally be a distorted oval in a projection of the globe. **a** Mean geocentric axial dipole field causes a simple geometric relation between magnetic inclination (I) at a site and the site's paleolatitude (λ). **b, c** Projection of the magnetic vector and its confidence cone produces an elliptical confidence region about the paleopole position that will appear as a distorted, oval-shape in plane-projection. **d** On spherical projections, the paleopole's confidence region is progressively more distorted toward the horizon. **e** On map projections the paleopole's confidence ellipse (shaded) only has a valid aspect ratio at the location of the paleopole; the actual confidence region must be distorted according to the map projection; in general it will not be elliptical and it may not be symmetrical

thing plots in its appropriate coordinate position on the map.

However, the projection problem is a quantum step more significant where we deal with information that involves direction. Whereas this will be more fully appreciated after reading Chapters 9 and 10, the essence of the problem may be grasped here. Let us illustrate this with a consideration of paleomagnetism. Over the long periods of time that some rocks require to magnetise, the geomagnetic field averages out to a geocentric axial dipole (Fig. 4.4a). At the time of magnetisation, the locked-in magnetism points everywhere to the North, vertically down at the north pole, horizontally to the North at the equator and vertically up at the south pole. At any latitude the northwards inclination of the magnetic vector I is related to the latitude λ by the following formula and illustrated in Fig. 4.4a.

$$\tan(I) = 2\tan(\lambda) \qquad (4.6)$$

Upwards inclinations and southerly latitudes are assigned negative signs by geophysical convention. Positive signs are for downwards inclinations and northern latitudes. When rock is translated and rotated on the earth's surface by plate tectonic motion, the rock's mean paleomagnetic vector points to the relative position of the paleopole responsible for its magnetisation. The first conclusion of interest, especially for paleoclimate studies, is that from the above equation we infer the paleolatitude of the rocks from the inclination of the paleomagnetic vector. In the context of this Chapter, it is of interest because the paleolatitude (λ) is related to the observation (magnetic inclination, I) by a nonlinear equation. Although the confidence limit on the mean paleomagnetic inclination is symmetrical (Fig. 4.4b; $\pm\,\alpha_{95}$), the confidence limits on the paleolatitude must be asymmetric with the exception of special cases ($\lambda = 0, -90$ or $+90$):

$$\lambda_{\max} = \arctan\left(\frac{\tan(I+\alpha_{95})}{2}\right)$$
$$\lambda_{\min} = \arctan\left(\frac{\tan(I-\alpha_{95})}{2}\right) \qquad (4.7)$$

The main use of paleomagnetic data goes much further than determining the paleolatitude. Much effort is expended to locate the paleopole responsible for the magnetisation at the site. It is very important to assess the confidence around each paleopole. After all, plate tectonic reconstructions rest on the interpretation of this data.

From a determination of the characteristic ancient magnetic vector preserved in a rock, one may calculate the relative position of the North Pole responsible for its magnetisation. Plate tectonics has moved continental rocks by ocean-floor spreading, continental rifting and by subduction. Consequently, for a coherent geological terrain, rocks of progressively greater age commonly give apparent pole positions that depart more and more from the present North pole (e.g. Tarling 1983; Butler 1992). The locus of these apparent pole positions is termed the *apparent polar wander path (APWP)*. Of course, it is not the North pole that moved significantly but rather the terrain. However, the APWP does show the relative changes in latitude between the rock and the pole responsible for its magnetisation. The precision with which the APWP is defined is critically important to paleogeographic, stratigraphic and Plate tectonic reconstructions of the World. Therefore, paleomagnetists go to great lengths to refine their estimate of each apparent pole position (Butler 1992).

In most paleomagnetic studies, the orientation precision of the mean paleomagnetic vector is described by a circular cone of confidence, of radius α_{95} (Fig. 4.4b), the *Fisher confidence cone* (more complicated situations may arise; Chap. 10, Fig. 10.11). However, when this circular confidence limit is passed through the calculation to fix the relative position of the paleopole, its cone of confidence is invariably elliptical on the surface of the sphere. The angular distance to the mean paleopole, the magnetic co-latitude cp, is fixed by the inclination of the vector (I), and its cone of confidence has radius dp in the azimuthal direction connecting the site and paleopole. The perpendicular radius is dm (Fig. 4.4c).

Thus, on the sphere, any confidence cone about the mean paleopole must be elliptical, except under extraordinarily rare circumstances. We must therefore take care in interpreting some older illustrations on which all confidence cones appear routinely circular.

The problem is now compounded when we plot the confidence limits around paleopoles on any plane map projection. Stereographic global projections reproduce the confidence limits most faithfully (Fig. 4.4d), but these suffer from the limitation that data near the horizon is badly distorted and any information beyond the horizon cannot be plotted at all. This frustrates presentation of results because magnetic vectors from a single coherent terrain may be attributable to paleopoles on the opposite hemisphere.

If we use a familiar plane projection, we have the advantage that all possible pole positions may be represented. However, the elliptical confidence cones for the poles are unreasonably distorted at high latitudes, so that the confidence limits are oval rather than elliptical (Fig. 4.4e). But should this really concern us? After all, the confidence ovals truthfully convey the ground-positions on which the paleopole could lie with 95% confidence. This is all a matter of judgement that must balance the relative needs to see undistorted *ellipses* of confidence, the entire World surface, and a familiar geographical representation. More seriously, whatever projection system is used, the relative positions of successive paleopoles (APWP) cannot be represented so that they are linearly related to time. This is a significant drawback because it follows that it will be difficult to visualise estimates of rates of plate movement from the APWP. Despite gross longitudinal distortion at high latitudes, a simple projection such as Gall's projection permits ready visualisation, and the equal linear scales for latitude and longitude permit simple recalculation of the apparent shape of confidence regions (Fig. 4.4e). Although the geographical distortion is somewhat different from popular projections such as Mercator, conic, etc., the coverage is global and the geography is still readily recognisable to everyone. Sophisticated software that manipulates a World database for paleomagnetism, with many plotting and projection options, is available from the website of the Norwegian Geological Survey.

4.3
Hypothesis Testing

Estimating confidence in observations and their means is a common and welcome addition to geological reports. A logical and not too complex extension arises when we ponder on the similarity or differences between the means of different samples. One of the most common inquiries in any research is; "Are these samples the same?" A straightforward yes or no answer is usually optimistic. More formally, we are asking whether these samples have been drawn from the same population, based on the sample statistics available, usually the mean and standard deviation. Moreover, the answer is always qualified with a confidence level. For example, the popular 95% confidence level means that there is a one-in-twenty chance (5%) that the same result could have arisen due to unfortunate sampling.

The most usual application compares the means of two samples, but we may also compare a sample mean with the population mean, usually the mean of the Normal distribution. Popular questions are posed on the comparison of a specific statistic. Are the sample means significantly different? Are the sample variances

significantly different? If the difference is not significant, we may infer that the two samples could be drawn from the same population. Later we may even address seemingly vague but important issues such as, are two distributions different *in any way at all* (Chap. 5)? Any of these types of comparison require the determination of a ratio that expresses the relationship between the compared two sample statistics. This dimensionless quantity is called the *test statistic*, given by some simple algebraic expression. The test statistic may follow various kinds of theoretical distributions whose shape may vary with sample size and the *degrees of freedom* (v or *d.f.*). Degrees of freedom may be regarded as the minimum amount of information required to regenerate the sample distribution.

We introduce the topic by considering the test statistic z that follows the standardised Normal distribution and is in units of standard deviation (σ). Consider that the horizontal axis represents the test statistic's values so that the curve represents its probability distribution, following the standardised Normal distribution (ND: 0, 1), which has mean $\mu = 0$, and standard deviation $\sigma = 1$. The area enclosed by this curve is unity, representing the 100% probability that the z-value must lie somewhere under the curve (Fig. 4.5a).

The probability of the test statistic lying between any two values of the test statistic, z_1 and z_2, is represented by the area under the curve between z_1 and z_2. It will become clear that z_1 and z_2 correspond to actual observations x_1 and x_2 with the help of simple arithmetic that we term "standardisation".

There is a probability of 1.0 (or 100%) that the test statistic z lies somewhere in the range from $-\infty$ to $+\infty$. We can also determine z-ranges that enclose more useful probabilities, such as that corresponding to 95% of the observations. It can be shown that 95% of all possible values of the test statistic lie within 1.96 standard deviations of the mean value $z = 0$. Alternatively, rare values of z that occur with a probability of $\leq 5\%$ lie in the *critical regions where $z < -1.96$ or where $z > +1.96$*. When we recall that each z-value corresponds to an actual observed value x, the value of this procedure becomes apparent. There is, at most, a 5% probability that our test statistic might lie further than 1.96σ from the mean. The 5% area under the extreme tails of the curve is designated as a probability region with area, $\alpha = 0.05$. α is the *significance level* and defines the rejection region of the distribution in which improbable values of z occur. If sampling is random, they should occur with a probability of $\leq 5\%$, that is to say, in not more than one sample out of every twenty. Consequently, if the z-value calculated from our data lies in the shaded, critical regions of Fig. 4.5a, we *reject the hy-*

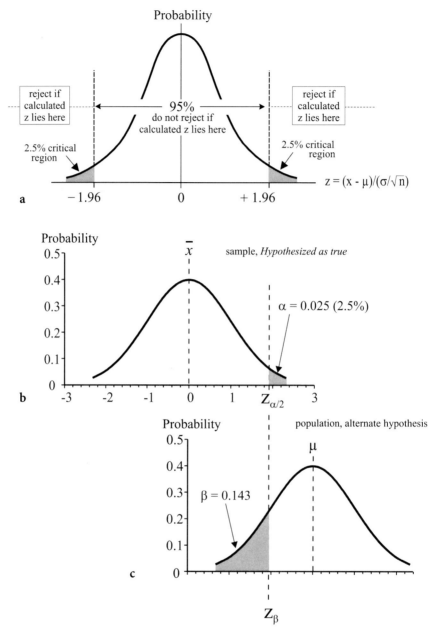

Fig. 4.5. a Standardised Normal distribution showing two-tailed rejection region at 95% confidence ($\alpha = 0.05$): the shaded areas represent the probability of rejecting values that lie further than 1.96σ from the mean. 95% is a suitable and common level of confidence used in earth science. **b** Sample shows a rejection region for a 95% confidence level, two-tailed test. If the test statistic (z_0) calculated from some value (x) lies in the *unshaded region*, we do not reject the null hypothesis that $\bar{x}$ represents the population mean. **c** Suppose, however, that the sample is unrepresentative and the population is shown as here. We see that our measurement actually lies in the rejection region of the true distribution. We have wrongly accepted a false hypothesis. The probability with which this may occur is 14.3%, in this example. However, in general, the probability of this β-error is unknown because the population distribution is unknown

Table 4.1. Outcomes of hypothesis tests

Possible outcomes	Postulate is true	Postulate is false	Significance
Reject H₀	Type I error	**Good**	α, chosen by researcher, optimum usually 0.05 Confidence = $100\alpha\%$ [test statistic lies in tail(s) of distribution with area = α]
Do not reject H₀ ($\alpha \approx$ *accept*)	**Good**	Type II error	β, generally unknown power of test = $(1-\beta)$

Since rejection is unequivocal, the null hypothesis H_0 is chosen such that its non-rejection will achieve the researcher's goal. H_0 is commonly an equality statement, one of no differences, to be most effective.

pothesis that $z = 0$ at the 95% confidence level. If the value of z calculated from our data lies in the unshaded region, we do not reject the hypothesis that $z = 0$ with a 95% confidence. Any corresponding observation x would lie in the range $(\pm 1.96\sigma)$, and we should *not reject* the possibility that it is equivalent to $\bar{x}$. The term *acceptance* is too strong.

This is a convenient point at which to expand upon the awkward phrase "*not be rejected*". The implication is that if a value x corresponding to a test statistic z is acceptable at the 95% level, it lies less than 1.96σ from the mean. However, the tails under the edge of the probability distribution deal with the negative side of the argument that our value may be unrepresentative because it is further than 1.96σ from the mean. Hence, the formal phrasing is that we *reject* values if their test statistic lies in the shaded region of Fig. 4.5 a. Conversely, if the value lies within 1.96 standard deviations of the mean, we *do not reject* the possibility that the value is representative. Typically, conclusions are phrased less formally. When comparing the means of two samples, authors may write, with increasing generosity:

"*We do not reject the hypothesis that two samples are drawn from the same population.*"
"*The two samples are drawn from the same population.*"
"*The samples are not different.*"
"*The samples are the same.*"

However, based on probabilities and visual considerations of the probability distribution, we can really do no better than to *reject* or *not reject* that a value lies within a given range defined by a chosen confidence level that has the value $100\,(1-\alpha)\%$ (Table 4.1).

4.4
Mechanics of Hypothesis Testing

The principle behind hypothesis testing was outlined in general terms above. Statistics texts formalise the procedure into six steps which are condensed for convenience in Table 4.2. Let us consider an example of a sample of coal seam thicknesses produced in a rhythmic (cyclothymic) sequence; there are good reasons to

Table 4.2. Steps in testing a hypothesis

Step	Example
(1) Formulate null hypothesis (H₀) that may be rejected	that $\bar{x} = \mu$, a sample mean = population mean (needs *two-tailed test* because when H₀ is untrue $\bar{x} > \mu$ or $\bar{x} < \mu$)
(2) Select probability α of rejecting H₀ when it is true (type I error) (α is area under tail of probability distribution)	commonly significance level = 0.05, as chosen below (= 5% confidence level)
(3) Select appropriate distribution and test statistic	Normal distribution, $z = (\bar{x} - \mu)/(\sigma/\alpha\sqrt{n})$
(4) Determine appropriate sampling distribution if H₀ is true	Consult table of percentage points for the appropriate distribution
(5) Determine test-statistic value(s) that define edge of critical region(s); only $100\alpha\%$ of values lie in critical region	e.g. For Normal distribution with $\alpha = 0.05$, and *two-tailed test* these are $+1.96$ and -1.96 (For a one-tailed test there is only one critical value.)
(6) Calculate test statistic from data. Reject H₀ if test statistic lies in critical region	*If calculation from data should yield:* (a) $z = +2.0$ this is $> +1.96$, **reject** H₀, that $\bar{x} = \mu$ (b) $z = -2.4$ this is < -1.96, **reject** H₀, that $\bar{x} = \mu$ (c) $z = +1.1$ this is $< +1.96$, **do not reject** H₀, that $\bar{x} = \mu$ (d) $z = -1.5$ this is > -1.96, **do not reject** H₀, that $\bar{x} = \mu$

suspect a sensitive control of sedimentation and compaction on average bed thickness. For this reason it may be necessary to determine if a sample of beds with mean thickness $\bar{x} = 1.91$ m ($n = 50$) is typical of a population with a mean $\mu = 1.95$ m and standard deviation $\sigma = 0.2$ m? Consider the example, step-by-step.

Step (1) Initially, we must formulate a testable hypothesis, the *null hypothesis, H_0*. We arrange our program of testing to make this a sensitive test which is often best achieved with *a test of no differences*, e.g.

$H_0: \mu = 1.95$

A null hypothesis expressing equality is a *sharp* or *point* test. We need not really believe that true equality is possible. It is merely a device to enable us to proceed in testing. If H_0 is untrue then an *alternative hypothesis H_A* must be accepted:

$H_A: \mu \neq 1.95$

This accounts for possibilities that $\mu > 1.95$ or $\mu < 1.95$, constituting a two-tailed test. Clearly, we should design H_0 so that we can accept something of interest to us.

Step (2) Select a significance level (α), noting the sample size which may also affect the next step. Most commonly in earth sciences, $\alpha = 0.05$ (= 95% confidence level). α is the area of the shaded rejection region (Fig. 4.5b). In our example, $n = 50$.

Step (3) Choose a test statistic appropriate to the items being considered. Are we comparing means, standard deviations or proportions? Some test statistics have special requirements such as a known distribution for the population, certain sample sizes and knowledge of certain population parameters. Suitable test statistics and details are indicated for popular situations later. In this example, to compare a value with the mean, the z-variate is appropriate:

$$z = \frac{\bar{x} - \mu}{\left(\sigma / \sqrt{n} \right)} \qquad (4.8)$$

Step (4) Consider the appropriate sampling distribution when H_0 is true. For the comparison of means, the Normal distribution is used. The standardised test statistic z may thus be compared with the percentage points of the Normal distribution (Table 3.5).

Step (5) Determine the critical regions, sometimes noted by the variate with a subscript indicat-

ing the significance level in question, e.g. $z_{\alpha/2}$ or $z_{\pm 0.05/2}$. There is a 0.05 or 5% probability that only $100\alpha\%$ of observations should take such extreme values and $\alpha/2$ indicates that the rejection region is divided into the two extreme tails (shaded in Fig. 4.5a). We reject H_0 if z lies in the low probability tails. For a two-tailed test with $\alpha = 0.05$ using the Normal distribution, we know that the critical regions are defined by <-1.96 and $>+1.96$ (Fig. 4.5a; Table 3.6).

Step (6) For the sample ($n = 50$), with $\bar{x} = 1.91$ compute the z-value, knowing $\mu = 1.95$ and $\sigma = 0.2$.

$$z = \frac{1.91 - 1.95}{\left(0.2 / \sqrt{50} \right)} = -1.41 \qquad (4.9)$$

Clearly, the test statistic's value, $z = -1.41$, lies between the critical limits -1.96 and $+1.96$ so that we do not reject H_0. Fig. 4.5a shows that z lies *between* the shaded rejection regions.

4.5
Further Considerations in Hypothesis Testing

So far we have outlined and tabulated the procedures with a common type of example. The principal variations on this concern choices in steps (1), (2), (3) and (6) above. For example, in step (1) the test may be two-tailed or one tailed; in step (2) α may be varied dependent on subject-specific conditions and the desired power of the test; and in step (3) an appropriate test statistic and probability distribution must be chosen. In the final step (6), there is no further choice in the formal statistical sense; H_0 has been rejected or has not been rejected. However, we will discuss how that decision is relayed in a subject-specific context; sloppy reporting could devalue the rigour of the analysis. The room to manoeuvre within hypothesis testing is outlined in Table 4.3.

4.5.1
Selecting H_0; One- and Two-Tailed Tests

At the outset, we must realise that the compared distributions are of measurements that are *mathematically independent, not correlated in any way*, and that the *sample sizes are adequate*. Some branches of earth science simply cannot provide sufficient data for which a distribution can be inferred. This is not laziness or carelessness of the workers; some natural phenomena are intrinsically rare.

Table 4.3. Principal choices available during application of a hypothesis test

Step (1): choices		Step (2): e.g., $\alpha = 0.05, 0.025$ or 0.01		Step (3): choice: more later	Step (6): decision
H_0 $n = ?$	Nature of decision	α	Rejection regions	Test-statistic sampling dist.	H_0 rejected where:
$\bar{x} = \mu$	Two-tailed	0.05	2.5% per tail	z, Normal	$z > +1.96$ or $z < -1.96$
$\bar{x} > \mu$	One-tailed	0.05	5% Upper tail	z, Normal	$z < +1.64$
$\bar{x} < \mu$	One tailed	0.05	5% Lower tail	z, Normal	$z > -1.64$

However, we shall see later that different tests are available, sometimes dependent on sample size, which compare two means, two variances or which simultaneously compare means or variances of many samples. There are even non-parametric tests that compare all aspects of a sample distribution without assuming anything about the population from which the samples were drawn (Chap. 5). In our examples, we shall assume that the sample has been drawn from a population that follows the Normal distribution (Fig. 4.5), which is a very common situation. For other situations, different test statistics from other sampling distributions may be used.

The null hypothesis must be carefully chosen to yield the most stringent decision. If we wish to test whether the means of two samples are equal (H_0: $\bar{x}_1 = \bar{x}_2$), our decision will be clearly rejected if the test statistic lies in the critical region. For this reason, *hypotheses of no differences are favoured*. Some critical values of x, that we call $\pm x_C$, define the edges of the tail regions beyond which only 5% of observations are found. The z-values corresponding to x_C for $\alpha = 0.05$ are $+1.96$ and -1.96. In other words, if $z > +1.96$ or $z < -1.96$ we must reject H_0. $\alpha = 0.05$ defines the 5% probability of the extreme situations in which such values could only rarely occur. These are distributed equally in the two extreme tails, 2.5% in each. In other situations, we may be concerned with an asymmetric distribution of critical areas, for example, H_0: $\bar{x} > \mu$. In this case, the 5% probability of extreme and unlikely values lies in the lower tail defined by $z = -1.64$ The *magnitude* of z is different because all of the critical area lies in one tail, corresponding to 5% of the probability distribution on one side of the curve. Critical values of z for one-tailed and two-tailed tests at $\alpha = 0.05$ and $\alpha = 0.01$ are presented in Table 4.4.

Table 4.4. Normal-distribution confidence level and critical limits

Significance level (rejection area under tails of distribution)	Confidence level	Two-tailed critical z-values	One-tailed critical z-value
$\alpha = 0.05$	95%	-1.96 and $+1.96$	-1.64 or $+1.64$
$\alpha = 0.01$	99%	-2.58 and $+2.58$	-2.33 or $+2.33$

bility distribution in which unlikely, extreme values of the test statistic are found. They correspond to confidence levels of 95, 97.5 and 99% given by $100(1-\alpha)$%. The 95% confidence level ($\alpha = 0.05$) is most common in the earth sciences, especially field studies. Higher confidence levels, for example 99% ($\alpha = 0.01$) are usually appropriate in high-precision laboratory work. The significance level may be chosen to suit the required rigour and that is justified by data quality, but there are theoretical limitations to be discussed below. The significance level should be selected on objective scientific criteria related to the subject at hand, before statistical testing. Otherwise, it might be tempting to select subsequently a probability level at which the desired comparison would become acceptable.

4.5.3
Choosing a Test Statistic and a Sampling Distribution

Test statistics and sampling distributions are summarised briefly below for commonly occurring situations. The z-variate follows the Normal distribution with which we are already familiar. Other useful test statistics follow the t-, F- and χ^2-distributions which are derived by sampling the Normal distribution in special ways. They are briefly described and tabled towards the end of this chapter. One test uses the χ^2-statistic, a discussion of which is postponed until Chapter 5 where it is best illustrated with applications. At the end of Chapter 5, the theoretical forms of the distributions are compared in a table. Note that certain situa-

4.5.2
Selecting α

Our hypothesis test may be made at different significance levels (α), usually 0.05, 0.025 or 0.01. These represent the fractional areas under the tails of a probations require different test statistics for *large samples* ($n \geq 30$) and for *small samples* ($n < 30$).

1. *Compare sample mean and population mean (n ≥ 30)*: use Normal distribution, σ known

$$z = \frac{\bar{x} - \mu}{\sigma / \sqrt{n}} \qquad (4.10)$$

2. *Compare sample mean and population mean (n <30)*: use *t*-distribution, small samples or σ unknown, $(n-1)$ degrees of freedom

$$t = \frac{\bar{x} - \mu}{s / \sqrt{n}} \qquad (4.11)$$

3. *Compare two sample means (n ≥ 30)*: use Normal distribution, σ_1 and σ_2 are known, or sample standard deviations are good proxies because sample size is adequate.

$$z = \frac{\bar{x}_1 - \bar{x}_2}{\sqrt{\left(\dfrac{\sigma_1^2}{n_1} + \dfrac{\sigma_2^2}{n_2} \right)}} \qquad (4.12)$$

4. *Compare two sample means (n< 30)*: use *t*-distribution, σ unknown

$$t = \frac{\bar{x}_1 - \bar{x}_2}{\sqrt{\left(\dfrac{s_1^2}{n_1} + \dfrac{s_2^2}{n_2} \right)}} \qquad (4.13)$$

The degrees of freedom required for use of the *t*-tables are calculated with:

$$d.f. = \frac{\left(\dfrac{s_1^2}{n_1} + \dfrac{s_2^2}{n_2} \right)^2}{\left(\dfrac{(s_1^2 / n_1)^2}{n_1 + 1} + \dfrac{(s_2^2 / n_2)^2}{n_2 + 1} \right)} - 2 \qquad (4.14)$$

5. *Compare sample proportion (p) with hypothesised proportion (p$_0$) (n ≥ 30)*: proportions $q = (1-p)$; use Normal distribution

$$z = \frac{p - p_0}{\sqrt{\dfrac{p_0(1 - p_0)}{n}}} \qquad (4.15)$$

6. *Compare sample variance with population variance*: use χ^2-distribution with $(n-1)$ degrees of freedom to compare sample standard deviation with that of population (see Chap. 5 for application of χ^2-test).

$$\chi^2 = \frac{(n-1)s^2}{\sigma^2} \qquad (4.16)$$

7. *Compare two independent variances, s$_1$ > s$_2$*: use F-distribution; degrees of freedom are $v_1 = (n_1 - 1)$ and $v_2 = (n_2 - 1)$. See applications in connection with comparison of variances of multiple samples

(ANOVA) and of coefficients of determination in regression analysis (Chaps. 6 and 7).

$$F = \left(\frac{s_1}{s_2} \right)^2 \qquad (4.17)$$

8. *Compare sample's correlation coefficient (r) with population correlation coefficient (ϱ)*: discussed in Chapters 6 and 7, but mentioned for completeness here, the test statistics are different for the standard regression line of *y* on *x* (all uncertainty in *y*) and for the bivariate-regression line where both *x* and *y* contribute to the uncertainty.

for simple regression of *y* on *x*:

$$t = \left| \frac{R\sqrt{(n-2)}}{\sqrt{(1 - R^2)}} \right|$$

uses *t*-distribtuion with $(n-2)$ degrees of freedom. For bivariate (x, y) simple regression:

$$z = \left(\frac{\sqrt{(n-3)}}{2} \right) \log_e \left(\frac{1 + R}{1 - R} \right) \qquad (4.18, 4.19)$$

follows the Normal distribution.

4.5.4
Interpreting the Decision

The formulation of the hypothesis test may appear discouraging. Primarily, we establish some stringent criterion (H_0) that we hope to reject in favour of some alternate hypothesis. The alternate hypothesis expresses the concept in which we are interested. If we reject the null hypothesis, we must accept the alternate hypothesis. The shaded areas ($= \alpha$) under the probability distribution and the critical values that limit it permit no other possibilities (Fig. 4.5a). However, there is a small probability ($100\alpha\%$) that the value we rejected lay in the rejection region due to the misfortunate effects of random sampling. In that case, we wrongly rejected a true postulate simply because the sample was unrepresentative for reasons of randomness.

This is apparent from consideration of the shaded areas under the probability distribution. The *false rejection of a true statement is referred to as a type I error* with a probability of occurrence denoted by the significance level, here $\alpha = 0.05$ (Table 4.1). In most common earth science applications, $\alpha = 0.05$, we risk a 5% or one-in-twenty chance that we may wrongly reject H_0. *It cannot be emphasised too strongly that this routine approach does no more than establish the probability of rejecting a correct hypothesis.* It is for this reason that the null hypothesis is best designed to be rejected, a matter that seems paradoxical upon first acquaintance.

The *alternative hypothesis* H_A is accepted when we *must reject* the null hypothesis. As one infers from the previous paragraph, "acceptance" might be too strong or perhaps slightly misleading. The test statistic for the sample should not lie in the shaded, critical region in this case (Fig. 4.5a). Let us further consider H_A: our sample's test statistic lies in the unshaded portion of the probability distribution (Fig. 4.5b). However, we are dealing simply with likelihood. Have we accepted (failed to reject) an incorrect hypothesis? After all, we are dealing merely with a random sample. Playing devil's advocate, suppose that the true distribution is different, as shown in Fig. 4.5c. Clearly, any test statistic that we fail to reject on the basis of our sample, lying in the unshaded portion of Fig. 4.5b, actually lies in a low-probability area of the true distribution (Fig. 4.5c). In other words, we would have accepted a false value or rejected a true value with a probability of 85.7%. This more serious error is classified as *type II* (Table 4.1). The shaded area in Fig. 4.5c represents the probability of accepting (failing to reject) an invalid value; its probability is β (here 14.3%). It is said that the hypothesis test has a *power* $(1-\beta)$; in this case, the *power of the test* is 85.7%. Usually, the population or true distribution (Fig. 4.5c) is unknown so that the probability of this type of error (β) and the power of the test is also unknown. In natural sciences, it is usually unknowable. In most situations, it is prudent to set $\alpha = 0.05$. Even $\alpha = 0.01$ may unduly increase the risk of a type II error. Laboratory experiments and manufacturing processes with a high degree of control may provide the opportunity to optimise α and β, improving the decision-making process.

Failing to reject the hypothesis may imply that it is true. However, it may be that the test was not sensitive enough. Therefore, we may attempt to manage our data collection and hypothesis testing to minimise the type II error and increase the test's power. There are basically two ways to achieve this.

The first option is to increase the sample size (n), recalling that the standard deviation of possible sample means $(\sigma/\sqrt{n})$ decreases with increasing n. This will decrease the sample variance, making the frequency distribution of the sample narrower, reducing its overlap with the true distribution (Fig. 4.6a). The overlap area, representing β, is therefore reduced and the power of the test is increased. Unfortunately, in earth science the option to increase sample size is not always available, given the limited access or availability of certain geological phenomena, events, specimens, sites and outcrops.

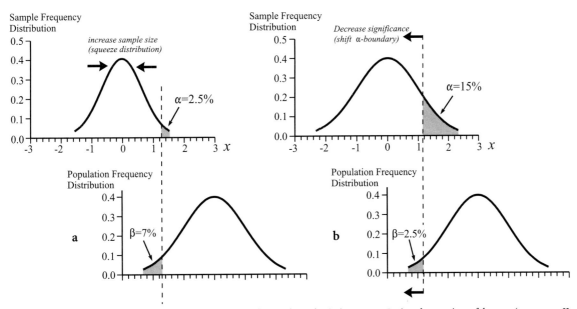

Fig. 4.6. Consider the balance between risking rejection of a true hypothesis (*error type I*, α) and accepting a false one (*error type II*, β), as in Fig. 4.5. Two options exist to increase the power of the hypothesis test. **a** Increasing sample size tightens the sample distribution and reduces the β-error. **b** Decreasing the confidence of the test, or increasing α reduces β, but this increases the possibility of rejecting a true hypothesis, rendering the test less powerful. In general, $\alpha = 0.05$ or $\alpha = 0.01$ are satisfactory compromises when dealing with natural observations

The optimum sample size required to maximise the power of the tests may be predetermined where β is known or where it may be estimated. This option is not normally available in the natural sciences. However, let us proceed. Consider Fig. 4.5 b, c, where x_C is some critical value common to both distributions, standardised as $z_{\alpha/2}$ and z_β respectively in the two distributions. Concerning α and β, we may write the equations for the test statistic, z as follows.

$$z = \frac{\bar{x} - \mu}{\left(\sigma / \sqrt{n}\right)} \quad \text{and for the difference}$$

$$z_{\alpha/2} - z_\beta = \frac{\bar{x} - \mu}{\left(\sigma / \sqrt{n}\right)} \tag{4.20}$$

Equating and rearranging, we see that the minimum sample size, n, needed to optimise the power of the hypothesis test is given by the following equation:

$$n = \left(\frac{\sigma(z_{\alpha/2} - z_\beta)}{\bar{x} - \mu} \right)^2 \tag{4.21}$$

The second option is to manipulate the test's significance level to reduce the overlap of the sample's critical region with the ideal distribution. If we reduce the confidence level to say 85%, i.e. a significance level of $\alpha = 0.15$, the overlap will clearly be smaller, reducing the probability that a false hypothesis will be accepted to 2.5% (Fig. 4.6 b). Whereas this reduces the chance of accepting a false hypothesis (type II error), it increases the probability of rejecting a true value of the test statistic (type I error). The most common compromise between these two conflicting requirements is usually to set $\alpha = 0.05$, to adopt a 95% confidence level. In comparison to some other scientists, we may feel disadvantaged by the small sizes of some samples and the complexity of multi-variable natural systems. However, medical researchers deal routinely with small samples and more complex systems. In clinical studies, ethical and humanitarian considerations are so great that sample sizes are very small, even by geological standards. For example, the median number of experimental subjects was between 16 and 36, for research published in three leading clinical medical Journals over a seventeen-year period (Fletcher and Fletcher 1979). Consequently, researchers may reduce their confidence level to satisfy a hypothesis test (Glantz 1997). Unfortunately, this commonly results in tests whose power is unjustifiably low. Over a four-year period in the same Journals, less than 36% of tests had adequate power

(Mohler et al. 1994). Some common applications of hypothesis tests in geological examples follow in subsequent chapters with presentation of the appropriate test statistics, mainly in connection with the comparison of distributions or the significance of correlations and use of regression lines.

4.5.5
Communicating the Result

Consider the statement, resulting from a carefully applied statistical test using the Normal distribution (Fig. 4.7 a). "We do not reject the hypothesis that the two sample means are equal at the $\alpha = 0.05$ significance level." Whereas this would suffice for a statistician, it would not fall favourably on the ears of most geologists. The question posed was probably, "are these two samples the same?". To satisfy the question, the answer may be rephrased but some caution is necessary so that careful work is not undervalued by overstated conclusions or exaggerated claims. It is never possible to say that two independent samples are *the same*. We have simply *not rejected* the hypothesis that the means are equal but there is a 5% chance that we are incorrect due to the effects of random sampling. The concept of 5% probability derives from the small shaded region under the tails of the probability distribution that represents a fractional area $\alpha = 0.05$ which represents the significance level. The confidence level is $100(1 - \alpha)$%.

At this point it is tempting to say that we "accept" the hypothesis that $\bar{x}_1 = \bar{x}_2$. However, if we consider the procedure of manipulating test statistics and critical values (Fig. 4.7 a), what we are really asking is "are these two samples drawn from the same population?" Therefore, it is logical to consider other ways in which to compare the samples. Why not use another characteristic as well as the mean? The similarity of the two samples is more convincing if another statistic (e.g. σ) also compares favourably. If we do not reject the hypotheses, $H_0: \bar{x}_1 = \bar{x}_2$ and $H_0: \sigma_1 = \sigma_2$, it is much more reasonable to suggest that the two samples were drawn from the same population. Informally, we could conclude that the samples are "the same". Finally, we can always make our result more convincing by using a large sample and by the best quality of measurements. The latter two items are probably most convincing to the readers of geological reports. We shall see that these arguments apply to the comparison of other statistics than means, shown by other distributions in Fig. 4.7 and discussed below.

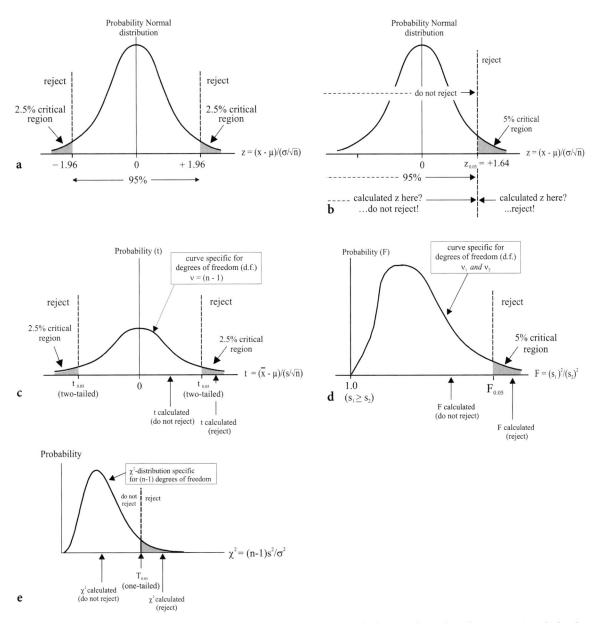

Fig. 4.7. Rejection limits for hypothesis tests with commonly used distributions. The horizontal axes show the test statistics calculated from the sample. In each test, the shaded regions indicate the $\alpha = 0.05$ significance level: this is the probability of rejecting a true hypothesis. $\alpha = 0.05$ corresponds to a 95% confidence level. The null hypothesis is rejected when the test statistic lies in the critical shaded region. Tests using the Normal distribution are independent of sample size whereas the t-, F- and χ^2-distributions change form with sample size, which controls the degrees of freedom (ν). For the F-distribution, two sample sizes and two different degrees of freedom control the shape of the function. **a** Two-tailed test for the Normal distribution. $H_0: x = \mu$; this hypothesis is rejected if the calculated value for $z > +1.96$ or < -1.96 **b** One-tailed test for a sample drawn from a Normal distribution. $H_0 : x > \mu$; this hypothesis is rejected if the calculated $z > 1.64$. **c** Two tailed-test for two means, using the t-distribution. The hypothesis that the two samples are drawn from the same population (i.e. sample means are the "same") is rejected if the calculated t-value lies in either shaded region. (**d**) Comparison of sample variances ($s_1 > s_2$) using the F-distribution. Because the larger variance is in the numerator, $F \geq 1$ and only a one-tailed test is possible. The hypothesis that the variances are equal is rejected if the calculated value of F exceeds $F_{0.05}$ which must be obtained from Tabled values, requiring two degrees of freedom. **e** Comparison of single-sample variance with population variance using χ^2-distribution: only one-tailed tests are possible

4.6
Sampling Distributions Used in Hypothesis Tests

4.6.1
The *t*-distribution (Comparing Means)

To compare a small sample ($n < 30$) with the population, or to compare two samples, it is inadvisable to use the z-variate from the Normal distribution to make significance tests. Instead, we need a distribution *whose form depends on n* (Fig. 4.7c). For this purpose, the *t*-distribution was developed by William Gossett (1876–1937) who modestly wrote under the nom-de-plume "student", hence the designation "Student's *t*-test". The test may be applied to small samples with a humped distribution that need not closely follow the Normal distribution. Its shape differs from the Normal distribution for small samples, but as sample size increases, the differences become less noticeable. For most practical applications, there is very little difference between the *t*-distribution and Normal distribution when the sample size approaches 30. Thus, $n = 30$ is generally considered to provide an upper limit on "small samples" from normally distributed data. As a corollary, a sample must have $n \geq 30$ if there is to be a reasonable prospect of comparing it with the Normal distribution.

The use of the *t*-value is identical to that of the z-variate and may be applied either as a one-tailed or two-tailed test (Fig. 4.7c). The value calculated for t (e.g. equations above) is compared with that in the appropriate table and may be used like z in previous explanations. The difference is that since t depends on the sample size n, we must construct a different table for each sample size. In practice we simply tabulate critical values of t for different sample sizes at the most commonly used significance levels $\alpha = 0.05$ and 0.01 (95 and 99%) as shown in Table 4.5.

This test introduces the notion of *degrees of freedom* (v or *d.f.*). In many statistical tests this is one less than the sample size ($n-1$). It reflects the minimum number of items sampled that are *independent*. For example, we know the mean so we need only know ($n-1$) of the x_i values in order to derive all values of x_i. Otherwise stated, the degrees of freedom is the minimum number of observations that are required to fully specify the sample, after some arithmetic manipulation, using the statistic we know. Thus, where we know both the sample mean and standard deviation, the degrees of freedom would be ($n-2$). For most simple statistical tests, $v = (n-1)$ or ($n-2$). In the *t*-tables we seek t under the

Table 4.5. Critical values of $|t|$ that define the edge of the rejection region which represents a probability α

v, degrees of freedom; often ($n-1$) or ($n-2$)	Two-tailed		One-tailed	
	$\alpha = 0.05$ 95%	$\alpha = 0.01$ 99% conf.	$\alpha = 0.05$ 95% conf.	$\alpha = 0.01$ 99% conf.
1	12.706	63.660	6.314	31.821
2	4.303	9.925	2.920	6.965
3	3.182	5.841	2.353	4.541
4	2.776	4.604	2.132	3.747
5	2.571	4.032	2.015	3.365
6	2.447	3.707	1.943	3.143
7	2.365	3.499	1.895	2.998
8	2.306	3.355	1.860	2.896
9	2.262	3.250	1.833	2.821
10	2.228	3.169	1.812	2.764
15	2.131	2.947	1.753	2.602
20	2.086	2.845	1.725	2.528
25	2.060	2.787	1.708	2.485
30	2.042	2.750	1.697	2.457
40	2.021	2.704	1.684	2.423
60	2.000	2.660	1.671	2.390
120	1.980	2.617	1.658	2.358
∞	1.960	2.576	1.645	2.326

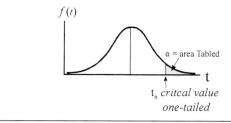

$f(t)$

$\alpha = $ area Tabled

t

t_α *critcal value one-tailed*

appropriate α-probability level for the appropriate degrees of freedom, v. Table 4.5 gives information only for common confidence levels of both one-tail and two-tailed tests. One-tailed tests are used where we are only concerned if a value is greater *or* less than a certain value. A two-tailed test is sensitive to extreme values regardless of their sign. For example, we may wish to determine whether or not two samples (A, B) are drawn from the same population, based on a comparison of their means ($H_0: \bar{x}_A = \bar{x}_B$). Our conditions are $v = 5$, a 95% confidence level and a two-tailed test. Thus, we expect t to lie inside the range $-2.57 < t < +2.57$, for non-rejection of H_0 (see Table 4.5).

For speedy reference, a graphical illustration of the critical *t*-values at different significance levels, for both one-tailed and two-tailed tests, is shown in Fig. 4.8a. If the value of t calculated from the data lies above the appropriate curve, we reject the hypothesis that the means could be drawn from the same population.

4.6.2
The F-distribution (Comparing Variances s_1, s_2)

The comparison of variances is an end in itself, but it may also be a supplement to a test comparing means. After all, if both means and variances compare favourably for two samples, it strengthens our conviction that the samples are drawn from the same popula-

tion. The appropriate test statistic is $F = (s_1)^2/(s_2)^2$, where $s_1 > s_2$. The F-test takes its name from the initial of its creator, Sir Ronald Fisher (1890–1962). The probability distribution of F ranges from unity to ∞ and is asymmetric, thus one-tailed tests are appropriate (Fig. 4.7d). The form of the distribution depends on both sample sizes n_1 and n_2, and upon the significance level, α. Normally earth scientists use the tabled critical

critical values of t-statistic

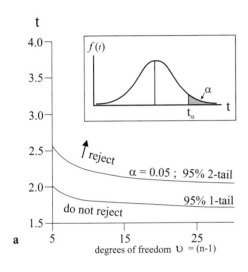

a

critical values of F-statistic
$(\alpha = 0.05, 95\% \text{ conf.})$

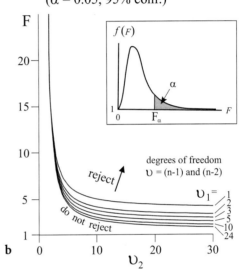

b

critical values of χ^2-statistic

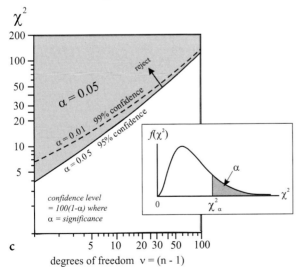

c

Fig. 4.8. Estimates of rejection limits defining critical regions for the common statistical tests shown in Fig. 4.7. In the earth sciences, the significance level is usually $\alpha = 0.05$ (95% confidence) but this may be assigned by the researcher, taking into account the subject matter. The degrees of freedom (v) are dictated by the sample size

Table 4.6. Critical values of F for $\alpha = 0.05$ (95% confidence). Degrees of freedom for denominator are $v_2 = n_2 - 1$, in rows

v_2	Degrees of freedom for numerator, $v_1 = n_1 - 1$														
	2	3	4	5	6	7	8	9	10	12	15	20	24	30
1	161	200	216	225	230	234	237	239	241	242	244	246	248	249	250
2	18.5	19	19.2	19	19.3	19.3	19.4	19.4	19.4	19.4	19.4	19.4	19.5	19.5	19.5
3	10.1	9.55	9.28	9.1	9.01	8.94	8.89	8.85	8.81	8.79	8.74	8.7	8.66	8.64	8.62
4	7.71	6.94	6.59	6.4	6.26	6.16	6.09	6.04	6	5.96	5.91	5.86	5.8	5.77	5.75
5	6.61	5.79	5.41	5.2	5.05	4.95	4.88	4.82	4.77	4.74	4.68	4.62	4.56	4.53	4.50
6	5.99	5.14	4.76	4.5	4.39	4.28	4.21	4.15	4.1	4.06	4	3.94	3.87	3.84	3.81
7	5.59	4.74	4.35	4.1	3.97	3.87	3.79	3.73	3.68	3.64	3.57	3.51	3.44	3.41	3.38
8	5.32	4.46	4.07	3.8	3.69	3.58	3.5	3.44	3.39	3.35	3.28	3.22	3.15	3.12	3.08
9	5.12	4.26	3.86	3.6	3.48	3.37	3.29	3.23	3.18	3.14	3.07	3.01	2.94	2.9	2.86
10	4.96	4.1	3.71	3.5	3.33	3.22	3.14	3.07	3.02	2.98	2.91	2.84	2.77	2.74	2.70
11	4.84	3.98	3.59	3.4	3.2	3.09	3.01	2.95	2.9	2.85	2.79	2.72	2.65	2.61	2.57
12	4.75	3.89	3.49	3.3	3.11	3	2.91	2.85	2.8	2.75	2.69	2.62	2.54	2.51	2.47
13	4.67	3.81	3.41	3.2	3.03	2.92	2.83	2.77	2.71	2.67	2.6	2.53	2.46	2.42	2.38
14	4.6	3.74	3.34	3.1	2.96	2.85	2.76	2.7	2.65	2.6	2.53	2.46	2.39	2.35	2.31
15	4.54	3.68	3.29	3.1	2.9	2.79	2.71	2.64	2.59	2.54	2.48	2.4	2.33	2.29	2.25
16	4.49	3.63	3.24	3	2.85	2.74	2.66	2.59	2.54	2.49	2.42	2.35	2.28	2.24	2.19
17	4.45	3.59	3.2	3	2.81	2.7	2.61	2.55	2.49	2.45	2.38	2.31	2.23	2.19	2.15
18	4.41	3.55	3.16	2.9	2.77	2.66	2.58	2.51	2.46	2.41	2.34	2.27	2.19	2.15	2.11
19	4.38	3.52	3.13	2.9	2.74	2.63	2.54	2.48	2.42	2.38	2.31	2.23	2.16	2.11	2.07
20	4.35	3.49	3.1	2.9	2.71	2.6	2.51	2.45	2.39	2.35	2.28	2.2	2.12	2.08	2.04
21	4.32	3.47	3.07	2.8	2.68	2.57	2.49	2.42	2.37	2.32	2.25	2.18	2.1	2.05	2.01
22	4.3	3.44	3.05	2.8	2.66	2.55	2.46	2.4	2.34	2.3	2.23	2.15	2.07	2.03	1.98
23	4.28	3.42	3.03	2.8	2.64	2.53	2.44	2.37	2.32	2.27	2.2	2.13	2.05	2.01	1.96
24	4.26	3.4	3.01	2.8	2.62	2.51	2.42	2.36	2.3	2.25	2.18	2.11	2.03	1.98	1.94
25	4.24	3.39	2.99	2.8	2.6	2.49	2.4	2.34	2.28	2.24	2.16	2.09	2.01	1.96	1.92
26	4.23	3.37	2.98	2.7	2.59	2.47	2.39	2.32	2.27	2.22	2.15	2.07	1.99	1.95	1.90
27	4.21	3.35	2.96	2.7	2.57	2.46	2.37	2.31	2.25	2.2	2.13	2.06	1.97	1.93	1.88
28	4.2	3.34	2.95	2.7	2.56	2.45	2.36	2.29	2.24	2.19	2.12	2.04	1.96	1.91	1.87
29	4.18	3.33	2.93	2.7	2.55	2.43	2.35	2.28	2.22	2.18	2.1	2.03	1.94	1.9	1.85
30	4.17	3.32	2.92	2.7	2.53	2.42	2.33	2.27	2.21	2.16	2.09	2.01	1.93	1.89	1.84
40	4.08	3.23	2.84	2.6	2.45	2.34	2.25	2.18	2.12	2.08	2	1.92	1.84	1.79	1.74
60	4	3.15	2.76	2.5	2.37	2.25	2.17	2.1	2.04	1.99	1.92	1.84	1.75	1.7	1.65
120	3.92	3.07	2.68	2.5	2.29	2.18	2.09	2.02	1.96	1.91	1.83	1.75	1.66	1.61	1.55
∞	3.84	3	2.6	2.4	2.21	2.1	2.01	1.94	1.88	1.83	1.75	1.67	1.57	1.52	1.46

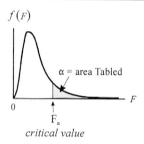

values for $\alpha = 0.05$ (95% confidence). The critical value must be selected using the degrees of freedom, $v_1 = (n_1 - 1)$ and $v_2 = (n_2 - 1)$, for each of the sample distributions with sample sizes n_1 and n_2. If the value for F exceeds the appropriate critical value for the degrees of freedom of our data (Table 4.6), we reject the hypothesis that the samples are sufficiently similar to have come from the same population at the 95% confidence level (Fig. 4.7 d). Of course, there is a 1 in 20 chance that we erred on the side of caution, due to selection of an unusual sample. Nevertheless, informally, one may accept the equivalence of the standard deviations. The one-tailed nature of the test restricts us to considering either $s_1 = s_2$ or $s_1 \neq s_2$. F has a very sensitive dependence on the samples' degrees of freedom ($v_1 = n_1 - 1$, $v_2 = n_2 - 1$) as shown in Fig. 4.8 b, for $\alpha = 0.05$ or 95% confidence level.

The F-distribution is also used to compare standard deviations of multiple samples by a technique known as the <u>an</u>alysis <u>of</u> <u>va</u>riance (ANOVA), discussed below.

Present in many spreadsheet programs, this enables the homogeneity of a phenomenon to be studied by determining whether sample means are equal (i.e. the null hypothesis $H_0: \bar{x}_1 = \bar{x}_2 = \ldots \bar{x}_n$) against the alternative hypothesis that at least one sample mean is different. The variance of a single sample may be compared to the population variance using a χ^2-test (Eq. 4.16), described more fully in Chapter 5.

4.7
Multiple Independent Variables: Paired Samples and Randomised Block Design

This subject was approached from a different perspective under spatial-sampling strategies in the first chapter (Fig. 1.8e), but it affects all aspects of data collection where multiple independent variables affected the observations. Independent variables may often be viewed as controls as they are the chief factors influencing the observations. In assessing confidence, in hypothesis testing, and also in correlation and regression (Chap. 6), it is particularly important to determine which is the *control* or *independent variable* and which is the *response* or *dependent variable*. In the simplest case, we normally see *y* as dependent on the (single) controlling variable *x*, and there is relatively little cause for concern. A familiar geological example would be the melting temperature of feldspar. The higher the Ca-content, the higher is the temperature required to melt it. Ca content is the independent control (*x*-axis) on which melting temperature is dependent (*y*-axis). Unfortunately, natural systems are usually more complex, and several controls may influence the dependent variable that we are investigating. It is advisable to reduce the complications as far as possible if we hope to detect relationships in which we can express some confidence, perhaps with the ultimate goal of establishing a theoretical basis for the process.

Consider the colour index used to classify igneous or plutonic rocks. When working in the appropriate terrains, geologists may employ colour codes for certain rocks as an approximate field-guide to their composition. The topic is simplified here, but two prominent controls are the total content of (Fe + Mg) and, to a lesser extent, grain size. High (Fe + Mg) contents are associated with dark silicates and dark rocks. However, very fine-grained rocks may be darker because there are so many grain boundaries, that a light crystalline appearance is masked and the rock appears darker. If we sample in a completely random fashion *with regard to colour* (white samples, Fig. 4.9a), the contributions of grain size and (Fe + Mg) content are confounded.

How can we be sure that a reasonably dark sample owes its hue to a high (Fe + Mg) content if it is also fine-grained? Where there are two controls the solution is relatively easy. We sample randomly, but from a sub-selection or *block* at constant grain size, so that the effects of varying grain size are excluded (black sample-symbols, Fig. 4.9a). Such samples are referred to as *paired samples* because each sample is selected so that its random (Fe + Mg) content is paired with a fixed grain size. This may sound rather demanding, but in a surprising number of earth science situations it is not as difficult as it seems because the less important control may be very nonlinear in its influence. For example, with this colour index example, only very fine grain sizes darken the rock's colour.

How do we proceed if there are several controlling variables? With many natural processes, it is not possible to isolate the contribution of the independent variable which interests us, because the others cannot be held constant. (Our colleagues in the laboratory may be more fortunate.) The best we can do is to design the data collection so that the effects of the primary control are maximised while the contributions of the secondary independent variables that do not interest us are kept "similar". Statisticians refer to this as a *randomised block design*. It is difficult to illustrate on paper; even three dimensions rarely suffice. However, let us attempt to do this by extending the rock-colour index example. Our intention is still to investigate (Fe + Mg) as the primary control on colour, but now we address the fact that secondary, independent variables include Ca/Al ratio, SiO_2 content, as well as grain size. Completely random sampling across the colour spectrum mingles contributions of all four independent variables (Fig. 4.9c). However, subject-specific knowledge would enable us to isolate *blocks of samples* in which one or more of the secondary controls was absent, or at least almost invariant. This would greatly increase our chances of determining whether (Fe + Mg) was the chief control on colour and just how strong was that relationship. Similar blocks of samples are shown by the black-sample dots in Fig. 4.9c. Such sampling is described as a randomised block design.

In some natural systems, the variables are confounded in a more intricately interwoven manner and it is impossible to remove any of the independent controls. Consider the ductile strain of rocks. Experimental and field research in connection with inferences from metallurgy, crystallography and geotechnics show that the rate at which rocks change shape under ductile conditions (ductile strain-rate) is controlled by several factors simultaneously. For a monomineralic rock at a fixed depth, these include temperature (*T*),

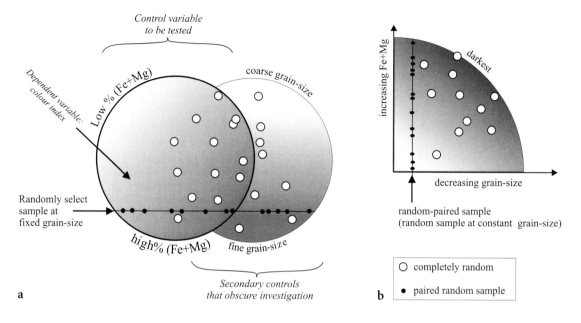

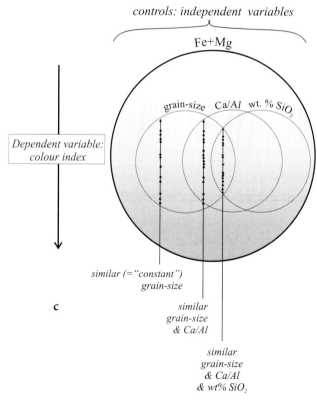

Fig. 4.9. The effects of the primary control (independent variable, *x*) may be obscured by some secondary control. For example, primarily, the colour index of igneous rocks may be a function of (Fe + Mg) content. However, grain size may also influence colour. **a** Completely random samples confound contributions of (Fe + Mg) content and grain size so that the significance of any relationship between the supposed control (Fe + Mg) and colour may be suspect. This is overcome by sampling randomly with respect to colour but at constant grain size, as shown by the samples along the line. Such samples are called *random paired samples*. **b** The same argument considered graphically. **c** Where there are more than two control variables (Fe/Mg, grain size, Ca/Al, wt%, SiO₂), paired sampling must be performed in campaigns that attempt to isolate or emphasise the effect of one control at a time; at best, this is a compromise

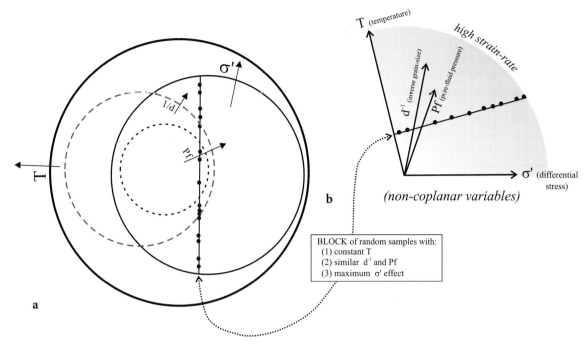

Fig. 4.10. Interacting multiple variables complicate sampling further. Attempts must be made to isolate contributions from one control variable at a time. It will not normally be possible to keep all secondary controls constant, but samples can be selected randomly with respect to each of the other controls in that sample block as similar as possible. **a** Four controls influence ductile strain rate of rocks; temperature (T), differential stress (σ'), inverse grain size (d^{-1}) and pore-fluid pressure (Pf). To investigate the relationship between temperature and strain rate, specimens (or experiments) should be taken so that the effects of as many other variables as possible are minimised. **b** An attempt to represent the same concept graphically

differential stress (σ', the difference between maximum and minimum stress), grain size (d) and the pressure of fluid in the rock's pores (Pf). Suppose that we wish to detect the influence of σ' on strain rate. Since all factors interact simultaneously, the best attempts at paired sampling can only keep *blocks of samples similar* with respect to one or more secondary controls (Fig. 4.10 a). We select samples to maximise the variation in σ'; it may be possible to sample so that T-effects are removed completely (sample at constant-T) but at best we may only minimise the compounded contributions of secondary controls by d and Pf (Fig. 4.10 a). The line of black sample dots schematically depicts the randomised block sample conceptually (Fig. 4.10 a) and graphically (Fig. 4.10 b).

4.8
Comparing Multiple Samples: ANOVA

Geologists in particular should note that in the following, "sample" is used in the statistical sense to refer to a set of observations, as elsewhere in this book. "Sample" does not refer to an individual geological specimen.

Previously, we compared pairs of samples, each with a unique variance and mean. Very commonly, earth science deals with groups of observations from different locations or resulting from different processes, so it is logical to consider the comparison of multiple samples. Now, our intention is not to consider the variance within the sample, but rather the variance between the samples, e.g. the degree to which sample means differ. ANOVA was developed originally to study quality control in manufacturing, where it is desirable to keep similar small variances and means in each production campaign. Consequently, the terminology refers to multiple "trials", yielding different samples, whose differences from the populations are regarded as "errors".

A logical application in experimental science is to repeat an experiment with a multivariable system, allowing only one variable to change in each experiment. Thus, each experiment represents a different trial, and observations from that experiment constitute the sample. The relative variances of samples from different trials then indicate the relative influence of the different variables. Where the variation between samples is attributable to a single independent variable, we may compare their variances by *single-factor ANOVA*.

A useful application of ANOVA in earth sciences is the comparison of field data from different sites. For example, we may wish to compare structural measurements or sedimentary facies or petrological characteristics from different locations. For a meaningful detection and evaluation of any regional trend, the variance of measurements within outcrops should be less than that between outcrops. The general phrasing would be that we require *within-sample variance (ISV)* to be less than *between-sample variance (BSV)*. These are sometimes termed the *intra-sample variance* and *inter-sample variance* respectively. In most common applications, we prefer ISV < BSV. For example, there is little reason to conduct a field study of spatial variation of some measurement if the variation within sample sites (ISV) exceeds that between sample sites (BSV); this is rarely assessed formally, but it is grasped intuitively by geologists. If ISV ≥ BSV, the area may be homogeneous with respect to the variable observed.

In practice the calculations deal with the sums of squares of deviations which, in view of the simple relationship above, require only the sums of

1. the squares of disparities of sample means from the grand mean of all samples
   BSS = *sum of **between-samples** squared deviations*
   (*SST* = sum of *between-treatments squared deviations in many statistics texts*)
2. the variation within sample squares of disparities from sample means
   ISS = sum of with**in**-sample squared deviations
   (*SSE* = sum of *within-sample squares of "errors" in statistics texts*).

The fundamental identity for ANOVA is a surprisingly simple relationship that defines the sum of the squares of deviations for all samples or the total variation in the system of all observations (SSTot) as the sum of between-samples-squared deviations and within-samples-squared deviations, weighted by sample-size:

$$\text{SSTot} = \text{BSS} + \text{ISS} \qquad (4.22)$$

ANOVA requires that:

1. each of the k samples ("treatments" of statistics) is drawn from a Normal population
2. standard deviations of the populations from which samples are drawn are equal ($\sigma_1 = \sigma_2 = \sigma_3 = \ldots = \sigma_k$)
3. the observations in each sample are independent of those in any other sample.

In practice, these criteria are not too stringent. The first criterion is most commonly violated in geological ANOVA with observations of weak concentrations that commonly follow something close to the log-Normal

distribution. This problem may be circumvented by re-expressing the variable of interest as a dummy variable, which does more closely follow the Normal distribution. Several transformations will be mentioned in Chapter 6, but the most common one is a logarithmic re-expression. For example, the data of Fig. 4.11 appear to follow a log-Normal distribution. A suitable arbitrary transformation of the concentrations x to $x' = (a + b\ln(x))$ can be devised by trial and error until the dummy variable x' more closely follows a Normal distribution (Table 5.8). The frequency axis (y), may be transformed also if that has not already been achieved during standardisation. Statistical testing then proceeds with the dummy variable. Of course, where results are required in original units they must be back-transformed.

The second criterion is most easily violated, but the procedures seem to work even when the largest standard deviation is twice the smallest, and techniques exist for dealing with less cooperative data sets (Miller 1986). The third criterion is usually easily satisfied when the variables and the system are evaluated with the benefit of subject-specific knowledge.

The notation in dealing with single-factor ANOVA can become cumbersome when formalised, but a simplified version is presented below.

Where k samples have:

sample sizes $n_1, n_2, n_3 \ldots \ldots n_k$

sample means $\bar{x}_1, \bar{x}_2, \bar{x}_3 \ldots \ldots \bar{x}_k$

sample variances $s_1^2, s_2^2, s_3^2 \ldots \ldots s_k^2$

total count $\quad N = n_1 + n_2 + n_3 + \ldots \ldots + n_k$

grand total observations

$$T = n_1\bar{x}_1 + n_2\bar{x}_2 + n_3\bar{x}_3 + \ldots \ldots + n_k\bar{x}_k$$

grand mean of all observations $\quad \bar{X} = \dfrac{T}{N} \qquad (4.23)$

(a) the total variation in the system is given by:

$$\text{SSTot} = \text{BSS} + \text{ISS} \ = \ \sum_{i=1}^{N}(x_i - \bar{X})^2$$

where x_i are all x-values

(b) BSS or SST $= n_1(\bar{x}_1 - \bar{X})^2 + n_2(\bar{x}_2 - \bar{X})^2$
$+ \ldots \quad \ldots + n_k(\bar{x}_k - \bar{X})^2$

(c) ISS or SSE $= (n_1 - 1)s_1^2 + (n_2 - 1)s_2^2 + \ldots \ldots$
$+ (n_k - 1)s_k^2$

... which in practice is easily determined by rearranging (a).

We then proceed to establish the logical and sensitive hypothesis that the samples all have the same mean. Of

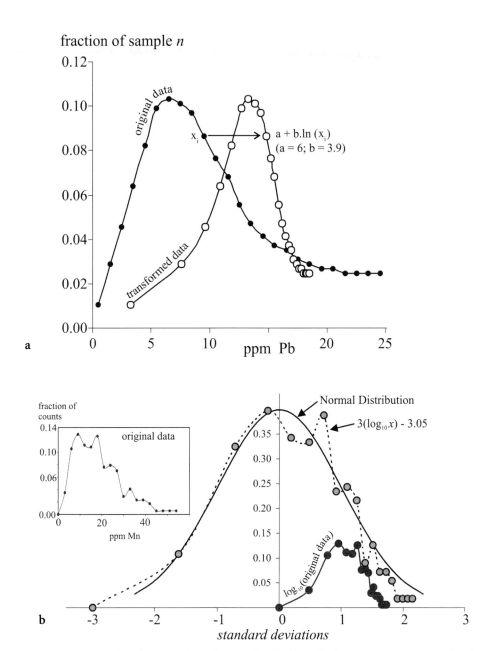

Fig. 4.11. For many purposes, such as the comparison of means of multiple samples by ANOVA, it is necessary that the samples are drawn from normally distributed populations. Where that is not possible, samples that are not Normally distributed may be transformed into one suitably close to the Normal distribution for the purposes of testing. **a** A common distribution that is amenable to treatment in this way is one similar to the log-Normal distribution. Re-expressing the original observations, (x) with logarithms produces a distribution that is similar to the Normal distribution. **b** Example of logarithmic transformation to a Normal distribution. Inset: Mn from a New Zealand soil sample approximately follows a log-Normal frequency distribution, like many low-abundance elements in rocks. Steps in the transformations show $\log_{10}(x)$ producing a humped form and a further empirical transformation $3\log_{10}(x) - 3.05$ which produces a transformed sample distribution that corresponds closely to the Normal distribution

course, this is always desired or even expected, but it permits us to define a statement that can be rejected formally at a certain confidence level. In some cases, we may wish that all our samples be "the same". For example, if we analyse ground-water samples for suitability for human consumption, we may wish the samples to be equally pure with respect to some contaminant. On the other hand, if we wish to study regional variations, we must be sure that the variance at individual sites is small enough that it does not disguise the differences between sites. Only then is it meaningful to compare site means. If we are satisfied that this is the case, we can establish the null hypothesis that the samples are drawn from a population with the same mean:

$$H_0: \mu_1 = \mu_2 = \mu_3 = \dots \dots = \mu_k \quad (4.24)$$

If the samples are considered "the same", it is implicit that they are drawn from populations with the same standard deviations $\sigma_1 = \sigma_2 = \sigma_3 = \dots \dots = \sigma_k$. The alternate hypothesis (H_A) is that at least one mean differs from the others.

The test statistic we require is a ratio of variances between samples (BSV), to that within samples (ISV) which follows the F-distribution:

$$F = \frac{\dfrac{BSS}{(k-1)}}{\dfrac{ISS}{(N-k)}} \quad (4.25)$$

with degrees of freedom $v_1 = (k-1); \ v_2 = (N-k)$

If $F > F_\alpha$, the test statistic lies in the rejection region of the upper tail of the F-distribution for the associated degrees of freedom. As we saw earlier, because the F-probability distribution depends on two degrees of freedom, tables are presented usually for selected common significance levels (α); our simplified Table 4.6 is only valid for $\alpha = 0.05$ or a 95% confidence level. Where F exceeds the critical value, we must reject the null hypothesis that the samples are drawn from the same population. In simple terms, the samples are dif-

ferent at that confidence level. This would be bad news where we want all our water samples to be of "the same" high quality. Where we wish to detect regional variations, it would be good news because our samples are measurably distinct from different sites.

The results of single-factor ANOVA are summarised conventionally as shown in Table 4.7. This is illustrated with a comparison of five drill cores ($k = 5$) from which between 24 and 55 gold assays had sample means (μ_1 to μ_k) of 0.5, 0.52, 0.69, 2.19, and 2.85 ounces per ton respectively (Koch and Link, 1970), for which the ANOVA calculation and original data are summarised in Table 4.8. Suppose we pose the question, "Are these five cores equally mineralised?" Assuming the necessary criteria for ANOVA are satisfied, we set the null hypothesis $H_0: \mu_1 = \mu_2 = \mu_3 = \mu_4 = \mu_5$ and proceed with single factor ANOVA as set out in Table 4.7. The calculated value of F is 0.89 whereas at a 95% confidence level, the critical rejection region for $F \geq 2.65$. Since F_{CALC} does not lie in the critical region, we do not reject the hypothesis (H_0) that the sample means are all equal. Less conservatively, the geologist might conclude that the drill cores are "equally mineralised". However, that makes considerable assumptions about the manner in which the rocks were sampled and assayed, the spacing of assays in the drill core and rock homogeneity. None of these factors is considered in the terse statistical result of the hypothesis test.

4.9
ANOVA: Identifying Important Variables in a Multivariable System

In the preceding section we compared samples whose variance was caused by a single factor. That approach may be possible even if the system under investigation has a secondary control, because its effects may be avoided by taking paired samples (Fig. 4.9), illustrated for this example in Fig. 4.12a. The example of gold mineralisation of five drill cores (Table 4.8) concerned a single variable, Au concentration, which represents the variable, "mineralisation". A common increment of

Table 4.7. Tabulation of results for single-factor ANOVA has following form

Source of variation	Degrees of freedom	Sum of squares of deviations	Mean square	F_{CALC}	$F\alpha$
Between samples (treatments)	$k - 1$	BSS (= SSTr)	BSS/$(k - 1)$	= BSV/ISV	e.g., $F_{0.05}$ from tables
Within samples (errors)	$N - k$	ISS (= SSE)	ISS/$(N - k)$	$H_0: \mu_1 = \mu_2 = \mu_3 \dots \mu_k$	
Grand total variation from all samples	$N - 1$	SSTot		Decision: reject H_0 if $F_{CALC} > F\alpha$	

Table 4.8. Example of single-factor ANOVA to compare gold assays in five drill cores from Homestake mine (Koch and Link 1970)

Source of variation	Degrees of freedom	Sum of squares of deviations	Mean square	F_{CALC}	$F_{0.05}$
Between samples (Treatments)	4	177.9	44.48	0.89	2.65 (from tables)
Within samples (errors)	51	2536.1	49.73	$F_{CALC} < F_{0.05}$, thus do not reject H_0; "accept" that means are equivalent for the five samples at the 95% confidence level	
Grand total observations from all samples		219	2714		

Data for the five Homestake drill cores

Drill core	Number of assays	Mean assay values $\bar{x}_k$ ounces/ton	Assay standard deviation
1	58	0.52	1.45
2	58	2.19	7.6
3	24	2.85	8.27
4	39	0.5	1.16
5	40	0.69	1.57

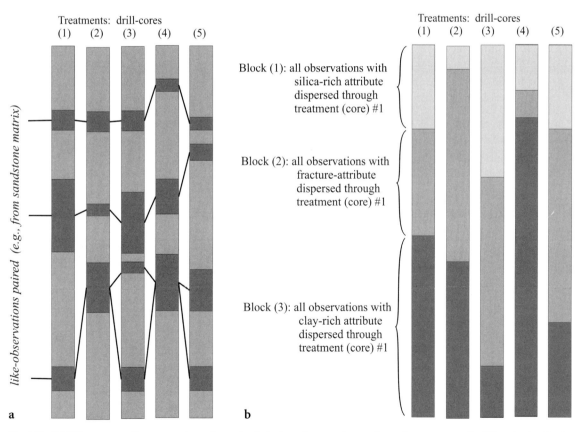

Fig. 4.12. ANOVA compares the mean values of samples, but where the samples are drawn from groups with different subordinate controlling influences, they should be arranged to make the comparisons fair. This is illustrated with respect to five treatments represented by five drill cores in which the observations are gold assays. **a** Where the drill core comprises two rock types, the observations should be compared only if the secondary attribute (rock type) is similar. Thus *paired samples* are used to determine sample means for each drill core; assays for each rock type are grouped together for each core. **b** Where the drill core comprises multiple secondary influences that may co-exist in the material providing the same observation, we must assemble the observations in blocks that are as similar as possible in each drill core. Here, three attributes are shaded differently and separated into separate sections; in reality, the observations that they represent are dispersed and intermingled throughout the core

complexity is introduced where the host rocks respond differently to mineralisation, e.g., sandstone is porous and lava is impermeable. How then do we sensibly compare mineralisation? Earth and life scientists deal with this sort of situation routinely and intuitively; in statistical terms, we pair samples: *observations* (assays) for each *sample* (drill core) are *grouped* according to some secondary variable or attribute (porosity). Thus, paired assays for sandstone-host rocks are formed into a "similar group" and could be compared by single-factor ANOVA as in the previous section. That procedure works reasonably well and may provide suitable answers for our investigations where the secondary variable is discrete; i.e. the groups are sandstone *or* lava, with no possible gradation (e.g. Fig. 4.12a). The situation is more complex where the secondary attribute by which we group observations is a continuous variable, such as clay content which reduces porosity and permeability and reduces mineralisation in a gradual fashion or where there are multiple discrete secondary influences. Then, "groups" must be assembled on the basis of "similarity" with respect to the secondary attributes (Fig. 4.12b). For example, on petrological grounds we may have reason to believe that mineralisation is similar in sandstone samples with 5–15% clay content. We should then group assays from such sandstones in different samples (= drill cores) in order to make comparisons of mineralisation on an equal basis. As this example shows, the designation of "similar groups" almost always requires subject-specific knowledge.

Advancing further towards the realities of earth science, let us now consider the comparison of multiple samples with multiple treatments and multiple blocks. Let us extend our example:

(a) let the treatments be the drill cores (5)
(b) let "similar" samples from *within each treatment* be characterised as:

　　block (1) = sample comprising assays from silica-rich rock
　　block (2) = sample comprising assays from fractured rock

block (3) = sample comprising assays from clay-rich rock

Figure 4.12b illustrates this arrangement, although assays (observations) with different attributes (blocks) are dispersed throughout the drill core (= treatment); in the diagram they are clustered for ease of illustration.

The general formulation of the randomised block design ANOVA then follows using the terminology:

number of treatments = k

number of blocks = L

mean for all measurements in treatment $i = \bar{x}_i$

mean for all measurements in block $i = \bar{b}_i$

grand mean for all measurements (kL)
　　in all treatments and all blocks $i = X$　　(4.26)

The sums of squared deviation for observations must now be assessed for three arrangements, treatments (SSTr), blocks (SSBl) and errors (SSE). The appropriate expressions are presented in Table 4.9. But in practice, SSE are calculated more easily from the identity SSTot = SSTr + SSBl + SSE which expresses that the total variation in the system comprises contributions from treatments (e.g. drill cores), blocks (grouped observations according to similar attribute) and from all individual observations (e.g. assays).

The calculation would be presented in the form shown in Table 4.10. If the calculated value of F exceeds F_α we reject the null hypothesis that all treatment means are equal. That would imply that at least one treatment mean was different, with a probability of $100(1-\alpha)$%. Our simplified F-table is only presented for a 95% confidence level ($\alpha = 0.05$, Table 4.6). Erikson et al. (1966) provide an example in which methods of gold assaying were compared. The randomised block ANOVA compared:

(a) three treatments: the analytical techniques
(b) twenty-two blocks: assays grouped for 22 "similar" rocks

Table 4.9. Sums of squares for randomised block experiment

Source of variation	Degrees of freedom	Sums of squares of deviations
Treatments	$(k-1)$	$SSTr = L\{(\bar{x}_1 - \bar{X})^2 + (\bar{x}_2 - \bar{X})^2 \ldots \ldots (\bar{x}_k - \bar{X})^2\}$
Blocks	$(L-1)$	$SSBl = k\{(\bar{b}_1 - \bar{X})^2 + (\bar{b}_2 - \bar{X})^2 \ldots \ldots (\bar{b}_k - \bar{X})^2\}$
Errors	$(k-1)(L-1)$	$SSE = SSTot - (SSTr + SSBl)$
Grand total (every measurement)	$(kL-1)$	$SSTot = \sum_{i=1}^{n=kL} (x_i - \bar{X})^2$

Table 4.10. Tabulation results for a randomized block design; ANOVA has following form

Source of variation	Degrees of freedom	Sum of squares of deviations	Mean square	F_{CALC}	$F\alpha$
Treatments	$k-1$	SSTr	$MSTr = SSTr/(k-1)$	$= MSTr/MSE$	e.g., $F_{0.05}$ from tables
Blocks	$L-1$	SSBl	$MSBl = SSBl/(L-1)$	$H_0: \mu_1 = \mu_2 = \mu_3 \dots \mu_k$	
Errors	$(k-1)(L-1)$	SSE	$MSE = \dfrac{SSE}{(k-1)(L-1)}$		
Grand total	$kL-1$	SSTot			

ANOVA requires that the observations be drawn from normally distributed populations but Au-abundances follow something like a log-Normal distribution. For this reason, the analysis was based on the natural logarithms of the Au-concentrations to achieve a standardisation of the sample distributions into that of the Normal distribution. The results are presented in Table 4.11, for which $F_{CALC} = 0.899$. For the degrees of freedom relevant in this example, $F_{0.05} = 3.20$ from Table 4.6 so that their calculated test statistic does not fall in the rejection region (≥ 3.20). Thus, we do not reject the hypothesis that the means of the three treatments are equal. This could be interpreted as meaning the three analytical methods agree, at the 95% confidence level.

Geochemical analysis was an area in which earth scientists were sensitive to precision and reproducibility in the early days of quantified geology. In the 1950s and 1960s, geologists were so concerned about geochemical analyses that inter-laboratory comparisons of techniques and research teams were conducted using material drawn from the same large homogenised batches of granite and basalt (e.g. Flanagan 1960). Since the early 1980s, analytical techniques have changed significantly and precision and reproducibility have improved. However, the problems of replication of results will always be with us because our curiosity inevitably drives us to work close to the contemporary levels of detection, sensitivity and accuracy (see Chap. 2, Table 2.1).

4.10
Informal Uses of Probability to Assess "Importance"

Science in general has many examples where probability is used informally to assign a loose "significance" to the frequency or intensity of some event. The statistician's use of the word significance would be too strong in this context since rigorous hypothesis testing is not involved. It is preferable to consider such usage as a qualitative or semi-quantitative assessment of the importance of some process.

One may read that on the basis of some natural data or experiments, the probability of some event is estimated dependent upon the value of one or more variables and assumptions which should be ascertained. A graph of probability ($0-1$ or $0-100\%$), or some value that proxies for probability, is then plotted against an independent variable. This is used to infer the importance of some controlling process. Lovelock's important study of the equilibrium of the earth provides an interesting example concerning the atmosphere's oxygen content (Lovelock 2000).

Table 4.11. Example of randomised block design; ANOVA: comparison of mean $ln(Au)$ for 3 different methods on 22 blocks of grouped specimens (Erikson et al. 1966)

Source of variation	Degrees of freedom	Sum of squares of deviations	Mean square	F_{CALC}	$F_{0.05}$
Treatments (analytical methods)	2	0.456	0.228	0.899	3.2
Blocks (similar samples)	21	229.3	10.92	$F_{CALC} < F_{0.05}$, *thus do not reject H_0: "accept" that means are equivalent for the five samples at the 95% confidence level and infer that analytical techniques do not differ*	
Errors (contents)	42	10.69	0.254		
Grand total	65	240.45			

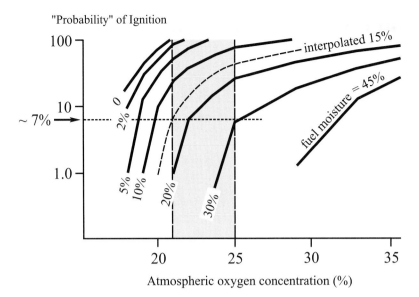

Fig. 4.13. An example of the informal use of probabilities to assess the "importance" of some phenomenon. Here, we see the probability of ignition of plant material due to natural causes (lightning) at different atmospheric oxygen concentrations and for different moisture contents of the organic fuel. Oxygen forms 21 % of the present-day atmosphere (Lovelock 2000). The shaded band indicates atmospheric oxygen concentrations between 21 and 25 % that can tolerate a 7 % ignition probability for fuel moisture contents between 15 and 30 %

Lovelock discussed the interesting, natural self-regulation of the Earth's atmosphere such that the oxygen content of the atmosphere has remained at a value advantageous for biological activity for hundreds of millions of years. However, oxygen concentrations have remained below the level at which spontaneous conflagration would consume the planet's grasslands and forests due to the ignition caused by lightning. We are not concerned here with the elegance and complexity of the Gaia hypothesis, but rather with an interesting application of ignition probabilities of natural organic fuels at different oxygen concentrations. Lovelock's graph presents these variations for different moisture contents of the organic material (Fig. 4.13). The author states *"At the present oxygen content (21 %) fires do not start at more than 15 % moisture content. At 25 % oxygen content even the damp twigs and grass of a rain forest would ignite…"*.

By interpolation we see that the 15 % moisture-content line on the graph would intersect the present atmospheric concentration (21 % vertical line) corresponding to an ignition probability of ~7 % which is presumably the present tolerable limit. More explicitly, this may imply that there is a 7 % probability of natural fire due to lightning strikes and that this affects any grasses or wood with < 15 % moisture content. Thus, we may infer that the author considers < 15 % moisture content typical, and that this does not exceed the tolerable 7 % ignition probability at a 21 % oxygen content. The implication appears to be that higher oxygen contents would lead to uncontrollable conflagrations. For example, fuels with up to 30 % moisture content would exceed the tolerable 7 % ignition probability. However, if the atmosphere had too much oxygen, say 25 %, there is a 7 % probability that even plants with as much as 30 % moisture would ignite. Whereas Lovelock's conclusions are not stated in a formal statistical sense, they illustrate an informal use of probabilities to express a complex and important concept. Explanations in this style are common in earth science, but the concepts of this chapter may permit us to evaluate them more effectively.

CHAPTER 5

Comparing Frequency-Distribution Curves 5

In this chapter, we consider the form of simple frequency distributions and the means by which they may be compared. In most earth-science applications, our observations will not correspond closely to the Normal distribution, nor to any other theoretical distribution, but that does not necessarily prevent description and characterisation with a few statistics. However, sometimes a direct comparison with the Normal distribution may be helpful and, in a few instances, it may be important, for example, to satisfy the requirements for ANOVA (Chap. 4). Therefore, this chapter commences with methods that compare observations to a Normal distribution. The discussion commences with simple arithmetical and graphic methods and then continues with the process of transforming observations by standardisation to facilitate direct comparison with the Normal distribution. The cumulative version of the frequency distribution sometimes expedites a cursory visual comparison of samples, and cumulative frequency distributions of grain size may reveal underlying processes. Finally, it is shown that any two distributions may be compared without making any assumptions about the form of the population that was sampled with the χ^2 and Kolmogorov-Smirnov techniques.

5.1
The Empirical Rule to Compare with the Normal Distribution

If we believe that the frequency distribution of our sample follows the Normal distribution, we may utilize a few specific and simply applied probability values of the theoretical distribution by which to gauge our sample. From Chapter 3, in particular Table 3.6, we know that a normally distributed population has 99.7, 95 and 68% of values within $\pm 3\sigma$, $\pm 2\sigma$ and $\pm\sigma$ of the mean value, where σ is the population standard deviation. If our sample follows the Normal distribution, similar proportions of observations should lie within $\pm 3s$, $\pm 2s$ and $\pm s$ of the sample mean, where s is the sample's standard deviation. This empirical rule is also known as the *68–95–99.7 rule*.

5.2
Linearisation Using Normal-Probability Paper to Compare with the Normal Distribution

The next approach uses the entire sample and provides some visual assurance of the validity of any comparison. A common device in managing geological data is to rearrange the scales of graphs so that the pattern falls into the simplest possible geometric pattern, i.e. a line. A special graph exists with a nonlinear vertical scale, distorted so that normally distributed observations fall on a straight line when their cumulative frequencies are plotted against their x. A cumulative frequency is the fraction of observations that have values $\leq x$. The distorted y-axis uses a Normal probability scale, a sample of which is shown in Fig. 5.1. Normal probability paper is available with a finer grid than the example shown, but for the precision possible with this approach the example may suffice. A high degree of linearity indicates similarity to the Normal distribution and is usually estimated visually, although it could be quantified by regression (Chap. 6). Of course, it is a subjective approach in essence, especially when there is much scatter about a line.

If the data plot as a straight line, the mean of their Normal distribution, $\bar{x}$, will be found at the cumulative frequency of 50% and the standard deviation is represented by $(x_{84} - x_{50})$, where the two values represent the x-coordinates of the 50 and 84% cumulative frequencies. Thus, the plot provides a simple shortcut for the estimation of Normal-distribution statistics. An example is shown in Fig. 5.4d, where grain sizes of sediment fractions approximately follow a line on the Normal-probability plot, suggesting that the grain sizes may approximately follow the Normal distribution.

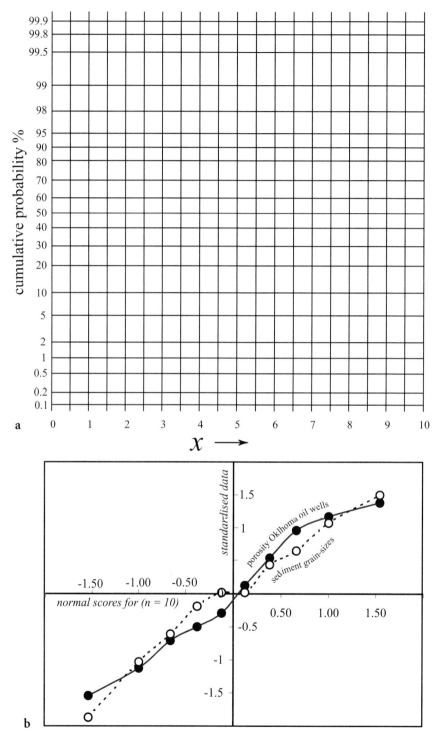

Fig. 5.1. a Normal-distribution probability paper: when one plots cumulative probabilities of a sample distribution on this graph, they will fall on a straight line of positive slope if they followed the Normal distribution. An example is shown in Fig. 5.4d. **b** Normal scores also indicate the degree to which a sample follows the Normal distribution. Standardised data (z-variate) for oil reservoir porosities (n = 10) and for sediment grain sizes (n = 10) are plotted against the *normal scores for a sample with n* = 10. The linearity of the plots indicates that each of these different samples was drawn from a population similar to the Normal distribution. The same data may be viewed in conventional frequency-distribution graphs in Figs. 5.2 and 5.4

5.3
Normal Scores to Compare with the Normal Distribution

A procedure superficially similar to linearisation can be performed more rapidly, by using tabled values which are now provided painlessly by spreadsheets and statistical software. Suppose we have ten observations (x_i where $i = 1$ to 10). Since we wish to make a comparison with the Normal distribution, the next step is predictable; we standardise the observations as the z-variate to create dummy variables $z_i = (x_i - \bar{x})/s$. We now rank our transformed observations in ascending order of magnitude. For example, the z_i rearranged in order of ascending magnitude may be:

$$-1.51, -1.00, -0.71, -0.32, 0.11, 0.20, 0.41, 0.62,$$
$$1.08, 1.53$$

Fortunately, based on the Normal distribution, statistical tables are available showing that for a sample of ten measurements, the smallest, next-smallest, second-to-smallest, etc. values should ideally have the following values:

$$-1.54, -1.00, -0.66, -0.38, -0.12, 0.12, 0.38, 0.66,$$
$$1.00, 1.54$$

These are the *Normal scores*, valid only for the sample with $n = 10$.

We may now plot z_i against the Normal score on the horizontal axis. Clearly, the more linear the plot, the closer is our sample to a Normal distribution. Examples are shown in Fig. 5.1b, for the porosity of a petroleum reservoir and for grain-size fractions from sieved sediment; these data may also be viewed in conventional frequency distribution graphs in Figs. 5.2 and 5.4 respectively.

Normal scores must be consulted for each sample size (n). They are obtained by repeatedly drawing a random sample of n values from the Normal distribution, with replacement. This is best performed as a Monte-Carlo operation, using a computer to select randomly, e.g. 1000 samples of ten observations ($n = 10$) from the Normal distribution. The 1000 smallest values from each sample are then averaged to give the smallest Normal score, the 1000 next-smallest values are averaged to give the second-smallest normal score, and so on. Again, the linearity of the Normal-score plot may be quantified by regression (Chap. 6).

5.4
Standardisation of Observations to Compare with the Normal Distribution

In many aspects of natural science we wish to compare distributions that have different means, different variances and even different units of measurement. In other words, how does one group of measurements compare to another in terms of fundamental statistics? This introduces the essential and elementary concept of standardisation: we must devise a method by which a distribution can take on a universally comparable mask.

This exercise compares the porosity of samples derived from petroleum-bearing strata in Texas with those from Oklahoma, from data listed by Davis (1973). The frequency is represented by n, the number of well sites, from which estimates of reservoir porosities have been obtained. The x-axis plots percentage porosity, which is naturally confined to the range zero to 100%. Strictly speaking, this can never be similar to the Normal distribution, which requires an infinite range for x ($-\infty$ to $+\infty$). Therefore, we can expect some discrepancies in comparing the tails of the curves near the limits of the range (0 and 100%). The data will be converted to *standard normal form*, which has been briefly mentioned previously (Chap. 2). This is explained in steps whose effects are illustrated in Fig. 5.2.

Fig. 5.2a Normalise the frequencies (y-axes) of the two distributions by using percentages or relative fractions. Thus, both distributions will have equal peaks at 100% or at a relative fraction = 1. This now permits some basis for reliable comparison. The relative fraction is also the probability of observing a porosity of the associated value.

Fig. 5.2b Now centralise the two distributions by bringing their means ($\bar{x}$) together at zero, shifting one distribution x_i to be coaxial with the other, x_i'

$$x_i' = x_i - \bar{x} \qquad (5.1)$$

The new distribution of x_i' still has the same units as the original one. However, the relative dispersions about their means are now very clear.

Fig. 5.2c Finally, to assess the conformity of the two distributions and to compare their dispersions as closely as possible, we must standardise the two distributions by dividing by

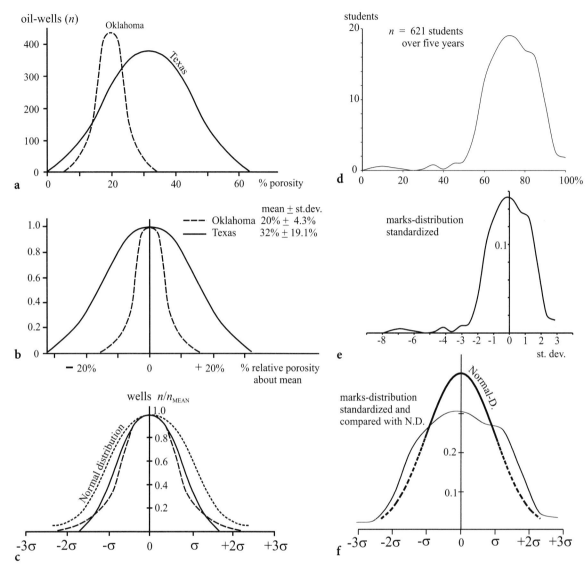

Fig. 5.2. a–c Standardisation of a sample distribution facilitates comparison with the Normal distribution using an example of petroleum reservoir porosities from Davis (1973). Steps: **a** Initial sample frequency distributions. **b** Centre sample distributions about zero by subtracting the mean value from each distribution. **c** Give each sample distribution the same *x*-scale by dividing by their sample standard deviations (*s*). The resulting standardised distributions are now scaled without dimensions, in numbers of standard deviations. They may now be compared easily with the standardised Normal distribution, sometimes recorded with the shorthand [ND (0, 1)]. That notation signifies a Normal distribution with mean $\mu = 0$ and standard deviation $\sigma = 1$. Neither sample follows the theoretical Normal distribution exactly because the data lie within its 2σ range. This is not surprising because porosities may only range from 0 to 100%. The Oklahoma reservoirs have a narrower range of porosities. **d–f** Student marks distribution over 5 years of a first-year geology course; **d** Frequency distribution of marks almost fills range giving rise to unusually broad distribution with long tails. **e** In standardised form shows measurable frequencies over a range of eleven standard deviations. **f** Compares unfavourably with the standard Normal distribution

the sample's standard deviation (s). Thus, we create a dummy variable, the z-variate:

$$z_i = \frac{x_i'}{s} \quad \text{more usually written as} \quad z = \frac{x_i - \bar{x}}{s}$$

$$(5.2)$$

The horizontal axis is now dimensionless in multiples of sample standard deviations (s). The theoretical Normal distribution is scaled in this way and has a mean of zero and standard deviation of one unit, sometimes abbreviated to ND(0, 1). When sample distributions are standardised, as described above, they may be readily compared with the theoretical Normal distribution and with one another.

If the standardised curves for two samples are very similar, we may conclude that they were drawn from the same population. If they were similar to the Normal distribution, one might further conclude that the two samples were drawn from a normally distributed population. At this stage, a visual comparison is still somewhat subjective (Fig. 5.2c). But at least we are able to make comparisons on a level playing-field. Objective conclusions might be that both samples show a tighter distribution than the Normal distribution. The Texan oil reservoirs show the greatest departure from the Normal distribution because there are no data beyond $\pm 2\sigma$; if the sample distributions were normally distributed, slightly more than 5% should lie further than that from the mean. The Oklahoma reservoirs have a narrower range of porosities. Both sample distributions are more peaked (larger *kurtosis*) than the Normal distribution, which is only to be expected because the maximum permissible range of porosities is constrained to a 100% range.

A second example in Fig. 5.2d–f shows the comparison of a distribution of a large sample ($n = 621$) of student marks in a first-year university geology program. This shows the problems that arise where a distribution has a finite x-range and occupies all of that range. When standardised, the data spread over 11σ (Fig. 5.2e), whereas a Normal distribution has 99.8% of its values in a $\pm 3\sigma$ range. The standardised sample distribution compares unfavourably to the standard Normal distribution (Fig. 5.2f).

5.5
Cumulative Frequency Distributions

The frequency distributions discussed so far plot the individual frequencies (n), or their normalised values (n/n_{MAX}) to represent probabilities, against the associated x-values. The conventional frequency distribution shows a *spectrum of frequencies* for x. An alternative presentation that sometimes improves visual comparisons is the *cumulative frequency distribution* or *cumulative probability distribution*. It is also a good idea to become familiar with *cumulative probabilities* because spreadsheets and statistical tables return cumulative probabilities for theoretical distributions.

The basic concept behind the cumulative frequency distributions (*cfd*) is that, instead of plotting the frequency of x_i on the y-axis, we plot the *sum of frequencies with values* $\leq x_i$, i.e.

$$\text{cumulative frequency at } x = \sum_{1}^{i} x_i \qquad (5.3)$$

The cumulative frequency distribution provides some useful advantages when comparing samples or a sample with a theoretical distribution, most commonly the Normal distribution. Comparisons are further expedited by standardising the x-axis in units of standard deviations. This is illustrated with an initial simple example of a large sample of measurements of gravity anomalies. Free-air anomalies (FAA) are deviations of the earth's gravitational field from the expected value. They involve relatively little data reduction and no consideration of local geology or density variation of the rocks, and are generally used to survey large-scale tectonic problems such as isostasy, mean land elevation or seafloor bathymetry. The frequency distribution of worldwide FAA are given in Fig. 5.3a (Rapp 1989); each sample represents the mean FAA value in gravity units (gu) from each of 64,800 segments bounded by 1^0 latitude and 1^0 longitude. The conventional frequency distribution is mound shaped but differs from the Normal distribution in that it is not symmetrical about the mean, its kurtosis (peakedness) is greater, and its range is clearly finite (Fig. 5.3a). We have already seen various techniques of making comparisons with the Normal distribution. Now let us compare the sample *cfd* with the *cfd* for the standardised Normal distribution (Fig. 5.3b). The difference between the mean of the sample cfd and the mean of the Normal distribution are evident, and the percentage of observations with $x < 0$ is immediately legible. The sample-distribution asymmetry is somewhat clearer by visual comparison of the curves' "S"-shapes. Since the probability scale is in cumulative %, percentage points of the Normal distribution may be readily compared with the data (Fig. 5.3b). The central portion of the sample *cfd* shows its variance is smaller than expected for a normally distributed sample. The sample's small range is shown by the fact that the sample *cfd* is enclosed within the standardised Normal distribution. The sample mean is not at zero

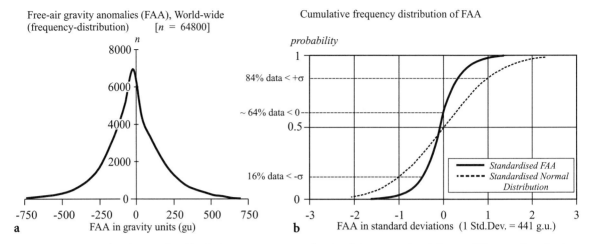

Fig. 5.3. a Frequency distribution of worldwide free-air gravity anomalies (FAA) from Rapp (1989). **b** Some differences from the theoretical Normal distribution may be more obvious when the data are plotted as a cumulative frequency distribution (*cfd*). Percentage points of the Normal distribution, for example corresponding to the $\pm\sigma$ limits, are readily visualised and compared with the standardised Normal distribution, ND (0, 1). Statistics other than standard deviation may also be read more easily from the *cfd*, for example skewness and kurtosis; see Table 5.2

but at – 10 gu which is ~ 2 % of one standard deviation. The sample's peakedness (kurtosis) is 2.06 and its skewness is 1.59; the Normal distribution has a kurtosis of 1.0 and is symmetrical.

We should pause to consider why we are making such comparisons. The misfit with the Normal distribution in no way reflects negatively on the quality of data. In this procedure, the data-acquisition errors are trivial and the data set is large ($n = 64,800$). We deduce only that the data cannot be described by the Normal distribution. In this instance, the large sample size and well-behaved distribution indicate an underlying physical control that has a strong tendency towards small non-zero values, with a sharper tighter peak than the Normal distribution. The differences from the Normal distribution in no way imply abnormality of the observations. It is just that the Normal distribution is an inappropriate model. In fact, the geophysical consequences are significant and relate to the different mean elevations of continents and ocean floors and the degree of isostatic compensation of continental crust.

5.5.1
Application of Cumulative Frequency Plots in Sedimentology

The *cumulative frequency distribution* finds extensive use in sedimentology, but requires a somewhat complicated *x*-scale that requires explanation. When sieving sediment for progressively finer grain-size fractions, successive sieve fractions are conveniently tallied as sums, leading to the convenient use of a cumulative plot. However, sieve-mesh sizes increment by a factor of two, and grain-size variation covers several orders of magnitude. Therefore, a logarithmic scale was adopted using base-2 logarithms. The Wentworth grain-size scale is explained in Table 5.1, with typical particle de-

Table 5.1. Wentworth grain-size scale (phi scale = ϕ-scale). $\phi = -\log_2 d$ and $d = 2^{-\phi}$

Sieve-limiting diameter d (in mm)	ϕ-Units	Size	Description
> 2048	– 11	Very large	Boulders
> 1024	– 10	Large	
> 512	– 9	Medium	
> 256	– 8	Small	
> 128	– 7	Large	Cobbles
> 64	– 6	Small	
> 32	– 5	Very coarse	Pebbles
> 16	– 4	Coarse	
> 8	– 3	Medium	
> 4	– 2	Fine	
> 2	– 1	Very fine	
> 1	0	Very coarse	Sand
> 0.5 (500 µm)	1	Coarse	
> 250 µm	2	Medium	
> 125 µm	3	Fine	
> 62 µm	4	Very fine	
> 31 µm	5	Very coarse	Silt
> 16 µm	6	Coarse	
> 8 µm	7	Medium	
> 4 µm	8	Fine	
> 2 µm	9	Very fine	

scriptions. It uses ϕ-*units* to define grain sizes for sieves where the mesh-spacing or *limiting grain-diameter d* is in millimetres, thus:

$$\phi = -\log_2 d \quad \text{and} \quad d = 2^{-\phi} \qquad (5.4)$$

Therefore, grain sizes >1 mm have negative ϕ-values (e.g. –11 for boulders) and grain sizes < 1 mm have positive ϕ-values (e.g. +5 for silt). Thus, one sieve passes grains half the diameter of the preceding sieve and twice the diameter of those of the next finer sieve. Cumulative plots of grain-size distributions are therefore easily compiled by plotting the weights of successively finer sieve fractions, with decreasing grain size to the right. One should note that "grain size" is actually the *limiting grain size* passed by the sieve mesh.

An example of a *cfd* for grain size is shown in (Fig. 5.4), after Blatt and Tracy (1996). The conceptual steps in formulating a cumulative frequency distribution start with the histogram, binning the weights of sieve fractions in each size range (1 to 2ϕ; 2 to 4ϕ, etc.; Fig. 5.4a). These are regarded as frequencies in each size range. The bar chart is converted into a continuous frequency distribution by joining column midpoints (Fig. 5.4b). The peak is readily identified in both cases (P). At each value of the horizontal axis (ϕ units), we plot the cumulative percentage of grains that have smaller grain sizes on the y-axis (Fig. 5.4c). This produces a cumulative frequency distribution curve that shows an "S"-shape, characteristic of any observations that possess a *humped frequency distribution*. Note that the peak, P, of the hump in the conventional frequency distribution is found at the inflexion point of the cumulative distribution curve (Fig. 5.4b, c).

There also exists an option to convert the scale of the vertical frequency, or probability axis of the *cfd* to a Normal-distribution probability scale. We learned that the Normal distribution forms a straight line if the y-axis has a Normal-probability scale (Fig. 5.1a). Plotting a *cfd* in this way makes it easier to compare with the Normal distribution. Thus, the hump-shaped frequency distribution (Fig. 5.4b) becomes fairly linear using the Normal-probability scale (Fig. 5.4d) and encourages comparisons and the further use of Normal-distribution statistics. Here, we see another advantage of this version of a cumulative frequency graph: we may readily read off percentage points of the Normal distribution. Most simply, 50% of data lie on either side of the mean and the 16 and 84% cumulative frequencies compare to the $\pm\sigma$ limits of the Normal distribution (Fig. 5.5a, b; Table 5.2). Other statistics such as skewness and kurtosis may also be read from the *cfd*.

The sedimentologist's use of the *cfd* provides some instructive insights into the generality of the Normal distribution as a guide to understanding any mound-shaped frequency distribution: the advantages of standardisation of the x-scale. Moreover, if we plot a cumulative frequency distribution, almost any humped frequency distribution may be profitably compared with the Normal distribution. This is possible because of the ease with which critical percentage points, corresponding to $\pm\sigma$ limits, may be read from the graph (Fig. 5.5a, b; Table 5.2).

The use of the ϕ-scale in sedimentology, like many special numerical and statistical devices in applied and earth science, is based on the initial convenience of some technique, in this case sieving. A convenience in one sense shows disadvantages elsewhere. Counter-intuitively, ϕ is negative for coarse grain sizes and positive for fine grain sizes: grain size decreases as ϕ increases. Further caution is needed in considering the *range* of grain size, or *sorting*, which may indicate useful depositional information such as transport distance. The actual size ranges (sieve-limited diameter ranges) are obtained from separated sieve fractions. In contrast, the *sorting texture* observed from a two-dimensional section under a petrographic microscope is very difficult to associate with the actual range of three-dimensional grain sizes. Views of sections may underestimate true grain diameters and make sorting appear worse than it is because grains are rarely sectioned through their centres (Fig. 5.5c). This is another manifestation of the *cut-effect* explained in Chapter 1 (Fig. 1.16). Moreover, this effect is exaggerated as the sorting becomes worse.

Assuming that the data conform closely to the theoretical Normal model, it is possible to use some shortcuts to estimate statistics directly from the cumulative frequency distribution. For sedimentologists, this requires that the horizontal scale be in ϕ-units (e.g. Fig. 5.5b), but the approach is valid for any *cfd* (e.g. Fig. 5.3b). One must record x-values corresponding to 5, 16, 50, 84 and 95% points of the cumulative frequency distribution. You may recognise that these percentages correspond to the mean, $\pm\sigma$ and $\pm2\sigma$ limits of the Normal distribution from which even estimates of skewness and kurtosis may be rapidly estimated (Table 5.2).

5.5.2
Interpreting Natural Processes from Frequency Distributions

With cumulative frequency distributions, it is sometimes particularly easy to use percentage points of the Normal distribution as descriptors. One may either standardise the x-scale in units of standard deviation

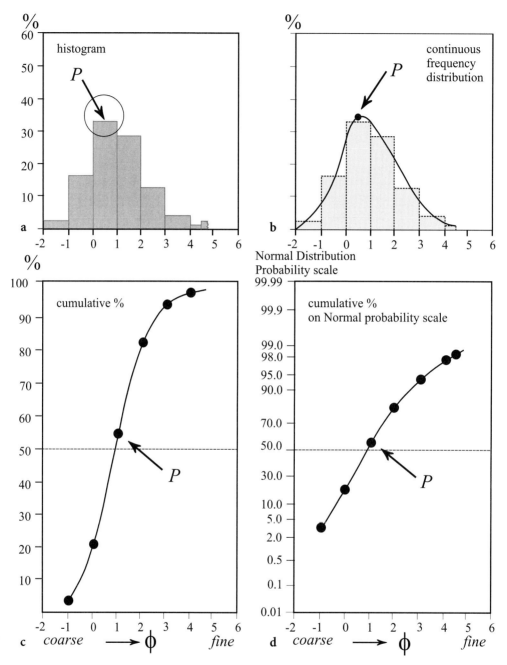

Fig. 5.4. After Blatt and Tracy (1996): **a** An example of a histogram of grain sizes in sediment with the frequencies in each grain-size interval represented as percentages of the total; the peak value is P. The horizontal scale is in ϕ-units ($-\log_2 d$ where d = sieve-limiting grain diameter in millimetres). ϕ increases with decreasing grain size. **b** The same data as a continuous frequency distribution by joining the tops of histogram columns; the percentages now could represent the probability of observing a grain of that size. The peak (P) might be slightly differently located according to the manner in which the continuous frequency curve is fitted to the histogram columns. **c** The same data plotted as cumulative percentages, so that at each grain size ϕ, the vertical axis sums all the coarser (smaller ϕ) grain sizes. The peak value, P, is now located in the centre of the curve, which always has positive slopes and which has an S-symmetry for any hump-shaped frequency distribution. **d** When the percentage-frequency (vertical) axis is plotted on a "Normal-probability" scale, it is possible to compare the original frequency distribution of **b** directly with a Normal distribution. If the data lie on a straight line, the original frequency distribution may be drawn from a Normally distributed population. This type of plot permits rapid estimation of statistical moments (variance, skewness, kurtosis). To expedite plotting data manually, as in former times, special Normal probability graph paper exists (e.g. see Fig. 5.1)

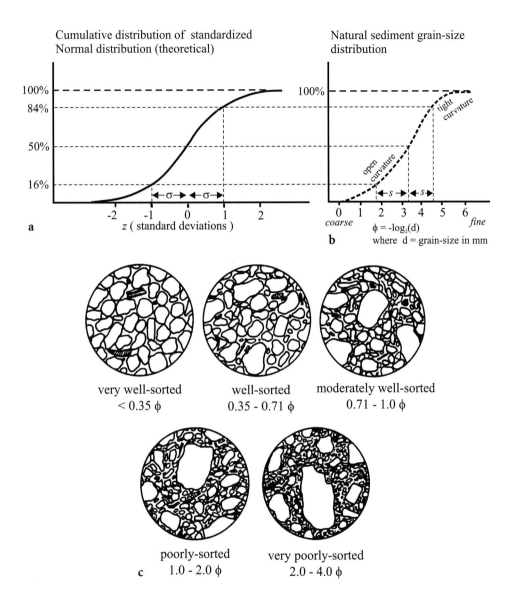

Fig. 5.5. a The cumulative version of the standardised Normal distribution (ND (0, 1)) with its *x*-scale in units of standard deviation. The asymmetrical "S" form has an inflexion point at the centre of the distribution (50% cumulative probability) that corresponds to the peak of the ND. Standard deviation (σ) is recognised very simply on a *cfd* using percentage points of the Normal distribution (see Table 5.2). Skewness and kurtosis may also be read from the *cfd* (see Table 5.2). **b** A cumulative frequency distribution of sediment grain sizes in millimetres. The grain-size (*d*) scale is in –log₂*d* or *φ-units*. Thus, 2φ is a grain size half of *φ*, and 4φ is half that of 2φ. Many sediments reveal a humped frequency distribution of grain sizes that may be profitably compared with the Normal distribution by using a cumulative plot. Here, the asymmetry is revealed by conflicting estimates of the sample standard deviation (*s*). **c** Appearance of different grain-size *ranges* in φ-units. Note that the range of true diameters is exaggerated by the *cut-effect* (Chap. 1, Fig. 1.16), since any sectional view cannot pass through the centre of every grain. The exaggeration also increases as the sorting gets worse

Table 5.2. Statistics estimated from percentage points of cumulative frequency distributions, mainly used with ϕ-scale grain-size plots. (Folk 1975)

Statistic	Estimate
Mean	$(\phi_{16} + \phi_{50} + \phi_{84})/3$
Standard deviation	$(\phi_{84} - \phi_{16})/2$
Skewness	$(\phi_{16} + \phi_{84} - 2\phi_{50})/(\phi_{84} - \phi_{16})$
Kurtosis	$(\phi_{95} - \phi_{5})/(2.44\,(\phi_{75} - \phi_{25}))$

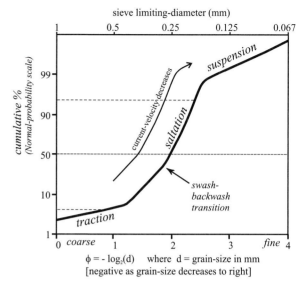

$$\phi = -\log_2(d) \quad \text{where} \quad d = \text{grain-size in mm}$$
[negative as grain-size decreases to right]

Fig. 5.6. The cumulative frequency distribution of grain sizes, using the ϕ-scale, may reveal segments with distinct slopes that correspond to different processes of grain transport according to Visher (1969). This characterisation of process by geometrical features of the frequency distribution is most easily recognised using a cumulative frequency distribution

and estimate associated percentage points from the y-axis (Figs. 5.3b, 5.5b), or use a Normal-probability scale for the y-axis to pick out the x-limits associated with $\pm\sigma$ and $\pm2\sigma$ (Fig. 5.4d). This approach makes it easy to describe complex natural frequency distributions and associate their features with a process or different components of a process. This does not require any profound theoretical understanding of the process, which may defy mathematical formulation. Normal-distribution statistics may be a useful tool, but it is not believed nor implied that the Normal distribution is in any way a description of a causative process. It is possible to handle any humped frequency distribution in this manner.

The petroleum-reservoir porosities described earlier were more tightly distributed than a Normal distribution (Fig. 5.2c). We know that the finite range for the variable is partly responsible. Porosity of 100% is impossible; there would be no rock. However, zero porosity is possible. Nevertheless, the distributions are symmetrical. This tells us that the process responsible for the porosity distribution has some random component, like the distribution of measurement errors for which the Normal distribution was devised. The steep flanks and flat-topped peaks of the distributions also imply some discrete control on the range of most frequent porosities. Geologists would attribute this immediately to strong lithological control.

Simply changing the format of presentation of data can be particularly helpful in revealing processes or differences in processes of complex natural phenomena. The use of the cumulative frequency distribution (*cfd*) simplified the comparison of worldwide free-air gravity anomalies with the Normal distribution. The tight-distribution, non-zero mean and peakedness (kurtosis) apparent from the frequency distribution (Fig. 5.3a) may be readily benchmarked against the Normal distribution on the *cfd* (Fig. 5.3b). The distribution's well-defined structure indicates the role of effective underlying controls. The distribution relates to isostasy and global distributions of elevation and continental versus oceanic crust.

Grain-size distribution plots in sedimentology provide common examples of process-related, if not process-controlled frequency distributions. Despite the complexities of hydraulics and sedimentation, differences in the shape of *cfd*-plots are commonly associated with different processes and environmental conditions. The percentage points of the Normal distribution may be used to describe or to discriminate between different processes or environments, although it is not suggested that the theoretical statistical model provides a direct physical description of the processes involved (Fig. 5.6).

5.6
Log-Frequency Distributions; the Logarithmic and Other Transformations

Some geological distributions cannot be compared to the Normal distribution even at a cursory level. One of the most common examples is a positively skewed distribution, one that has a long tail to the right, towards high values of the measurement. Most observations are of small values, commonly with a peak near the lower limit of zero or unity that is fixed by the nature of the variable. However, the long tail to the right indicates that observations with unusually large x-values occur less often as x increases. One can impose numerical symmetry on a distribution with a range $0 < x < \infty$ since logarithms of numbers in this range are stretched towards the range from $-\infty < \log x < \infty$. This is the basic idea behind the *log-Normal distribution*; if we plot logarithms of the x-variable, such distributions become more symmetrical. Very often, the hump of log x-frequencies appears sufficiently similar to that of the Normal distribution to permit its characterisation with Normal-distribution percentage points and statistics. We shall restrict this discussion to the use of logarithms to make a skewed distribution more symmetrical, and perhaps quite similar to the Normal distribution with the advantage that the re-expressed variable may take full benefit of statistical procedures based on the well-understood Normal distribution. However, there are less popular transformations, which may be equally successful in standardising other sample distributions towards Normality. They are summarized in Table 5.8. For quite different purposes, the next table (Table 6.1) in the following chapter describes transformations used to re-express data so that they follow a linear distribution on an x-y graph.

Asymmetric, positively skewed properties with a potential range of $0 < x < \infty$ include features such as grain size and the flow discharge of streams. Commonly, phenomena without an upper limit are associated with work or energy input into a system. For example, earthquake magnitudes, energy released in volcanic eruptions and meteorite impacts have no theoretical maximum limit. In a quite different sense, and with a suitable approximation, the frequency distributions of low-abundance items such as trace-element concentrations are very compressed at the lower limit of their scale. Although the upper limit of their x-ranges is fixed at 100%, abundances are usually so low (parts per million) that the upper limit does not cause a problem. A more worrisome issue with low-abundance analyses is the *detection limit* of the analysis. Such information should always accompany data descriptions because many chemical analyses report abundances very close to the detection limit, which varies with the element analysed, the matrix and the technique of analysis (see Chap. 2, Table 2.1). The detection limit may suppress a low-concentration peak or cause it to be inaccurately estimated. Other types of data produce positively skewed distributions with a hump close to the lower limit of their range. For example, shape ratios, strain ratios and many other ratios fall in a range $1 < x < \infty$ with the mode near the lower limit. These too may be amenable to a *logarithmic transformation*, which has the further advantage that in some applications, such as strain analysis, ratios combine as products that simplify to summations on the logarithmic scale.

Taking logarithms of a positively skewed distribution produces a symmetrical hump which may be similar to the Normal distribution. We may therefore wonder what is the form of the Normal distribution when its x-values are plotted with a logarithmic scale. The form is an idealised positively skewed distribution with zero frequency at $x = 0$ with mean α and standard deviation β. Its theoretical definition, where $x > 0$, $\alpha > 0$ and $\beta > 0$, is:

$$y = \frac{1}{x\beta\sqrt{2\pi}} \cdot \exp\left\{-\frac{(\ln x - \alpha)^2}{2\beta^2}\right\} \qquad (5.5)$$

where α is the mean of the natural logarithms of x, and β is their standard deviation. In turn, the mean of x is given by

$$\bar{x} = \exp\left(\alpha + \frac{\beta^2}{2}\right) \qquad (5.6)$$

An example of a log-Normal distribution is shown in Fig. 5.7a, revealing the characteristic positive skewness shown by many frequency distributions of low concentrations, such as accessory minerals or trace elements. These data record the volume percentage of opaque minerals in granitoids; the modal frequency is close to 0.1% but there is a long tail, with specimen contents rarely exceeding 1%. Transforming the x-axis to a logarithmic one, i.e. simply plotting $\log_{10}$ (volume %), produces a more symmetrical curve, similar to a Normal distribution (Fig. 5.7b). Thus, one might entertain the possibility that the frequency distribution is log-Normal. Failing that, one can at least more readily assess the mean value and variance from visual inspection. Although log-Normal distributions are discussed quite commonly in connection with petrographic and geochemical data, the limitations imposed by a fixed linear scale from zero to 100% should not be forgotten. Physical properties may also show log-Normal distributions, especially where they are due to some low-concentration source that also has a log-Normal distribution. Magnetic susceptibility may show this where it is due to low concentrations of magnetite (Fig. 5.7c). Since sedimentologists use a logarithmic scale for grain size (albeit to base 2), called the phi scale or Wentworth scale, any positively skewed nearly log-Normal concentration automatically appears symmetrically humped like a Normal distribution (Fig. 5.7d). It does not matter that sedimentologists use logarithms to base 2, nor that elsewhere $\log_e$ and $\log_{10}$ are equally common. This is because the *transformation of logarithmic-bases* is linear. In general, if we

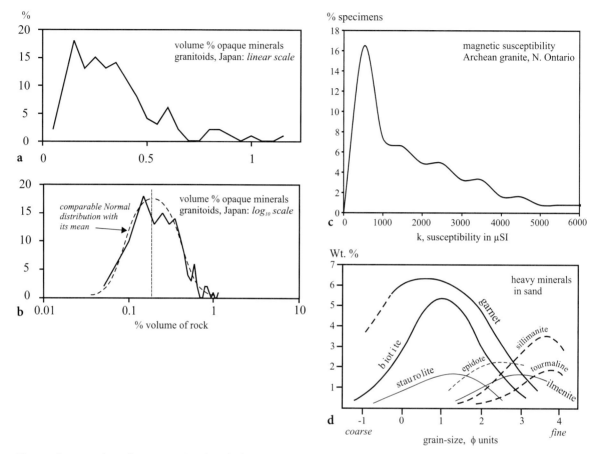

Fig. 5.7. Concentrations of accessory minerals and other rare components, such as trace elements, may show a positively skewed frequency distribution with a mode at a low concentration. The long tail to the right indicates the rapidly diminishing probability of recognising samples with large abundances. Such data may follow the theoretical log-Normal distribution, as shown by these examples. **a** Conventional frequency distribution of accessory magnetite content in granite (Ishihara 1979). **b** The same data using a logarithmic x-scale makes the humped distribution more symmetrical and comparable to a Normal distribution. The original data may therefore follow a log-Normal distribution. **c** Physical properties of minerals whose abundances are log-normally distributed may also follow a log-Normal distribution. Here, magnetic susceptibility, mainly due to accessory magnetite, shows an asymmetric hump at low values and a strong positive skew toward rare high values. **d** In sedimentology, the ϕ-scale used for grain size (d) uses a grain-size axis that has already been transformed logarithmically ($-\log_2 d$). Therefore, abundances that were log-normally distributed appear as symmetrical hump-shaped distributions comparable to the Normal distribution in shape. Accessory heavy minerals in clastic sediments show this effect, as here

transform logarithms from base b to base a, we multiply by $\log_a b$, thus:

$$(\log_a b) \cdot (\log_b x) = \log_a x \qquad (5.7)$$

Since the bases a and b are constants, the scale factor between different logarithms is also constant. For example, to convert to base 10 from base e we multiply by $\log_e 10 = 2.3026$; to base 2 from base 10 we multiply by $\log_{10} 2 = 0.3010$. Regardless of the base of the logarithms, the distribution will appear more Normal-like if it had a positively skewed hump, like the theoretical log-Normal distribution.

The log-Normal relationship provides a reasonable descriptive basis for distributions of many natural observations, as shown in the examples of Fig. 5.7. Furthermore, in a few special circumstances, an actual process or theoretical model may be causally associated with the log-Normal distribution. The log-Normal function in this case describes a causal process and is not restricted to characterisation of the observations. For example, joints and faults may have spacings or displacements that follow a log-Normal relationship (Epstein 1947, 1948). Theoretical models for fractional magma crystallisation predict log-Normal crystal-size distributions, and channel hydrology may produce log-Normal distributions for grain size or channel dimensions (Kolmogoroff 1941; Kottler 1950). Upon reflection, many natural processes tend to produce a common value, but the lower part of the objects' magnitude range is favoured, resulting in a positively skewed distribution. This is true for the objects of many "growth processes" and propagation processes, such as the growth of crystals, fractures and stream channels as well as for the abundances of rare items.

5.7
Quantified Comparison of Frequency Distributions: Non-parametric Statistics

The preceding sections of this chapter provided some simple methods to compare frequency distributions visually and somewhat subjectively. Of course, statisticians provide much more rigorous approaches to this topic. A simple introduction follows that suffices for most common applications in the natural sciences. The techniques discussed below are termed *non-parametric statistics* because they are not tied to population parameters. Other authors use the easily understood term, *distribution-free statistics*, because no assumptions are made concerning the distribution of the population that was sampled. Obviously, this is valuable in nature when we wish to compare natural distributions that may have no theoretical basis. Thus, the distributions being compared need not follow any particular theoretical model. Any empirical distribution may be used and compared with any other empirical (or theoretical) distribution. Moreover, the methods are applicable to variables that may be quantified in various ways. Apart from the usual numeric real or integer values, they are applicable to ordinal values describing position in a sequence (1st, 2nd, etc.), to differences in rank or order, to enumerative values (i.e. counts or tallies in classes), and even to qualitative observations (e.g. rock type) to which integer labels have been assigned. This gives considerable flexibility although the various methods each have specific limitations and ranges of conditions for their reliable use. Two easily applied and popular tests are described: the χ^2-test and the Kolmogorov-Smirnov test. They possess the potential to summarise *all and any differences* between samples represented by their frequency distributions. For example, the samples' frequency distributions may differ in central tendency, variance or skewness, but their differences are summarised by a single test statistic whose value may be used to decide whether the samples are drawn from the same population, at a chosen confidence level.

5.7.1
χ^2-Distribution and Goodness-of-Fit

The χ^2-distribution (chi-squared, pronounced more properly as kigh-squared) is a positively skewed theoretical distribution ranging from zero to $+\infty$. Its form depends on the number of classes into which the data is grouped (k), and degrees of freedom usually defined by $v = k - 1$ (Fig. 5.8c). As k increases, the distribution takes on the shape of the Normal distribution. Here we will concern ourselves with a pragmatic rather than rigorous explanation of the formulation of this distribution.

We will show that a simple statistic may be devised to compare two frequency distributions, permitting a test of the hypothesis that the two distributions were drawn from the same population (i.e. the two samples are "the same"), by reference to the probabilities represented by areas under the χ^2-distribution curve (Table 5.6). The χ^2-distribution is commonly used with experimental data where one curve closely follows another. It may provide more information where the two distributions compare the same variable, e.g. from two experiments with the same conditions or a comparison of natural measurements from different environments but influenced by some well-constrained physical law. Hence, it is sometimes considered as a test of the *goodness-of-fit* of curves or frequency distributions.

However, because this test is *distribution-independent*, it may be applied to compare distributions of natural observations that do not follow any simple law or curve. Since it is somewhat intuitive, we introduced it informally at an early stage to compare site distributions on maps with hypothesised uniform distributions (Chap. 1, Figs. 1.6, 1.7). Where the sample size is large, or where at least one of the two compared distributions is well-constrained, the χ^2-approach is very successful. For example, in computer experiments one may remove the effects of tectonic strain of geological structures or stratigraphic sections by incrementally matching the progressively de-strained structure with its known pre-strain form. In such cases, the comparison at each stage of the de-straining procedure may be performed with a χ^2-test (Borradaile 1976). Unfortunately, natural frequency distributions usually show some considerable mismatch with a hypothesised distribution, even where we have some good reason to expect a certain model for the behaviour (e.g. trace-element concentrations or grain-size distributions). The problem of comparing distributions is worse when we attempt to compare quite different variables, in different units of measurement, related by some natural process or phenomenon, e.g. crustal heat-flow and the concentration of radioac-

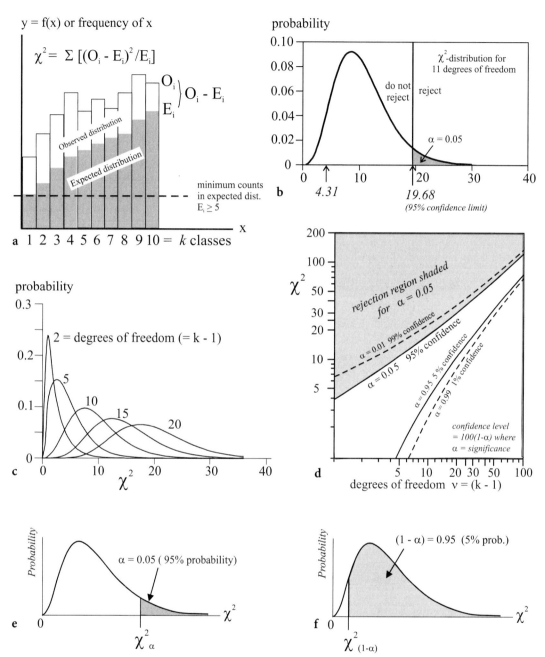

Fig. 5.8. χ^2 (chi-squared) is a distribution-free or non-parametric statistic. We may use it without any assumptions or knowledge of the distribution of the population. **a** χ^2-tests goodness-of-fit of an observed (O) with an expected (E) distribution. The distributions are divided into k equivalent classes and the value of χ^2 is calculated for the differences between the observed and expected counts in each of the k classes. No class should have an expected count $E_i < 5$. χ^2 has small values for distributions that are similar: large values indicate dissimilarity. **b** An example of a χ^2-test that compares two distributions with 12 classes (thus, $v = 11$ degrees of freedom). The calculated value of $\chi^2 = 4.31$, whereas the rejection region at 95% confidence has a lower bound of 19.68. The calculated value does not lie in the rejection region. Therefore, we do not reject the null hypothesis H_0 that the two distributions are drawn from the same population. **c** Examples of χ^2-distributions for some different degrees of freedom ($v = k - 1$); the mean of each distribution is equal to its degrees of freedom. **d** Critical values of χ^2 for various degrees of freedom (v) at the 95 and 99% confidence levels. If χ^2 lies above the appropriate curve, we reject the null hypothesis that two distributions are drawn from the same population, at that confidence level. The shaded rejection region is for the upper or right-hand tail of the distribution within which there is a 0.05 probability (95% confidence level) of rejection. This is illustrated in **e**. **e** Right-hand tail rejection region with $\alpha = 0.05$ significance level. **f** Left-hand tail with 0.05 probability; note that its critical χ^2-value corresponds to that of the *right-hand tail's $\alpha = 0.95$ probability rejection region*

tive elements. In contrast, laboratory experiments are more likely to produce sample distributions in close agreement with a theoretical model. For example, in petrological experiments, trace elements may show sympathetic variations with major elements on a variation diagram whereas rock samples from the field may be much less convincing. In other cases, the looseness of the relationship between distributions may be complicated by some offset or delay between cause and effect (*hysteresis*). Where the variables are time-dependent, such as rainfall or stream discharge on a hydrograph, they are not reported as frequency distributions but rather as *time series* that require special attention (Chap. 8). Earth science generally tolerates imperfect agreement between theoretical models and observations more readily than the physical sciences.

The χ^2-test statistic may be introduced in the comparison of distributions, as follows: Let us count the total number of observations (n) and consider their distribution over k intervals, also known as classes or bins. The number of classes, k, must be chosen so that more than 5 counts could be expected in any class, if the counts were evenly distributed, i.e. $(n/k) \geq 5$. The expected number of counts in any cell is $E_i \geq 5$. The number of observed counts, O_i, is recorded in the same class. The boundaries of classes may be adjusted freely, and the intervals need not be equal in size. At least ten classes are preferred so that the total count of observations (n) should be ≥ 50. The χ^2-statistic is then defined, intuitively, as the mismatch between the observed and expected distributions:

$$\chi^2 = \sum_{i=1}^{k} \frac{(O_i - E_i)^2}{E_i} \tag{5.8}$$

It has $(k-1)$ degrees of freedom, where k is the number of intervals, not total counts. This statistic may be used to compare two distributions, as shown in Fig. 5.8a. If χ^2 is large, the fit between the observed and expected distributions is poor; if $\chi^2 = 0$, the fit is perfect. Hence, the phrase *goodness-of-fit* is sometimes used in connection with this statistic.

This statistic follows a well-known distribution that permits hypothesis testing by comparing the value of χ^2 obtained from the observations with a critical tabled value. For completeness, the theoretical form of the probability density distribution of χ^2 is given by the following equation, which uses the gamma-function (Γ), mentioned at the end of Chapter 3 (Table 3.7):

$$\frac{1}{\Gamma\left(\frac{\mu}{2}\right) \cdot 2^{\frac{\mu}{2}}} \cdot x^{\left(\frac{\mu}{2}-1\right)} \cdot \exp\left(\frac{-x}{2}\right) \tag{5.9}$$

This formulation is rather intimidating but shows a dependence only on the variable x and its mean (μ). Moreover, 2μ is the variance. This distribution is represented by a positively skewed curve, which tends to the form of a Normal distribution as the number of degrees of freedom (ν) increases (Fig. 5.8c). ν is the number of independent comparisons that could be made between the two distributions given by $(k-m)$. The number of classes is k, and m is the number of quantities calculated from the observed data that we need in order to calculate the expected frequencies from the observations. Since we need only the mean, $m = 1$ and thus, $\nu = (k - 1)$.

The positively skewed χ^2-distribution ranges from zero upwards so that all tests are one-tailed, and the curve is less peaked and progressively more symmetrical for higher degrees of freedom. Figure 5.8c shows χ^2-distributions for 2, 5, 10, 15 and 20 degrees of freedom and values of χ^2 up to 35. Note that the standard deviation increases with the mean, in accordance with the simple relationship from the theoretical equation given above. Values of χ^2 corresponding to specific cumulative percentages of its probability distribution are given in Table 5.6. As with parametric statistics, the χ^2-distribution provides a basis for rigorous hypothesis testing. Thus, we can do better than to say that a large χ^2-value indicates dissimilar distributions.

Let us set the null hypothesis (H_0) such that the observed and expected counts in each of the k classes are equal, in other words, that the two distributions are identical:

$$H_0: (O_1/E_1) = (O_2/E_2) = (O_3/E_3) = \ldots \ldots = (O_k/E_k) = 1$$

against which there must be an alternative hypothesis which must be accepted if H_0 is rejected.

$$H_A: \text{that one or more } (O_i/E_i) \neq 1$$

Clearly, if H_0 is true, the calculated value of χ^2 will be zero. Alternatively, the greater the mismatch between the distributions, the greater will be χ^2. Therefore, in order to reject the null hypothesis, a certain critical value of χ^2 must be exceeded. We will use an acceptance level $\alpha = 0.05$ (95% confidence level) and compare our χ^2 with the tabled critical value $\chi^2_{0.05}$ for the appropriate number of degrees of freedom $\nu = (k - 1)$. As an example, consider two hypothetical stratigraphic sequences (A, B) in which the numbers of species of fossil are tallied at corresponding levels (Table 5.3).

The summed value records the mismatch between the two sequences, so $\chi^2 = 3.74$. There are 12 horizons (classes) in which the tallies were recorded, so $\nu = 11$. If we are prepared to use a 95% confidence level, we find the extreme right-hand tail of the χ^2 probability occurs

Table 5.3. Example of numbers of fossil species by stratigraphic horizon

Horizons ($k = 12$)	A: No. species per horizon	B: No. species per horizon	$(A-B)^2/B$
12	7	6	0.17
11	8	8	0.00
10	5	6	0.17
9	6	7	0.14
8	9	8	0.13
7	7	9	0.44
6	12	10	0.40
5	7	9	0.44
4	8	9	0.11
3	9	6	1.50
2	11	10	0.10
1	8	7	0.14
			$\Sigma = 3.74$

above the critical value $\chi^2_{(0.05,\, v=11)}$, which Table 5.6 gives as 19.67. The calculated value (3.74) is much less than the critical value, therefore the mismatch between the frequency distributions is not extreme. We do not reject the null hypothesis that the two distributions are similar (i.e. each horizon has a consistent number of species in each section).

Consulting critical values from tables is the preferred method of comparing a calculated statistic. However, nomograms such as Fig. 5.8d permit visual evaluation of hypothesis tests and promote a better understanding of the relationships between the statistic's value, sample size (related to v) and confidence level given by $100(1-\alpha)\%$, where $\alpha =$ significance. Where the coordinates (v, χ^2), determined from the data, lie above the curve appropriate for the chosen confidence level, we must reject the hypothesis that the distributions could be drawn from the same population. The shaded area indicates the rejection region for a 95% confidence level appropriate to the application of the χ^2-test in which two distributions were compared (Fig. 5.8d).

One additional feature of the χ^2-test is that it may be used with *ordinal data*, not only conventional values. For example, we could assign integers to certain attributes, *1= sandstone, 2 = mudstone, 3 = limestone,* etc., and use them to generate a synthetic frequency distribution of ordinals illustrating the relative importance of the different lithologies. If one applies this approach to an extensive sequence, the similarity between the two numerically coded stratigraphic sections may be quantified (part of such a sequence is shown in Fig. 8.1b). For the complete sequence, one calculates $\chi^2 = 12.8$ for 21 classes. Thus, $v = 20$ and Table 5.6 gives $\chi^2_{(0.05,\, v=20)} = 31.4$ as the critical value that defines the lower limit of the rejection region. Our calculated value

is less, so we do not reject the hypothesis that the sequences are different at the 95% confidence level.

χ^2 takes advantage merely of the enumerative aspects of the observations, i.e. the number of counts in classes. Where the counted items represent observations in a fixed order or sequence, other methods of comparison that use the structure of the sequence (e.g. rhythm, pattern) may be more effective or informative. This is discussed under *sequences* and *time series* that are introduced in Chapter 8.

5.7.2
χ^2-Distribution and Contingency Tables

In goodness-of-fit, discussed above and intuitively in Chapter 1, we have seen that χ^2 provides a simple mechanism for comparing the discrepancies between two-frequency distributions of counts of observations in the same units of measurement, or of two sets of ordinal data. It may also be applied to two distributions of probabilities (or proportions from 0 to 1), as long as the probabilities in each class were based on 5 or more expected observations. This data manipulation uses *contingency tables* in which the probabilities of certain observations are recorded in a series of table cells, which form the classes in a χ^2 analysis. Commonly, the variables are not numerical values but attributes or properties.

For example, let us compare the counts of mineral grains in a sediment with the source rock from which it is derived. Each type of mineral grain constitutes a variable in the form of a property or attribute. We have "observed" counts (O_i) of each mineral (i) in a sediment sample, but how do we compare it with those in the source rock? From the source rock we divide the counts of each mineral by the total number of mineral grains to obtain their proportions, p_i (Table 5.4).

The proportions of mineral grains in the source rock may be considered as the expected distribution of attributes, whereas the proportions in the sediment are observed values that may differ from the source according to the degree of chemical weathering and alteration during transport. In the unlikely event that the two distributions of mineral grains matched perfectly, χ^2 would have a zero value. That would indicate complete ineffectiveness of the erosion and transport process in the breakdown of the minerals. In the sediment the expected number of grains of a type i is given by $E_i = p_i(\Sigma O_i)$. The interesting null hypothesis is that the proportions of minerals in each category are not different after erosion and transport:

$$H_0: p_1 = p_2 = p_3 = p_4 = p_5$$

Table 5.4. Example of one-way contingency table comparing mineral content of source and sediment

Mineral grain	Source counts of grains	Sediment counts of grains = O	Source proportion of grains p	Counts expected in sediment E = p × **98**	$(O–E)^2/E$
Quartz	54	**48**	0.383	37.532	2.9197
Feldspar	45	37	0.319	31.277	1.0473
Mica	22	5	0.156	15.291	6.9258
Hornblende	15	2	0.106	10.426	6.8092
Magnetite	5	6	0.035	3.475	1.8344
Sums =	141	**98**	1	98	$\chi^2 = 19.54; \nu = 4$

Here, the proportions' subscripts indicate the five different minerals. If the hypothesis of no differences is rejected, we must accept the alternative hypothesis that the proportion of at least one mineral has changed. Because proportions must sum to unity, this requires that at the very least, two proportions change. This introduces us to the *closure* problem, discussed further in Chapter 7.

For the source-sediment problem of Table 5.4, the five mineral categories provide 4 degrees of freedom. From Table 5.6, the calculated value of χ^2 must exceed $\chi^2_{(0.05,4)} = 9.49$ to reject the null hypothesis at a 95% confidence level. In fact, our calculated value is 19.54, lies in the rejection region under the right tail of the χ^2-distribution; we reject the null hypothesis that the proportions of mineral grains remained constant during erosion and transport.

The preceding example used χ^2 to compare two distributions where only one variable (mineral type) was compared. One may compare observations comprising two attributes, in a *two-way contingency table*. For example, consider counts of pebbles classified according to two attributes, distance from the source of the stream, and their roundness. Both could be quantified precisely, but it is certainly easier and quite sufficient to use qualitative terms for roundness. Such semi-quantitative data is usual in field studies. The upper part of Table 5.5 shows the observed counts. A useful hypothesis to test is that there is no association between the pebble roundness and transport distance:

$$H_0: n_{11} = n_{12} = n_{13} = \ldots p_{RC}$$

Here, the subscripts R and C refer to the rows and columns of the table. In that case, the proportion of counts in every class should be equal to the total number of counts (all pebbles in all cells), divided by the number of cells. For the same size sample with no preferential rounding due to proximity, the mean expected number of counts in each cell should be 246/20 or 12.3. Therefore, the null hypothesis is extended:

$$H_0: n_{11} = n_{12} = n_{13} = \ldots p_{RC} = 12.3$$

There are 20 cells or classes in the table so that there are 19 degrees of freedom. We use χ^2 to compare the

Table 5.5. Two-way contingency table for pebble roundness and source distance (association indicated by arrangement of shaded cells) 246 counts in specimens at distance from source if no preference for rounding with distance, mean counts expected (246/20) = 12.3

Attribute	0–5 km proximal	5 to 10	10 to 20	20 to 30	30 to 40 distal
Angular	28	21	11	1	0
Sub-angular	23	23	14	8	9
Sub-rounded	5	8	12	14	18
Rounded	0	2	8	12	29
Totals	56	54	45	35	56

Proportions in specimens at distance from source

Attribute	0–5 km proximal	5 to 10	10 to 20	20 to 30	30 to 40 distal
Angular	0.114	0.085	0.045	0.004	0.000
Sub-angular	0.093	0.093	0.057	0.033	0.037
Sub-rounded	0.02	0.033	0.049	0.057	0.073
Rounded	0.000	0.008	0.033	0.049	0.118
Totals	1	1	1	1	1

Table 5.6. Table of χ^2-values (α = significance, confidence = $100(1-\alpha)\%$, ν = degrees of freedom), α = area (probability) under the curve's right-hand tail

ν	0.01	0.025	0.05	0.10	0.25	0.5	0.75	0.90	0.95	0.975	0.99	$=\alpha$
1	6.63	5.02	3.84	2.71	1.32	0.455	0.102	0.0158	0.0039	0.0010	0.0002	
2	9.21	7.38	5.99	4.61	2.77	1.39	0.575	0.211	0.103	0.0506	0.0201	
3	11.3	9.35	7.81	6.25	4.11	2.37	1.21	0.584	0.352	0.216	0.115	
4	13.3	11.1	9.49	7.78	5.39	3.36	1.92	1.06	0.711	0.484	0.297	
5	15.1	12.8	11.1	9.24	6.63	4.35	2.67	1.61	1.15	0.831	0.554	
6	16.8	14.4	12.6	10.6	7.84	5.35	3.45	2.20	1.64	1.24	0.872	
7	18.5	16.0	14.1	12.0	9.04	6.35	4.25	2.83	2.17	1.69	1.24	
8	20.1	17.5	15.5	13.4	10.2	7.34	5.07	3.49	2.73	2.18	1.65	
9	21.7	19.0	16.9	14.7	11.4	8.34	5.90	4.17	3.33	2.70	2.09	
10	23.2	20.5	18.3	16.0	12.5	9.34	6.74	4.87	3.94	3.25	2.56	
11	24.7	21.9	19.7	17.3	13.7	10.3	7.58	5.58	4.57	3.82	3.05	
12	26.2	23.3	21.0	18.5	14.8	11.3	8.44	6.30	5.23	4.40	3.57	
13	27.7	24.7	22.4	19.8	16.0	12.3	9.30	7.04	5.89	5.01	4.11	
14	29.1	26.1	23.7	21.1	17.1	13.3	10.2	7.79	6.57	5.63	4.66	
15	30.6	27.5	25.0	22.3	18.2	14.3	11.0	8.55	7.26	6.26	5.23	
16	32.0	28.8	26.3	23.5	19.4	15.3	11.9	9.31	7.96	6.91	5.81	
17	33.4	30.2	27.6	24.8	20.5	16.3	12.8	10.1	8.67	7.56	6.41	
18	34.8	31.5	28.9	26.0	21.6	17.3	13.7	10.9	9.39	8.23	7.01	
19	36.2	32.9	30.1	27.2	22.7	18.3	14.6	11.7	10.1	8.91	7.63	
20	37.6	34.2	31.4	28.4	23.8	19.3	15.5	12.4	10.9	9.59	8.26	
21	38.9	35.5	32.7	29.6	24.9	20.3	16.3	13.2	11.6	10.3	8.90	
22	40.3	36.8	33.9	30.8	26.0	21.3	17.2	14.0	12.3	11.0	9.54	
23	41.6	38.1	35.2	32.0	27.1	22.3	18.1	14.8	13.1	11.7	10.2	
24	43.0	39.4	36.4	33.2	28.2	23.3	19.0	15.7	13.8	12.4	10.9	
25	44.3	40.6	37.7	34.4	29.3	24.3	19.9	16.5	14.6	13.1	11.5	
26	45.6	41.9	38.9	35.6	30.4	25.3	20.8	17.3	15.4	13.8	12.2	
27	47.0	43.2	40.1	36.7	31.5	26.3	21.7	18.1	16.2	14.6	12.9	
28	48.3	44.5	41.3	37.9	32.6	27.3	22.7	18.9	16.9	15.3	13.6	
29	49.6	45.7	42.6	39.1	33.7	28.3	23.6	19.8	17.7	16.0	14.3	
30	50.9	47.0	43.8	40.3	34.8	29.3	24.5	20.6	18.5	16.8	15.0	
40	63.7	59.3	55.8	51.8	45.6	39.3	33.7	29.1	26.5	24.4	22.2	
50	76.2	71.4	67.5	63.2	56.3	49.3	42.9	37.7	34.8	32.4	29.7	
60	88.4	83.3	79.1	74.4	67.0	59.3	52.3	46.5	43.2	40.5	37.5	
70	100.4	95.0	90.5	85.5	77.6	69.3	61.7	55.3	51.7	48.8	45.4	
80	112.3	106.6	101.9	96.6	88.1	79.3	71.1	64.3	60.4	57.2	53.5	
90	124.1	118.1	113.1	107.6	98.6	89.3	80.6	73.3	69.1	65.6	61.8	
100	135.8	129.6	124.3	118.5	109.1	99.3	90.1	82.4	77.9	74.2	70.1	

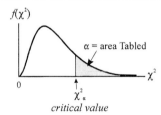

observed counts (O_i) with the expected mean counts (E_i = 12.3) on the hypothesis of no differences. Using the usual formulae (5.6) yields χ^2 = 124.8. The lower limit of the rejection region at a 95% confidence level for 19 degrees of freedom is given in Table 5.6 as $\chi^2_{(0.05, 19)}$ = 30.14. The calculated value greatly exceeds the critical value, placing the statistic well in the rejection region. We must therefore accept the alternative hypothesis that the roundness of pebbles is dependent on transport-distance.

The example serves to introduce another aspect of bivariate data, the degree of association between the two variables. Where each of the variables is placed in some meaningful order, as in Table 5.5, any symmetry to the numbers of counts reveals association or correlation between the two variables, or attributes. This is clarified if the counts are presented as proportions of the total number of counts in all cells, as in the lower part of Table 5.5. The shaded higher values clearly indicate an association: rounding increases with transport distance. This introduces the concept of *correlation* that is discussed in Chapter 7, using bivariate data in which both variables are numeric values. However, the two-way contingency table may reveal association qualitatively. This example is a fairly typical application of two-way contingency in natural science. Generally, the degree of association will be loosely defined for two reasons. First, the variables are classified discretely, even with qualitative or semi-quantitative criteria; this yields classes (table cells) that may be rather imprecise. Secondly, the problems to which two-way contingency is applied may defy precise correlation of the variables due to intrinsic natural complexity. In the example of pebble roundness, many factors were ignored, including: the rock type of the pebbles and the fact that not all pebbles originate at the source, e.g. tributaries may introduce pebbles whose shapes were controlled by other channels.

The *confidence on any proportion* in a cell of a contingency table comes from an appreciation of binomial probabilities (Chap. 3). If the proportion (probability) in a certain cell is p then its 95% confidence limits are given by:

$$p \pm 1.96 \sqrt{\frac{p(1-p)}{n}} \tag{5.10}$$

where n is the total number of counts from all classes.

5.7.3
χ^2-Distribution and Single-Sample Variance

At the close of Chapter 3, reference was made to the interrelationships of several commonly used frequency distributions, in particular their relation to the Normal distribution. A practical benefit of this may be illustrated using the χ^2-distribution. For example, consider a normally distributed population, whose standard deviation is σ. If we draw numerous random samples from it, whose standard deviations are s and sample sizes are n, it can be shown that the following quantity follows the χ^2-distribution:

$$\frac{(n-1)s^2}{\sigma^2} \tag{5.11}$$

This provides a useful test statistic with which to *compare the variance (s^2) of a single sample with variance σ^2 of a normally distributed population*. Note that if $s^2 = \sigma^2$, the value given by the above expression is equal to the degrees of freedom $(n-1)$. This statistic is used to test the null hypothesis H_0 : sample variance = population variance. Therefore, the greater the difference between the value calculated in the preceding expression and $(n-1)$, the more likely it is that the sample and population variances are different. Although this provides another method of comparing distributions, this particular application of the χ^2-test is only valid if the population is normally distributed.

Consider the assessment of ore passing through a separation facility where the mean long-term concentration of Au has a variance $\sigma^2 = 49$ ppm. Suppose, on one day the actual variance of ten batches sampled was 40 ppm; i.e. $s^2 = 40$ and $n = 10$. Does this day's production show significantly less variance than the long-term production variance? This information could be valuable as it may reveal concerns about the consistent supply of ore and consistent ore-processing techniques. We will use a 95% confidence level, i.e. $\alpha = 0.05$.

We set the null hypothesis of no-differences:

$$H_0: s^2 = \sigma^2 \text{ against the alternative } H_A: s^2 < \sigma^2$$

Using Eq. (5.11), the value calculated from the mill's production data gives $\chi^2_{CALC} = ((10-1) \times 40/49) = 7.34$. Some care is needed here: the formulation shows that we must reject the null hypothesis for *small* values of χ^2_{CALC}. Therefore, the rejection region is the *lower tail* of the distribution to the left of $\chi^2_{(0.95,9)} = 3.33$ (Table 5.6, Fig. 5.8f). χ^2_{CALC} exceeds this value and does not lie in the rejection region. Therefore, we do not reject H_0. We may infer that the day's production variance is not different from the long-term variance.

The formulation of χ^2 in Eq. (5.8) leads to the definition of *confidence intervals for sample variance*, and thus *confidence intervals for sample standard deviation*. For $(n-1)$ degrees of freedom and a significance level α, the lower and upper limits are given by:

$$\frac{(n-1)s^2}{\chi^2_{(\alpha/2)}} < \sigma^2 < \frac{(n-1)s^2}{\chi^2_{(1-\alpha/2)}} \tag{5.12}$$

5.8
Kolmogorov-Smirnov Two-Sample Test

This test permits the comparison of two frequency distributions without making any assumptions about their form, i.e. it is a non-parametric or distribution-free test. It may be applied to conventional numeric (nominal) or ordinal data and is most effectively applied where both sample distributions have $n \geq 40$. If the sample sizes, n_1 and n_2, each exceed 40, they need not be equal. This procedure is somewhat easier to apply than the χ^2-test due to its relaxed constraints concerning sample size.

The frequency distributions of the two samples are plotted cumulatively to achieve a sensitive detection of a unique maximum discrepancy, because the test relies on a "maximum difference" between the distributions. Moreover, to facilitate the comparison of different sample sizes, the two samples are plotted as proportions (relative frequencies). This is simply achieved by dividing each frequency by the peak frequency in that distribution. Thus, the relative frequencies (y) range from zero to unity for each sample, and the two distributions are directly comparable due to standardisation in the same "units".

One notes the maximum y-difference between the curves as D. The test compares D with a test statistic D_α for the sample size variable, N:

$$D_\alpha = f_\alpha \sqrt{\frac{n_1 + n_2}{n_1 \cdot n_2}} \quad \text{and} \quad N = \frac{n_1 + n_2}{n_1 \cdot n_2} \quad (5.13)$$

where $f_{0.05} = 1.36$ for a two-tailed test

The f_α-coefficient depends on the significance (α) or confidence level ($1 - \alpha$) and whether the test is one-tailed or two-tailed. Other useful values are presented

Table 5.7. f_α-coefficients for the Kolmogorov-Smirnov test statistic

	95% Confidence Significance $\alpha = 0.05$	99% Confidence Significance $\alpha = 0.01$
Two-tail test	1.36	1.63
One-tail test	1.22	1.51

in Table 5.7. Two-tailed tests are usual because we are normally interested in *any* difference between the curves, not merely whether one "exceeds" the other. An example is shown in Fig. 5.9a, the cumulative version of data shown in Fig. 7.6. The data are from a study that noted an association between the injection of fluid-wastes in deep boreholes and microseisms induced by this process. The fluid pressure opposed the confining pressure on faults, thereby allowing them to slip and produce small earthquakes. (This is an effective-stress principle; it has nothing to do with lubrication.) The cumulative relative frequencies show a maximum discrepancy $D = 0.21$; there were 40 measurements of each variable; $n_1 = n_2 = 40$ and thus $N = 0.05$. The (D, N) coordinates define a point on the graph that does

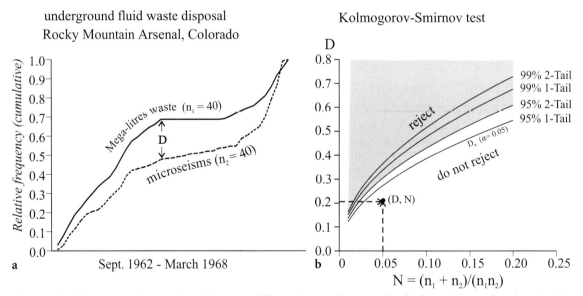

Fig. 5.9. The Kolmogorov-Smirnov test is sensitive to any differences between frequency distributions of two samples, whose sizes (n_1, n_2) may be unequal. The test is non-parametric; it does not assume any knowledge of the population from which the two samples are drawn. **a** The distributions are plotted cumulatively and as proportions, so that unequal sample sizes may be compared. The maximum difference between the two sample distributions (D), and the term $N = (n_1 + n_2)/(n_1 n_2)$ are used to test the hypothesis that the samples were drawn from the same population. **b** Values of the test statistic D_α for common confidence levels and one-tailed and two-tailed tests. If the determined value D for the sample size term N, exceeds the test statistic D_α given by the appropriate curve, we reject the hypothesis that the samples are drawn from the same population. The *shaded area* shows (D, N)-combinations that require rejection of the null hypothesis that the samples are drawn from the same population, for the most common application, a two-tailed test at a 95% confidence level

Table 5.8. Transformations that make skewed frequency distributions of x more symmetrical or perhaps closer to the Normal distribution in appearance

Transformation	Dummy variable (a, b = discretionary constants)	Conditions
Logarithmic	$a + b \, log(x)$	Logarithms to any base, $x > 0$
Square root	$a + b \, \sqrt{x}$	$x \geq 0$
Arcsine	$arcsin\,(p)$ or $arcsin\,(\sqrt{p})$	$0 \leq p \leq 1$; first standardise x to this range by dividing by a maximum value, $p = x/x_{MAX}$
Logit	$ln\,(P/(1-P))$	P may be a proportion or concentration $0 < P < 1$ $P \neq 0$ $P \neq 1$

not exceed the test statistic D_α. Thus, we should not reject the hypothesis that these distributions are similar at the 95% level (Fig. 5.9b). (Informally, we would "accept" some correspondence between the phenomena.)

The usual formulation of the test rejects the hypothesis that the distributions represent samples drawn from the same population when D exceeds the critical value appropriate for our chosen level of significance. The shaded region of Fig. 5.9b indicates (D, N) values for which the null hypothesis must be rejected at a 95% confidence level for a two-tailed-test; these are the conditions for the most common application of this test in elementary earth science problems.

CHAPTER 6

Regression: Linear, Curvilinear and Multilinear   6

Previously, we mainly dealt with *univariate statistics*, for observations that constitute a single measurement. They were presented in frequency-distribution graphs. The next step is to understand systems where each observation comprises measurements of two different variables, *x* and *y* using approaches of *bivariate statistics*. Bivariate data are presented on an *x-y* graph, sometimes referred to as a *scatter plot*. In the earth sciences, some *x-y* plots are so unstructured that "scatter" may be a deserved description but it usually does not carry a pejorative connotation. Some useful information can be obtained from scatter plots by contouring the density of points or by simple visual inspection. However, even quite complex geological systems sometimes show *x-y* data organised along a line or curve. The latter may be described by a mathematical relationship that may be used for predictive purposes. The simplest relationship is a straight line and the (*x, y*) relationship is then said to be linear. The straight line is defined by an equation of the form, $y = m\,x + c$ where *m* is the slope and *c* is the intercept of the line on the *y*-axis (Fig. 6.1a). Even in carefully controlled laboratory experiments there may be some departure of the data points from a line, and observations of natural phenomena may include more *noise*, or spurious variation. We can reduce the influence of this uncertainty on our interpretation of the scatter plot by calculating a *regression line* that provides the best predictions and description. The line regressing *y* on *x* is most useful, in which the response, *y*, contains all the uncertainty due to a well-defined control, *x*. The approach may be extended to non-linear (*x-y*) distributions by first transforming *x* or *y*, or both, so that the plot appears linear. The line fitted to (*x-y*) or to their transformations need not imply cause and effect; regression is primarily an objective predictive tool. Even if *y* is indeed caused by *x*, the regression equation is probably not a mathematical description of the actual physical response mechanism.

6.1
Control and Response Variables

Conventionally, the horizontal or *x*-axis represents the independent variable, which is usually more precisely defined than the dependent *y*-variable. Thus:

> *x = independent variable = control variable*
> *y = dependent variable = response variable*

Regression analysis provides methods for predicting the *y*-response from *x*, despite scatter that usually affects the dependent variable most, i.e. most of the noise will be represented by deviations parallel to the *y*-axis. Of course, casual observation of scattered data points will not reveal that, but an understanding of the system and of which variable is dependent allows us to appreciate which variable should contribute the most dispersion. Prediction is mainly within the range of observations, but modest extrapolation beyond the range may be justified. Regression takes its name from the pioneer study by Sir Francis Galton (1822–1911) who noted that taller-than-average fathers produced taller-than-average sons, and shorter-than-average fathers produced shorter-than-average sons. However, the tall sons were not as tall as their fathers, and the short sons were not as short as their fathers. The regressive tendencies of progeny toward a mean height produced scatter in the father-son (*x-y*) heights that would otherwise directly correspond. Galton referred to this as regression toward mediocrity.

Clearly defined control and response variables occur in laboratory experiments. For example, consider measurements of the Fe/Mg ratios of olivine crystals grown in the laboratory at different temperatures. Because temperature (T) is controlled and subject to little error, it is the *independent or control variable* and T would be plotted as the horizontal *x*-axis. In contrast, the measurements of Fe/Mg ratio involve analytical error as well as real differences due to varying levels of disequilibrium in the chemical system. The Fe/Mg ratio would be plotted as *y*, the *dependent* or *response variable*. The observations may show some linear pat-

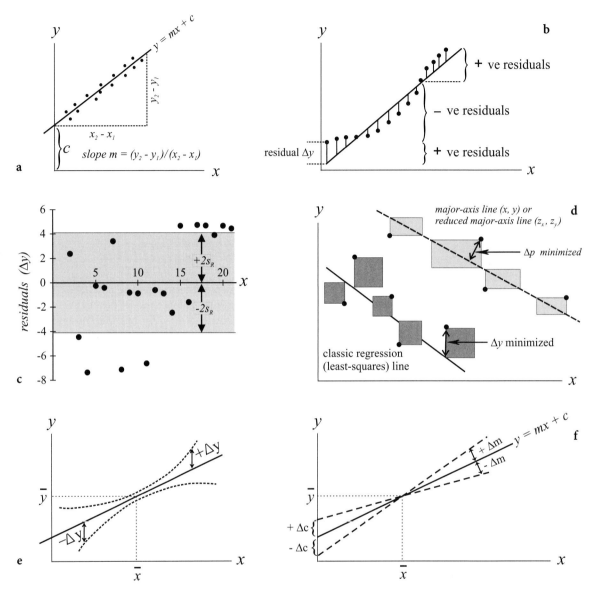

Fig. 6.1. **a** Scatter plot with regression line that has slope = m, and y-intercept = c. **b** The systematic pattern shown by the Δy-residuals shows that linear regression is unwise. **c** Δy-residuals plotted against x. This residual plot shows no obvious pattern with a Normal distribution of residuals about their means; this favours the regression approach. One may be justified in rejecting observations with large residuals, e.g. beyond the s_R or $2s_R$ confidence band where s_R is the standard deviation of the residuals. **d** *Classic (least squares) regression line*: for the case where all uncertainty is in y, the line is the one for which the Δy-residuals are a minimum, least-square areas *shaded*. *Major axis line*: for the case where x and y contribute to uncertainty, line minimises the perpendicular distances Δp between points and the line, least areas *shaded*. Fitted to exactly the same data points, these different "best-fit" lines would have different slopes and intercepts unless the points were perfectly aligned. **e** Confidence band about the regression line defined by the magnitudes of residuals Δy. The confidence band is pinched near the centroid $(\bar{x}, \bar{y})$. **f** Confidence limits on the slope and intercept of the regression line, usually at the 95% confidence level

tern. How can we use this pattern to predict the Fe/Mg ratio for a given T, and how may we express our confidence in it? Such questions may be answered by linear regression. The *regression line* is the most favourable trend line through the (*x-y*) points which are calculated on the assumption that scatter is due primarily to noise in the *y*-measurements. This is achieved by choosing a line for which the Δy *residuals* are smallest (Fig. 6.1 e). The equation for the regression line combines residuals (Δy), making due allowance for the fact that some residuals are positive and others negative by using the sum of their squares, $\Sigma(\Delta y)^2$. The residuals also provide a useful means of assessing the validity of attempting regression. One may verify the assumption of linearity by calculating or estimating a suitable line (e.g. Fig. 6.1 a) and inspecting the residuals, which are the discrepancies between the line and the data points (Fig. 6.1 c). If the residuals show any systematic pattern, it is unwise to entertain the possibility of linear behaviour in the system (e.g. Fig. 6.1 b). It is more objective to detect any non-random pattern from a separate graph of the Δy-residuals against *x*. Ideally, regression requires that the residuals are normally distributed (Fig. 6.1 c).

Systematic patterns to the residuals may be apparent from the raw data (e.g. Fig. 6.1 b). In this case, the data may be described mathematically by a non-linear equation, requiring a more complicated procedure to give a close fit to the data, with suitably small residuals (Fig. 6.1 c). Fitting curves, for example by curvilinear regression, has become a simple task using spreadsheets such as *Microsoft Excel*, or at a more advanced level with scientific graphing programs such as *SPSS SigmaPlot* and *Origin*.

6.2 Associated Variables Without Clear Control

In some instances, any consideration of dependence between *x* and *y* may be irrelevant, but it may still be reasonable to attempt predictions from the data set or merely quantify the description. For example, the concentrations of two heavy accessory minerals in beach sand, such as magnetite and hornblende, may appear in similar proportions in different environments. However, the abundance of magnetite does not "control" the abundance of hornblende. It is simply that heavy minerals respond similarly with regard to hydrology and sedimentation. In this case, a linear relationship between the two concentrations may exist, but we should not feel obliged to regress *y* on *x* (e.g. Fig. 6.1 e). Now both variables are equally subject to scatter. Therefore, for predictive purposes, the *regression line* which only minimises the Δy residuals, may not be suitable. Instead, we could minimise the residuals of both *x* and *y* (i.e. Δx, Δy). This produces the *major axis line*, obtained by minimising the perpendicular distances from the points to the line (Fig. 6.1 d). The major axis line may sometimes be loosely referred to as the *line of best fit* but this really is a matter of opinion. The regression line is a better fit if the entire scatter is due to variation in *y*; the major axis line is better where *x* and *y* contribute equally to the uncertainty.

In fact, there are several equally valid ways to fit a line through a scatter plot depending on our needs. The two most common are the *least-squares regression line* and the *major axis line* (Fig. 6.1 d). Applications of the major axis line are widened if it is plotted using standardised

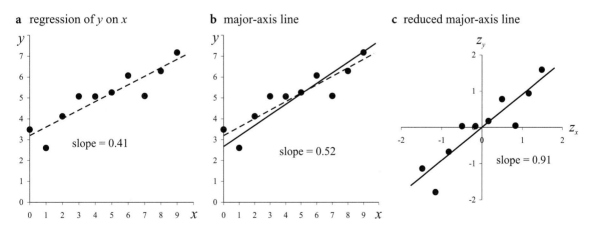

a regression of *y* on *x*   **b** major-axis line   **c** reduced major-axis line

Fig. 6.2. Comparison of the classic least squares **a** regression line of *y* on *x* with the **b** major axis line and the **c** reduced major axis line, for the same ten observations. The reduced major axis line uses standardised versions of *x* and *y* (e.g. $z_x = (x - \bar{x})/s_x$) so that it is scale-independent. Note the difference in slopes of the lines, and the different intercepts for **a** and **b**

x- and y-variables (i.e. $z_x = (x - \bar{x})/s_x$). This gives the *reduced major axis line,* which has the advantage of being independent of scale units. To review, the following are all possible linear fitting mechanisms, compared with geological examples by Agterberg (1974):

1. **regression line** (usual; regress y on x, minimise y-residuals, uncertainty is in y)
2. regression line (regress x on y, minimise x-residuals, uncertainty is in x)
3. **major axis line** (simultaneously minimise x and y-residuals; uncertainty is in x and y)
4. reduced major axis line (as above, but standardise x, y as z_x, z_y; thus, scale-independent)

Unless the points lie precisely along a line, these techniques will normally yield lines of different slope, and the differences will be greater for less linear distributions as should be apparent from Fig. 6.2 where all three lines are calculated for the same data. Most of the following discussion concerns the classic regression line and the major axis line. However, before we understand the mechanics of fitting lines to earth science data, it is important to identify the true nature of the variables and any interdependency.

6.3
Lurking Variables

Relationships may not always be what they seem, in life and in statistics. This may be true whether y truly depends on x, or whether the x-y association is looser with noise in both observations, like much earth science data. Well-constrained linear relationships are unlikely to be entirely coincidental, but they may be entirely illusory. For example, it is commonly found that deeper mines (y) exploit richer ore deposits (x). It would be unwise to conclude that the richness of ores increases with depth. The real issue here is that a deep deposit must have high ore concentration to justify the expense of its exploitation. What we are recognising is that the value-expense ratio is a more fundamental control than ore grade, and there is a much better behaved relationship between the mine depth and value-expense ratio. The more fundamental *lurking* variable is camouflaged by another variable that is usually deceptively appealing. The lurking variable usually acts in the same sense as the simpler variable over the range of the entire data set and may possess a stronger or a weaker association with the response variable than does the illusory control.

Two well-known examples from mineral resource economics illustrate the complexity of earth-science associations and the role of lurking variables that perhaps enhance an apparent relationship. Such data commonly have great intrinsic value but lurking variables may cause their over-interpretation or misinterpretation. One such relationship is commonly misinterpreted from its original source. The Lasky relationship shows an apparently strong inverse relationship between logarithmic cumulative tonnage of copper deposits, and their grade (De Young 1981), reproduced here as Fig. 6.3 a. Grade represents the concentration of economically useful material, most simply given as a percentage. While we cannot deny the simplicity of the graph, we should note first that the response variable is *cumulative* tonnage, each y-value includes the y-values of all richer grade deposits. For statistical treatment such as regression, we require that individual observations be independent. When the actual tonnages are plotted, the scatter is a little greater, and the regression line explains 91 % of the distribution rather than 99 % that would be claimed using cumulative tonnages. Also, strictly from a scientific perspective, the relationship is deceptive because the tonnage of deposits is not *controlled* by grade. High-grade deposits may be considered as reserves despite a low tonnage simply because the value-cost ratio is favourable. Regardless of the concentration, the deposit will only be exploited or even assessed as a potential reserve if there is sufficient tonnage to justify potential exploitation. Not only is the value-cost ratio an important lurking variable, many other nonlinear influences could affect the data set either in its entirety or over just part of the range. For example, certain political and tax jurisdictions encourage the publication of information on reserves to different degrees and, at different times in the economic history of the world, reporting may be biased differently. It is unfortunate that such diagrams could be misinterpreted by spurious phrasing, e.g. "it (= *the Lasky relationship*) provides a comforting indication that the volume of mineral material available will increase greatly for even a small decrease in grade. (Kesler 1994)". First, the scarcity or abundance of ore deposits of a certain tonnage is not *controlled* by grade. It is the rarity of one or more specialised geological processes acting as lurking variables that control the available tonnage. There are many such processes and many types of Cu deposit, e.g. massive sulphides, porphyry coppers, or native copper. Secondly, the cumulative nature of the available tonnage is not controlled by grade. Rich deposits (> 4 % Cu) collectively represent < 0.1 % of the estimated world reserves, so any predictive value of accessible tonnage from grade appears optimistic.

A further fascinating diagram in mineral economics relates reserves of economically important metals to

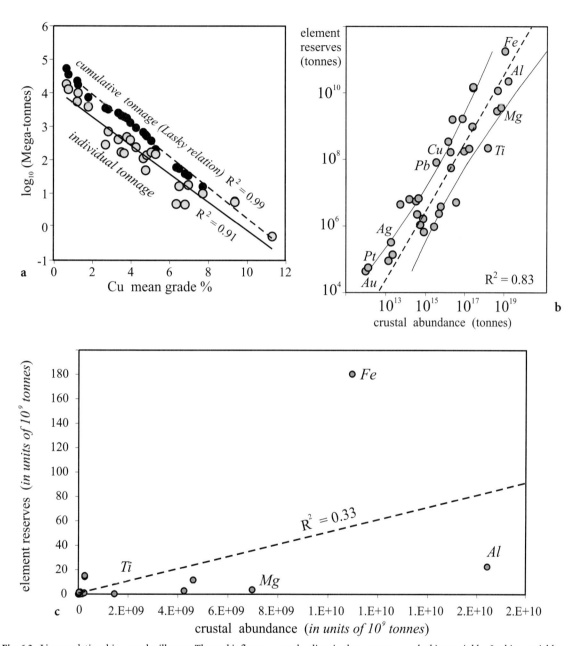

Fig. 6.3. Linear relationships may be illusory. The real influences may be disguised as one or more *lurking variables*. Lurking variables control the relationship at a more fundamental level than the more obvious ones chosen for the axes of the scatter plot. **a** The Lasky relationship could be interpreted as showing a strong inverse relation between the tonnage of ore deposits and their richness or grade (De Young 1981). However, grade does not control tonnage: rather we only mine weakly concentrated ore if there is sufficient tonnage. The lurking variable is the value-cost ratio of the mining activity. Moreover, the data are usually presented with the tonnage of ore deposits plotted *cumulatively*, each point including the values of all ore deposits of higher grades. In that case, the x-values are not independent and the strength of the association is exaggerated; Lasky's relation would claim to explain 99 % of the scatter. When the individual tonnage is plotted (*solid line*), the relationship is a little weaker. **b** The McKelvey (1960) relation usually graphs crustal abundances of economically important metals as *y* against their known economic reserves as *x*. Here, the axes are exchanged since if there is any valid dependency it could only be of reserves upon abundance. That any relationship appears at all is remarkable, but there are many lurking variables in the assessment reserves that far outweigh crustal abundance. **c** In part, the McKelvey relationship is an artefact of logarithmic scaling of variables that have too large a range. On linear axes, as here, large abundances are clearly outliers whereas in log-log space as in **b**, they give a false impression of a linear relation. Note that reserves are generally < one hundred-millionth of crustal abundance (the same units are used for both axes)

their crustal abundance (McKelvey 1960). Non-geologists may be surprised to learn that crustal abundances of elements, and even minerals, may be predicted more accurately than economic reserves because the crust's chemical constitution is well constrained with the aid of various well-founded geological and geochemical principles. As but one example, continental-glacial sediments represent chemically unaltered rock specimens from vast areas of the continental crust and thus provide an average crustal composition more effectively than could be devised by any human sampling scheme.

The McKelvey diagram usually shows crustal abundance as y and reserves as x. However, reserves are dependent upon abundance, so we plot reserves (y) against crustal abundance (x) in Fig. 6.3b. With even rudimentary understanding of the complexities of earth science, it is wondrous that there is any relationship. This surprise may be moderated by appreciating that the logarithmic axes encompass large values more than 1 million times greater than the smallest, leading to a massive suppression of scatter. Given even small errors in the estimates of reserves, the scatter will increase dramatically for more abundant elements. "Reserves" is a subjective variable, depending on economics, fiscal policy and politics, as well as Nature. For example, Fe is at least 100,000,000-fold more abundant than Au. A conservative 1% error in reserve estimates of Fe corresponds to ~100,000,000 tonnes. That uncertainty is larger than the magnitude of the reserves of most other metals (Fig. 6.3b). Although a regression line and its 95% confidence band are shown, its interpretation and use require caution. The association is real, but weak due to logarithmic-scaling and large errors in reserve-estimates. When the data are plotted with linear axes, we see that the association is weaker (Fig. 6.3c). Logarithmic axes with an inappropriately large range suppress the erratic variation in large abundance outliers (Fig. 6.3c). Moreover, more fundamental difficulties are the meaningfulness of making such inferences from one metal-reserve to another, and the influence of lurking variables on reserves.

We have mentioned political and economic factors that suppress reliable information about mineral commodities. However, geology also provides lurking variables. For example, reserves of Al would mostly be considered in the form of bauxite, an easily extracted superficial deposit produced by tropical weathering. Where that is not available for political or geographic reasons, as in the former Soviet Union, hard-rock sources (e.g. feldspathoids) may provide an alternative source. How may these two vastly different reserves be merged in the diagram? Although aluminium is the third-most abundant element in the crust, the amounts

present in economically extractable forms are negligible, and any relationship that falls in the same confidence band on a log-log graph is almost fortuitous (e.g. Fig. 6.3b). In turn, we could apply similar arguments to each metal although the actual economic and scientific factors would differ from element to element. In summary, there are so many possible lurking variables that the association shown must be evaluated carefully, despite its general interest. It is unwise to use it as a predictive tool for other elements along the lines suggested by one author: "reserves for less-explored commodities can be estimated from reserves for well-explored ones."

It should now be clear that it is advisable to ponder the complexities of the variables before attempting regression. It may change our choice of what should be regressed upon what, and whether by simple regression or by major axis line. We may even decide against regression or at least view the results with extra caution.

6.4
Basis of Linear Regression

To explain the details of regression, we will limit the discussion to the common and simple case of *linear regression*, where y responds to x alone, and the entire scatter is in the y-values. It is said that we regress y on x. Since all of the discrepancy between the observations and the proposed line is due to Δy residuals, we must find a way to choose a line so that their combined effect is minimal. However, observations lie both above and below the regression line so Δy deviations must be squared before the summation, to ensure that the contributions of negative and positive residuals do not cancel one another. Thus, from the values of the independent variable $x_1, x_2, \ldots x_n$ and of the associated values of the dependent variable $y_1, y_2, \ldots y_n$ that are related by:

$$y = mx + c \qquad (6.1)$$

We must minimise the sum of the deviations Δy_i between the observation y_i and the value $(mx + c)$ predicted by the regression line from Eq. (6.1) to permit a *least-squares linear regression*. The sum of all the deviations is given by the following equation for the *residual sum of squares*:

$$\sum_{i=1}^{n} (\Delta y_i)^2 = \sum_{i=1}^{n} [y_i - (mx_i + c)]^2 \qquad (6.2)$$

The minimum value is obtained by partial differentiation of the right-hand term and equating to zero, leading to relatively simple definitions of the slope (m) and intercept (c) of the regression line where, $\bar{x}, \bar{y}$ are mean values:

$$m = \frac{\sum\limits_{i=1}^{n}(x_i - \bar{x})(y_i - \bar{y})}{\sum\limits_{i=1}^{n}(x_i - \bar{x})^2} \qquad (6.3)$$

The intercept (c) is obtained by substituting the mean values back into Eq. (6.1):

$$c = \bar{y} - m\bar{x} \qquad (6.4)$$

The regression line must pass through the data centroid $(\bar{x}, \bar{y})$ so that we may predict any (x, y) values that lie on the regression line by using the following equation.

$$y - \bar{y} = m\,(x - \bar{x}) \qquad (6.5)$$

Linear regression may be for one of two main purposes. First, it may be merely for description, permitting estimates of values within the data range (*interpolation*) or modest estimates beyond the data range (*extrapolation*). No assumption is made about the relationship between the variables because the line is merely descriptive and does not assign cause. For example, examine the change in mean global temperatures between the years 1850 and 1990 (Fig. 6.4a). The data are normalised arbitrarily to a zero value at 1971. There is general acceptance of 20th century global warming, and we may point to a few clear culprits, exhaust from fossil-fuel combustion and the production of other greenhouse gases. However, any relation between cause and effect must be very complex, even assuming "mean global temperature" and observations of some "causal" phenomenon may be measured accurately. There is no reason to believe that there is an underlying linear control on the system. Nevertheless, linear regression provides a reasonable approximation to the data, and the individual points are fairly equally scattered about the regression line, indicating that the distribution of residuals is at least balanced (Fig. 6.4b). The regression line characterises a generalised temperature trend, without compromising our appreciation that the actual response of global temperature is due to multiple complex variables and is not controlled linearly by a single, time-dependent phenomenon.

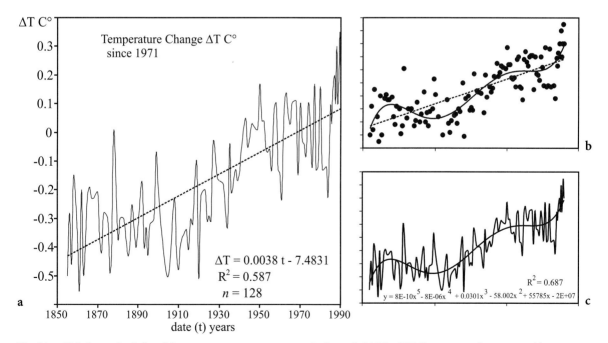

Fig. 6.4. **a** Global warming inferred from temperature measurements in the period 1854–1982. Temperature changes are with respect to the mean value for 1971. Our interest is in natural and industrial causes such as volcanicity, variation in solar activity, emissions from combustion of fossil fuels and changes in other greenhouse gases, but the scatter also includes errors in determining a mean annual global temperature and changes in methods of data collection. Linear regression would predict temperatures increasing at 0.0038 °C/year. **b** Scatter plot compares linear regression with a fifth-order polynomial fit. **c** The polynomial fit compared with the line graph of the original data

The linear regression equation predicts temperature change (ΔT) with time (t) by the equation $\Delta T = 0.0038(t) - 7.4831$. The average annual temperature increase has been $0.0038\,°C$ in this period, merely as a matter of observation.

6.5
An Introduction to the Correlation Coefficient

Correlation will be discussed in more detail in the next chapter but some background details are pertinent here. *Regression* provides a simple linear description of empirical data mainly for predictive purposes by interpolation or extrapolation. However, the regression line may possess interpretive significance if it represents some physical association between the variables, even in the absence of direct causality. The degree of association and linearity is quantified by a statistic known as the *correlation coefficient* (R), which is different from the slope of the regression line but has the same sign. Its value varies between -1 and $+1$. The *coefficient of determination*, R^2, expresses the proportion of y-variation explainable by the regression line and it is sometimes quoted as a percentage.

A simple grasp of the relationship between R and the data distribution is possible by considering the Normal-standardised forms of the variables. We recall that the z-variate or z-score is a standardised form of a variable. To standardise x, subtract the mean value ($\bar{x}$) from each of the x-values, and divide by the standard deviation (s_x) (Chap. 3). This proxy variable centres on zero and is in scale-independent units of standard deviation. In general, the variability could be equally due to deviations in y-values so we must consider z-values for both x and y:

$$z_x = \frac{x - \bar{x}}{s_x} \quad \text{and} \quad z_y = \frac{y - \bar{y}}{s_y} \tag{6.6}$$

Now consider the sum, $\Sigma(z_x, z_y)$ which combines the variations due to scatter in the system, referred to as the *x-y covariance* in Chapter 7. If the scatter is completely random in x and y, there should be equal contributions from positive and negative z-values, and this term should be close to zero. When the z-values have the same sign, they reinforce the summed product and give a large positive value. If y increases with x, the sum has a large positive value and if y decreases with x, the sum has a large negative value. This may be made universal for all sample sizes by dividing by the degrees of freedom $(n - 1)$ to give the correlation coefficient:

$$R = \frac{\sum_{i=1}^{n}(z_x z_y)}{(n-1)} \tag{6.7}$$

Thus:

R = + 1 for perfect positive correlation (*y increases with x, no scatter*)

R = 0 for no correlation (*random uncorrelated x, y values*)

R = – 1 for perfect negative correlation (*y decreases with x, no scatter*)

For the linear regression study of changes in mean global temperature, R = 0.77 (Fig. 6.4a, b), is quite reasonable in view of the sample size ($n=128$), and even significant at the 95% level (see Chap. 7). However, $R^2 = 0.58$, indicating that the regression line explains only 58% of the variance in y. Clearly, it would be unwise to extrapolate the line to predict future temperatures, or temperatures before 1850. Indeed, from other sources, we know that the rate was lower before 1850 (e.g. Fig. 7.10a), and it is believed that the rate could continue to increase in the future. As a general note, there is a general loose reference to "time" being a control variable in earth science processes. In most cases, the control variable is actually a geological or environmental factor that changes with time, and for which time may be a rather inaccurate proxy.

6.6
Confidence Interval About the Regression Line

Although the correlation coefficient gives us a sense of the strength of the *x-y* association, we may need to quantify the regression line's precision with a confidence band (Fig. 6.1e), especially since the regression line may be used for prediction. We saw that the total of the y-deviations from the line was given by the *residual sum of squares* (Eq. 6.2). Its right-hand term also yields an estimate of the *residual variance*, $(s_R)^2$, using:

$$(s_R)^2 = \frac{1}{(n-2)} \sum_{i=1}^{n} [y_i - (mx_i + c)]^2 \tag{6.8}$$

The term $(n-2)$ shows that two degrees of freedom have been lost since m and c are already known.

Since the residuals are normally distributed, we may then define the envelope about the regression line at the confidence level of $100(1-\alpha)\%$ using the appropriate tabled value of the t-statistic for a two-tailed test with $(n-2)$ degrees of freedom (Table 4.5). Usually, $\alpha = 0.05$ for a confidence level of 95 %. The coordinates of the upper and lower confidence envelope may then be calculated at any (x, y) point using the following expression:

$$y \pm t_{\alpha/2}s_R \left[\frac{1}{n} + \frac{(x - \bar{x})^2}{\sum (x_i - \bar{x})^2} \right]^{\frac{1}{2}} \qquad (6.9)$$

The confidence band is always pinched most at $\bar{x}$. An example is shown in Fig. 6.3 b.

6.7
The Major Axis Line and the Reduced Major Axis Line

The regression line is very popular but is limited to applications where the uncertainty resides in only one of the variables. On the other hand, the *major axis* line is chosen so that the sum of the perpendicular distances to the points is a minimum. The Δx and Δy residuals are simultaneously selected to give the minimum total deviation from the line (Fig. 6.1 d). The major axis line has the form $y = mx + c$ but the slope, $m = \tan \alpha$, is given by the following equation which, however, uses the double angle, 2α (Agterberg 1974):

$$\tan 2\alpha = \frac{2 \sum_{i=1}^{n} (x_i - \bar{x})(y_i - \bar{y})}{\sum_{i=1}^{n} [(x_i - \bar{x})^2 - (y_i - \bar{y})^2]} \qquad (6.10)$$

The intercept (c) for the major axis line may then be determined by substituting m into $\bar{y} = m\bar{x} + c$.

The confidence intervals about the slope and intercept of the major axis line may be determined from their standard deviations, using the following equations:

$$s_m = m \sqrt{\frac{1 - R^2}{n}} \quad \text{and} \quad s_c = s_y \sqrt{\left(\frac{1 - R^2}{n} \left(1 + \frac{\bar{x}^2}{s_x^2} \right) \right)} \qquad (6.11)$$

Confidence limits may use multiples of standard deviation directly, the one-sigma, two-sigma limits, etc. However, most likely, one would use 95 % confidence limits given by $\pm 1.96 s_m$ for the slope, and by $\pm 1.96 s_c$ for the intercept.

Manipulation of the major axis line is facilitated when the variables are standardised according to

Eq. (6.7). This results in the *reduced major axis line*, which plots in z_x–z_y space using scale units of standard deviations, s_x and s_y. The line passes through the origin, which corresponds of course to $(\bar{x}, \bar{y})$. The standardised scale is independent of units and therefore very useful in comparing different data. Moreover, the definition of its slope is greatly simplified to the following equation:

$$m = \frac{s_y}{s_x} \qquad (6.12)$$

We should not forget that for the same data, the *regression line*, *major axis line* and *reduced major axis line* all have different slopes unless the points are perfectly aligned. The discrepancy between the slopes increases with the scatter in the x-y distribution. The major axis line is a special two-dimensional case of a more general fitting procedure called *principal-component analysis* that may be applied to data in three dimensions (x, y, z) or more. Such multivariate procedures define the principal components as eigenvectors of the variance-covariance matrix. These multivariate procedures are mostly beyond our scope and lead into *factor analysis*, so the reader is referred to Agterberg (1974) and Davis (1973, 2002).

6.8
Linearisation to Facilitate Regression Analysis

Occasionally, we may wish to predict values from an (x, y) distribution that clearly curves across the scatter plot rather than following a line (e.g. Fig. 6.1b). It is possible to fit a regression curve directly to the observations, but it is much simpler if the x-axis, y-axis or both axes of the scatter plot can be transformed so that the scatter of points is linearised. Linear regression may then be performed on the transformed variables (x', y').

For example, consider the way in which y changes with x. Consider a small unit increment Δx in the control variable x. If x increases by progressively greater amounts as successive Δx increments accumulate, we could apply a transformation that *progressively reduces* larger x-values, e.g. replacing x with the dummy variable, such as $x' = \sqrt{(x)}$. It may be desirable to add a constant (a) to the dummy variable to shift the origin and a coefficient (b) to rescale the values. The transformed data (x', y) can be treated as (x, y) in regression analysis. Subsequently, if any x' results are required in original units, they must be fed in reverse through the transformation. It may also be desirable to apply some transformation to y.

Table 6.1. Power ladder for selection of linearisation transformation (applicable to y or x or both)

Distord x with increasing x	Power	Transformation
Progressively stretch initial variable	3	$x' = a + bx^3$
	2	$x' = a + bx^2$
	1	no transformation
$\updownarrow$	$^1/_2$	$x' = a + bx^{1/2}$
	$^1/_3$	$x' = a + b(x)^{1/3}$
		$x' = a + ln(y)$ or $x' = a + log_{10}(y)$
Progressively compress initial variable	-1	$x' = a + b\,(x^{-1})$

A suitable transformation is selected by inspecting the initial (x, y) distribution. A qualitative assessment permits us to decide how severely the x-values must be "stretched" or "compressed" to straighten the pattern of data points. The more the x-values have to be distorted with increasing magnitude of x, the higher the power of transformation should be employed (Table 6.1). The power-transformations ladder ranks transformations in their relative effectiveness in stretching or compressing the x-scale. We should note that the stretching and compression are progressive, being more noticeable for larger magnitudes of x, and that a transformation of y-values may also be required. Linearisation is a useful trick to simplify more complex systems for the purpose of analysis by linear regression, or merely for the purposes of graphical presentation, description and interpretation.

Where a non-linear association is evident from the scatter plot, a more direct approach finds the best curve to fit the data. The curve's equation then provides a mathematical description of the observations but probably not of their cause. Statistical software makes that direct approach simple to implement, but it is sometimes more instructive to re-express the initial observations so that linear regression may be applied. A few simple linearisation techniques satisfy the needs of almost all our problems in the earth sciences. All are easily applied in spreadsheets, if not automatically available in statistical programs.

The most commonly successful transformation of non-linear data takes logarithms of the variables, usually to base 10 or base e, although base 2 may be useful in sedimentology where it forms the basis of the Wentworth grain-size scale (see Chap. 5). The choice for the base of the logarithms is not too important as it affects only the slope of the line, if linearisation is successful. Logarithmic linearisation is achieved similarly by plotting the variables on axes with logarithmic scales. The data distribution is sheared and straightened from some curved initial pattern into a linear form in log-log space. If this is successful, the linear form in log-log space has a slope (n) which defines the index of a power-law relationship of the original variables in the form, $y = a\,x^n$. Most often, the linearised fit does not explain the data distribution; it serves as a first-order approximation and as an empirical description. Care must be taken not to apply logarithmic transformation of data whose initial range is too large. That may suppress the role of aberrant values (*outliers*) and produce an illusory linear association in log-log space (e.g. Fig. 6.3b, cf. c).

Plotting one or both axes with logarithmic scales is an instantaneous option in computer spreadsheets. However, historically, it was achieved at a less sophisticated level by plotting the values on graph paper with logarithmic-scaled axes. If the observations are constrained to lie on a line by this process, the *logarithms* of the variables may be treated like dummy linear variables. Another advantage of logarithmic plots is that they tend to disperse low data values as long as the data range is not unreasonably large, for example using base-10 logarithms, a cautious data range might be ≤ 1000 units of the linear scale. Also, the degree of scatter at low values may reveal the importance of errors of observation or natural noise where the fractional error is larger. A familiar non-linear, but very well-behaved, relationship is the gas law due to the 17th century physicist Robert Boyle. Boyle's law shows that the volume of a gas is inversely proportional to its pressure, demonstrated by gas trapped beneath a column of mercury in a J-shaped tube, sealed at the short end (Fig. 6.5d). Using his original data, we see that the air volume appears to decrease inversely with pressure from the raw data. This is confirmed by three approaches. First, directly fitting a power law to the raw data yields an exponent -1.0015, agreeably close to the expected -1. Second, if we did not share Boyle's intuition, or we were dealing with less well-behaved data typical of the earth sciences, we could attempt to linearise the data by taking logarithms. The experiment clearly showed some relationship of the general form $P.V^k = c$ where k and c are as-yet undetermined constants. Using logarithms, we may rewrite:

$$log_{10} P = - k\,log_{10} V + log_{10} c \qquad (6.13)$$

which is analogous to the straight-line equation $y = mx + c$. We see indeed that a plot of $log_{10}P$ against $log_{10}V$ yields a straight line, shown in the inset in Fig. 6.5d. Third, having confirmed an inverse relationship, our most effective test would be to express the variables di-

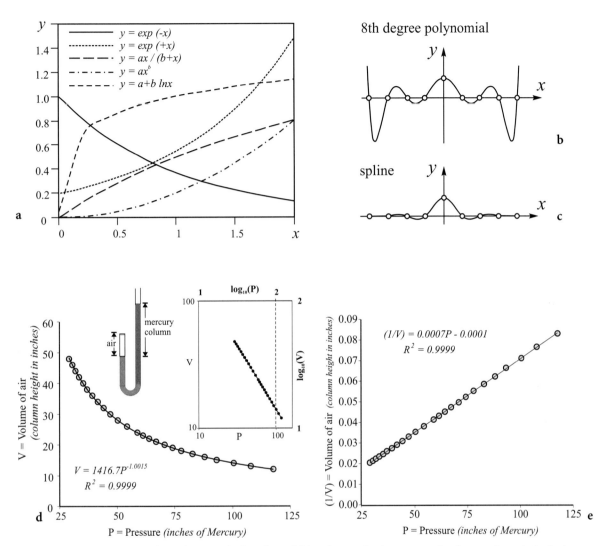

Fig. 6.5. **a** Some common curve-fitting functions. **b** The polynomial function may be chosen at varying degrees of complexity (*orders* defined by the highest power of x). The higher the order, the more closely the curve fits the data until the curve simply passes through every data point and provides no useful generalisation at all, as shown for an eighth-degree polynomial that is forced to pass directly through a sample of nine points. **c** The spline, fitted to the same data as in **b**, minimises the curvature of a curve that joins the points. It is a descriptive tool whose equation is unlikely to have any physical significance. **d** Original gas-law data of the 17th century physicist Robert Boyle showed an inverse relationship of volume to pressure. This is confirmed by the best power-law fit, which has an exponent of -1.0015 (≈ 1), or by the linearisation achieved by taking logarithms (*inset*). **e** Having confirmed the suspicion of an inverse relationship in **d**, the variable $(1/V)$ may be linearly regressed upon (P). The coefficients in the equations are due to the unusual original Imperial units, but the linearity of the graph is convincing confirmation of the inverse relationship

rectly in a form that could be tested by linear regression. In other words, we may now compare $(1/V)$ directly with P; the resulting linear regression is shown in Fig. 6.5e. Boyle's original data clearly demonstrate a well-behaved inverse relationship with $(1/V)$ proportional to P from all three aspects of this simple investigation. The correlation coefficient is very high and R^2 shows that the proposed gas law explains 99.99% of the volume variation due to pressure. We will rarely see such strong correlations outside the laboratory or in earth science, due to the intrinsic looseness and poorly constrained behaviour of environmental and geological processes.

A geological example of the use of linearisation is provided by a simple hydrological empirical relationship between the mass (m) of a particle and the mean velocity (v) of a stream required to move it by bed traction. The relationship is $m = k\,v^5$ where k is a constant summarising characteristics of the channel (slope, cross-sectional area, bed roughness) and of the particle's shape. Clearly, those factors together with errors in measuring average stream velocity and particle mass make this an imprecise relation, more of an approximation. We expect considerable scatter of v and m. For these reasons it would be difficult to compare a fitted power-law curve directly with the $(m - v)$ scatter plot, due to the scatter $v \pm \Delta v$ and $m \pm \Delta m$. It is easier to compare the scatter plot of $log_{10}(m)$ against $log_{10}(v)$ with their regression line; if the relationship is valid, the line should have a slope of 5.

6.9
Directly Fitting Curves to the Data

Regression permits us to detect the presence and strength of a linear relationship in order to make predictions from scattered data. That goal is thwarted where there is an underlying non-linear relationship. Empirically, one may straighten out the data by the linearisation transformations suggested above (Table 6.1). However, a different approach is possible where we may reasonably expect the relationship to follow a certain mathematical form, or where we make an educated guess from its scatter plot. This has been rendered practical by the personal computer, spreadsheets and statistical-graphing programs. (*Excel, SigmaPlot, SigmaStat* and *Origin* have been found very useful.) These permit us to directly compare common nonlinear functions with the data. Where necessary, we may examine the residuals for any systematic pattern (Fig. 6.1b, c) or quantify the goodness-of-fit in some more refined manner (χ^2 test, Chap. 5).

Computer programs permit us to attempt a large number of functions to explain the data distribution by *curvilinear regression*. However, the combination of random variation in the data and the complexity of the true curve may obscure the choice of an appropriate fit for earth science observations. It is tempting to go on a fishing expedition and find many different potential fits to scattered data; some or all of these fits may be misleading or unnecessarily complicated. It is therefore important that the geologist has some external scientific reason for choosing what kind of function to fit, e.g. exponential as opposed to a power law. Common forms are shown in Table 6.2 and Fig. 6.5a. If the choice is to facilitate extrapolation or interpolation, this may not be too critical. However, the selection of the appropriate function is very important if there is some hope of inferring a mathematical relationship. As with simple linear regression, the more complex forms also require that the variables act independently and that the y-residuals are normally distributed.

Two special curve fits deserve mention. The polynomial can be used to describe many data distributions. The order of the polynomial is given by the highest power of x, and the best-fitting curve is described by the coefficients a_0, a_1, etc. (Table 6.2). Curvilinear regression of the polynomial achieves the curve for which the sum-of-squares of the y-residuals are a minimum. The general polynomial of Nth order may be rewritten for each observation (x_i, y_i) and summed to give the total deviation as:

$$\sum_{i=1}^{n} [y_i - (a_0 + a_1 x_i + a_2 x_i^2 \dots \ \dots a_N x_i^N)]^2 \qquad (6.14)$$

Partial differentiation with respect to coefficients and rearrangement yields $N + 1$ equations from which the $N + 1$ unknown coefficients may be determined:

$$\sum y = na_0 + a_1 \sum x + \dots \ \dots + a_N \sum x^n$$
$$\sum xy = a_0 \sum x + a_1 \sum x^2 + \dots \ \dots + a_N \sum x^{N+1}$$
etc.
$$\sum x^N y = a_0 \sum x^N + a_1 \sum x^{N+1} + \dots \ \dots + a_N \sum x^{2N}$$

$$(6.15)$$

These are termed the *normal equations for polynomial regression*. Computer programs permit one to choose as high a polynomial as one desires, but the curve may be unnecessarily contorted for higher orders. Eventually, as the order of the polynomial approaches the number of data points, the curve simply passes through all points and shows many spurious maxima and minima (Fig. 6.5b). A polynomial of $(n-1)$th order will always pass through n points. There is usually no theoretical reason to assume a particular order for the

Table 6.2. Common forms of equations to fit directly to $(x$-$y)$ plots

Form	Example
Linear	$y = a + bx$
Hyperbolic	$y = ax/(b + x)$
Exponential Growth	$y = a\,e^{bx}$
Exponential Decay	$y = a\,e^{-bx}$
Power-law	$y = a\,x^b$
Polynomial of degree n	$y = a_0 + a_1 x + a_2 x^2 + a_3 x^3 \dots + a_n x^n$
Logarithmic	$y = a_0 + a_1 \ln x$ or $y = a_0 + a_1 \log_{10} x$
Spline (flexible curve-fitting function where f_n are cubic functions of x)	$y = a_1 f_1 + a_2 f_2 + a_3 f_3 \dots$

polynomial, so we should choose the lowest order polynomial that satisfies the observations. A simple starting criterion relates the polynomial's order to the number of turning points (crests and troughs) in the data set. An initial approach may be to fit a polynomial whose order exceeds the number of turning points by one. Consider the global temperature change data (Fig. 6.4a): there appear to be four major turning points, starting with a temperature crest ~1860, followed by the first trough at ~1910, etc. Thus, a fifth-order polynomial was chosen that explains 69% of the variation in the scatter ($R^2 = 0.69$). Visually, it appears more agreeable when compared with a line graph (Fig. 6.4c) than with the scatter plot (Fig. 6.4b). Scatter plots are more objective for the visual assessment of trends and curve fits. If our goal is to provide a predictive tool, it is difficult to justify the sophistication of the polynomial over the simplicity of linear regression when we compare them against the spurious background variation of the scatter plot (Fig. 6.4b).

Fitting polynomials may be approached more methodically by comparing the fits achieved with different orders of polynomial. One might start with linear regression (coefficients a_2 to a_n, set to zero) and then repeat the regression for a_3 to a_n, set to zero, etc. In this way the quality of the regression, judged by the coefficient of determination (R^2) may be used to assess the correlation. For example, if the coefficient of determination is highest when a_3 is non-zero, but reduced for a_2 and a_4, we may assume that a cubic (third-order) polynomial is a good fit.

Polynomials higher than sixth order are rarely used for the description or prediction of geological phenomena that relate observations, unless we are evaluating a single variable (y) against time (t); such *time series* are discussed in Chapter 8. Generally, interrelationships between common variables such as temperature and composition of a mineral, particle mass and stream velocity, pore-fluid pressure and slope stability are related by polynomials of less than sixth order and commonly only third order (cubic) or less. This is even true of time series where y undergoes few cyclic changes, as in the case of continental latitude change discussed in the paleomagnetic example of Fig. 6.6. Even if the relationship truly follows a high-order polynomial, the scatter in most earth science data precludes a convincing verification.

The *spline* also attempts to use much detail from the data but, unlike the polynomial, its number of turning points are not so restricted. It takes its name from a simple, flexible mechanical drafting tool, usually some sort of chain encased in plastic which a draftsman could flex to pass as smoothly as possible through the points. The mathematical spline is a series of polynomial functions, usually cubic, fitted to sections of the data and then usually knotted to join at points of equal gradient. A spline and polynomial are fitted to the same data in Fig. 6.5b, c. The spline curve can take any arbitrary curvatures, minimising unnecessary oscillations and producing short arcs. It is therefore valuable as a description but has no value as a model for causal relationships. The spline is rarely used in earth science. Apart from the linear relationship, which is included here under "curve" fitting for completeness, the most commonly used fits are polynomial, power law, logarithmic and exponential (Table 6.2).

The polynomial may easily be overextended as shown by the paleomagnetic example of Fig. 6.6a. Paleomagnetic analysis of rocks of ages, from 550 to 40 Ma, at sites in Oklahoma reveal a variety of inclinations of the ancient geomagnetic field. The ancient inclination I is related to the paleolatitude λ, by $tan (I) = 2 tan (\lambda)$. Paleolatitudes vary considerably over almost 550 Ma, but there are only eight data points, a reasonable situation in paleomagnetism as suitable data are scarce and data-rejection criteria are stringent. The data clearly support some plate motion because paleolatitudes vary from 30°S to 40°N. Moreover, continents must move in some gradual fashion, so it must be possible to connect the data of Fig. 6.6a in a smooth curve to reveal the actual North-South motion. The published source of this data shows a curve that is more complicated than those presented here, with some sharp turning points that are highly improbable on tectonic grounds (Fig. 6.6b). It even implies a brief northward motion at about 450 Ma ago, unsupported by any data, perhaps due to fitting a spline. The simpler possibilities attempted here show polynomials ranging in degree from linear to sixth order (Fig. 6.6a). The coefficients for the equations and the success with which they explain the data scatter ($R^2 = 1$ for a perfect explanation) are given in Table 6.3. The original published interpretation for the Oklahoma paleomagnetic data shows a contorted curve, passing through every data point with sharp turning points. This is unlikely, firstly because the data are subject to errors in paleo-inclination so that the curve describing the actual motion would be unlikely to pass precisely through the data. Secondly, continents cannot undergo jerky movements. Selecting an appropriate curve must balance an appreciation of possible errors, mainly in inclination but possibly also in age, with the limitations on erratic motion caused by the mass and inertia of the tectonic plates. For the Oklahoma data, the sixth-order polynomial suggests at least four changes in latitude (longitudinal changes pass undetected in paleomagnetism).

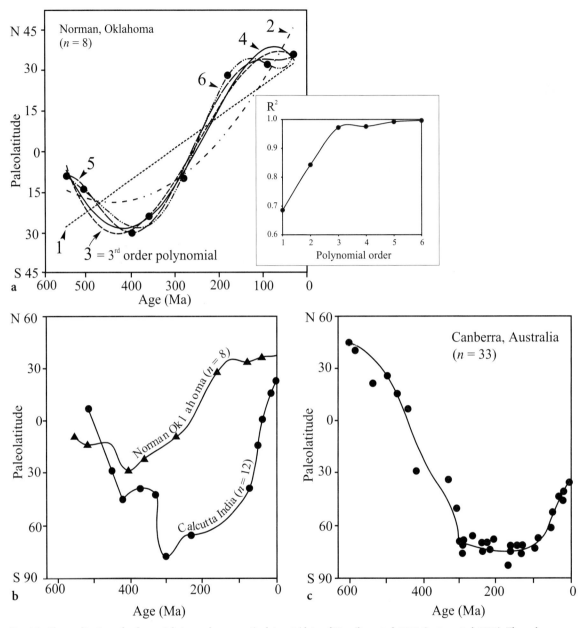

Fig. 6.6. The application of polynomials to a paleomagnetic data set (data of Hamilton et al. 1979; Scotese et al. 1979). The paleomagnetic vector can fix the paleolatitude at which a rock formed (see Chap. 4 or 10). Consequently, changes in paleolatitude recorded in rocks at a site may reveal their plate-tectonic motion. **a** Eight sites in Oklahoma yield discrete paleolatitudes varying from 30°S to 35°N. The linear fit (*1*) or first-order polynomial is clearly too smooth. It ignores all the important variations in plate motion. The second-order polynomial (*2*) misleadingly suggests slight southward motion followed by steady northward movement. The third-order polynomial (*3*) better approximates trend and retains the main features. Moreover, some deviations from the curve are expected due to observational errors. **b** Published curves of paleomagnetic data for Norman (Oklahoma) and Calcutta (India). The curve cited for Oklahoma has maxima at 450 and 150 Ma that shows bends unsupported by data points. The curve for India passes precisely through all points, implying there was no consideration for inherent uncertainty in those data. **c** Some Australian data appear to comprise two data sets joined by separate curve fits that are disjunctive at ∼310 Ma

Table 6.3. Example of polynomial coefficients for Oklahoma paleolatitude data ($n = 8$). $y = a_0 + a_1x + a_2x^2 + a_3x^3 + a_4x^4 + a_5x^5 + a_6x^6$ (y = paleolatitude, x = time)

Order	a_0	a_1	a_2	a_3	a_4	a_5	a_6	R^2
1 (= linear)	39.523	−0.116						0.685
2 (= quadratic)	56.065	−0.337	0					0.843
3 (= cubic)	30.071	0.225	0	3×10^{-6}				0.972
4	23.99	0.426	0	7×10^{-6}	-4×10^{-9}			0.976
5	41.79	−0.335	0.01	-3×10^{-5}	7×10^{-8}	-5×10^{-11}		0.992
6	59.27	−1.246	0.019	-1×10^{-4}	4×10^{-7}	-5×10^{-10}	2×10^{-13}	0.996

The selection of a sixth-order polynomial is unreasonable, given only eight data points. At the other extreme, a first-order (linear) or second-order polynomial indicates gradual northward motion, ignoring well-documented low- and high-latitude locations at 400 and 100 Ma respectively. A reasonable compromise would be a third-order polynomial that favours southerly drift to 30°S at 400 Ma, northward to 40°N at 100 Ma with the possibility of slight southward drift thereafter. Clearly, the most reasonable graphical explanation should account also for geological reality and measurement errors. It need not necessarily provide the best mathematical fit; Table 6.3 shows that $R^2 = 0.972$ for the third-order polynomial. Subsequently we shall show that $R^2 = 1$ for a perfect mathematical fit of the curve to the data. A published curve that claims to describe the motion of the rocks near Calcutta, India, was also forced to pass through each data point, suggesting the original authors may have used a spline function (Fig. 6.6 b). Apart from measurement uncertainty in the data, it is highly unlikely that there were abrupt changes from southward to northward motion, and vice versa, at 430 and 300 Ma respectively. These are probably due to an unwise choice of curve-fitting function. The data for rocks near Canberra, Australia, are also interesting (Fig. 6.6 c). The curve does not pass through all points, so that we may assume some sort of best-fit curve was selected, but probably in two sections according to the very different frequency with which data is found at various ages. Before 300 Ma, there were only nine observations, approximately one per 30 Ma. The published curve is an agreeable, sensible approximation to the results, given the errors of determination of paleolatitude and age. After 300 Ma, there are 24 data points, almost one per 10 Ma. The best-fit curve is also agreeable and has a smooth form, which would be appropriate for the description of changes in latitude of a tectonic plate. However, where the two curves join, there is an inflexion point, perhaps due to fitting separate functions. The inflexion point at 310 Ma could be an artefact of data processing; such a sharp change in direction of plate motion is unlikely,

and a smooth curve linking the two segments may better represent actual plate motion. Of course, tectonics may cause sudden changes in plate motion, following plate collision. However, that involves massive crustal deformation, perhaps creating mountain belts over tens of millions of years. They also produce more significant changes in paleomagnetic records.

Examples of curve fits to bivariate data are quite common in earth science and are illustrated here by the relationship between magnetic susceptibility and the wt % Fe in igneous rocks. Formerly, it was commonly regarded that the magnetic susceptibility of a rock was directly and simply related to the Fe content, sometimes with the implication that the contribution was exclusively from iron in magnetite. In fact, all the minerals contribute to susceptibility with ferromagnetic, paramagnetic or diamagnetic responses. The ferromagnetic response of high-susceptibility magnetite would be significant but, due to the low concentration of such accessory minerals, rock-forming matrix minerals may contribute more to the bulk susceptibility from their paramagnetic response. Fe content is not the sole control; it depends on the way it is incorporated into a mineral lattice, and other transition metals (e.g. Mn, Cr, Ti) also make strong contributions to susceptibility. It is therefore encouraging that Fe content so clearly relates to susceptibility with relatively little scatter (Fig. 6.7 a). An initial reasonable approximation to the raw data has an exponential form. However, it does not explain the variation as well as other models, and visual comparison is difficult (Fig. 6.7 a). Other reasonable possible models include power law, polynomial, and logarithmic (Fig. 6.7 b–d).

Inspection shows that exponential and logarithmic fits are inadequate; R^2 explains only 75 and 67 % of the variation in susceptibility (Fig. 6.7 a, d). The power-law fit seems satisfactory, as it explains 84 % of the variation ($R^2 = 0.84$, Fig. 6.7 b). However, simply plotting the data with logarithmic axes and performing regression on the log-log plot achieves the same result more easily (Fig. 6.7 e). This has the added advantage that the low precision of the low Fe-content data is revealed

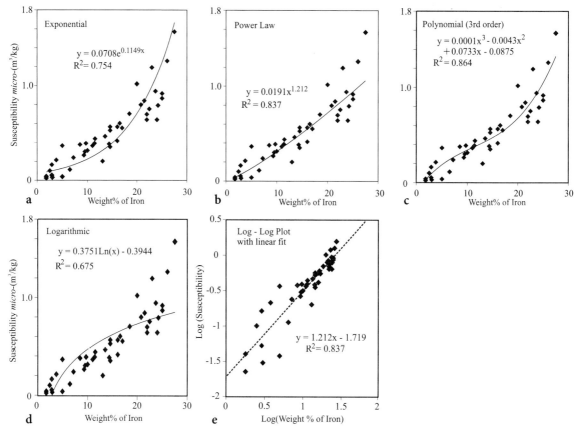

Fig. 6.7. a–d Different curves fitted to the same data in an attempt to characterise the variation of magnetic susceptibility with iron content of some rocks. Of the different fits, a polynomial (third-order) in **c** gives the highest correlation with the data shown by the highest value of the coefficient of determination (R^2). $R^2 = 0.864$ means that the regression line explains 86.4% of the variance in susceptibility. **e** Plotting the original data on logarithmic axes is a simple method that linearises the data and helps to detect the strength of the relationship between susceptibility and Fe content without proposing a mathematical model

from its wider scatter. Lower susceptibilities have larger proportional measurement errors and also represent specimens in which there is competition from multiple mineralogical sources. Further analysis might suggest omitting the scattered low-Fe% data from Fig. 6.7e to refine the mathematical description of the data set, or breaking it into two subsets of different degrees of scatter. The third-order polynomial provides a marginally better approximation, explaining 86% of the variation (Fig. 6.7c).

The advantages of linearisation with a log-log plot seem to increase with the complexity of the topic. In earth science, the response variable may be influenced by many secondary variables as well as the independent control variable considered in linear regression. Scatter is commonly due more to this effect than to errors of measurement. Regression is still our most useful predictor, even with small sample sizes if those

measurements have relatively high intrinsic precision, as with some laboratory experiments and chemical analyses. An archeomagnetic example illustrates this (Fig. 6.8). This is not the conventional archeomagnetic dating that matches precise orientations of the past geomagnetic field with a paleosecular calibration curve (e.g. Chap. 8, Figs. 8.13, 8.14). Instead, the method used the acquisition of viscous remanent magnetization by masonry since its installation. Upon installation, a new piece of masonry may acquire magnetism parallel to the ambient field, overprinting the intrinsic paleomagnetic record acquired through geological time. The *unblocking temperature* required to erase the viscous magnetism is clearly related to the acquisition period, although the actual causes are many and have differing influence, including grain size, shape and domain structure of magnetite. Progressively higher unblocking temperatures are needed to erase the viscous rema-

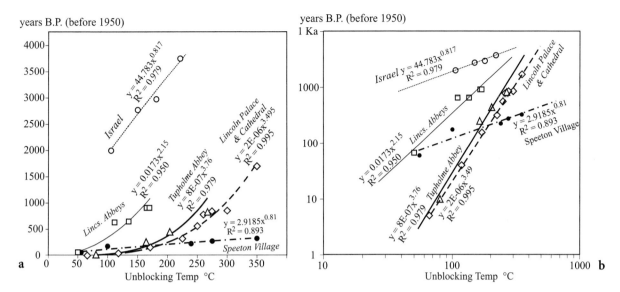

Fig. 6.8. Archeological ages of stone monuments determined from the unblocking temperatures of viscous remanent magnetic components acquired since the masonry was installed (Borradaile 1996b; Borradaile and Brann 1997; and work in progress). **a** On a graph with linear-scale axes, it is difficult to characterise the data using power-law curves. Although the empirical curves are significant at the 95% level, it is difficult for archeologists to use them for interpolation. **b** Simply plotting the data on a graph with logarithmic axes makes it easier to predict and interpolate ages of enigmatic masonry

nence of older masonry. The measurements appear to lie on power-law curves that do not readily lend themselves to extrapolation or interpolation (Fig. 6.8a). Linear regression is our main tool for prediction, so it is logical to linearise the plot with logarithmic scales; the regression lines then provide a magnetic "clock" that archeologists may employ to date enigmatic monuments of the same building stone (Fig. 6.8b). Moreover, the distinctly different slopes of the regression lines indicate different sensitivities of different rocks for age determinations in different parts of the archeological time scale. Subsequently, we will learn that the regression lines are significant statistically, despite the few observations (Chap. 7).

Two further examples show other advantages of linearisation from the use of logarithmic plots. Earthquakes show an enormous variation in their energy output. The common measure of earthquake intensity, the Richter Scale, is already logarithmic for that reason, using base-10 logarithms. Thus, each integer increment on the Richter Scale represents a ten-fold increase in ground-motion amplitude. A study in California shows a fairly common distribution of earthquake intensities: many smaller earthquakes and microseisms, but scarce powerful events. From the scatter plot of raw data in Fig. 6.9a, one may conjecture that the distribution may be log-Normal or it could be part of a Poisson distribution (e.g. see Chap. 3). By applying a logarithmic scale

to the frequency axis (n), the data are somewhat linearised, excluding the counts for the weakest events (Fig. 6.9b). A possible conclusion is that the distribution of Fig. 6.9a was Poisson, but that many weak events go undetected and are under-represented in the frequency distribution.

The distinction between the seismic signature of earthquakes and those of remote nuclear tests is of strategic importance and newsworthy. Nuclear detonations are rated in terms of the equivalent energy release of a conventional explosion, rated in thousands of tonnes of TNT. Nuclear tests are readily detectable by the seismic station network designed to record earthquakes. The scatter plot uses a logarithmic scale for nuclear yield; the P-magnitudes for earthquakes are already logarithmic. Regression provides a satisfactory line on the log-log plot, representing an underlying power-law relationship (Fig. 6.9c). One important difference between earthquakes and nuclear explosions is the way in which the energy release is partitioned between surface waves and body (P) waves. The log-log scatter plot distinguishes the power-law relation for nuclear tests from earthquakes (Fig. 6.9d). Here the linearisation shows that the natural and man-made events produce data along lines of similar slopes but with different intercepts (Mb = 1.76 and 2.36 respectively).

A demanding environmental problem arises from the prediction of flooding, and river-basin manage-

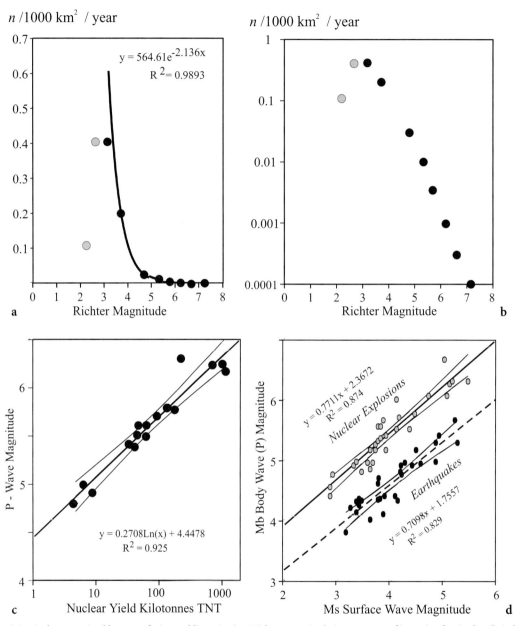

Fig. 6.9. Seismic data examined by curve fitting and linearisation. Richter magnitude is a measure of intensity that is already in $\log_{10}$ units; a scale-3 event is ten times more energetic than a scale-2 event. **a** Frequency distribution of earthquake magnitudes in California. **b** Same data with a logarithmic frequency axis. **c** Nuclear detonations cause seismic P-waves whose magnitudes are linearly related to nuclear yield. **d** Nuclear explosions may be distinguished from earthquakes by the intercept of their linear distributions on a plot of body-wave versus surface-wave magnitudes

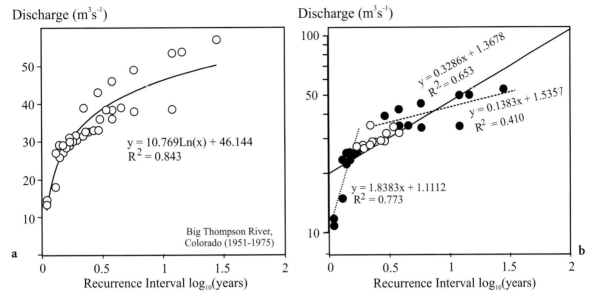

Fig. 6.10. Flood-frequency curve for the Big Thompson River, Colorado, 1951–1975 (data in Montgomery 1979 from USGS Open File 79-681). **a** Channel discharge versus its recurrence interval. The total data set shows an acceptable logarithmic relationship with a coefficient of determination $R^2 = 0.843$, meaning that the curve explains 84.3% of the variation in discharge. **b** Inspection of the raw data reveals heterogeneous subsets, with particular differences between the first and last ten years. Individual linear regression for these subsets may be more representative of natural trends, despite their weaker statistical significance ($R^2 = 0.77$ and 0.41 respectively)

ment in general. In order to recommend suitable land use, to minimise flood hazard or even to determine potential insurance risks in actuarial studies, flood frequencies should be assessed. Clearly, regression of historical records is our best predictive tool. It is worth noting that historical weather patterns cannot be projected forwards, in view of incontestable global warming and increased variability of meteorological patterns. This must moderate any inferences based on massive floods that are very rare and for which there is proportionately less historical data. For example, reviewing past records of stream discharge (or water levels), we may determine how commonly a certain flood level (discharge rate) has been recorded in the past. Usually, this is cited in terms of recurrence intervals, e.g. a flood with a discharge of 45 m³/s has occurred every 10 years on average. This is sometimes referred to as the "10-year flood", and it is considered to have therefore a 10% probability of occurrence in *any* given year, including successive years. Two 10-year floods could even occur in 1 year, but the probability would be 0.1×0.1 or 1%. Data are given for a river basin in Colorado over a 24-year period (Fig. 6.10). Unfortunately, the record covers a short time period. A more reliable record would cover several centuries, and in Europe and China such historical records are available. However, in view of contemporary climate

change, records before 1950 may be of limited predictive value. The raw data are very difficult to interpret, although a moderately satisfactory power-law description can be provided from a spreadsheet program (Fig. 6.10a). The equation in no way implies an underlying physical law that is explained by that mathematical equation. However, the curve might be used to predict discharge expected from a low-probability flood, one with a long recurrence interval. For example, what discharge could be expected of a 100-year flood? Such information would be useful to determine the area of the flood basin that might be submerged, but such events are rare and poorly documented in the record. Therefore, the larger scatter for the recurrence intervals of catastrophic floods (right-side of graph) makes prediction difficult (Fig. 6.10a). Linear regression on a bi-logarithmic plot facilitates prediction (Fig. 6.10b). Furthermore, it clarifies the different degrees of scatter for common and for rare large floods, suggesting the data set might be subdivided, and separate regression lines may be used for prediction in different recurrence interval ranges (Fig. 6.10b). In this example, prediction is all but impossible without a plot linearised by logarithmic scaling. Although the three different linear components have moderately low R^2-values, the lines are significant at the 95% level, from a consideration of the magnitudes of R and *n*. Acceptance at the

95% level requires R > 0.58 for $n = 10$, which is achieved here as shown in Chapter 7, Table 7.3.

6.10
Multiple Regression

Bivariate statistics characterise observations that require two values (x, y) for their description. Multivariate statistics are needed for observations that are defined by three or more values. Physics and statistics students are not confronted by multivariate observations until they are quite competent with bivariate systems amenable to simple linear regression. Unfortunately, from their very first lectures, earth science and life science students are confronted with observations that comprise more than two variables. One cannot even begin to start teaching mineralogy or petrology by simplifying compositions or physical conditions to two variables. Consequently, earth scientists intuitively and successfully manipulate *multivariate observations* into readily visualised problems. Scientists have always been adept at collapsing multivariate situations into two dimensions for ease of calculation. Although modern software removes those obstacles, it is still convenient to visualise relationships in two dimensions.

A convenient illustration is provided by the data of Hutchison (1975) on the relation between cationic %K in lava and the depth from the volcanic vent to the underlying Benioff zone, in Java. Le Maitre (1982) nicely illustrated regression calculations with this example. It has been widely observed that subducted lithosphere yields magma for which the K-content increases with depth. The magma is then vented at different distances from an oceanic trench, its K-content being proportional to the depth of its source. The observation is valuable for detecting subduction direction in ancient terrains. But how good is the correlation between K% and depth? If one assumes that the Benioff zone is perfectly planar and that magma production occurs along a fixed-plane surface, the regression line of K% versus depth should reveal a perfect association. Unfortunately, the geometry is much more loosely constrained, but projecting the data onto a two-dimensional vertical section produces quite a good simple-linear regression coefficient ($R = 0.7$; $R^2 = 0.49$, explaining 49% of vari-

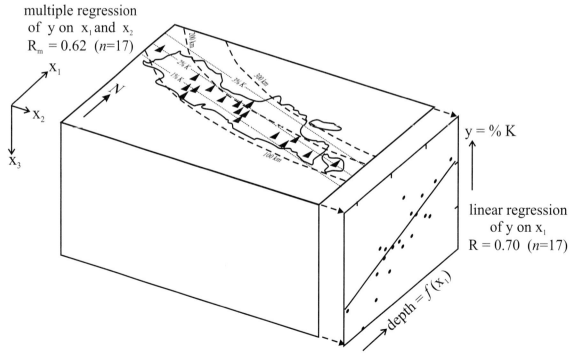

Fig. 6.11. As the depth from the volcanic vent to the underlying Benioff zone increases, so does the potassium content of the lava increase. This example uses data from Hutchison (1975) and a regression analysis by Le Maitre (1982). Projecting the data onto one plane simplifies the three-dimensional regression problem of vent position to a simpler two-dimensional one (depth is really not just a function of x_1 but is influenced by x_2 also). However, the two-dimensional regression appears stronger ($R = 0.70$) than that for multiple-regression ($R_m = 0.62$)

ation). This would be representative of the true relationship if the volcanoes lay along a line perpendicular to the Benioff zone, but their scatter on map coordinates ($x_1 - x_2$ in Fig. 6.11) indicates that the vents probably sample the subduction zone in a less perfect manner. The principal control variable, depth, is a simple function of vent-trench distance (this trend is almost parallel to x_1) on which K% was regressed. When multiple regression is performed, regressing K% upon the coordinates of the vents ($x_1 - x_2$), x_1 is the main control variable and x_2 is a secondary control, somewhat obfuscating the hypothesised two-dimensional process. Both regressions are significant at the 95% level, using the techniques of Chapter 7, but it is important to realise that reducing a multivariate situation to a bivariate one may falsely enhance relationships. Let us consider the multiple regression process in general.

Where three variables are concerned (*y dependent on x_1 and x_2*), a surface may be defined to visualise the regression of *y* on x_1 and x_2, which is a plane in the simplest case (Fig. 6.12a). In a controlled laboratory environment, one would hope to define the control variables (*x*) precisely, perhaps by instrumentation. Thus, *y* would be regressed on the *x*-variables, which are controls. An example might be a chemical petrology experiment that synthesises minerals of differing chemical composition (*y*) in response to control variables: temperature (x_1) and pressure (x_2). More commonly in geology, multiple regression concerns relationships between variables, each of which is subject to considerable scatter, and none may be clearly and uniquely dependent on another. For example, consider the following variables measured from outcrops of sandstone: grain size, grain roundness and the concentration of heavy accessory minerals. In this case, the observations are all independent variables resulting from some depositional process. A multiple-regression surface would permit us to quantify the description of the interrelationships, to interpolate, and to make confidence estimates on predicted values. In this, and most cases, the regression equation describing the surface would not define a physical law or imply a genetic relationship between the variables. Causality is rarely a consequence of regression. However, when we introduce a physical property (e.g. permeability) into a multiple-regression study, it usually is considered the dependent or response variable. On the other hand, the geological observations (e.g. frequency distribution of fractures, variance of grain size or fracture width) would be considered as independent and possibly controlling variables. In multiple regression concerning natural phenomenon, it may be difficult to identify the dominant control variable.

Multiple regression investigates associations between three or more related properties. For the moment, consider just three related variables where *y* depends on x_1 and x_2. Standard statistical procedures assume that x_1 and x_2 are controls, not correlated with one another, free of "error" so that *y* may be regressed simultaneously on x_1 and x_2. In the case of *linear multiple regression*, the response surface is a plane from which the *y*-residuals are solely responsible for the deviations (Fig. 6.12a). Unfortunately, few natural variables show a planar regression surface, and it may be very difficult to re-express the variables to produce a planar regression surface. Although modern graphing programs readily permit the determination of curving regression surfaces, some understanding of the variables may aid linearisation of the function so that a planar regression surface may be defined between *y*, x_1 and x_2 (Fig. 6.12b). Using logarithmic scales is the most commonly successful procedure to achieve linearisation. As with simple linear regression, the variables must act independently and the *y*-residuals must be normally distributed about the regression surface. The major problem in geological examples is that the departure from the best relationship may be attributable to scatter in all variables (*y*, x_1, x_2) and, as in the following examples, one or more of the *x*-variables may not be a control in the strict sense. Even worse, it may be very difficult to isolate x_1 from x_2 as truly independent factors in the system. This is very clear in geomorphology, petrology, geochemistry, sedimentology, palaeoclimatology, etc., where the observed phenomenon (*y*) is influenced by many interwoven variables (x_i) that may show some degree of *co-linearity*. For maximum effectiveness, we should select just two completely uncorrelated controls (x_1 and x_2) and investigate both their influence on the *y*-variation, and the degree to which their residuals are normally distributed about the proposed regression surface. Probably the worst problem is caused by a *lurking variable*, one that is not recorded, but controls or strongly influences the *y*-response either directly or indirectly, via one of the x_i-variables. Although we mentioned these earlier, they are more heavily disguised in multivariate situations.

Multiple regression involving more than three variables cannot be visualised but the results may be tabulated. A classic study in geology investigated the dependence of the stability of beach sand (*y*) on observations of grain size (x_1), moisture content (x_2) and porosity (x_3) (Krumbein and Graybill 1965). Here, we consider an example in the literature that establishes a relationship between the number of water sources supplying a drainage basin and several geomorphological variables (x_i) (Krumbein and Shreve 1970; as discussed

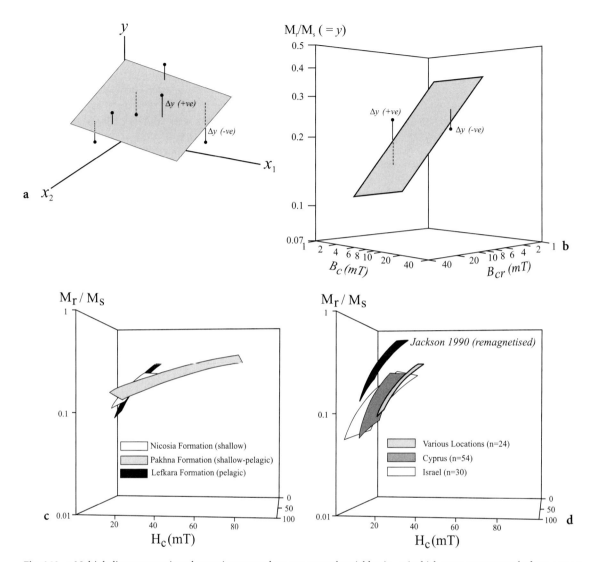

Fig. 6.12. a Multiple linear regression plane; y is regressed on two control variables (x_1, x_2) which must act separately from one another. The combined contributions of all Δy-residuals from the plane are minimised to define the regression plane. **b** Using suitably scaled axes, there is a planar regression surface relating magnetic-hysteresis variables in a certain limestone. **c** Non-planar regression surfaces for limestone hysteresis variables. Different regression surfaces correspond to parts of a limestone sequence that were deposited in different environments (Borradaile and Lagroix 2000). **d** Non-planar regression surfaces for limestone hysteresis showing contrast between remagnetised limestones (Jackson 1990) and primary depositional magnetic characteristics of different environments. (Borradaile and Lagroix 2000)

Table 6.4. Example of multiple regression of six variables influencing number of water sources supplying drainage basin $(n = 92, R^2 = 0.61)$

	Multiple linear regression coefficients						
Coefficient	a_0	a_1	a_2	a_3	a_4	a_5	a_6
Normalized value	−2.24	0.01	0.02	−23.28	6.26	−0.2	−11.66
Associated variable	*Deviations from surface*	Height of basin outlet	Basin relief	Basin area (BA)	Total stream length (Σ sl)	Σ sl/BA	Basin shape
Units (L = length)	Ratio	L	L	L²	L	Ratio	Ratio

by Davis 1973; Table 6.4). The variables include total stream length and basin area, but these cannot be separated entirely; a very wide basin must have long streams if they are to reach the outlet. Similarly, basin relief must influence stream length, since mountainous relief would favour short, straight mountain streams whereas gentler topography favours long, meandering channels. However, this is the essence and challenge of earth science problems. Our observations commonly force us to integrate more than one effect in a single value. Finally, one of the variables is a ratio of two others (total stream length divided by basin area) so it cannot be independent of the variables "stream length" and "basin area".

Multiple regression for studies with more than three variables defies visualisation. Of course, plotting the ratios of some values may permit plotting, but it is generally inadvisable because it confounds the contributions of two controls on one graph axis. The details of calculating multiple-regression surfaces need not concern us here, but we should be prepared to comprehend the results of multiple regression that are readily provided by commercial software and to appreciate the process where more than two control variables are at work. An equation comparable to those used previously for simple linear regression and multiple linear regression will be given for continuity of style (Eq. 6.18), but matrix algebra summarises the procedure more succinctly by rewriting the *normal equations for linear regression* (Eq. 6.15) as follows:

$$\mathbf{X} = \begin{bmatrix} 1 & x_{11} & x_{12} \\ 1 & x_{21} & x_{22} \\ \vdots & & \\ 1 & x_{n1} & x_{n2} \end{bmatrix} \quad \mathbf{y} = \begin{bmatrix} y_1 \\ y_2 \\ \vdots \\ y_n \end{bmatrix} \quad \mathbf{a} = \begin{bmatrix} a_0 \\ a_1 \\ a_2 \end{bmatrix} \quad (6.14)$$

which give matrices for the measurements of the control variables (x_{1i} and x_{2i}), measurements of the response variable (y_i) and estimates of the regression surface coefficients (a_0, a_1 and a_2). The regression coefficients needed for tabulation, from which we might identify the strongest controls, are obtained from:

$$\mathbf{a} = (\mathbf{X}^t \mathbf{X})^{-1} \mathbf{X}^t \mathbf{y} \quad (6.17)$$

where $\mathbf{X}^t$ is the transpose of $\mathbf{X}$.

Now let us return to the drainage-basin example, which considers six possible, quantifiable geomorphological controls (x_i) on the number of water sources for a drainage basin (y). The successful multiple regression yields an equation of the form:

$$y = a_0 + a_1 x_1 + a_2 x_2 + a_3 x_3 + a_4 x_4 + a_5 x_5 + a_6 x_6 \quad (6.18)$$

The coefficients (a_1 to a_6) and a term a_0 represent deviations of y from the regression surface. However, the largest coefficient cannot be used immediately to identify the most effective dependent variable because the controlling variables (x_i) have different units and magnitudes dependent on their nature. In order to compare the coefficients fairly, they are standardised by dividing them by their standard deviations. This renders all coefficients dimensionless and in multiples of their standard deviations. The relative importance of each variable is then indicated by the relative magnitude of its standardised coefficient. Table 6.4 presents the coefficients associated with the six variables influencing the number of water sources feeding drainage basins.

Magnetic hysteresis properties of limestone provide an example of multiple regression and show that regression surfaces are different for limestones of different sedimentary environments. This is interesting to paleomagnetists who wish to know how long magnetic memory is retained and how quickly it was acquired. It is also interesting to sedimentologists who wish to evaluate the biological or climatic controls on biomineralisation. Limestones are of particular interest because the sedimentary environment may have a sensitive control on the magnetic mineralogy and thus on their paleomagnetic potential. Three magnetic parameters are plotted in Fig. 6.12b, although their logarithms are plotted to achieve linearisation of the relationship. In this case, a planar multiple linear-regression surface suffices to predict relationships among the three variables. One might argue that the three hysteresis variables are not entirely independent at the most fundamental crystallographic level. However, the rocks provide variation in source, domain structure and dislocation density, and it is worthwhile to investigate the relationships, treating the variables as if they were independent. In the case of Fig. 6.12b, the regression surface is statistically significant at the 95% confidence level and is represented by a plane. More commonly, the distribution of data in three dimensions is best described by a curving multiple regression surface with linear graph axes (Fig. 6.12c, d), rather than logarithmic axes (Fig. 6.12b). Magnetic hysteresis regression surfaces differ from one sedimentary environment to another. In other words, the potential of limestones for paleomagnetic use is controlled by depositional environment of limestones, especially water depth (Borradaile and Lagroix 2000).

A more complicated natural situation reveals sensible relationships between the variables describing bulk magnetic susceptibility (k_{MEAN}), and the anisotropy of susceptibility that are statistically significant at the 95% level (Fig. 6.13a; Borradaile and Nakamura, in

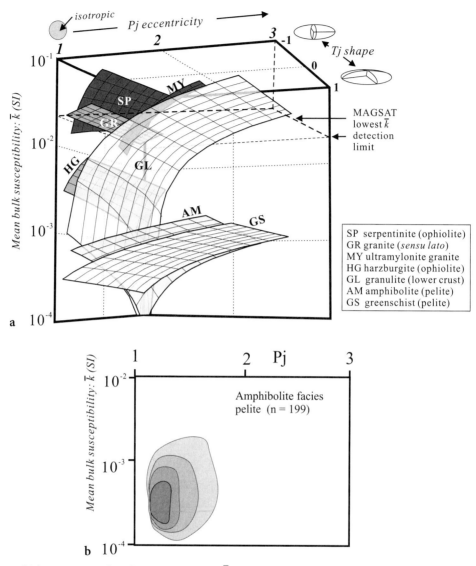

Fig. 6.13. a Multiple regression surfaces for bulk susceptibility $\bar{k}$ and anisotropy of susceptibility may characterise selected metamorphic and plutonic environments (Nakamura and Borradaile, in press). Anisotropy is documented by the eccentricity (Pj) and shape (Tj) of an ellipsoid that represents the magnitudes of the susceptibility tensor (see Chap. 11). **b** Two-dimensional projections of just two variables may fail to reveal any relationship, as shown here for the data shown in **a** for pelites in the amphibolite facies

press). The anisotropy may be represented by an ellipsoid whose eccentricity (intensity) is defined by Pj. (These parameters are defined in Chapter 10, Section 10.11.) Its shape may range from cigar-like to a disc and is recorded by the value of Tj. (Some other anisotropy parameters, such as ratios of principal susceptibilities, are mathematically interdependent and therefore unsuitable for regression or correlation.) The three variables, k_{MEAN}, Pj and Tj, are physically independent. Nevertheless, in different metamorphic or plutonic environments, statistically significant, different multiple-regression surfaces characterise the association of mean susceptibility and anisotropy of susceptibility. Attempts to project the same data onto an x-y plane suitable for simple linear regression fail if the three-dimensional multiple-regression surfaces are strongly curved, obscuring the two-dimensional view due to the scatter of points (e.g. Fig. 6.13b).

CHAPTER 7

Correlation and Comparison of Variables 7

Correlation concerns techniques that are mostly used to assess the association or degree of dependence between the dependent response variable (y) and the independent or control variable (x). This uses the concept of the regression line from the previous chapter, but it is rarely a mathematical model of underlying theoretical relationship or causal association. The value of the correlation coefficient and its significance qualify the strength of the association. This may be illustrated by confidence limits on the slope and intercept of the regression line, and confidence regions about the regression line that will be discussed below. However, a different approach is required where the scatter is influenced by contributions from both x and y, which in the natural sciences is usually due to a lack of direct dependence of y on x. In that case, the *reduced major-axis line* is used to quantify the association; it minimizes the scatter in x and y simultaneously (Fig. 6.1d). Initially, and through most of this chapter, we use the regression line to detect the degree of association between variables.

7.1
Covariance and the Correlation Coefficient

The *centroid* of a data set is located at the mean x and mean y, $\bar{x}, \bar{y}$. For randomly scattered (x, y) points, we may expect equal numbers dispersed above and below the centroid, and to the left and right of the centroid, as in the example of Fig. 7.1a. If the points show any linear association, they may only collect in diagonally opposed quadrants about the centroid. This *association* is reflected in the value of the product of the deviations from their means. When summed, they convey the intensity of the association:

$$\sum_{i=1}^{n}(x_i - \bar{x})(y_i - \bar{y}) \tag{7.1}$$

If the points lie in the lower-left and upper-right quadrants, as in the example of Fig. 7.1b, both terms in the

product are of the same sign and association or correlation would be positive. On the other hand, should the points lie in the upper-left and lower-right quadrants, the terms in the product have opposite signs, and the result is a large negative value for the association. This associative term may be standardised by dividing by the sample size, n to yield the covariance:

$$s_{xy} = \frac{\sum_{i=1}^{n}(x_i - \bar{x})(y_i - \bar{y})}{n-1} \tag{7.2}$$

Clearly, the covariance preserves the sense in which x and y correlate: if the data lie along a line of positive slope, covariance will be positive. If the (x, y) define a line of negative slope, the covariance is negative. To improve the universality of this association, we standardise it further, dividing by the standard deviations of x and y (respectively, s_x and s_y), which yields a dimensionless number between -1 and $+1$. This statistic is *Pearson's product-moment correlation coefficient* (R), devised by Karl Pearson (1857–1936):

$$R = \frac{\text{covariance }(x, y)}{\sqrt{\text{variance }(x) \cdot \text{variance }(y)}} \quad \text{or} \quad R = \frac{s_{xy}}{s_x s_y}$$

and the general form for the population is given by

$$\rho = \frac{\sigma_{xy}}{\sigma_x \sigma_y} \tag{7.3}$$

The value of R is different from the slope of the regression line. However, its sign is the same and its magnitude indicates the degree of correlation, thus:

R = -1 for perfect negative correlation (y decreases with x)

R = 0 for no correlation

R = $+1$ for perfect positive correlation (y increases with x).

Examples of regression lines are shown in Fig. 7.2.

The existence of a correlation between variables is commonly of interest. Therefore, one may set the null hypothesis $\rho = 0$, that no correlation exists in the population. If the sample yields a value for R that is signif-

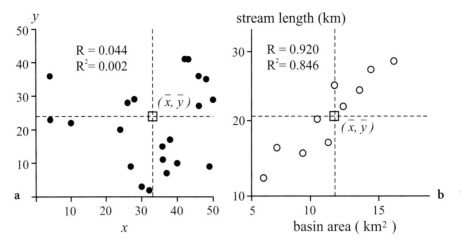

Fig. 7.1. a Covariance concept illustrated with a scattergram. The centroid is the data's centre of mass at $(\bar{x}, \bar{y})$. For this disordered distribution, the data do not cluster in two diametrically opposed quadrants on either side of the centroid. Thus, their covariance is near zero, and subsequently we will see this yields a small value for the correlation coefficient (R). **b** Consider variables between which it is reasonable to suspect a simple systematic relation such as stream length and drainage basin area (Doornkamp and King 1971). The data lie systematically in opposed quadrants on either side of the centroid. Covariance is positive for data lying below-left and above-right of the centroid. Where data lie above-left and below-right of the centroid, covariance is negative, yielding a negative correlation coefficient. This yields a regression coefficient of positive sign, R = + 0.92. It will be shown that this correlation is significant at the 95% level; for this sample size any correlation with $|R| > 0.636$ would be significant at the 95% level (e.g. Table 7.3, Fig. 7.5b)

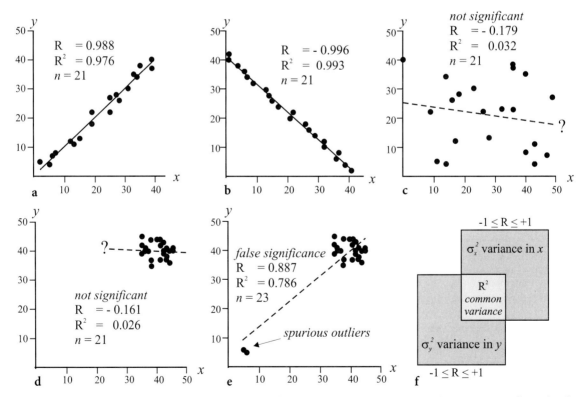

Fig. 7.2 a–f. Examples of regression lines and correlation coefficients, in which y is regressed on x. The regression coefficient has the range $(-1 \leq R \leq +1)$; its sign gives the sense of slope of the regression line. The significance of R at the 95% level may be assessed from Table 7.3 and Fig. 7.5, knowing the number of observations, n. The coefficient of determination, R^2, gives the fraction of y-variance explained by the regression line. **a** Strong positive correlation significant at the 95% level. **b** Strong negative correlation significant at the 95% level. **c, d** Insignificant correlations; R is not significantly different from zero at the 95% level. **e** Two spurious outliers added to the data in **d** yield a misleading positive correlation. **f** Conceptual model attempting to show that the coefficient of determination R^2 represents the variance in common between x and y, i.e. the degree to which the variance in x explains the variance in y

icantly different from zero, one may accept the alternate hypothesis that an acceptable correlation exists at the chosen confidence level. A test statistic that follows the t-distribution will be calculated from R and n to permit this. Examples will be given using Table 7.3 and Fig. 7.5 to decide whether a correlation is acceptable or not using this principle, for the 95% confidence level ($\alpha = 0.05$).

Other approaches exist to detect association or correlation in a broader or more qualitative sense. For example, an alternative expression for the correlation coefficient based on Normal scores (z-variate values) was mentioned briefly in Chapter 6. Also comparisons of data in two-way contingency tables may detect association between two variables even though the variables' relation may be very complex (Chap. 5, Table 5.5). A more rigorous detection of complex or poorly linear associations use a generalisation of the Pearson correlation coefficient. Therefore, we will next discuss *Spearman's rank correlation coefficient* (R_S) that may quantify association for noisy or even nonlinear x–y relationships.

7.2
Spearman's Rank Correlation Coefficient

Although Pearson's correlation coefficient (R) quantifies the strength of an identifiable linear relationship, it is very sensitive to outliers and it cannot be used directly to detect a nonlinear relationship. Of course, one may re-express the original data using a linearisation transformation (Chap. 6) and then analyse the transformed data. However, that may fail if the sample size is too small to establish a nonlinear relationship in the first place.

An alternative direct approach uses *Spearman's rank correlation coefficient* (R_S), devised by Charles Spearman (1863–1945). Its main use is to reconnoitre associations in complex nonlinear data sets and, upon reflection, it therefore also may be used with non-numerical data to which an ordinal or rank may be assigned (1, 2, 3...). For example, rock types may be designated with an arbitrary number code to detect correlation between stratigraphic sequences (e.g. Chap. 8, Fig. 8.1 b, c).

For numerical data, the actual values of x and y are replaced by their ranks; the smallest value of x has rank 1, next-larger has rank 2, etc. Similarly, ranks are assigned to y. Table 7.1 shows an example of rank assignments for five observations. Where two or more x values have the same rank, their average y rank is recorded.

Table 7.1. Example of ranking of x, y for Spearman's rank correlation

i	x_i	xr_i rank of x_i	y_i	yr_i rank of y_i	$d_i = xr_i - yr_i$
1	4.3	1	4.1	2	–1
2	5.7	2	3.8	1	1
3	6.2	3	5.9	3	0
4	8.1	4	8.6	5	–1
5	9.3	5	8.5	4	1

Rank correlation requires the differences in rank of the two variables, x and y, recorded as d_i. The definition of Spearman's rank correlation coefficient is then:

$$R_S = 1 - \frac{6\sum_{i=1}^{n} d_i^2}{n(n^2-1)} \tag{7.4}$$

For example, consider the example of the suspected relationship between fluid waste disposal in deep wells and the minor earthquakes apparently triggered by an effective stress-effect (Fig. 5.9 a). Visual inspection leads us to suspect a relationship but linear regression yields a correlation coefficient R = 0.59, which is not very encouraging. Since the data set is noisy and we do not know the form of the actual distribution, it is appropriate to use Spearman's rank correlation coefficient to detect the presence of any association, even nonlinear. Spearman's correlation coefficient has the value $R_S = 0.87$, encouraging our suspicion that an association exists. The success of this coefficient is essentially due to its insensitivity to the individual contribution of outliers.

7.3
Coefficient of Determination, R^2

It is useful to formulate an expression which can explain the success of the regression in a single value. Large or nonrandom y residuals ($y_{OBSERVED} - y_{PREDICTED}$) would indicate a weak correlation. Logically and intuitively, the fraction of the variance of y explained by the proposed linear $-y$ relation seems an appropriate measure of the success of a correlation (e.g. Fig. 7.2 f). This statistic is termed the *coefficient of determination*. We will need to understand how the y_i values differ from their mean, and also from each associated predicted value $\hat{y}$. Thus, we formulate terms for:

(SS_{yy}) = sums of the squares of deviations from mean-y

(*SSE*) = sums of the squares of deviations from predicted value $\hat{y}$ ("errors").

Now the coefficient of determination, or sample variability that is explained by the regression, R^2 is:

$$\frac{\text{variability explained by regression line}}{\text{total sample variability}} = \frac{SS_{yy} - SSE}{SS_{yy}}$$

(7.5)

This value ranges from zero to unity and, in the case of *linear* regression, it is given directly by the square of the correlation coefficient, R^2. Simple graphic examples showing different coefficients of determination and their relationship to R may be viewed in Fig. 7.2. Recall that since $|R| \leq 1$, the variance-in-common must be a fraction of the x-variance and the y-variance (Fig. 7.2f).

For the association between fluid-waste disposal and microseisms, the correlation coefficient R = + 0.59. However, the coefficient of determination, $R^2 = 0.35$, indicates that only 35% of the y-variance is explained by the regression line (Fig. 5.9a). An example of a strong linear dependence is shown by stream lengths on their catchment areas (Fig. 7.1b). The strong positive correlation, R = + 0.92 confirms that bigger drainage basins have longer streams. However, the degree to which the regression line explains the variance in stream length is given by $R^2 = 0.846$. In other words, 84.6% of the variance in stream length is attributable to the linear relationship with basin area.

The classic regression line of y on x is sensitive only to the y-residuals (Fig. 6.1b, d). Uneven data distribution, probably beyond our control, may therefore affect the outcome of correlation. Regression is usually facilitated if the independent variable (x) has evenly spaced values. However, there may be good reasons to discard certain extreme points that lie far from the regression line. Such *outliers* may simply be rejected because of an improbable statistical value, but it is more satisfactory to find a scientific reason for the rejection of an outlier. Outliers may then be excluded and the correlation repeated to yield a more sensible result.

A more rigorous assessment of outliers can be adopted by considering the distribution of y-residuals. We expect the y-residuals to be normally distributed, so that their standard deviation (s_R) may be used to establish confidence limits for the isolation of outliers (Table 7.2). In a laboratory environment, it may be wise to discard results whose residuals lie more than $3s_R$ from the regression line and then recalculate the regression line. In field studies, where noisier observations must be tolerated, a two-sigma or one-sigma rejection level may be more appropriate.

Table 7.2. Ranges of normally distributed y-residuals from regression line

Lower limit of y-residual	Percentage of residuals	Upper limit of y-residual
$-3s_R$	99.7	$+3s_R$
$-2s_R$	95.4	$+2s_R$
$-s_R$	68.3	$+s_R$

Where the dependent variable is influenced by two or more dependent variables simultaneously (e.g. y controlled by x_1 and x_2), we speak of multiple regression, and a surface defines the relationship. Such surfaces are rarely planes for natural data; some examples were given in Chapter 6 (Figs. 6.12, 6.13). Multiple regression is a complex procedure readily performed by commercial software packages and described briefly at the end of this Chapter. It should not be confused with much more complex multivariate statistical situations in which each of several variables mutually interact.

Scatter-plots that show differing degrees of correlation are shown in Fig. 7.2; they will remind us also that an inspection of the raw data can save some wasted effort. Strong positive and negative correlations are shown in Fig. 7.2a, b, in which 97.6 and 99.3% of the y-variance is explained by the respective regression lines. The absence of meaningful correlation in Fig. 7.2c is obvious, but for the purpose of demonstration, a regression line is shown, which accounts for <3% of the variance in y. Preliminary, visual inspection of the data set can avoid the misinterpretation possible in Fig. 7.2d, e. In Fig. 7.2d, an isotropic cluster of points has no correlation and its regression line only accounts for 2.5% of the y-variance. However, if the same data includes a few spurious outliers (Fig. 7.2e), the illusion of an apparently strong positive correlation appears. Regression now apparently accounts for 78.6% of the y-variance. Obviously, it would be wise to investigate the observations that give the outlying (x, y) values. Despite their rarity, *outliers* bias the distribution because they are far from the data mass. They could be valid observations on scientific grounds, but if the sampling was unbiased and statistically valid, it is unreasonable to include them in the regression.

Outliers should always be considered with caution. The regression of geophysical data in Fig. 7.3a initially used all the points, including one in the upper right that might be considered an outlier on statistical grounds. Removing it from consideration, the regression is recalculated, yielding a steeper line, with reduced correlation coefficient and significance (Fig. 7.3b). If the observations were independent and the sampling free from bias, as demanded by statistical

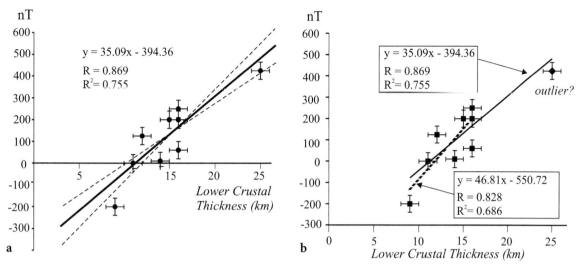

Fig. 7.3. Over long distances (~ 100 km), regional magnetic anomalies may be related to crustal thickness, as in this example from Manitoba (Hall 1974). **a** Regression line with 95% confidence limits, for all data ($R^2 = 0.755$). **b** Dashed regression line is appropriate if the upper-right data point is rejected as an outlier, although the regression coefficient is reduced ($R^2 = 0.686$). However, the outlying point should not be neglected without independent justification, especially since each point summarises a vast amount of geophysical data from a large area. Both correlations are significant (i.e. R is different from zero) at the 95% level (see Fig. 7.5b; Table 7.3)

theory, we may feel obliged to discard the "outlier". However, each data point corresponds to thousands of hours of work, combining extremely elaborate and expensive seismic surveys to estimate crustal thickness (x), and hundreds of hours of flying time to collect aeromagnetic data (y). It is unlikely that the outlying point is "wrong". We must accept that it is in the nature of earth sciences to impose obstacles to sampling, especially spatial sampling (Chap. 1). The sample is small, not ideally random in the statistical sense, but there is no independent geological reason to reject any point. Moreover, they have similarly small error-bars, so that they are all at least consistently precise. This simple example is merely illustrative; a deeper appreciation of the subject matter would lead one to suspect that any relationship would be more complex than the simple linear case.

7.4
Significance of Correlation (R)

How do we assess the significance of the correlation? Usually, we wish to prove the existence of some correlation. Therefore, we establish a null hypothesis that, upon rejection, may support our contention (Chap. 4). We set the null hypothesis that the population's correlation coefficient $\varrho = 0$, i.e. H_0: $\varrho = 0$. If some test statistic takes an extreme value beyond the range of 95%

of its possible values, we will reject H_0 and we may therefore accept the alternate hypothesis that a correlation exists with 95% confidence. Note that in this application, due to the formulation of H_0, we prefer to find a test statistic with an extreme value in the tail of its distribution.

To evaluate the significance of R, we use a test statistic that follows the t-distribution (Table 4.5). A confidence level of 95% ($\alpha = 0.05$) is mostly used in earth science, but the following definition of the statistic is generalised for any significance, corresponding to a confidence level of $100(1 - \alpha)\%$. The following expression compares the test statistic with a tabled t-value that we choose based on our desired confidence level, and on the degrees of freedom designated as ($n - 2$) in this test.

$$\left| \frac{R\sqrt{(n-2)}}{\sqrt{(1-R^2)}} \right| \ge t_{\alpha/2} \qquad (7.6)$$

Note that the use of the t-statistic for the correlation coefficient requires ($n - 2$) degrees of freedom. Also, a two-tailed test is generally appropriate because we wish to detect any significant departure of R from zero, extremely low or extremely high. The test statistic calculated from the data must exceed the tabled t-value if R is significant (Table 4.5). Minimum t-values for various sample sizes that support the alternate hypothesis

Table 7.3. If $|R|$ exceeds the tabled value, reject H_0: that population correlation coefficient = 0 (for the 95% confidence level or $\alpha = 0.05$)

Sample size n	$|R|$ Pearson's correlation coefficient ($= n - 2$)	$|R_s|$ Spearman's rank-correlation coefficient
5	0.878	1.000
6	0.811	0.886
7	0.754	0.786
8	0.707	0.715
9	0.666	0.700
10	0.632	0.649
11	0.602	0.619
12	0.576	0.588
13	0.553	0.561
14	0.532	0.539
15	0.514	0.522
16	0.497	0.503
17	0.482	0.488
18	0.468	0.474
19	0.456	0.460
20	0.444	0.447
25	0.396	0.399
30	0.361	0.363
40	0.308	
50	0.273	
60	0.251	
70	0.232	
80	0.217	
90	0.205	
100	0.195	

The F-statistic is defined in this case as:

$$F = \frac{\text{variability explained by regression/(d.f.)}}{\text{Summed Squares of Errors/(d.f.)}}$$

$$F = \frac{(SS_{yy} - SSE)/k}{SSE/(n-k-1)} \tag{7.7}$$

$$F = \frac{R^2/k}{(1-R^2)/(n-k-1)}$$

From this equation, F is calculated and may then be compared with a tabled critical value for F (see Table 4.6) at a suitable confidence level (usually 95%, i.e. significance $\alpha = 0.05$) for the respective degrees of freedom of numerator and denominator, k and $[n - k - 1]$ respectively. Usually, we are interested in detecting a significant coefficient of determination so the null hypothesis is set that it is zero in the population; H_0: $\varrho^2 = 0$. If the F-value calculated from the sample's R^2 exceeds the critical tabled value, we must reject the null hypothesis and therefore accept the alternate hypothesis that the coefficient of determination is significant at that level. Regression software performs this test routinely, also for the more involved case of multiple-regression involving more than two variables.

that a correlation exists with a 95% confidence level are illustrated in Fig. 7.5a. The same test may be used with Spearman's rank-correlation coefficient. Note, however, that where the uncertainty is not just in y, but in both x and y, each measurement is fully bivariate and a different test is required for the significance of the regression coefficient. This is discussed later in the section concerning the *bivariate Normal distribution*.

Rapid consultation tables and nomograms are presented that give the minimum value of R for a given sample-size, n, for a correlation to be significant at the 95% level. These are presented for Pearson's correlation coefficient as well as for Spearman's rank-correlation coefficient (Table 7.3; Fig. 7.5b, c).

We may test also the significance of the coefficient of determination, R^2, the statistic that explains the regression's success in explaining the variance in y. R^2 was defined in Eq. (7.5) as the ratio of terms that are sums of squares of deviations. This is an appropriate situation in which to use the F-statistic. It requires two degrees of freedom, the number of independent variables ($k = 1$ in the case of simple linear regression) and a term combining k with sample size as $(n - k - 1)$.

7.5
Confidence in the Regression Line

Where the regression line is used to investigate the correlation between two variables, it may be especially useful to determine the degree of confidence for the regression line. The regression line was discussed in the previous chapter and compared with other lines of "best fit", each of which is appropriate for a different purpose. For the most commonly used regression line, regressing y on x, the confidence limits on its slope (m) and intercept (c) are given as Δm and Δc respectively. Most often we use the 95% confidence limits that correspond to a significance level $\alpha = 0.05$ but the following equations are generalised. The t-distribution must be consulted for ($n - 2$) degrees of freedom at the appropriate confidence level defined by $100(1 - \alpha)$% using a two-tailed test (Table 4.5). The equations require the use of the *residual variance* (s_R^2) discussed in the previous chapter and defined by:

$$(s_R)^2 = \frac{1}{(n-2)} \sum_{i=1}^{n} [y_i - (mx_i + c)]^2 \tag{7.8}$$

Confidence limits on the slope are given by:

$$\Delta m = \pm t_{\alpha/2} \left[\frac{s_R^2}{\left(\dfrac{1}{n-1}\right)\sum(x-\bar{x})^2} \right]^{\frac{1}{2}} \qquad (7.9)$$

and for the intercept by:

$$\Delta c = \pm t_{\alpha/2} \left[\frac{s_R^2 \sum x^2}{\dfrac{n}{n-1}\sum(x-\bar{x})^2} \right]^{\frac{1}{2}} \qquad (7.10)$$

A range of reasonable regression lines, defined by maximum and minimum slopes, is shown in Fig. 7.3 a. The mean intensities of the geomagnetic field (y) above sections of the Canadian Shield are plotted against the thickness of the lower crust (x). The thicknesses are known from seismic profiles and are believed to be the control variable on which the magnetic anomalies depend. It is important to know what confidence we may place on the slope and intercept of a regression line if we plan to use it for predictions. Therefore, the range of slopes (m) and intercepts (c) that may be tolerated at the 95% level should be reviewed. Using the preceding equations, we can draw the steepest and shallowest possible regression lines on either side of the dashed regression line (Fig. 7.3 a). There is a one in twenty chance that the regression line could actually have a slope lying outside the limits shown.

However, confidence in the regression line may be presented in a more sophisticated manner that draws a confidence band encompassed by a convex curve above the regression line and by a concave curve below it. As explained in Chapter 6, the two curves are defined by the following terms:

$$y \pm t_{\alpha/2} s_R \left[\frac{1}{n} + \frac{(x-\bar{x})^2}{\sum(x_i-\bar{x})^2} \right]^{\frac{1}{2}} \qquad (7.11)$$

This expression may be used to locate the y-coordinates of the confidence envelope for a given x-value. The band is narrowest near the centre of the data cluster, with the smallest vertical separation at $\bar{x}$. An example of a confidence band is shown in Fig. 7.4.

The preceding expressions allow us to quantify the importance of a correlation statistically. The interpretation is more difficult and requires the examination of samples of different sizes and different degrees of or-

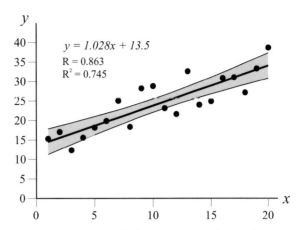

Fig. 7.4. Confidence bands for the regression line at the 95% level; this indicates acceptable variation of the regression line due to the scatter in y. The regression line passes through the mean values of x and y, where the confidence band is most narrow

dering. The importance of the association is not entirely conveyed by the magnitude of R even if we assume the sample is taken randomly, and there are no outliers or unusual clusters. Moreover, smaller samples require a stronger correlation, with R closer to ± 1, to be significant. Quite low values of R may be significant for large samples ($n > 100$). Correlation is not automatically significant just because $R > +0.5$ or $R < -0.5$. With a sensible data distribution devoid of clusters and outliers, scattergrams with $R > 0.8$ or $R < -0.8$ usually reveal some meaningful correlation. Of course, negative R-values do not signify an absence of correlation, they merely imply that y decreases as x increases. Table 7.3 and Fig. 7.5 b show minimum absolute values of Pearson's correlation coefficient (R) that are significant for certain sample-sizes at the 95% confidence level. The interpretation of R will be illustrated with some examples below.

Before we proceed further, we should note that different branches of science tend to have different tolerance levels for what is considered as an acceptable correlation. In laboratory physical sciences, such as physics, observations may be made with great accuracy and precision. The use of a higher confidence level translates into a higher minimum acceptable value for R. In earth sciences, loose underlying relationships between variables are confounded with observational errors, so that we adopt a pragmatic approach and rarely use better than the 95% confidence level ($\alpha = 0.05$).

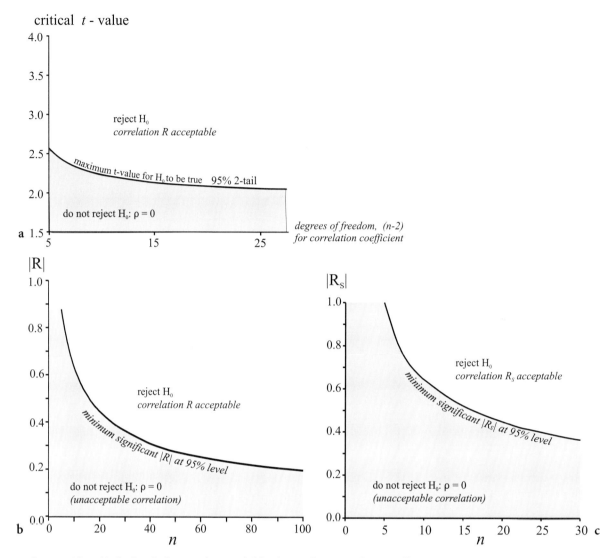

Fig. 7.5. a The critical value of t that must be exceeded for the population correlation coefficient (ϱ) to be significantly different from zero, at the 95% level for the degrees of freedom given $(n-2)$, where n = sample size. Satisfactory combinations of the sample correlation coefficient (R) and n must define a point that lies above the curve. **b** Minimum value of *Pearson's correlation coefficient* (R) that is significant at the 95% level for different sample sizes. **c** Minimum value of *Spearman's rank correlation coefficient* (R_S) that is significant at the 95% level for different sample sizes

7.6
Progress, Problems and Pitfalls with Simple Linear Regression

The advantages of quantifying correlation and estimating its significance far outweigh the pitfalls of misinterpretation. Modern spreadsheets remove any hardship and, with careful application, one may obtain a deeper understanding of the nature of the problem be-

ing investigated, as well as qualifying the conclusions statistically. An example of a much quoted data set in textbooks on environmental, physical and structural geology shows an association between fluid waste disposal and earthquakes. The data usually appear in the form of histograms; the reader is left to visualise and imagine the correlation (Fig. 7.6a). They document an ill-advised period of toxic-waste disposal, pumping noxious fluids into deep wells at the US Rocky Moun-

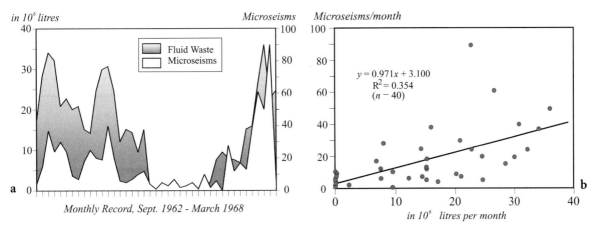

Fig. 7.6. a Fluid-waste disposal at the Rocky Mountain Arsenal, Colorado. The histogram gives a visual impression that the amount of waste disposal is related to the number of microseismic events. **b** Scatter plot of the same data; R² shows that the regression line accounts for only 35.6% of the variance in microseism frequency. Nevertheless, R differs significantly from zero at the 95% level. (The test statistic is 4.38, exceeding the critical value of $t = 2.02$ for $n = 40$ with $n - 2$ degrees of freedom; see Table 4.5; Fig. 7.5b). Thus we reject the hypothesis that the population correlation coefficient is zero and accept the alternate hypothesis that a correlation exists at the 95% level

tain arsenal, in the 1960s, before an awareness of potential environmental problems had developed. The fluid pressure caused an effective-stress effect, reducing the binding stress that acts across fault planes (friction is *not* reduced). Thus, the less tightly bound faults moved, producing small earthquakes or microseisms. This is of great interest in environmental geology as it presents an optimistic solution for preventing a large earthquake by inducing a series of microseisms that would gradually relieve dangerous levels of high stress. Of course, in a benign campaign, water would be the most suitable fluid. The toxic fluids accidentally used in this example could contaminate the groundwater. The visual comparison of histograms does seem to favour an association between the amount of military waste disposal and the frequency of microseismic events. Quantitative techniques exist for comparing raw data sets directly, without making assumptions about the data distribution (e.g. χ^2 and Kolmogorov-Smirnov tests, Chap. 5).

Here, we attempt correlation by plotting waste disposal as the control variable (x) and the numbers of seismic events as the response variable (y) to yield the scatter plot of Fig. 7.6b. Upon cursory inspection, the spatial distribution of points seems discouraging and the correlation, R = 0.597, is not strong. The coefficient of determination $R^2 = 0.356$, means that only 35.6% of the y-variance is explained by the regression line. However, the minimum significant value of R at the 95% confidence level for this sample size, is 0.26 (Table 7.3; Fig. 7.5b), so we may accept the correlation. Alterna-

tively, for the 40 data points we may calculate the test statistic of Eq. (7.6) and compare it with a tabled t-value. The calculated value for t = 4.38 which exceeds the critical value of t = 2.02 for $n = 40$ and ($n - 2$) degrees of freedom (see Table 4.5). We may reject the null hypothesis that the correlation coefficient is zero and accept the alternate hypothesis that the correlation is significant at the 95% level. The same data were discussed earlier in this chapter using Spearman's rank-correlation coefficient and in Chapter 5 where the data were compared using the Kolmogorov-Smirnov statistic (Fig. 5.9).

Most correlations do not imply causality or genetic significance. At any rate, in order to do so, the correlated variables must be mutually and mathematically independent. In complex earth-science systems, variables may be so intricately interwoven that it may be difficult to guarantee their independence. This leads to misapplications of regression and correlation that give statistics unfair bad publicity. Campbell (2000) quotes some amusing examples to illustrate this point. They concern the interaction of the geomagnetic field and the solar wind and their "effect" on human activity. For example, one prominent scientist was particularly prolific every 11 years, just after years of heightened solar activity. Was his brain stimulated by the electromagnetic atmosphere? Yes, but not directly; he researched solar and geomagnetic activity so that he just had more to write about after those natural phenomena provided him with new data. A second example quoted is of the correlation of geomagnetic-solar activity with the collapse and heart failure of agricultural workers. However, we need not attribute electromagnetic causes to the illnesses or injuries. The geomagnetic activity simply follows a similar seasonal cycle to those of agriculture. Correlations such as these may still be of some use, respectively in connection with the history of scientific publication and with socio-economic studies of the agricultural workforce. However, the correlated data have no causal link and there is not really any sensible basis for considering their formal correlation. Earth science provides many examples of correlation falsely implying *causality*. This is due to the complexity and number of variables involved in natural processes. We should not regard this in a negative light. It serves to keep us in awe of nature's complexity and stimulates us to a deeper understanding of the interrelationships among the variables.

Data that can truly be said to produce scattergrams are common in several branches of field geology. Some observational errors may be large but, more commonly, the underlying physical relationship may have a loosely defined nature. An earlier example relating stream length and drainage-basin area was affected by complex variables such as microclimate, terrain, rock type and tectonic stability of the land surface. Strictly speaking, we should only attempt to correlate stream lengths and drainage-basin areas for which all of these conditions are identical. Of course, it is unreasonable and impractical to make those requirements too stringent. Otherwise, there would not be enough data to consider. Nevertheless, some simple data confirm conventional wisdom: the longer the stream, the larger its

drainage basin (Fig. 7.1b). The correlation is significant at the 95% level, the test statistic is 6.63 according to Eq. (7.6) which exceeds the critical t-value of 2.306 for 8 degrees of freedom ($n - 2$) at the 95% confidence level. The significance of the correlation may also be accepted by rapid inspection of Fig. 7.5b and Table 7.3 which require only that R > 0.632 for a significant correlation at the 95% level.

The classic data correlating stream length and basin area revealed a surprisingly well-defined linear association, in comparison to data from many hydrological and climatic studies. The next example will show that scattered data and quite low values of |R| correlations may be significant for a suitably large sample size (n) (Table 7.3; Fig. 7.5b). Nevertheless, even where a correlation appears significant on the basis of a statistical calculation, some common sense and an appreciation of the subject matter must qualify any subsequent interpretations.

An example revealing significant scatter is shown from a hydrological study of spring-flow with the passage of time at a location in Italy (Fig. 7.7). A ubiquitous contemporary concern is the effect of global climate change on weather patterns, agriculture and less directly on groundwater supplies. Changes in climate may produce less precipitation, which lowers the water table and thus may reduce the yield of springs. From 1974 to 1991, spring-flow data are presented; time *proxies* as the independent variable (x) and is not responsible for introducing uncertainty. However, the measurement of spring flow is notoriously subject to error and uncertain behaviour. Furthermore, it is not influenced

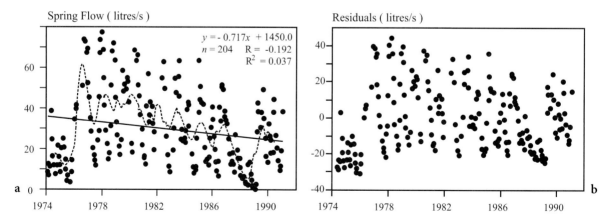

Fig. 7.7 a, b. Broadly scattered spring-flow data for an Italian source, 1974 to 1992 (Dragoni 1996). Caution is recommended in interpretation of their statistics, due to the patchy data distribution and broad scatter. **a** Moving average of ten observation groups is *dotted*. Linear regression yields a test statistic = 2.83, exceeding the critical value $t \sim 1.96$ for 202 degrees of freedom (Table 7.2) and implying that the weak association may be a significant correlation at the 95% level, despite broad scatter. **b** Residuals of the author's data from his regression line are not randomly scattered about the zero-residual value, due to some systematic departures from the broad trend shown in **a**

solely by precipitation, but also by temperature, vegetation cover and water use in the region, all of which may affect groundwater replenishment. Consequently, many factors affect spring flow, not just seasonal variation superimposed on a climatic trend. It is clear from the graph and suspected on grounds of common sense that some systematic seasonal variation may exist, despite enormous scatter. The moving average, represented by a curve, possibly indicates seasonal fluctuation (moving-averages are discussed further in Chap. 8). It brings out some highs and lows in the data set, more clearly than shown by the scattered points alone. However, lows in 1974–1975 and 1988 are obvious anyway. The suggestion of a seasonal fluctuation prevents a linear explanation of the data, even if a monotonic decrease in spring flow is hypothesised on grounds of climate change. Nevertheless, as a tool for extrapolation to future spring-flow levels, or to provide a crude indication of the rate of decline of water supply, linear regression was proposed. The regression line's negative slope suggests an annual decrease in supply of 0.71 l/s. The standard error on this rate of change is ±0.26 l/s (error on the slope of the regression line).

One may reasonably doubt the importance of any correlation in this data, although it is always possible to use the regression line for trend-estimation, as shown (Fig. 7.7a). The correlation coefficient R = 0.197 and the regression line only explains ~4% of the variance in spring flow ($R^2 = 0.037$). However, this yields a test statistic value of 2.83 due to the large sample size ($n = 204$), which exceeds the critical t-value of ~1.97 at this sample size (Table 4.5). Therefore, we may accept that this weak trend is significant at the 95% level. This correlation should not be taken at face value, in view of the complexity of the problem, the noise in the data set and the superimposed nonlinear seasonal fluctuations. Inspection of the residuals, the deviations Δy from the calculated regression line, indicates unsatisfactory aspects to the data (Fig. 7.7b). The residuals show some pattern: they are clearly negative around 1974–1975 and 1988–1989, and decrease after 1977 with smaller residual variance. It is difficult to place too much faith in the association suggested by the correlation, given the scatter of the data and the nature of the observations.

A further example of hydrological data shows the variation of precipitation with temperature at four Italian locations over 20 years (Fig. 7.8; Dragoni 1996). The four locations show data with varying degrees of scatter. Meteorological data such as rainfall and temperature are recorded accurately so that the scatter here reflects the looseness of the "controlling" processes rather than observational errors. Intuition tells us that rainfall should be less when the temperatures are higher so that temperature is selected as the independent variable, although it is not the only influence on rainfall. There could be many other factors such as wind direction and atmospheric pollution. Nevertheless, in all cases the regression line has a negative slope, suggesting that rainfall is lower at higher temperatures. However, the significance of the correlations at the 95% level varies. The test statistic has been calculated using Eq. (7.6) and is compared with the critical t-value, t_C, taken from Table 4.5 for a two-tailed test at the 95% level of confidence ($\alpha = 0.05$) with ($n - 2$) degrees of freedom. Setting the null hypothesis that *there is no correlation in the population* H_0: $\varrho = 0$, the results of the comparisons are:

(a) Arezzo   ($n = 90$) test-statistic 1.619 < t_C = 1.99
accept H_0; no correlation

(b) Rome   ($n = 90$) test-statistic 2.17 > t_C = 1.99
reject H_0; correlation

(c) Terni   ($n = 40$) test-statistic 2.399 > t_C = 2.02
reject H_0; correlation

(d) Cortona   ($n = 36$) test-statistic 2.17 > t_C = 1.99
reject H_0; correlation

At Arezzo, we accept the null hypothesis that the correlation coefficient is not significantly different from zero; no trend may be inferred. For Rome, Terni and Cortona, despite much scatter and low correlation coefficients, the correlations are significant at the 95% level. This may also be confirmed by inspecting the nomogram of Fig. 7.5b. For Cortona, the negative correlation is strongest, all 50 data have R = – 0.5, so that $R^2 = 0.25$, or the regression line explains 25% of the observed rainfall-variance. Previously, it was shown that it may be advisable to exclude outliers whose residuals from the regression line exceed a certain value, for example the one-sigma level. Here, however, there is a meteorological reason for considering that cooler precipitation represents aberrant data in the physical, not merely statistical, sense. Thus, excluding data for temperatures <14 °C improves Cortona's correlation to R = – 0.56. The new regression line provides a satisfactory explanation for 32% of the variation in rainfall ($R^2 = 0.32$). Now the regression line is steeper, showing a stronger dependence of rainfall decrease on temperature increase. Of course, we must define outliers and exclude them merely to support our hypothesis. We may reject them if they show unreasonably large residuals but, ideally, rejection should be at an early stage based on objective criteria such as instrumentation problems or inappropriate environmental

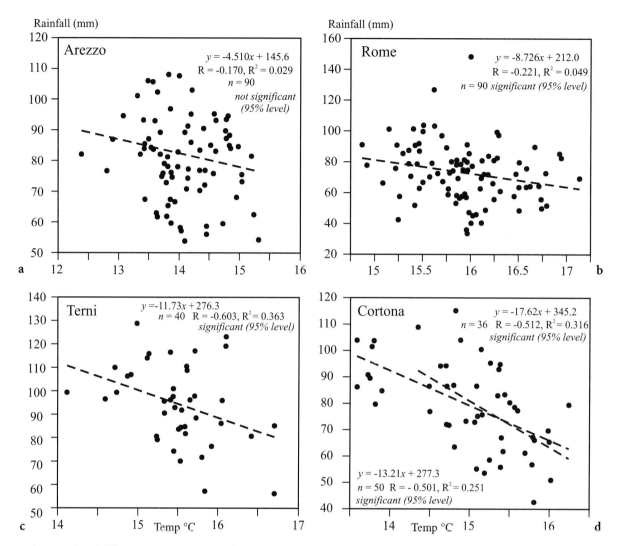

Fig. 7.8a–d. Rainfall-temperature scattergrams for monthly data over a 20-year period, Italy (from Dragoni 1996). **a** Nonsignificant correlation; **b, c, d** show weak negative correlations, all significant at the 95% level, despite scattered distributions that do not inspire much enthusiasm. **d** Exclusion of 14 outliers for temperatures <14°C, steepens negative slope and improves correlation (statistics at *upper right*)

factors. However, with such scattered *x-y* points and some intuition of the physical processes involved, it may be justifiable to exclude outliers.

More structured data sets are characterised with much more confidence and ease, but more importantly they permit rigorous comparison with other data. We have previously seen that the mean global temperature appears to have risen since the first reliable estimates of 1855. These are shown plotted with an arbitrary zero, i.e. the changes are meaningful but the absolute value of mean temperature is not represented (Fig. 7.9). In comparison, for Nicosia, Cyprus, the mean annual

temperatures are shown with accurate absolute values. Both data sets show positive correlations significant at the 95% level; the correlation coefficients are 0.709 and 0.766, whereas even a correlation coefficient of 0.2 would be significant for these sample sizes (Fig. 7.5b). Thus, we may compare the changes in temperature in Nicosia with those of the whole world. For this, we should compare the slopes of the regression lines. The global temperature appears to have increased by 0.0038 ± 0.0003°C/year. The error represents the 95% confidence limits on the regression-line slope. For Nicosia, the increase has been 0.0088 ± 0.0009°C/year. It would

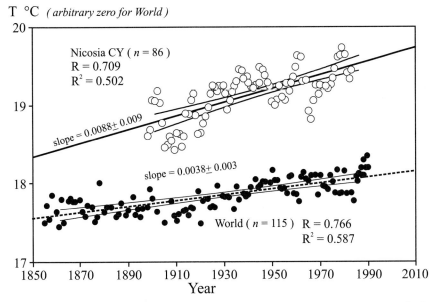

Fig. 7.9. Temperature increases at Nicosia, Cyprus, compared with global values. The Nicosia temperatures are absolute, but the global values represent changes only, relative to an arbitrary zero. This is not important since we are only concerned with rates of temperature change. Both regression lines are highly significant and indicate temperature increases of 0.0088 °C/year for Nicosia versus an estimated mean global temperature increase of 0.0038 °C/year. The 95 % confidence limits on the slopes of the two regression lines provide no opportunity for overlap so that we may be confident that the temperatures in Nicosia are increasing faster than those of the world. Cultural-economic effects combined with a special microclimate may explain this

appear that Nicosia has been warming more than twice as quickly as the entire world. How confident are we that this difference is significant? This is answered by comparing the slopes. With a 95 % confidence limit, the slope of 0.0038 for the Nicosia data could range actually from 0.0079 to 0.0097 °C/year. For the world, the comparable values are 0.0035 to 0.0041 °C/year. Since there is no possible overlap of the ranges of the two slopes, we are confident that the temperature increases differ significantly at the 95 % level. In fact, the calculation may be refined to the 3σ level to show the differences are significant at the 99 % level.

When we wish to investigate the variation in climatic conditions over historical and prehistorical time, the luxury of precise meteorological measurements eludes us. Then we are confronted by climatic measurements that must be determined by some scientific method, subject to assumptions and measurement errors. Typically, temperature is determined from some stable-isotope study, but relative temperatures may be estimated from studies of pollen, agricultural practices and estimated yields. Moreover, the controlled or independent variable, time, may not be well defined. In some ancient civilisations, the dates may be well determined from cultural material or even documentation, otherwise the very difficult radiocarbon-dating tech-

nique must be applied. An interesting archeological data set from the eastern Mediterranean is shown in Fig. 7.10a. The scatter in temperatures is considerable but remarkably manageable, considering that we are dealing with ancient events, and the correlations are significant at the 95 % level. Part of the success is because measurements of ancient temperatures (and other physical variables) are sometimes *averaged* by the very nature of the scientific measurement process. Thus, sampling incorporates a time-averaging effect which smoothes the data.

Within the data set there is some clear structure, evident from the numbers themselves, but also inferred from anthropological and archeological studies (Issar and Brown 1998). From B.C. 1000 to A.D. 1000, there was very little temperature variation, the slope of the regression line is practically zero, but the regression is not significant at the 95 % level. However, this does not mean that we should discard the data. It is simply intrinsically noisy to such a degree that we cannot infer a satisfactory, predictive linear relationship and we cannot detect any trend in temperature. After A.D. 1000 a clearer pattern emerges, temperatures decrease rapidly by 0.001 °C/year according to the regression line, with a correlation coefficient R = −0.72 which explains 51 % of the variation in temperature ($R^2 = 0.51$). The correlation is significant so that we may conclude that with a 95% confidence there is a progressive decrease in temperature after A.D. 1000. However, the dangers of extrapolation of regression lines become very clear in this instance. Since the industrial revolution, indisputable global warming has produced an enormous increase in temperature in a period of less than 200 years. These data, measured by the techniques used for the archeological data, appear as four outliers at the right-hand side of Fig. 7.10a. They were excluded in calculating the

regression line for the post-A.D.-1000 period. Their exclusion was not merely based on their disparate location on the graph; we know there are good reasons to expect different temperature regulation. The steep short regression line passing through the post-Industrial Revolution outliers is not fitted to those data but taken from real temperature measurements taken over the period 1860–1990 (Fig. 7.9). Since 1800, temperatures have mainly been influenced by industrial activity, before that, the effects were natural.

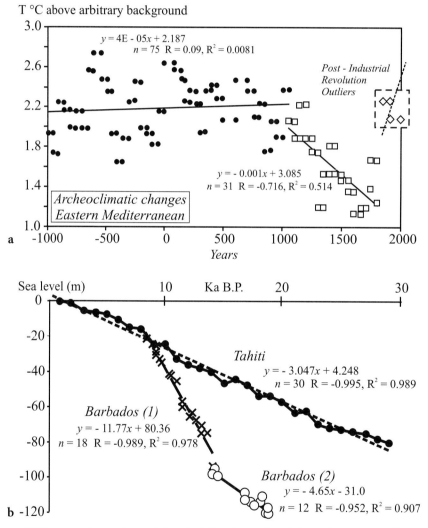

Fig. 7.10. a Eastern Mediterranean temperature variations in the last three millennia form a heterogeneous data set (Issar and Brown 1998). Using independent evidence, it is reasonable to exclude outliers for the post-Industrial Revolution period, which shows marked global warming. The remaining data set can be divided into two distinct groups by visual inspection. Before A.D. 1055, there was no significant change in temperature. From 1055 to 1855, temperature decreased at 0.001 °C/year, significant at the 95% level. For comparative purposes, the regression line for the post-Industrial Revolution period is shown; it is based on the detailed temperature measurements since the 1850s, taken from Fig. 7.9. **b** Global sea-level rise due to deglaciation; all regression lines are significant at the 95% level. The rise inferred from Barbados is complicated by local tectonic instability. A preliminary separation of the data in groups permits a sensible regression analysis. Data from Quinn and Mountain (2000)

Sea-level changes sensitively reflect mean global temperatures, mainly through variation in polar-ice volumes rather than thermal expansion of water. This is strikingly documented during the contemporary deglaciation shown in Fig. 7.10b (after Quinn and Mountain 2000), all significant at the 95% level. Around Tahiti, sea level appears to have risen steadily whereas relative to Barbados, sea level appears to have risen at three different rates. All of these rates (= correlations) are significant and they could only be realised by visually analysing and identifying subgroups of data before performing regression. The geological details are beyond the scope of this discussion, but local tectonic instability is the usual explanation for apparent differences in the changes in sea level along different coastlines.

Sometimes quite small samples yield strong correlations if there is some strong physical control on the response variable. A previous example showed the acquisition of viscous magnetism in archeological ma-

sonry revealed by a power-law growth in magnetisation with time (Fig. 6.8). The convex-downward curves are difficult to compare or to use for prediction. However, linearisation by plotting logarithms of the variables simplifies comparisons and predictions (Fig. 7.11). The regression lines may be treated like any other regression lines but their slopes now represent the indices of the power-law relations between the variables. Despite small sample sizes, the regression lines are significant at the 5% level, so the power-law relationships are similarly significant. The scarcity of data points for any line is disappointing for the application of statistical methods. However, each data point summarises a great number of experiments and measurements; their reliability is therefore considerable and reflected in the strong linearity. A few previous case studies in geochronology and geophysics show a similar picture; a few precise data warrant statistical treatment and permit conclusions to be formed (Figs. 4.2, 7.3).

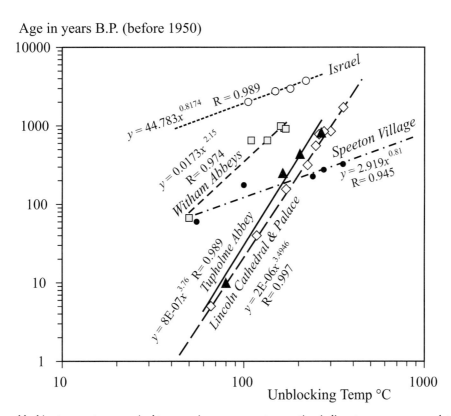

Fig. 7.11. The unblocking temperatures, required to erase viscous remanent magnetism in limestone masonry, are related by a power law to the age at which the masonry was stabilised in a building. This provides an archeological-dating method (Borradaile 1996; Borradaile and Brann 1997). The theoretical basis for this relationship is as yet unknown, so that the correlations are empirical. The correlations for several sites are all significant, but yield different chronometric curves, depending on the nature of the masonry. To be able to apply the chronometric curve from one site to another, we need to know how significant the differences are in slope between the lines. This is addressed in the text, particularly Table 7.4

The technique reliability determines the age of stabilisation of the masonry in a structure from the elevated temperature required to erase, or unblock, its viscous magnetisation in the earth's magnetic field. The viscous magnetic data provide chronometric curves from which the ages of otherwise undatable limestone buildings may be estimated from the blocking temperature. Data are presented for sites in England and Israel. Clearly, the graphs suggest that different limestones at different archeological sites acquired viscous magnetism at different rates. How significant are the different rates? In other words, are the slopes of the regression lines significantly different in log–log coordinates? The information is most simply presented in Table 7.4. The agreement or disagreement between slopes may be determined from the overlap, or absence of overlap, of the upper and lower limiting slopes defined at a 95 % confidence level. It is safe to draw a few simple conclusions. For example, the range of possible slopes for Tupholme Abbey encompasses the range for the sites at Lincoln City. One may therefore conclude that the behaviour of the limestones at the two sites is not distinguishable at the 95 % confidence level. This is reasonable, as the ancient masons quarried the stone from the same geological formation, albeit at different quarries and at different stratigraphic levels. Another clear conclusion is that the range of slopes for the limestones of Speeton Village and Israel lies outside the range of slopes for the limestones of the Abbeys and Cathedral of Lincolnshire. We can safely assume a different type of viscous magnetic response and realise that the chronometric curves must be used cautiously, recognising that they are calibrated for a specific type of limestone. In fact, the limestones of the Israel and Speeton Village sites are pelagic chalks, whereas those of Lincolnshire are shelf limestones. Depositional conditions may influence the proportion of biogenic versus clastic magnetite and thus affect magnetisation characteristics.

Table 7.4. Example of confidence limits for regression-line slopes of viscous-magnetisation data in log–log space

	Slope: upper 95% limit	Regression slope	Slope: lower 95% limit
Lincoln Cathedral and Bishop's Palace	3.712	3.494	3.276
Tupholme Abbey	5.440	3.758	2.068
Bardney and Barlings Abbey	3.050	2.149	1.240
Speeton Village	1.803	0.810	0.399
Israel sites	2.430	0.817	0.460

7.7
Special Cases in Correlation

7.7.1
Co-linear Variables

The complexity of natural-science data occasionally gives rise to special cases where completely false correlations appear and, conversely, where true correlations are suppressed or obscured. Correlation only makes sense where the properties correlated are mathematically independent. They must be free to vary according to some natural processes. The independent variable x may represent some control, process, or observation whose value may be subject to natural variation as well as measurement error. The x observations must be independent of each other. Moreover, whatever relationship is recognised between a dependent variable y and x must be due to some external process, if the correlation is to have any meaning. A trivial example would be an attempt to correlate the volume of pebbles with their mean diameter; clearly, the two are not mathematically independent, and any association would not have geological significance. Variables that are constrained mathematically, or otherwise, are said to be *co-linear*. Attempts to correlate them will not add any new or meaningful information. Such interdependence problems arise commonly and may be quite subtle, as in the following example:

In many aspects of geology, we deal with shapes of three-dimensional objects that may be represented by ellipsoids. Their shapes are expressed as ratios, e.g. $2 : 1 : 0.5$, such that their product is unity. Actual measurements of ellipsoid axes X, Y, Z may be converted to this form by dividing each of them by $(XYZ)^{1/3}$. The use of ellipsoidal shapes is a powerful and simple representation of the variation in shapes of objects that represent anisotropy, or variation in physical properties with direction. For example, the shape of a pebble represents the anisotropy of resistance of the pebble to abrasion; that of the optical indicatrix reveals the directional variation of velocity of light in a crystal. (The most easily manipulated anisotropies are represented by second-rank tensors but not all second-rank tensors may be represented by ellipsoids and not all ellipsoidal representations require description by second-rank tensors; see Chap. 11.)

For structural geologists, the shape of the strain ellipsoid representing the relative amounts of extension and shortening in different directions is very important. Unfortunately, few rocks contain suitable features from which this can be determined easily. However, the orientation-distribution of minerals controls the ease with which a rock may be magnetised temporarily in the presence of an applied magnetic field (e.g. Chaps. 10, 11). Therefore, the directional variation, or anisotropy, of magnetic susceptibility provides clues to the orientation-distribution of minerals and thus perhaps of the strain ellipsoid's shape. Clearly, if the easily determined magnetic-susceptibility ellipsoids (x) could be calibrated with strain markers in one location, they could be used elsewhere to define the more elusive strain ellipsoid (y). It was claimed that the regression line of y on x (strain on magnetic anisotropy) would provide a universal calibration. For nonstatistical reasons, this has been shown since to be highly optimistic (Borradaile and Henry 1997). However, let us discuss the statistical aspect of this problem.

In the quest for a universal strain-magnetic anisotropy calibration, authors would represent the three strain-ellipsoid axes with strain parameters N_1, N_2, N_3 and the axes of the magnetic-susceptibility ellipsoid were designated M_1, M_2, M_3. The (N_i, M_i) values were logarithmic expressions of strain and susceptibility respectively. Clearly, M_2 is not free to vary independently of M_1 or M_3. The same reasoning may be applied to any of the three N_i values. Each N_i value and each M_i value is not free to vary but is constrained by the others in its group via the ellipsoidal geometry. To take the most

obvious restriction, the intermediate value (e.g. N_2 or M_2) must lie between that of the minimum and maximum axes, by definition. Thus, the x-values are already correlated and not independent of one another. Despite this, an unfortunate global correlation of all values was attempted (Rathore 1979, 1980). Even the distinct clustering of maximum, intermediate and minimum values should raise suspicion (Fig. 7.12a). Quite different results were achieved in a separate study, in which magnetic anisotropy was measured from the same specimens as the strain, using improved strain-analysis methods (Borradaile and Mothersill 1984; Fig. 7.12b). The more appropriate treatment of the data yields insignificant correlation, between individual axes ($n = 17$) as:

Maximum susceptibility regressed on maximum extension

$$R = -0.223 \qquad R^2 = 0.049 \qquad (\sim 5\%)$$

Intermediate susceptibility regressed on intermediate extension

$$R = +0.032 \qquad R^2 = 0.001 \qquad (\sim 0.1\%)$$

Minimum susceptibility regressed on minimum extension

$$R = +0.430 \qquad R^2 = 0.018 \qquad (\sim 1.8\%)$$

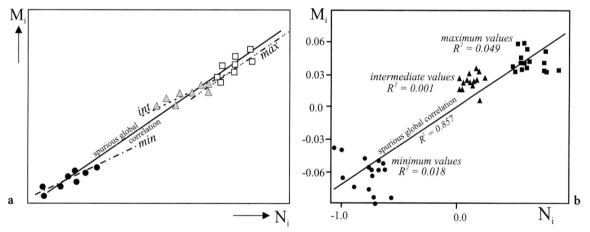

Fig. 7.12 a,b. Physically meaningful correlations only arise where the variables are independent, that is to say, they must not be interrelated. An ellipsoid is commonly used to represent the three principal strain axes, and another ellipsoid may be used to represent the three principal axes of magnetic anisotropy. However, in each ellipsoid, the three axes are not independent quantities; they are not completely free to vary independently with respect to one another. In this diagram, the strain axes and magnetic-susceptibility axes are represented by logarithmic quantities, N_i and M_i respectively for the maximum, intermediate and minimum axes ($i = 1, 2, 3$). **a** One author suggested that *all* strain axes could be regressed upon *all* susceptibility axes to produce a universal calibration function that would yield strain values from the more easily measured susceptibility anisotropy (Rathore 1980). This global correlation is apparently very strong but spurious. On the other hand, it may be more valid to compare individual pairs of axes (e.g. maximum strain with maximum susceptibility, etc.), but those associations are different and weaker as shown by the three dashed regression lines, evident here. **b** A quite different study on the same rocks used different strain-analysis methods and improved techniques (Borradaile and Mothersill 1984). It showed that the individual correlations (maximum strain with maximum susceptibility, etc.) are nonexistent. Moreover, it was clear that the global correlation of all axes of strain with all axes of susceptibility was entirely spurious with a meaningless but high value for R^2

The coefficient of determination, R^2, shows that no individual regression explained more than 5% of the variance. Moreover, the regression line showed a negative slope for the maximum axes, which would imply that maximum susceptibility decreased with strain, contrary to the simplified hypothesis of early workers.

In contrast, the inappropriate global correlation of all three groups of susceptibility magnitudes simultaneously against all extensions ($n = 3 \times 17$) yields a spurious correlation that falsely claims to explain 86% of the variance (Fig. 7.12a) with:

R = + 0.926 $R^2 = 0.857$ (~86%)

Clearly, this invalid global correlation gives an unreasonably strong and optimistic correlation. When collinearity is avoided by performing separate regressions for (N_1 versus M_1), (N_2 versus M_2), (N_3 versus M_3), the data show no simple relationship between magnetic susceptibility and strain. Indeed, there are subject-specific reasons for doubting the existence of any simple underlying relationship, in general (e.g. Borradaile 1991a; Borradaile and Henry 1997).

7.7.2
The Bivariate Normal Distribution

The Normal distribution is by now very familiar to us in univariate statistics; we understand that a single variable, x or y, may possess a Normal frequency distribution (Chap. 3). The form of the Normal distribution makes it particularly useful for many statistical tests, and its generality, through the Central Limit Theorem, makes its relevance almost universal for the estimation of confidence in results (Chap. 4). The Normal distribution may be extended to bivariate statistics. In that case, the observation requires two values, x and y, for its specification, but both x and y are subject to uncertainty that follows the Normal distribution. For example, one may make numerous measurements of a bivariate quantity, yielding numerous x–y value pairs. However, both x and y are normally distributed leading to scatter around the best estimate of the quantity given by the mean values $(\bar{x}, \bar{y})$. Intuitively, earth scientists appreciate this at an early stage in their training and the concept of error bars or a confidence ellipse around an $(x$-$y)$ point are well established both by calculation and directly from the scatter of measurements (Chap. 4, Fig. 4.2).

The bivariate Normal distribution is a generalised two-dimensional extension of the Normal distribution on a line, from Chapter 3. Recalling that is the population correlation coefficient, the probability distribu-

tion of a point in $(x$–$y)$ space is given most simply by:

$$P(x, y) = \left[\frac{1}{2\pi\sigma_x\sigma_y \sqrt{1-\rho^2}} \right]$$
$$\exp\left[\frac{-1}{2(1-\rho^2)} (z_x^2 - 2\rho z_x z_y + z_y^2) \right]$$

where z_x and z_y are standardised forms of x, y:

$$z_x = \frac{x - \mu_x}{\sigma_x} \qquad z_y = \frac{y - \mu_y}{\sigma_y} \tag{7.12}$$

The form of this distribution is indicated in Fig. 7.13. The surface encloses a volume of unity, which represents the 100% probability that any value in the distribution must lie beneath it. Any vertical section through the surface has the form of a two-dimensional Normal distribution. The contours of the probability density reflect the form of the confidence ellipses, which are familiar in earth science (Fig. 4.2). The shape of the ellipse is given by the term:

$$z_x^2 - 2\rho z_x z_y + z_y^2 \tag{7.13}$$

where $\varrho = 0$, the uncertainties in x and y are independent and the ellipses are parallel to the x-axis for $s_x > s_y$ and parallel to the y-axis for $s_x < s_y$. Where x and y are not independent, the ellipse's major axis is inclined with a slope $m = tan\theta$ where θ is given by the double-angle formula:

$$\tan 2\theta = \frac{2\rho\sigma_x\sigma_y}{\sigma_x^2 - \sigma_y^2} \tag{7.14}$$

Thus, for $\varrho \neq 0$, the slope of the ellipse with respect to the x-axis has the same sign as that of ϱ. Inclined confidence ellipses are familiar in systems with a complex interdependence of variables, for example in geochronology and petrology (Figs. 4.2, 7.15). Bivariate confidence zones may even be calculated for triangular diagrams, where they lose their elliptical symmetry and take on a crescent-shape (Fig. 7.16b; Weltje 2002).

Where we wish to detect the significance of the bivariate regression line, a more complicated statistic is required than that for classic linear regression of y on x, where all the uncertainty resides only in y. The test statistic follows the Normal distribution and used the probability table for the standard Normal distributions' z-variate (Table 3.5). The test-statistic for H_0: $\varrho = 0$ is:

$$z = \left(\frac{\sqrt{(n-3)}}{2} \right) \log_e \left(\frac{1+R}{1-R} \right) \tag{7.15}$$

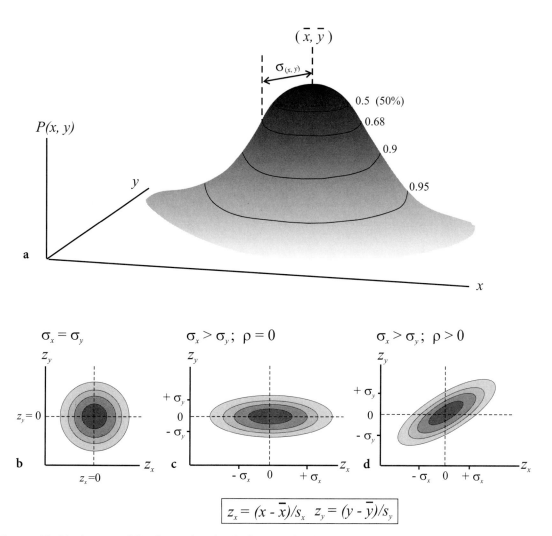

Fig. 7.13. The bivariate Normal distribution describes the frequency distribution of a bivariate quantity, described by x and y, where both x and y are normally distributed. **a** Probability distribution of a bivariate Normal distribution in x-y coordinates. The bivariate Normal distribution's surface is contoured to show the fraction of data enclosed within certain distances of the mean. These define confidence regions of elliptical shape. **b–d** The bivariate Normal distribution more simply represented using the standardised transformations of x and y to z_x and z_y. In **d** the confidence ellipses are not parallel to the coordinate axes because x and y are not independent

The most common null hypothesis is that the population correlation coefficient $\varrho = 0$; H_0: $\varrho = 0$. We calculate z using the sample correlation coefficient. If $z > |1.96|$, we reject the null hypothesis in favour of the alternate hypothesis that the correlation is significantly different from zero, with 95% confidence.

7.7.3
Constant Sums, Ratios and the Closure Restraint

In many geological data sets, the variables are not entirely independent but are constrained to form a con-stant sum. Chemical analyses or other contents are the most obvious example, requiring a sum of 100%. Any quantities that are ratios or percentages are subject to the *constant sum*, or *closure restraint* (Chayes 1971). Consider a hypothetical dolomitized limestone, initially composed of >95% calcite ($CaCO_3$). Due to progressive diagenetic alteration; it may become partially replaced by the mineral dolomite ($CaMg(CO_3)_2$). Clearly, the percentage volume occupied by calcite must be displaced by the dolomite that replaces it. Therefore, there must be a perfect inverse relationship between the two proportions (Fig. 7.14a), which should have been obvious from the nature of the problem, and

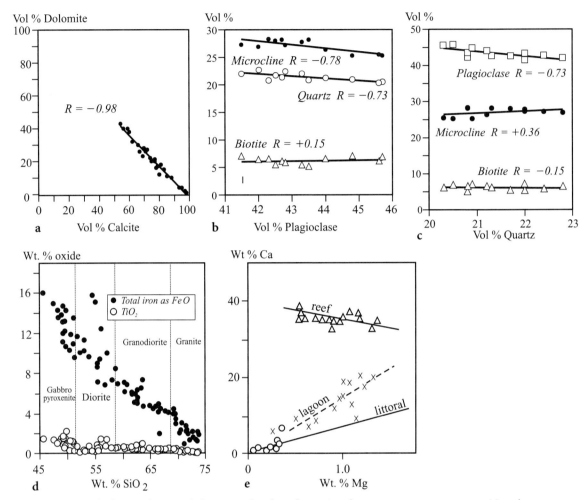

Fig. 7.14a–e. Meaningless, spurious correlations may arise where observations form a constant sum, e.g. compositions that must sum to 100 %, or proportions constrained in the zero-to-unity range (Chayes 1971; Till 1974). **a** Hypothetical calcite-dolomite contents for dolomitised limestones with < 5 % of other minerals. The variables are co-dependent and displace each other's volume forcing a spurious negative correlation. The meaningless correlation appears very strong because the two variables form more than 95 % of the volume. **b, c** Mineral content of the granite designated G-2, used for interlaboratory calibration of analytical techniques in the 1960s. Spurious negative correlations appear between the minerals that form a significant fraction of the rock because one displaces the other. Similarly, spurious positive correlations occur between major and minor components that displace each other. **d** Variation diagram of FeO and TiO_2 versus SiO_2 for plutonic rocks of the Canadian shield (Pilkington and Percival 1999). The decrease in FeO is not correlated with SiO_2 as the original study may have implied. The two contents are not independent and an increase in one displaces the other. **e** Ratio data for rock compositions may form clusters that should be identified on the basis of other attributes (e.g. here depositional environment) before even spurious correlations may be recognised. (Till 1974)

the regression line simply demonstrates that the variables are interdependent. The regression line does not provide any meaningful statistical information. This example is perhaps too obvious and too simple, having only two components. However, minor components, such as accessory minerals or sand contamination forming < 5 % of the volume, would not be expected to show any systematic variation with the major components. They may thus show some statistically meaningful trends. Unfortunately, the major components usu-

ally show spurious negative "correlations", especially where the number of major components is small (Fig. 7.14a, b). Unfortunately, the number of major components is necessarily small. For igneous and metamorphic rocks whose mineralogy is dictated by the rules of physical chemistry, the phase rule usually restricts the number of stable phases to ≤6 minerals. The proportions of as many as six major components usually show spurious negative correlations due to the closure restraint. Chayes (1971) uses the term "spurious correlation" rather cautiously and conservatively in this context. For most purposes, such "correlations" are meaningless if taken at face value.

A more complex example illustrates Chayes' (1971) arguments very clearly. It concerns a granite sample, designated G-2, that was used for comparison and standardisation of geochemical analytical techniques between different laboratories in the 1960s. As with most igneous (and metamorphic) rocks that have achieved a stable mineralogy and minimised chemical-free energy, the number of minerals is limited by the phase rule to ≤ 6. In G-2, the minerals are, in order of decreasing abundance, plagioclase feldspar (43%), microcline feldspar (27%), quartz (21%), biotite (6%) and some minor accessory minerals. The sympathetic variation between the percentage volumes of these minerals is shown in Fig. 7.14b, c. The results are predictable for quartz and plagioclase feldspar. They show a strong apparent inverse correlation since an increase in the proportion of one necessarily displaces the content of the other to some degree. The proviso "to some degree" is necessary because we are no longer dealing with a two-component system. As Chayes pointed out, minor components may show an apparently positive correlation with a major one, e.g. biotite with plagioclase in this example (Fig. 7.14b). Moreover, even modestly abundant components may show an apparent positive correlation, such as microcline with quartz (Fig. 7.14c). The spurious positive correlations are just as predictable and misleading as the negative ones expected for a small number of major components.

An example from a routine investigation of magnetic minerals in plutonic igneous rocks of the Canadian Shield is shown in Fig. 7.14d. This information is useful when considering the contribution of rocks to regional magnetic anomalies and also their potential for paleomagnetic study (Pilkington and Percival 1999). Magnetite, ilmenite and titanomagnetite are important in this respect, so that the *variation diagram* of FeO and TiO_2 versus SiO_2 is a logical first step in understanding the response of different plutonic rocks. A negative correlation of FeO against SiO_2 was implied. However, the decrease in FeO can be explained largely due to displacement by increasing SiO_2. Little can be concluded from the TiO_2 contents. Its slight decrease could also be a displacement phenomenon but one must always remember that elements like Fe and Ti may not behave independently. A substantial part of the Ti content could be in solid solution with Fe in certain minerals (e.g., ilmenite or titanomagnetite). Thus, TiO_2 may mimic FeO to some extent.

This introduces us to a special problem within the issue of closure restraint. The composition of most minerals is constrained to vary within certain prescribed limits. For example, olivine has a generic formula X_2SiO_4 where the X cation is controlled mineralogically to be some proportion of Fe and Mg, varying anywhere from 100% Fe to 100% Mg, e.g. $(Fe_{0.8}, Mg_{0.2})_2SiO_4$. Most rock-forming minerals show a more extensive *stoichiometry*, involving variable cation proportions in several lattice sites. Thus, statistics of mineral compositions, or of rocks dominated by a few minerals of strongly varying stoichiometry, constrain the freedom of compositional variation still further, exacerbating the closure restraint problem. Such variables pose intractable problems for conventional statistical treatment. However, in certain special applications in igneous geology, mineral stoichiometry promotes the understanding of certain magmatic processes from elemental ratios (see Pearce ratios, below).

Clearly, great care must be applied when inspecting constant-sum data, which is an inevitable and unavoidable form of data in any sort of compositional analysis. Only contents present as small traces, e.g., very rare minerals, trace elements, and noble metals are conducive to statistical treatment in this way, because dilute abundances occupy only a small part of the possible range, usually in parts per mil (‰), parts per million (ppm) or parts per billion (ppb). However, as with the example of Ti-Fe above, where a trace element consistently camouflages one major element in the rock, its response will not be independent but rather reflect the behaviour of the affiliated major component, which will be subject to the closure constraint. Components of constant-sum systems can only be considered free from closure problems where they are extremely diluted and independent from any other major component.

Constant-sum systems may become still more complex. Not only may there be spurious correlations between a few major components, but even that pattern may not be apparent until sub-samples are defined on the basis of some attribute that is not expressed as a ratio. For example, Till (1974) gives the example of Ca/Mg ratios in carbonates. The scattergram shows clear clusters but these are only apparent because the data within them are discriminated on the basis of a subsidiary attribute, a depositional environment (Fig. 7.14e). Although the individual correlations shown within groups are spurious, the differences between them indicate different proportion changes of Ca and Mg in different depositional environments.

For simplicity, we have introduced the reader to ratio correlation using concentrations expressed as percentages. This is familiar to geologists and many other earth scientists. However, forced pseudo-correlations with little physical significance may arise with any quantities expressed as a ratio or proportion. Examples include variation diagrams in petrology and ternary (triangular) diagrams in geochemistry. The most commonly overlooked examples concern orientations, which may affect many branches of earth science; for example, compass-azimuths must lie within a 360° range. Orientation statistics are discussed in Chapters 9 and 10.

7.7.4
Pearce Element Ratios

In petrology, Pearce element ratios provide a partial solution to some of the problems caused by closure in elemental analyses (Pearce 1968; Russell and Stanley 1990). The value of such diagrams is great but should not be overstated. They only circumvent the closure problem for specific petrological problems. Their success lies in the fact that during the crystallisation of certain magmatic rocks elements are taken up by crystallising minerals in fixed proportions dictated by mineral stoichiometry, and other *conserved elements* may not be involved at all. For example, every Si atom olivine takes up two atoms comprising some proportion of $(Fe+Mg)$ due to the stoichiometry of olivine, $(Fe, Mg)_2SiO_4$. We will use Pearce's original examples of basalt crystallisation, in which plagioclase feldspar provides a further useful example, taking up two Ca and three Na atoms for every Si atom. In the case of both olivine and plagioclase, K is conserved as it is not withdrawn from the magma by those minerals as they crystallise. Thus, in the case of olivine, we may plot $(Fe+Mg)/K$ against (Si/K). If olivine progressively depletes the magma, the rock's composition will follow a line of slope $= +2$ because $(Fe+Mg)$ is withdrawn at twice the rate of Si, both standardised against K which is a conserved element in this case (Fig. 7.15a). Similar reasoning shows that rock compositions follow a line of unit slope where fractionation is controlled by plagioclase sorting because $(2Ca+3Na)$ is withdrawn at the same rate as Si, while plagioclase crystallises (Fig. 7.15b). Pearce diagrams thus provide a very powerful test of the processes at work in well-defined magmatic systems where permitted by mineral stoichiometry and the presence of a conserved element for standardisation. However, the lines along which data are distributed are not regression lines and the variables plotted are clearly not independent, due to their standardisation to the same conserved element. This is evident objectively from the confidence ellipse limits (Fig. 7.15c) for the analyses that are oblique to the graph axes, showing the interdependence of the variables

(e.g., see Chap. 4, Fig. 4.2 b–d and the bivariate Normal distribution, previously in this chapter). Such mutual interaction of the variables hinders conventional statistical treatment, regression and hypothesis testing. It does not devalue the Pearce concept in solving specialised petrological problems, but the superficial similarity to scatter plots and regression lines of conventional statistics may cause confusion.

7.7.5
Triangular Diagrams

It is commonly possible in petrology and sedimentology to reduce the number of abundance variables to three, for which purpose a triangular diagram is ubiquitously used. First, we note that the data must be abundances subject to the constant-sum constraint. Second, reducing the system to three variables requires that there are no other significant components or that they are constant in abundance. The three components are represented by the sides of an equilateral triangle, ranging from 0 to 100% in an anticlockwise sense around the triangle. Each point within the triangle represents a distribution of components, i.e. a certain percentage of A, of B and of C (Fig. 7.15) according to the fractional distance of the point from the appropriate side. For example, points on the side opposite A represent distributions with 0% A and the percentage of A increases as points are found closer to the apex A. Compositions intermediate between any point within the triangle, x_1 and x_2, are found using a principle known as the *lever rule*, or *mass balance principle*. Thus, the composition at a point x, at fractional distance d from x_1 toward x_2 is given by:

$$d = \frac{x - x_1}{x_2 - x_1} \quad \text{or} \quad x = d(x_2 - x_1) + x_1 \quad (7.16)$$

x may represent the percentage of any of the three components: the simple case of the mean of x_1 and x_2 ($d = 0.5$) is shown in Fig. 7.16. The triangular diagram is probably the device that most obviously underlines the presence and problems of closure constraints. Some examples from sedimentology are shown in Fig. 7.17, in which weathering and provenance influence the composition of detritus. Inevitably, there is some compromise in reducing the components to three: quartz, feldspar and lithic fragments. Do these three components adequately describe the materials? How much of the lithic fragment is feldspar and therefore subject to degradation by weathering? Such details affect the validity of the triangular diagram and may weaken conclusions drawn from it. The distribution of data and confidence regions on triangular diagrams rarely

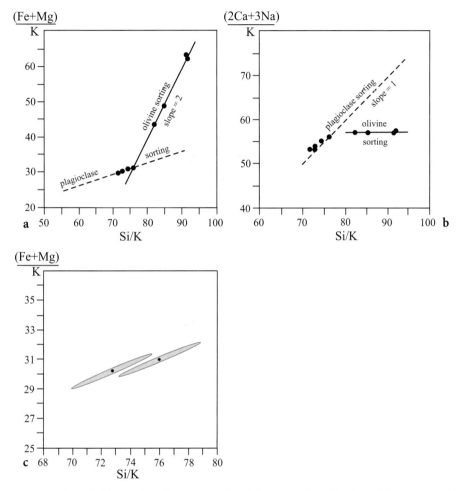

Fig. 7.15a–c. For restricted but valuable petrological purposes, ratios of elements and combinations of elements may yield information on the reassignment of elements between different minerals during the evolution of a magmatic rock, despite the problems caused by the constant-sum nature of compositional data. Pearce diagrams (Russell and Stanley 1997) take advantage of the fact that certain elements are conserved during magmatic evolution and that others are combined in mineralogically dictated proportions due to the stoichiometry of minerals. **a, b** For example, K is conserved as olivine crystallised from a basic magma, and the concentrations of other elements may be standardised to K for the purpose of evaluating magmatic-differentiation trends. Pearce diagrams may not be used for regression because the variables are not independent (each involves the same elemental concentration in the denominator), and slopes of the lines are fixed by mineral stoichiometry, not by independent correlation. Pearce element ratios are very useful in the study of magma evolution. They do not solve the closed-sum problem for statistical work; the variables are hopelessly confounded and the observations are not normally distributed about a regression line. **c** Error-confidence ellipses are inclined to both axes, indicating the interdependence of the variables and emphasising their uselessness for regression (Chap. 4, Fig. 4.2)

shows the elliptical symmetry of bivariate data. Crescent-shaped patches and curved "trends" may appear, but their shapes are artefacts of the triangular distribution (Weltje 2002; Fig. 7.16b).

7.7.6
Multiple and Partial Correlation

In Chapter 6, we saw that *multiple regression* enabled us to regress *y* simultaneously on several controls or independent variables, $x_1, x_2, \ldots x_n$ to define a regression surface. For three variables, the regression may be envisioned as a surface but it is rarely planar. Like simple regression, it is important that the variables, x_i, are not correlated or mutually dependent. However, collinearity may be more difficult to identify in the absence of graphic representation. The association among the variables may proceed by inspecting the coefficients of the equation for the regression surface, but there is no single correlation coefficient as with linear

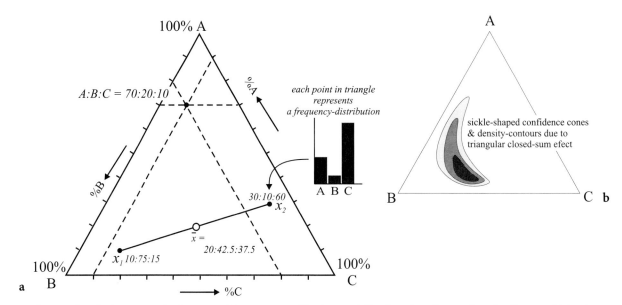

Fig. 7.16. Petrologists commonly plot compositions represented by three variables (here *A*, *B*, *C*) on a triangular diagram which epitomises the constant-sum problem. Each apex represents a composition comprising 100% of the constituent at that corner; each point within the triangle represents a distribution of the components as shown by the histogram for point x_2 with 30% *A*, 10% *B* and 60% *C*. As with other diagrams presenting ratios or percentage contents, the variables are nonstochastic. Strict limitations on the degree to which they may vary and the strong interactions between their values prohibit most statistical procedures. Compositions relate to the positions in the diagram (e.g., see mean of x_1 and x_2) referred to as a mass-balance principle

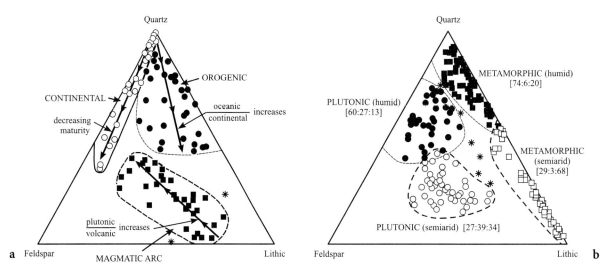

Fig. 7.17 a, b. Examples of ternary diagrams to illustrate compositions in terms of three components: Here, quartz, feldspar and lithic content (= rock fragment) may be used to identify depositional conditions or provenance. We must recall that the data points are constrained by the constant-sum issue unlike completely liberated stochastic variables. Therefore, trend lines cannot be determined by regression and cannot define correlations with the conventional objectivity of traditional statistics

regression as there are more than two variables. However, the quality of the association may be simply summarised by one statistic, the *multiple coefficient of determination*. Its value is represented by R_m^2, but it must be calculated similarly to Eq. (7.5):

$$R_m^2 = \frac{\text{variability explained by regression surface}}{\text{total sample variability}}$$
$$= \frac{SS_{yy} - SSE}{SS_{yy}} \tag{7.17}$$

Like its counterpart in linear regression, the multiple coefficient of determination ranges from zero to unity as the degree of association improves. A regression surface that explains perfectly all of the variation in the data set has $R_m^2 = 1$. The multiple coefficient of determination rarely has high values in earth-science data sets because our observations are subject to natural variation. Also, it is more difficult to regress satisfactorily as the number of variables increases. We should note that R_m, like the simple linear correlation coefficient (R), expresses correlation between two variables. Now, however, the "control variable" is really a blend of several variables. It is also possible to assign different weights to each of the control variables if there is a good scientific reason to believe that they contribute to the process to different degrees. For example, if x_1 and x_2, are believed to be half as effective as x_3 in controlling y, then weighting-factors of *0.25, 0.25* and *0.5* could be applied to each x_1, x_2, and x_3 value respectively. The sums of the weighting coefficients must be unity, of course, and they are sometimes called *beta-weighting coefficients (β-weights).*

Correlations within a system of more than two variables are quite involved. However, with multiple variables, one may also investigate *partial correlations* whose interpretation must be very carefully considered in the light of the specific subject. For example, consider a system where y is dependent on two variables x_1, x_2. If we determine the linear regression of y on x_1, we are performing a statistical experiment, regression in a plane perpendicular to x_2 in three-dimensional data space. This is comparable to a laboratory experiment in which one of two control-variables is held constant to permit an evaluation of the other control on the dependent variable. It should be clear that the absence of any correlation between the control variables, x_1 and x_2, is mandatory.

It is only possible to visualise this concept in a diagram where y depends on just x_1 and x_2, but it is valid also for more than two control variables. Laboratory scientists commonly use a phrase such as "*all other fac-*

tors being constant, y shows a strong dependence on x_2''. This means that x_1, x_3, etc., were kept constant. For example, the uniaxial strength (y) of sandstones may depend on grain size (x_1), quartz content (x_2), matrix silicification (x_3), and anisotropy (x_4) due to bedding fabric. In laboratory tests of rock strength, it is usual to test batches in which all but one of the independent variables (x_i) are held constant. *Partial correlation* provides a statistical analogue in which the correlation of y may be isolated with respect to just one control, e.g. x_2. However, most multivariate earth-science data are natural observations, sensitive to the whims of nature, and it is unlikely that we may isolate variation to just one control variable at a time.

Below, notation is simplified if all variables are indicated as x_i; in any case, it is not always meaningful or possible to define a single dependent variable as y. Thus, we define each observation in terms of its several values $x_1, x_2, \ldots x_n$ in n-dimensional space, without the notation bestowing dependence on any particular observations. The apparent linear correlation between x_j and x_k would be designated a partial correlation coefficient R_{jk}.

In three dimensions, $R_{12(3)}$ is the partial correlation coefficient between x_1 and x_2, simulating a situation where x_3 is constant. It is determined from the linear correlation coefficients by

$$R_{12(3)} = \frac{R_{12} - R_{13}R_{23}}{\sqrt{(1 - R_{13}^2)(1 - R_{23}^2)}} \tag{7.18}$$

The same equation may be extended up to any four of n-variables to determine, e.g. the partial correlation of x_2 and x_3 keeping x_1 and x_4 constant ($R_{23(14)}$). This is achieved by replacing the linear correlation coefficients on the right-hand side of the equation (e.g. R_{13}) with partial correlation coefficients such as $R_{23(4)}$. They, in turn, isolate the contribution of some other variable (x_4).

7.8 Multivariate Situations

Most of Chapters 6 and 7 have been concerned with bivariate situations in which the behaviour of the system may be described by two variables, dependent (y) and independent (x). The dependent variable could be influenced by more than one variable (e.g. y as a function of x_1 and x_2), permitting multiple linear regression (e.g. Fig. 6.11), but that does not constitute a full multivariate situation. In multivariate systems, several variables show varying degrees of dependence upon one another. For example, the simplest multivariate situation would involve three variables (x, y, z) but x would have an influence on y, y on z and x on z. With three variables, one (z) may be represented by contouring it in (x, y) space, similar to a topographic map. Specific three-variable plots in two dimensions have been devised for certain subdisciplines, most familiar of which is the triangular or ternary diagram in petrology that presents three values with a fixed range, for example percentages of minerals or oxides (Figs. 7.16, 7.17). Unfortunately, this necessarily awakens the constant-sum issue, and any procedures using the diagram are subject to closure-restraint problems. Triangular diagrams severely restrict statistical approaches. Graphic representations fail completely with larger numbers of variables $(x_1, x_2, ..., x_m)$, but the advanced field of *multivariate statistics* tackles these situations numerically. For example, fresh-water and marine sediments may be distinguished by their faunal assemblage of numerous species; the paleomagnetic potential of rocks may depend on many properties (mineralogy, domain structure, grain size and shape, coercivity, coercivity of remanence, and/or saturation remanence).

In multivariate systems, *co-linear variables* are particularly troublesome. Ideally and most simply, the independent variables $(x_1$ to $x_n)$ should behave without any artificial constraints. Thus, it is legitimate to investigate dependency of y on x_1 and on x_2, but x_1 must be an independent phenomenon from x_2. For example, atmospheric CO_2 may be influenced by volcanic activity, but we must choose measures of volcanic activity that are independent. The number of volcanoes (x_1) and the volume of volcanic rock (x_2) would be unsuitable, already correlated variables. Co-linearity problems are common with ratios or constant-sum data but they may be more subtle and real. For example, although a low-abundance trace element may be free from closure restraint in general, it may be hosted in a specific mineral. If the mineral is a major component of the rock, its abundance may be subject to the constant-sum problem, which may devalue the behaviour of the trace element as an independent variable. Identification of collinear variables may be difficult in multivariate situations, requiring detailed knowledge of the subject.

When manipulating systems of multivariate data, each quantity should be measured in similar units. This is self-evident and simple if, for example, all the quantities are concentrations measured in parts per million (ppm) or percentages. However, variables with very different magnitudes are not a deterrent. For example, we may compare trace elements in ppm and major oxides in weight percent by standardising each variable by its mean value.

In all of these cases, each variable, interacting to varying degrees with each other variable, possesses its own frequency distribution. Fortunately, in many multivariate situations, the variables may each approximately follow the Normal distribution. For m variables, in m-dimensional space, the multivariate mean is represented by a vector from the origin to the cluster's centre of mass. This is readily envisioned with just three variables (x_1, x_2, x_3) (Fig. 7.18a). The variances and covariances of each variable are represented by the elements of a matrix with m rows and m columns.

The two main goals of multivariate statistics concern either distinguishing or classifying clusters of observations according to the variables that define them. These are really two inverse applications of the same process. If we can successfully discriminate between groups of data as a corollary, we should be able to classify individual observations into each of the groups. Thus, first we outline the concepts of discriminant-function analysis and then of cluster analysis. For the purposes of illustration, we are restricted to three variables, but generally, there are more $(x_1$ to $x_n)$. In addition, it may be necessary to discriminate more than two data groups or to pigeonhole measurements into several clusters. Unlike other statistical procedures that occupy our attention throughout this book, there is no requirement for this type of data to follow some theoretical distribution. On the contrary, earth scientists specifically select objects such as geological processes, mechanisms, rock types, faunal assemblages, depositional conditions, etc., on the basis of a bias towards some attributes or variable values that interest them or that are imposed by Nature.

An important aspect of discrimination and cluster analysis is that there is an underlying assumption that although the variables (x_i) are continuous stochastic variables, the groups are not. We proceed on the assumption, or knowledge, that some underlying physical process separates the groups on the basis of the different weights of each variable, x_1 to x_n. This is different from previous techniques of bivariate analysis, such as linear regression that require a single homogeneous group of observations. Multivariate techniques that assume separate data groups are usually reserved for more complex situations. Geological intuition and circumstantial evidence may sow the seeds of interpretation, but multivariate statistical approaches may permit us to detect associations among any of many stochastic variables.

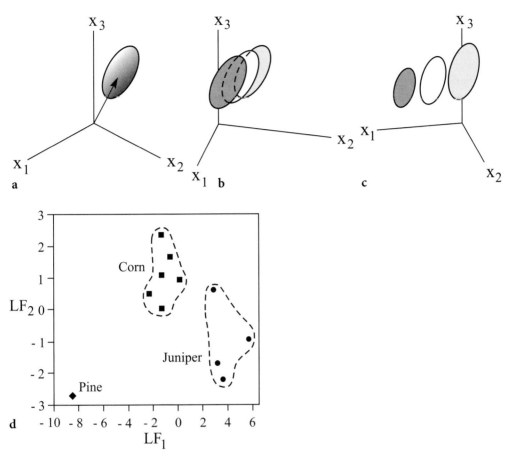

Fig. 7.18. Multivariate data represent observations for which several quantities may vary independently and stochastically. This can be visualised with three variables but is extendable to n variables in the general case. **a** A vector from the origin to the data centroid representing the mean and variance is intuitively indicated by the dispersion of the data cloud. **b** Linear discriminatory function analysis (LDFA) distinguishes groups of observations, represented here by different *shading*, on the basis of their distribution in n-dimensional space defined by stochastic, independent variables, x_1 to x_n. **c** LDFA determines a linear function, e.g. $LF = a_1x_1 + a_2x_2 + \dots + a_nx_n$ that defines a viewing axis that best discriminates between the groups. **d** Pierce et al. (1998) used this technique to discriminate the fuel sources used by ancient Native American cultures in the southwestern USA. Two linear discriminatory functions (LF_1, LF_2) distinguish discrete fuel sources on the basis of seven critical, non-interdependent elemental abundances obtained by chemical analysis. Great care is taken in this kind of study to select variables that are quite independent, i.e. not forced to vary collinearly

7.9
Linear Discriminant-Function Analysis (LDFA)

Complicated multivariate systems commonly beg the question, "What environmental or process factors distinguish between the main groups of data?" The problem is reduced essentially to finding a relationship amongst the variables $(x_1, \dots x_m)$ that clarifies the difference between the groups of observations most effectively. This can be appreciated visually where there are just three variables. Consider Fig. 7.18b, where the three clouds of data represent three rock types for

which measurements of the variables x_1, x_2 and x_3 are available. From any arbitrary direction, it is difficult to discriminate among the three rock types due to their overlap in that particular view through (x_1, x_2, x_3) space. However, one can appreciate that some particular view may discriminate more effectively between the rock types (Fig. 7.18c). LDFA finds the optimum viewing direction described by the coefficients of a line in the variables' space. For three variables, this linear function will be of the form:

$$LF = a_1x_1 + a_2x_2 + a_3x_3 \qquad (7.19)$$

For more than three variables, the visual analogies and "view" have no meaning but the linear discriminant function may still be calculated. The coefficients (a_i) of *LF* define the discriminatory function and act like weights that indicate the relative importance of each variable. However, the value of *LF* may suffice as a characteristic number to separate data groups because it defines a function that represents a view that reveals the minimum overlaps of data groups. For example, in geoarcheology, it can be important to know the fuel sources used by ancient civilisations in order to understand their economy and land-use patterns. Elemental analysis of ancient hearth ash reveals fuel sources several hundred years ago by Native Americans in the southwestern United States (Fig. 7.18d). From the hearth-ash analyses, seven elements were selected carefully to minimise collinearity that would otherwise produce spurious interdependence and thus a false correlation. That is always a potential risk with constant-sum compositional data. LDFA determined the most favourable coefficients (a_1 to a_7) of two linear functions LF_1 and LF_2, for which three clear data groups emerge. These three data groups discriminate and isolate three fuel sources which are attributed to pine, juniper and corn fuels by comparison with modern ash (Pierce at al. 1998).

7.10
Cluster Analysis

Geologists commonly wish to lump together many observations to define a complex concept, object or process in a particular category. For example, many observations of individual variables may characterise rock composition, faunal community, or depositional environment. Depositional environments from several different regions may fall into different categories depending on variables such as the absence or presence of several sedimentation structures, the number of stratigraphic breaks, the amounts of certain mineral grains, the relative abundances of clastic versus chemical sediments, etc. Clearly, the variables may be attributes, ratios or number values that must be normalised before further processing.

In the general case, we could study n objects affected by m variables. Objects may be plotted in *m-dimensional space* like any multivariable observation defined, for example by coordinates $x_1, x_2, x_3 \ldots x_m$. The similarity between objects would be difficult to recognise where several variables are involved; however, similar objects should cluster in m-dimensional space.

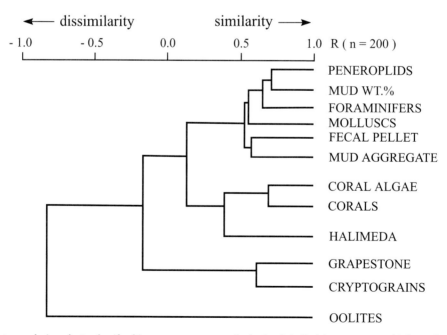

Fig. 7.19. Cluster analysis seeks to classify objects or processes on the basis of similarities amongst multiple stochastic variables. Philosophically, this could be considered as the inverse of the approach used in linear discriminatory function analysis. The affinity of data groups is indicated by small distances between group centroids in n-dimensional space, defined by the variables x_1 to x_n (e.g. see Fig. 7.17c). When the affinities between objects are established by small intercluster distances, one may calculate the correlation coefficients for appropriate object pairs. These provide a more useful scale for the cluster diagram: the example shown is from a classic study of the constitution of Bahamian carbonate sediments. (Parks 1966)

Table 7.5. Similarity matrix for cluster analysis for one measured variable (k)

	1	2	3	4	5	...j
1	0	d_{12}	d_{13}	d_{14}	d_{15}	d_{1j}
2		0	d_{23}	d_{24}	d_{25}	d_{2j}
3			0	d_{34}	d_{35}	d_{3j}
4				0	d_{45}	d_{4j}
5					0	d_{5j}
...i						0

A distance or separation coefficient can be determined on the basis of measurements of each of the m variables on n objects. The separation or distance coefficient between two objects i and j is given by the Euclidean distance, d_{ij} averaged by the number (m) of characters or variables measured from $k = 1$ to m.

$$d_{ij} = \left[\frac{\sum_{k=1}^{m} [x_{ik} - x_{jk}]^2}{m} \right]^{\frac{1}{2}} \tag{7.21}$$

For the total of n objects and m variables, a similarity table or matrix may be erected with values determined from the above equation. Inspection of the similarity-matrix table then reveals the closest similarities (smallest separations) between objects (e.g., Table 7.5).

From the individual values of this table one may construct a tree diagram that groups together samples into clusters of similarity, using a distance scale. For example, a study of clast type in Bahamian carbonates was based on 12 types of quantitative observation, which were normalised to unity (data of Imbrie and Purdie 1962, in Parks 1966). The observations were of abundances of: peneroplids, foraminifers, molluscs, Halimedas, coralline algae, coral, mud, mud-aggregate, fecal pellets, grapestones, cryptograins and oolites. Of course, these records were all normalised to a fraction of unity.

A matrix of inter-object distances is used to record similarities (Table 7.5) from which a cluster diagram may be constructed to reveal how samples with great affinity are grouped, based on the eight most important variables. However, it is of greater geological interest to inspect how the variables group with regard to the separation. The original authors also constructed a matrix of correlation coefficients of inter-object distances, notwithstanding the issues concerning the interdependence of some of the observations. Variables with a strong affinity have branches with high positive values of the correlation coefficient ($\sim +1$) and variables that do not show an affinity have strong negative correlation (~ -1). The interrelationship of the attributes used by Imbrie and Purdy (1962) to formulate a cluster diagram for the Bahamian carbonate sediments (Fig. 7.19).

CHAPTER 8

Sequences, Cycles and Time Series

Most earth-science data are *univariate*; single values suffice to describe the observations. Univariate statistics dominate introductory courses in statistics, and simple statistics determined from frequency distributions permit their characterisation (Chaps. 2–4). The next increment of complexity was introduced by bivariate data, where each observation required an *x*-value and *y*-value for its description (Chaps. 6, 7). In Chapter 7, the principles were extended to multivariate observations, each of which required several values (x_1, x_2, ... x_n) for its specification.

Nature provides another sort of dependency, one where the dependent variable (*y*) varies with time (*t*). The plot of *y* against *t* yields a *time-series*. Appropriate statistical procedures for *time-series analysis* are mentioned rarely in elementary statistics texts because applications are mostly restricted to specialised fields such as electrical engineering and physics. However, only rarely, for example in seismology and in certain analytical experiments, do earth scientists make precisely timed observations as would engineers and physicists. Although a time series plots an observed value (*y*) against time (*t*), time is not the control variable; it proxies for one or more interacting control variables that commonly change with time. Some examples are provided by stratigraphy (Fig. 8.1), measurements at geophysical observatories (Fig. 8.2), and climate change and dendrochronology (Fig. 8.2). At any one time, some unknown control variable(s) may take any one of a range of values, which is reflected in the value of y. All we know is that a single value of *y* was observed at time *t*. The time-series plot thus conveys a different meaning from a bivariate *x*–*y* plot.

Earth science is a unique combination of physical science and history that encourages us to observe some value or attribute against its age, which may be recorded directly or indirectly. Unfortunately, geological time is poorly and unevenly documented by comparison with other sciences. Where the absolute ages are fixed by radiometric-dating, magnetic-reversals, archeomagnetism, dendrochronology or archeology (Fig. 8.3), we have a time series amenable to treatment by the methods used in economics or electrical engineering. However, our timing errors are far greater; they vary with technique as well as absolute age. More seriously, we often overlook that the earth scientists' timed series include ages determined by different methods, which introduce errors due to internal inconsistency. This may do worse than to introduce non-linearity in the time scale; events may even be placed out of order. This is particularly the case if a dating method does not use a monotonic calibration curve but rather a complex cyclical one that introduces ambiguous possibilities for age, as we will see in the cases of archeomagnetic (Fig. 8.3 c) or radiocarbon (Fig. 8.3 d) age-determination.

We should not be discouraged that geologists have used *rate-dependent phenomena* to proxy for time. For example, varves (one per year), pelagic sedimentation rates (~ 10 mm/1000 a) and ocean-floor spreading rates (~ 50 mm/a) have been used as time proxies. To the surprise of our colleagues in more precise sciences, these rates are sufficiently linear and accurate for many purposes. Indeed, fundamental, independently verified discoveries about earth processes have been made in that way, for example in the field of plate tectonics. However, in the least accurately timed sequences, the order of observations may be merely relative, as for example, the types of rocks in a stratified sequence. These sequences may be investigated in terms of the randomness of their pattern or structure using *runs analysis* or *Markov chain analysis*. This has proven useful in certain aspects of sedimentology and structural geology where merely the order of recurring events is sufficient to characterise a process or phenomenon.

Time-series analysis may reveal a pattern with an identifiable renewal component, perhaps with a similar period for certain geological events but rarely with a reproducible waveform. For example, geochronologically dated reversals of the geomagnetic field throw light on the mechanism of the geodynamo, and variations in oxygen isotope ratios reveal climatic changes. In the simplest case, without involving any assumptions or theoretical models, periodicity may be identified by *autocorrelation*, which may permit some extrapolation beyond the studied time range. Natural

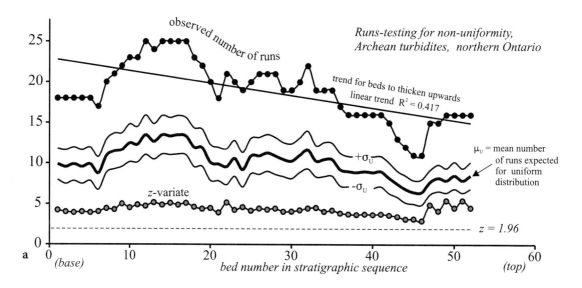

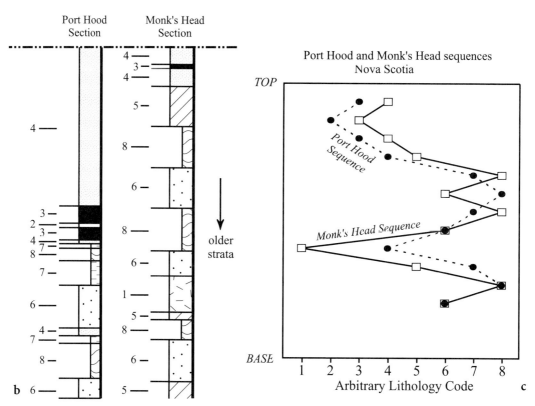

Fig. 8.1. a Runs testing is a simple technique to detect the tendency for patterns to persist in a sequence. Two states must be defined, and a *run* is defined as any cluster of like states from which the other state is excluded. This data, courtesy of P. Fralick (Lakehead University) defines two states as "next-bed thicker" and "next-bed thinner" in an 80-m stratigraphic section of turbidites. The observed number of runs exceeds that expected if the pattern was uniform and the z-variate exceeds 1.96, so we may reject the hypothesis that the sequence shows no pattern. In fact, there is a tendency for thicker beds to dominate towards the top of the sequence. **b** Two sections of a Middle Carboniferous cyclic sedimentary sequence from eastern Canada (Schenk 1969; simplified courtesy of P. Fralick, pers. comm.). Arbitrary code numbers are assigned to the different lithologies. **c** The graph of lithology code against stratigraphic depth reveals a pattern in the sequence. Stratigraphic position is commonly used as an inferior surrogate for time in geology; the stratigraphic (~time) scale is unlikely to be linear and it may not even be continuous. Nevertheless, it introduces the reader to the concept of a time series derived from the most fundamental record of geological history

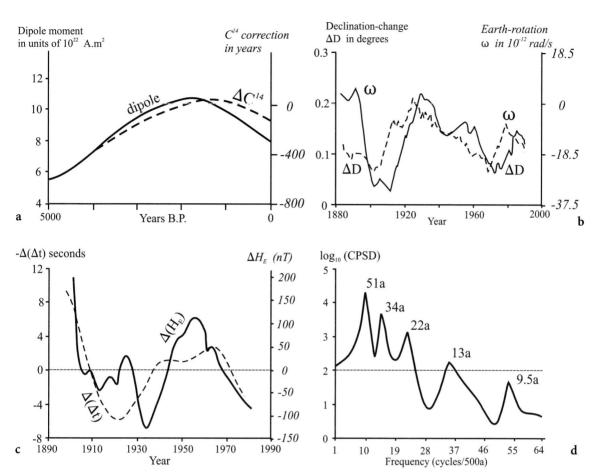

Fig. 8.2. Some geological time series may have very precise timing but may cover insufficient cycles to characterise the behaviour. This is common in observatory studies of recent geophysical and meteorological phenomenon. **a** The geomagnetic field's dipole-moment and the age correction required to adjust otherwise un-calibrated ^{14}C ages (Tarling 1983). These two phenomena are related as other considerations imply, although the curves are slightly out-of-phase. Unfortunately, the data do not even span one cycle. **b** Westward drift of the geomagnetic field and variations in the angular velocity of the earth have been precisely documented but barely span one cycle. Again, the phenomena are related but out-of-phase (Le Mouel et al. 1992). **c** Length-of-day fluctuations and pulsations in geo-magnetic-dipole moment are complexly inter-related cycles (Jin 1992). **d** Cross-power spectral density reveals characteristic mutual periods in the dataset of **c**

phenomena rarely produce time series comprising components of fixed waveform-geometry. The constituent waveforms, their periodicity, amplitude and phase may be rigorously defined by *Fourier analysis,* as used in electrical engineering. However, even where geological time series could satisfy the constraint that their constituent waveforms are precisely fixed, the data sequences may be too short and too noisy to give reliable results. In any case, component waveforms may have no genetic significance.

Earth-science time series illustrate the response of some complex phenomenon through time. However, we must always remember that time is not strictly a control or independent variable, it proxies for some environmental process in which there may be several lurking variables. For example, fluctuations in faunal assemblages of marine fossils may largely respond to the mean temperature of the oceans. Time is not a control and other controls lurk. They may include salinity, tectonically influenced oceanic-circulation patterns and volcanic "winters" that disrupt the food-chain. Thus, at best, *time* may be a vague umbrella proxying for multiple control variables.

Finally, in all kinds of sequences that are tied to an absolute geological time-scale, we must appreciate that the precision and accuracy of the ages in different parts of the sequence may differ. This should be obvious if, for example, geochronological ages are calculated using a mixture of U/Pb and Rb/Sr and K/Ar, according to the nature of material available. Still worse, one may use distinct techniques in different parts of the time scale, e.g. dendrochronology for more recent events and varves for older ones. The problems caused by such mixed procedures are commonly overlooked. Paradoxically, certain sophisticated laboratory-based methods may be deceptively heterogeneous in their precision and accuracy at different ages because they rely on a provisional calibration curve. Good examples of these are the radiocarbon method, archeomagnetic age determinations using the secular variation of the geomagnetic field, and the chronometric curves for viscous remanent magnetisation. Their elegance belies the variation of their sensitivity at different ages caused by the variable suitability of the calibration information used to make the master reference curve.

First, we will consider sequential events which may not be regularly spaced either in time or distance; they are merely observations that occur in a certain order. Geological field observations, such as samples along a traverse, provide a familiar example. Stratigraphy provides instructive examples where sequences of beds represent an order, which may have some meaningful pattern due to a controlling process. However, their occurrence in the sequence may not be related linearly to the progression of time or even to distance along a stratigraphic section. Notwithstanding the compression or dilation of time represented by the events (e.g. beds) in different parts of the sequence, their serial occurrence may be investigated as a quasi-time series. Limited analysis is possible by techniques such as *runs analysis* and *Markov chain analysis,* and they have the advantage that they are nonparametric; they do not make any assumptions about the form of the population from which the sample is drawn. Even for more quantified time series, it is usually safer to take a conservative nonparametric approach, e.g. using autocorrelation. However, the Fourier analysis approach, which assumes a constitution of component sine waves, is sometimes attempted because certain natural processes may possess some clear periodicity, e.g. those influenced by climatic, tidal or planetary cycles.

8.1
Runs Analysis

Some simple time-series are represented by the absence or presence of some phenomenon at different points in a sequence. Thus, the variable is *dichotomous,* having only two possible states, e.g. 1 and 0, and since time is relative, the duration of either state may be variable and is unimportant in this context. A run is defined as a group of adjacent *like-states* separated by the other states. The run may be of any length, including just one member. A possible sequence progressing with time could be read, left to right as:

| 1 | 0 | 0 | 0 | 1 | 1 | 0 | 1 | 0 | 0 | 1 | 1 | 1 |
| 0 | 1 | 0 | 1 | 0 | 0 | 1 | 1 | 1 | 1 | 1 |

(*time* →)

For example, at the end of this recorded sequence a run of five states of the same kind, represented by "1" occurred. A run is any group of like-states, no matter how short. The sequence possesses 13 runs that are separated in the following line for clarity:

1 **000** 11 **0** 1 **00** 111 **0** 1 **0** 1
00 11111

The usefulness of this approach to earth scientists is that events of uneven duration without absolute ages may be examined for randomness or order in their sequence. This is particularly useful in stratigraphy where only relative age is known. Runs may be devised by considering the observation in terms of two states. The main problem in applying runs testing in natural situations is to be sure that the record is complete. We must be sure that no states of one kind, or the other, have been overlooked or omitted in sampling. This may not only be a problem of the sampling strategy but an intrinsic feature of some geological process. For example, in a stratified sequence of beds the two states (1 and 0) may represent the presence or absence of some lithology or sedimentary structure of interest. However, intermittent erosion may have removed part of the record, changing the lengths of some runs and even selectively removing one state, such as an easily eroded rock type. In particular, careful observations must be made because a run will be omitted if one fails to observe a self-contact, i.e. two examples of the same state in contact.

In a sequence comprising just two kinds of observation, how can we be sure that there is any meaningful pattern? The simplest approach is to compare the number of runs of like kind with that which could arise by the random effects of nature. For simplicity, we shall model natural randomness with a uniform distribution. We need to record:

U = number of runs of either state (e.g. 13 in the example above)

n_1 = number of counts of one state (e.g. 14 counts of "1" in the example)

n_2 = number of counts of the other state (e.g. 10 counts of "0" in the example)

It can be shown that the mean (μ_U) and standard deviation (σ_U) of the number of runs (U) expected for a random distribution are given by the following equations:

$$\mu_U = \frac{2n_1n_2}{n_1+n_2}+1 \quad \sigma_U = \sqrt{\frac{2n_1n_2(2n_1n_2-n_1-n_2)}{(n_1+n_2)^2(n_1+n_2-1)}}$$

(8.1)

If $n_1 \geq 9$ and $n_2 \geq 9$, the sampling distribution of the total number of runs (U) approximately follows the Normal distribution. In that case, the observed number of runs may be standardised as the z-variate and compared with the percentage points of the Normal distribution (Table 3.5). The z-variate would be given as follows (from Chaps. 3, 4):

$$z = \frac{U-\mu_U}{\sigma_U}$$

(8.2)

The value calculated for the test statistic, z, is then compared with the tabled values of the Normal distribution of probabilities to decide if it is an extreme and improbable value that would indicate non-randomness. Usually, we wish to detect some order, i.e. non-uniformity or non-randomness, so we set the null hypothesis H_0: $U = \mu_U$, that the sequence is random. If $z > 1.96$ or $z < -1.96$, we must reject the null hypothesis at the 95% level, and accept the alternate hypothesis that the sequence shows some non-random pattern or structure. In other words, there are either too few runs of one state or too many runs of the other state for the sequence to be considered random. If the measured value of U yielded a z-value such that $-1.96 < z < +1.96$, it would not be possible to reject the hypothesis of randomness.

As an example, let us consider a sequence of Archean turbidite sediments studied by Barrett and Fralick (1989) in northern Ontario and data collected by one of Fralick's Lakehead graduate students (Purdon 1995, unpubl. data). It was suspected that the sequence showed some inherent pattern with a tendency for thicker beds to become more common with time. They carefully documented the thicknesses of beds and other sedimentation features through several hundred metres thickness of turbidity flows. Here we examine only ~70 m of the profile using just one of Fralick's observations, namely, whether a bed is followed by a thicker, or by a thinner, unit. Thus, two states are defined that may be analysed by the runs

method. We shall pose the null hypothesis that the sequence shows a random pattern of the two states. If this is rejected, we must accept the alternate hypothesis that the sequence is structured and shows some pattern.

The number of runs was recorded for the entire sequence. From the sequence, windows of data were selected that contained sufficient runs to validate use of the technique, as explained above ($n_1 \geq 9$ and $n_2 \geq 9$). For each window, the mean (μ_U) and standard deviation (σ_U) of the number of runs expected for a uniform distribution were calculated according to the above formulae. These statistics are plotted for many overlapping windows moving up the stratigraphic column (Fig. 8.1 a). The graph shows that the observed number of runs is much larger than could be reasonably expected from a random distribution. The z-variate was also calculated for each window subsample and its graphed values are clearly > 1.96. Therefore, it lies in the rejection region for a significance level $\alpha = 0.05$, and we must reject the hypothesis that the observed number of runs could occur in a uniform (~ random) distribution. Since we are 95% confident that the pattern is nonrandom, we must accept the alternative hypothesis that the sequence is structured. A regression line for the observed number of runs indicates the trend for thicker beds to dominate towards the top of the sequence. The trend is significant at the 95% level (see Chap. 7, Fig. 7.5).

The runs method is a simple robust method to identify patterns in sequences. However, the observations must be reduced to two possible states for its application. It is usually possible to arrange the actual observations in such a fashion and even to detect trends by applying runs analysis to moving windows through the sequence. Marsal (1987) shows some nice examples where runs testing may be applied to the study of petrographic textures, testing *fabric homogeneity* and *fabric isotropy* from grain counts or measurements along certain traverses through specimens. However, geology commonly requires tests of patterns with observations of multiple states and of the transitions from one state to another. Some elementary approaches are mentioned below.

8.2
Rank Correlation

Correlation in Chapter 7 mostly concerned the comparison of precisely quantified data, and the correlation coefficient was shown to be a simple and powerful statistic with which to establish the validity of a correlation between two sets of values. Unfortunately, most stratigraphic-sequence data are not easily described by precise quantities. However, one correlation technique was described that deliberately avoids the use of values but only compares their relative order, or rank, in a sequence. Spearman's rank-correlation coefficient assigns arbitrary ordinals (1, 2, 3...) to one set of observations and then determines the rank (= position in sequence) of the paired observation in the other set. With precisely quantified variables, the general value of Spearman's rank correlation is its insensitivity to outliers as only the rank of the observations is considered. Thus, one may detect any kind of association or correlation, however complex and nonlinear. This simple procedure may be applied to compare or correlate stratigraphic sequences. Consider the two sequences from Nova Scotia (Fig. 8.1b). An arbitrary lithology code has been assigned to the rock types which occur in both sequences, e.g. 3 = mudstone, 6 = sandstone. Starting at the bottom of the Port Hood section we write down a list of the lithology codes in ascending order. The corresponding list is assembled for the Monk's head sequence. The two lists are compared and the difference in code is taken to be the difference in rank. A sequence of 12 successive rock types was recorded from each list ($n = 12$) and the sum of the squared differences in rank between the two sequences was determined as 23. Using the formulae for Spearman's rank-correlation coefficient (Chap. 7) yields a modestly encouraging result, $R_S = 0.73$, which is significant at the 95% level (Table 7.3). One may conclude that there is some basis for a similarity between the two sequences, however complex that may be.

8.3
Markov Chain Analysis: Identifying Cyclicity

A method of sequence analysis that makes fewer demands on a theoretical model and makes no assumptions about the original pattern is Markov Chain analysis. This analysis focuses on the transition from one condition to another in a sequence and permits detection of stratigraphic cyclicity or preferred juxtapositions of lithologies. For example, if sandstone, limestone, mudstone and shale occur in various relative positions in a stratigraphic column, it may be possible to identify whether a sandstone-to-shale transition is more common than a sandstone-to-limestone transition. This would indicate if there is a preferred order to the lithological sequence. There can be good geological reasons to expect a certain sequential order. Progres-

sive evaporation precipitates salts in a certain order in evaporite sequences undisturbed by aqueous influx. Similarly, rhythmic deposits or cyclothems may be identified from coal deposits, turbidite deposits and seasonally controlled sediments such as glacial varves or loess. With care, we may be able to identify an ideal sequence, which natural processes try to maintain against the competition from spurious environmental disturbances that are considered to be randomising influences.

The incidences in which beds A, B, C and D follow one another might be recorded in a transition frequency matrix. Below we record the number of times there is a transition from rock type B to rock type D, etc. The leading diagonal (top left–bottom right) should be filled with zeroes if no discernable contact exists between two beds of the same rock type. However, that is not mandatory, as many cyclothymic coal sequences and turbiditic Bouma cycles may show the erosion or non-deposition of an expected rock type. Its omission from the sequence may therefore juxtapose two like beds and this may be identifiable from field observations, permitting a non-zero value to be entered into the leading diagonal. However, non-zero values in the leading diagonal may be artefacts of the spatial-sampling procedure, where records are made at predetermined stations, rather than observing individual transitions (see Chap. 1).

The observations are tallied in a frequency matrix which in turn is recalculated as a transition-probability matrix, with each entry expressed as a fraction of the grand total of transitions which is 50 in this example (Table 8.1). The probabilities will be expressed as a fraction of unity, and the grand total probability (bottom right table entry), should be the sum of the cells in the column above, or of the row to the left, and equal to one. Consider the top row of the transition-probability matrix (Table 8.2). It reveals that a transition upwards from A to B has a 6% probability, from A to C has an 8% probability and that 16% of the transitions are from A to D. The transition from A to D is most common, from the probabilities shown in Table 8.1.

Table 8.1. Frequency matrix: transitions recorded in sequence: hypothetical example

	To A	To B	To C	To D	Total
From A	0	3	4	8	15
From B	3	0	2	5	10
From C	4	2	0	5	11
From D	6	5	3	0	14
Total	13	10	9	18	50

Table 8.2. Transition probability matrix: hypothetical example

	To A	To B	To C	To D	Total
From A	0	0.06	0.08	0.16	*0.30*
From B	0.06	0	0.04	0.10	*0.20*
From C	0.08	0.04	0	0.10	*0.22*
From D	0.12	0.10	0.06	0	*0.28*
Total	*0.26*	*0.20*	*0.18*	*0.36*	1.00

Note: cell entries subject to rounding.

Some data for a real stratigraphic sequence in Nova Scotia has been provided by Dr. P. Fralick (pers. comm.), of which a part is illustrated in Fig. 8.1 b. Eight different lithologies are recognised, from which Fralick recorded how often one lithology passed upwards into another. For example, in how many instances was black mudstone succeeded by red mudstone? This was repeated for all combinations of the eight lithologies present. The tallies are recorded in a frequency matrix (Table 8.3), and these are converted into probabilities in Table 8.4 by dividing individual cell-entries by the grand total of 119 observed transitions.

The leading diagonal contains non-zero values because self-contacts are observed, records were made of every exposed transition. Some authors refer to a "sequence memory", where one lithology shows a strong tendency to be followed by another particular lithology. This anthropomorphic usage is justified in the studies of paleomagnetism, because there is indeed a physical retention of information in the form of magnetic ordering, directly analogous to human memory. However, in stratigraphy, repeatedly occurring lithological transitions point rather to a reproducible process, in which a sequence of lithologies is controlled by a recurring progression of physical changes, not by the physical retention of information that mimics memory. This sequence in Nova Scotia reveals that the more common transitions, neglecting self-contacts, in order of descending probability are:

1. $CaSO_4$ up to red mudstone (9.2% of transitions)
2. Microdolomite up to oosparite (6.7%)
3. Red mudstone to black mudstone (5.9%)
4. Oosparite to biomicrite and microdolomite (4.2%)
5. Biomicrite to oosparite (4.2%)
6. Red mudstone to microdolomite (4.2%)

Table 8.3. Frequency matrix for transitions in a stratigraphic sequence from Nova Scotia

Upward Transition to →	1	2	3	4	5	6	7	8	Row totals
From ↓									
1 Limestone	4	1			1			1	*7*
2 Biosparite	1		2	1			1	2	*7*
3 Oosparite	1	3	1	5	5			1	*16*
4 Biomicrite		2	5	3	1	2		2	*15*
5 Microdolomite	1		8	2	4	4		3	*22*
6 CaSO$_4$	1	1	1		4	3	1	11	*22*
7 Black mudstone		1	2	2	3			1	*9*
8 Red mudstone	1	2		2	5	3	7	1	*21*
Column totals	*9*	*10*	*19*	*15*	*23*	*12*	*9*	*22*	119

Table 8.4. Transition probability matrix for a stratigraphic section in Nova Scotia

Upward Transition to →	1	2	3	4	5	6	7	8	Row totals
From ↓									
1 Limestone	0.034	0.008			0.008			0.008	*0.059*
2 Biosparite	0.008		0.017	0.008			0.008	0.017	*0.059*
3 Oosparite	0.008	0.025	0.008	0.042	0.042			0.008	*0.134*
4 Biomicrite		0.017	0.042	0.025	0.008	0.017		0.017	*0.126*
5 Microdolomite	0.008		0.067	0.017	0.034	0.034		0.025	*0.185*
6 CaSO$_4$	0.008	0.008	0.008		0.034	0.025	0.008	0.092	*0.185*
7 Black mudstone		0.008	0.017	0.017	0.025			0.008	*0.076*
8 Red mudstone	0.008	0.017		0.017	0.042	0.025	0.059	0.008	*0.176*
Column totals	*0.076*	*0.084*	*0.160*	*0.126*	*0.193*	*0.101*	*0.076*	*0.185*	*1.000*

Note: cell entries subject to rounding.

Self-contacts of limestone (3.4%) and microdolomite (3.4%) are quite common. From the preceding information, commencing with the most common transition, one might conclude that two common transition sequences are as follows, established in the form of the following branching pattern or *tree*:

CaSO$_4$ → red mudstone → microdolomite →
oosparite → *EITHER* → biomicrite *OR* → microdolomite
red mudstone → black mudstone

This includes only the most prominent lithological transitions because there is insufficient evidence to include them all in a single idealised sequence from the available data.

Miall (1973) shows that the features of the transition probability matrix, above, may be enhanced as follows. One constructs a probability-difference matrix in which each cell contains the difference between the cell entry in the observed transition-probability matrix and the cell entry expected *if the transitions were completely random*.

8.4
Time Series

Data attached to a chronological scale can be treated more profoundly and yields more information for further study than observations simply arranged in some relative order. If one event occurs every second, its period (T) is 1 s and its frequency ($f = 1/T$) is 1.0 s^{-1}. Frequencies are commonly recorded in units of Ma^{-1} for pre-Pleistocene history, and in ka^{-1} for Pleistocene geology and prehistorical archeology. Renewable processes in earth science do not have precise periods unless they are related to solar activity or astronomical events, for example. We may still estimate useful parameters like frequency and period, but the actual graph of the observations with time may not have a constant waveform. This makes such natural time series intractable to many techniques developed to understand time-series for communications engineering and electronics (e.g. Wiener 1966; Spall 1988). Physicists and engineers would regard the terms "frequency" and "wavelength" as very flattering descriptions for the renewal processes found in the earth's prehistory.

One might characterise geology as a combination of physical science and history, so our fascination with time-series is understandable. Their study may reveal something about periodic controlling processes and permit extrapolation of events to earlier unrecorded events or even to future events. Where the time scale is

precisely fixed, for example by geochronology, interpretation is most reliable. However, even semi-quantitative chronologies based, for example, on constant pelagic sedimentation rates, permit some appreciation of the time-dependence of certain geological events.

Chronological errors may severely reduce the usefulness of geological time series, and this problem is sometimes overlooked. Fluctuations in the geomagnetic field, either secular variations of direction (0.6 to 1 per ka) or their complete reversal (1 to 3 per Ma) are one of the most clearly repetitive time series. However, like most geological renewal processes, they are not always suitable to establish periodicity by well-known methods applied to rigorously defined waveforms, e.g. in electrical engineering (fast Fourier transform; FFT). Instead, the short, more irregular waveforms may be studied by simple methods such as autocorrelation or the more sophisticated maximum-entropy method (MEM; Barton 1983). Systems driven by paleoclimatic variations, such as the time series for oxygen isotopes, abundances of certain marine fossils, and gas compositions of bubbles in glacial ice, capture the student's imagination more readily. However, in view of the imprecision of timing, one has the general impression that paleoclimatic time series tend to be over-interpreted in comparison to others, such as geomagnetic time series.

The initial accumulation of such sequential data includes errors of observation, but as we have seen before in geological examples, the natural variance of the phenomenon may dwarf observational errors. Nevertheless, quite complicated geological processes yield time series that show clear indications of trend or periodicity even if the time scale is poorly defined. In part, this paradox may be because the observations involve natural processes or laboratory experiments that average out, either singly or in combination, the intrinsic variation of the natural process over a time interval. This plots as a single point on the time axis but actually smoothes out some scatter. Consequently, the noise in the original measurement may be much less serious than the error in the age. This contrasts with other applications of time series, e.g. in economics, meteorology or physical experiments where timing is precisely known but the measurement error could be more important. Where time is reliably determined, the very techniques used for studying and characterising the time-dependent variable may help to suppress noise in its measurement.

A frustrating stumbling block in some of the most interesting earth-science time series is that the data run is too short to establish the repetition pattern with certainty. For example, adequate seismological records span only decades, the archeological dated record of secular changes in the earth's magnetic field spans less than two millennia. Still more tantalising are variations for which theory predicts some well-defined renewal process but where the existing database barely spans one cycle. The variation in the relation between dipole moment and C^{14}-correction, and the westward drift of the geomagnetic field are good examples (Tarling 1983; Le Mouel et al. 1992). Each of these series provides less than one complete cycle in the last 5 ka due to the limited archeological and dendrochronology record (Fig. 8.2; Table 8.5). Although observatory time scale geophysical methods are continuous and of high precision, they may reveal only a few cycles also, e.g. the relationships between length-of-day fluctuations and geomagnetic intensity (Jin 1992). From Jin's records, collected over less than 100 years, a few cycles permit determination of fundamental periodic behaviour at 51, 34, 22, 13 and 9.5 years (Fig. 8.2c, d). This requires sophisticated analyses beyond the scope of this book, but it is introduced here to show just how much can be achieved with the advantage of precisely timed sampling. Unfortunately, this chronological precision is unobtainable for most repeating geological phenomena.

Other time series concern phenomena that produce an intrinsically low number of cycles so that it is difficult to characterise the behaviour even though the dating may be adequate. Historical climate fluctuations, mining disasters, landslides, glacially controlled climate changes and geomagnetic behaviour may all provide examples of this (Figs. 8.2 to 8.12). Either the process under investigation is limited to an intrinsi-

cally small number of cycles or the combination of record length and precision is unfavorable. However, for the younger geological record and archeological time scale, it is not all bad news; loess, varves, and growth-rings in corals or trees may produce sequences that yield more than enough seasonally controlled cycles. Precipitation and temperature records may show trends and changes that correlate interestingly with cultural changes (Fig. 8.3a, b). Complex oscillations and cyclic behaviour may be detected in climate, magnetic field and excess radiocarbon (Fig. 8.3c–e). The radiocarbon record is particularly interesting since it forms the most common chronological basis used for Quaternary and prehistorical studies. A long-term trend is clear for $\Delta^{14}C$, despite poorly regulated oscillations and some spectacular spikes that make it difficult to calibrate radiocarbon ages. Periodicity is difficult to recognise with the compressed time-scale shown (Fig. 8.3c), but the details of the oscillations are sufficiently well documented to determine ages quite precisely (Fig. 4.3). Secular variation of the geomagnetic field is recognisably periodic, both in declination and inclination, and has an easily comprehensible underlying cause, at least at an elementary level (Fig. 8.3d). The westward drift of the geomagnetic field axis causes differently inclined (and declined) field orientations to sweep past each location, with a period of the order of 1 ka (Runcorn 1959). The pulsation of declination is not perfectly periodic, as would be expected from the precession of a perfect dipole. On average, 20% of the geomagnetic field is attributed to non-dipolar anomalies caused by turbulence near the surface of the core, superimposing noise on the secular variation record.

Table 8.5. Comparison of some absolute age determination techniques in geology and archeology. (Aitken 1990; Bull 1996a, b; Faure 1986; Stuiver et al. 1998; Taylor and Aitken 1997)

	Method	Range or top limit	±
G[a]	$^{206}Pb/^{238}U$ – $^{207}Pb/^{235}U$	Geological	~2%
G	$^{87}Sr/^{86}Sr$ – $^{87}Rb/^{86}Sr$	Geological	~2%
G	$^{40}Ar/^{39}Ar$ – $^{40}K/^{36}Ar$	Geological	~2%
G	$^{143}Nd/^{144}Nd$ – $^{146}Sm/^{144}Nd$	Geological	~2%
G	*, Magnetic polarity[b]	< 600 Ma	0.1–5 Ma
A and G	fission-track	< 1 Ga	≤10%
A and Q	*, ** Radiocarbon	< 25 ka	≥25 a
A and Q	Electron-spin resonance	10 Ka – 1 Ma	≥10%
A and Q	Obsidian hydration	< 200 ka	~10%
A and Q	Luminescence	< 0.5 Ma	~5%
A and Q	*, ** Archaeomagnetic	< 10 ka	50–100 a
A	** Viscous-magnetic	< 5 ka	25–200 a
A	* Dendrochronology	< 12 ka	1–20 a
A	** Lichenometry	< 300 a	10–50 a

a, year; ka, 1000 years; Ma, million (10^6) years; Ga, billion (10^9) years.
A, Archeological use; Q, Quaternary use; G, use over a large part of the geological time scale.
* Cyclical variation superimposed giving non-unique solutions; ** requires calibration, site-, material- or lithology-dependent.

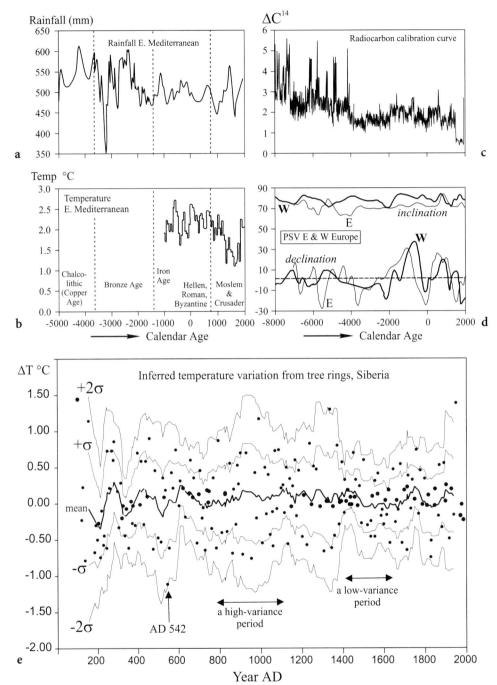

Fig. 8.3. Trend, stationarity and periodicity shown in Holocene time-series. **a, b** Historical rainfall and temperature records for Cyprus show negative trends, irregular periodicity and changes in variance with time. Note their influence on civilisation. **c** The radiocarbon calibration curve permits $\Delta^{14}C$ to be used to determine the age of woody material, and it is essentially globally valid (Stuiver et al. 1998). The general negative trend relates to magnetosheath variation (see Fig. 8.2a). The variation is periodic and also changes in variance with time. **d** Periodic secular variation (PSV) of geomagnetic field orientation for Europe; no trend, but periodic. PSV curves are valid for regions of about 1000 km in diameter. **e** Mean annual temperature inferred from tree rings (centre line = moving average for each decade); no trend but the σ and 2σ confidence limits reveal variance changes. A global climatic catastrophe suggested at ~A.D. 542 (Keys 2000) may not have caused anomalous cooling; the datum is within the 2σ limit; some scientists may consider this acceptable normal variation. Moreover, the following years were not cool. *Note: Radiocarbon, secular geomagnetic variation and dendrochronology require calibration curves, but their cyclicity may introduce ambiguities for age*

Of course, a deeper understanding of the geodynamo and the spatial and temporal anomalies of periodic secular variation tax leading scientists and are certainly beyond our scope here.

Absolute age determinations are essential to the understanding of earth history and its major processes. It is commonly understood that radioisotope methods applied to minerals provide a unique age determination from an isotopic ratio (Table 8.5). Geochronological methods, particularly the isochron technique using two related decay series, provide better than 1% precision in some cases, although conservative values are presented in Table 8.5.

However unwise it may seem, natural time series are sometimes used as the basis of a chronology, their "period" being analogous to the tick of some hypothetical geological clock. At this point, the reader is justified in wondering if it is not better to fix the time scale in order to study the phenomenon. Nevertheless, we are sometimes forced to resort to use some repetitive phenomenon as a clue to the passage of time. For example, the radiocarbon calibration curve yields many possible ages for a given $\Delta^{14}C$ value, due to large oscillations superimposed on a less notable long-term trend (Fig. 8.3b), but when calibrated against dendrochronology, this provides a powerful technique of age determination. Periodic secular variation of the geomagnetic field provides another natural time series that may be used as a dating method (Fig. 8.3d). Although the renewal component is less frequent and less well cross-calibrated by other Quaternary dating techniques, these time series form the basis of archeomagnetic dating and may be applied to a wide variety of materials. Unfortunately, their calibration curves are only valid for relatively small regions ~1000 km in diameter.

Time series are usually studied to understand the phenomenon rather than estimate ages, so their time scale must be fixed by some scientific method. The choice of dating technique depends on the type of materials available, the approximate age-range in question, and the desired precision. Any given method will yield progressively lower precision as the absolute age increases. For example, in geochronology, each unstable isotopic decay series is characterised by a half-life, after which time only half of the parent isotope remains. The determination of the age depends on the exponentially decreasing ratio between the parent isotope and the stable daughter isotope; this becomes more difficult to measure as time passes. After the passage of time equal to each half-life, half the preceding mass of parent isotope remains; after five half-lives the mass of the remaining parent isotope is reduced to 2^{-5} or ~3%, which starts to impose a theoretical limit on the precision of the age to be determined. Of course, all the practical geological and experimental sources of error are confounded, making matters worse (Appendix I for confounding of errors). Although it is not normally a concern to us, each technique has a minimum age limit also, usually a few percent of the half-life. In practice, the most accurate geochronological methods for very ancient rocks use isotope pairs. Their ratios are combined on an isochron graph that permits the age determination, sometimes with remarkable precision, e. g. ~ 0.1% for some U-Pb methods using zircons in rocks as much as 3,000 Ma old (Fig. 4.2b, c).

Unfortunately, it may be necessary to merge or overlap time series for which ages were determined by different techniques. This is common for combined Quaternary (≤ 1.6 Ma) and pre-Quaternary observations in which quite different techniques such as dendrochronology, archeomagnetism and radiocarbon are used to determine ages at different sites or in different materials. However, it is also true over any long geological interval, where it becomes necessary to use ages determined from different isotopic decay series. Although it may be necessary, mixing ages determined by two or more different methods is likely to introduce systematic differences in both absolute age and precision between the parts of the data sequence which are dated differently. Table 8.5 summarises in a cursory manner some details of the principal scientific methods of determining the ages of processes and events over geological and prehistorical time scales.

8.5 Stationary and Nonstationary Behaviour

Where a time series possesses a renewal component, it is usually inappropriate to characterise the raw data directly, e. g. by regression or curve-fitting. It is therefore more difficult to identify and describe a controlling process than in bivariate ($x-y$) systems. This is because time marches on and we have only one measurement of the response variable (y) at each instant (t). How do we know whether the process behaves similarly through time? In other words, is the process *stationary*, providing the same degree of association between control and response throughout time? If it is stationary, the response variable's mean ($\bar{y}$) and variance (σ_y^2) must be constant with time. For true *stationarity*, all the statistical characteristics of y should be independent of age.

The essence of the problem is that y is a response to some *stochastic* process. That process evolves according to some probability controlled phenomenon. For example, the temperature (y) measured at any time could vary about the mean value, $\bar{y}$, with a certain variance, σ_y^2. If time could stand still, the sample of temperatures might follow some well-behaved pattern, for example as if drawn from a Normal distribution. In reality, only one y-value may be obtained at any instant in time (t_n). If the stochastic process is stationary, the probability distribution of y is independent of time and can be treated like a sample of an ordinary univariate frequency distribution; the same values for $\bar{y}$ and σ_y should be obtainable at different times. A small sample of y over a small time interval ($t_2 - t_1$) may yield a meaningful estimate of $\bar{y}$ and σ_y.

Suppose we measure the pH of a lake with time; the pH records would scatter, presumably about some mean value with the passage of time. However, at any instant, we have only one pH value. How can we calculate statistics such as mean and variance at that point in time, from only one measurement? Suppose that time could be made to stand still but pH could show its stochastic behaviour. Frozen in time, we could measure pH many times and from the random sample calculate mean and variance.

In the earlier chapters, we saw how important it was to have enough measurements to determine reliable means, variances and confidence estimates. However, the progression of time restricts us to one observation of a stochastic process at one instant. How much variation is attributable to the instantaneous stochastic frequency distribution of the variable and how much is due to temporal changes in the controlling process? Changes in the mean or variance with time may alert us to nonstationary behaviour. Stationarity requires that all statistics for the variable are constant with time. Constant mean value is recognisable from the absence of a trend. For example, world temperatures from 1850 to 1910 appear stationary but there is a strong positive trend after 1910 (Fig. 8.3 d). However, constancy of variance is also desirable if we wish to proceed with simple analyses.

The first step in examining a time series, is to isolate trends so that anomalies or renewal features are more clearly identified. An overall increase or decrease in y indicates a trend and is an important type of *nonstationary* behaviour. Sequences without trend, in which y oscillates about a constant value, show one important property of *stationary* behaviour; their *mean* is constant. For example, the variation in mean annual temperature in Nicosia, Cyprus, shows a positive trend, for simplicity identified with the help of a regression line (Fig. 8.5 b). However, superimposed on this is a residual, cyclic variation. In comparison, global tempera-

tures show stationary behaviour to about 1910, and nonstationary behaviour thereafter (Fig. 8.5 c, d). If necessary, the trend may be subtracted from the time series to yield a *stationary residual signal* from which anomalies, changes in variance or renewal processes may be more clearly read.

Let us examine an example where it is difficult to decide whether the instantaneous potential for variability is large enough to disguise true time-dependent changes. A popular book on the effects of catastrophic climatic change on historical events provides evidence for global cooling between A.D 536 and A.D 542, perhaps with lingering effects for two decades (Keys 2000). This fascinating account discusses various possibilities, which lead the author to favour a "volcanic winter" caused by eruptions in Southeast Asia. One piece of evidence comes from tree-ring thickness variations in Siberia. Thicknesses vary with the success of the growing season and thus proxy for mean temperature, although many other factors can influence individual tree-ring records so we should not expect a very clear picture. The record for the last two millennia is shown in Fig. 8.3 e. The mean is shown by the central line and is "smoothed" as a moving average for each decade. Moving averages are explained below (Fig. 8.5 a, b). Mean temperature does not show any significant trend, but the temperature's variance (σ^2) is far from constant. Thus we may suggest that the series is not stationary and infer that some important changes occurred in climate control. For example, if we plot the $\pm \sigma$ and $\pm 2\sigma$ confidence band around the mean, we see that there are periods of high variance, e.g. A.D 800 to A.D 1100 as well as periods of low variance, e.g. A.D 1400 – A.D 1650. Let us return to the author's hypothesis that A.D 542 was an unusually cold year. It is the lowest data point in the series, but it also occurs in a period of high variability between A.D 400 and A.D 600. In fact, the data point in question lies on the -2σ limit, which would not be considered unusually cold by many earth scientists, given the intrinsic variability at any point in time, not to mention the errors in measurement and interpretation in terms of relative temperature. A.D 542 may just have been a random fluctuation. Two items seem to validate this. First, the following years were not similarly cold. Second, temperatures actually exceeded the $+2\sigma$ limit in $\sim$ A.D 1300 and $\sim$ A.D 1920 but there is no suggestion of catastrophic climatic events at those times. It is unfair, of course, to reject the author's hypothesis with this data set alone; he presented other independent evidence. However, it is an instructive example for our purposes.

Some natural time series are much more complex. Consider the geomagnetic field's westward precession around the earth's rotation axis (Figs. 8.3 d, 8.12). This causes a cyclic variation in magnetic inclination and declination with time at every location. Periodic secular variation (PSV) varies with location and has renewal periods of 600 to 1000 years. There is as yet no suggestion of an underlying mathematical law that can predict or explain the PSV at a given locality. Even if the fluid mechanics of the geodynamo in the outer core were predictable, its translation into localised surface effects would probably be severely camouflaged by random turbulent components near the surface of the outer core and the interaction of the magnetic field with the heterogeneous crust of the earth. The PSV time series have clear anomalies and renewal components but there are also trends on different time scales. In such examples, it may be very difficult to discriminate the renewal component from the short-term trend.

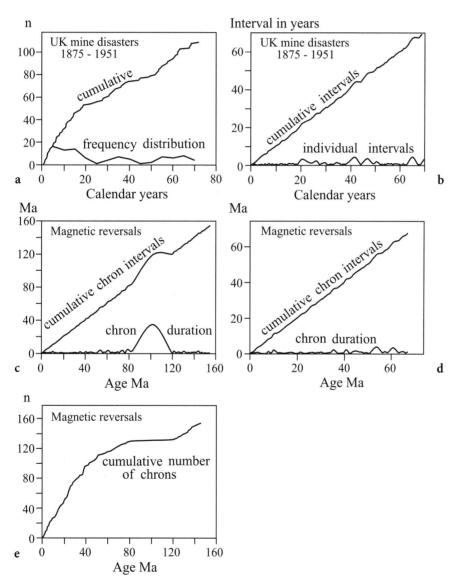

Fig. 8.4. Some time series are clarified by the use of cumulative plots in which all previous values are summed up to that time. Straight cumulative sum curves (CUSUM) show an absence of trend. **a** CUSUM reveals nonstationarity more clearly for countable events, like coal mine disasters (Maguire et al. 1952) than a frequency distribution. **b** Intervals between such events may also be investigated in this way. **c** Dichotomous variables such as normal/reversed polarities of the geomagnetic field can be CUSUM-plotted as intervals or **d** as durations of chrons. **e** Such data may also be CUSUM-plotted as counts

For some geological time series, y is not a variable value but an *event*. This may be a countable item, like the number of mine-accidents, or a dichotomous phenomenon like the normal/reversed states of the geomagnetic field. Neither can be treated as a stochastic variable that took its value at each time, t, from a probability distribution ranging from $-\infty$ to $+\infty$. However, a stochastic aspect may be identified by re-expressing the data. Countable events may be replotted as the number of events in successive time intervals, and dichotomous states may be replotted in terms of the time intervals between switches (Fig. 8.4). The data are then amenable to processing like other time series.

In the following sections, we discuss some methods that may isolate the trends of time series and investigate any underlying features. They are not equally useful in all situations. Firstly, to separate trends from anomalies:

1. *plot the data cumulatively against time, as the cumulative sum (CUSUM) of the variable*
2. *calculate moving averages, perhaps with filters*
3. *fit a regression line or a curve, usually a polynomial*

When any trend has been removed from the data, renewal processes may be identified via repetitive patterns using:

4. *data-stacking, in special cases where the period is known*
5. *autocorrelation*
6. *isolation of fixed waveforms by Fourier analysis*

8.6
Cumulative Sum Plot (*CUSUM*)

One is accustomed to the use of frequency distributions as a means of quickly evaluating the behaviour of a data set. In most statistical applications, the frequency of an observation (n) against the particular value (x) gives the simplest picture. In the case of time series, however, we are dealing with the behaviour of events against time. Replacing the x-coordinate with time is straightforward, so that the magnitude of the variable may now be plotted as the vertical axis, against (t) as the horizontal axis. Most time series appear in this way in the earth sciences (Figs. 8.2, 8.3).

However, time-sampled data may commonly fall into another category. The measurements may be *countable events*. An example of a countable event might be the number of coal-mine accidents involving ten or more fatalities, over a certain time period. Our interest in such data would be geological and environmental. The patterns could reflect geological conditions, economic activity and improvements in mine-safety. The number of disasters plotted against time does not produce a particularly revealing histogram (Fig. 8.4a). A variant is to plot the cumulative data, i.e. the sum of y-values up to time t:

$$\sum_{t=0}^{t} y_t \quad \text{against time, } t$$

This yields the cumulative summation plot (*CUSUM*). Its slope between t_1 and t_2 represents the mean number of events over that interval. This plot is thus very sensitive to any systematic changes in the trend. For example, the gentle slope between 1900 and 1940 may lead us to infer that the fewer serious accidents were associated with the reduced economic activity at that time (Fig. 8.4a, *cumulative curve*). For this data set, the CUSUM plot is probably most helpful. However, countable events provide another possibility for graphic representation. One may plot the *interval between events* as the variable, against time. The mine-accident data are illustrated in this way in Fig. 8.4b.

Another type of countable data is an *attribute*. In its simplest form, this may be a *dichotomous variable*, e.g. the presence or absence of some rock type or sedimentary structure. Such variables have only two values such as present/absent, yes/no, one/zero, etc. Nevertheless, they are amenable to graphic representation in a time series. Consider, for example, the reversals of the geomagnetic field. Only two stable states exist, normal or reversed polarity, and these states persist for the order of 1 million years between rapid switches. This data set is very well documented for the last 160 Ma from ocean floor magnetostratigraphy, and much less completely from continental paleomagnetic records to approximately 600 Ma. Simply summarised, there are rare long periods (20–80 Ma) of fixed polarity, e.g. the Cretaceous quiet-period normal *superchron* (85–115 Ma) and the Permo-Carboniferous reversed superchron (250–320 Ma). Mixed-polarity periods comprise *chrons* of normal polarity lasting ~1 Ma and *subchrons* of reversed polarity lasting ~0.1 Ma. For the purpose of discussion here, all of these intervals will be referred to generically as *chrons*. Their duration plotted against time is not particularly helpful, apart from identifying an anomalously quiet period of consistent normal polarity during a large portion of the Cretaceous superchron (Fig. 8.4c). The cumulative chron-interval version of the same graph is more revealing, for it shows a fairly constant slope, for the most part, indicating that ~160 chrons occurred in ~160 Ma, giving a crude estimate of chron duration of approximately 1 Ma. The details of this graph are shown in Fig. 8.4d, focusing on the last 60 Ma, for which behaviour is most consistent and best recorded. Finally, countable attributes, like reversals, can be tallied in terms of the cumulative number of chrons. This is perhaps the simplest and most helpful initial approach. The slope and slope changes show nicely the regularity of polarity switches over the last 60 Ma, and clearly reveal the polarity stability in the quiet period between 80 and 120 Ma (Fig. 8.4e).

The study of magnetic-polarity switching draws attention to some considerations in geological time series. The first concerns the time coordinate. In other sciences, accuracy of the time measurement would be taken for granted and it would be equally precise through the data set, being measured by a conventional clock. The absolute ages for the magnetic-polarity scale are mostly fixed by K-Ar geochronology. Since its resolution is not better than $\pm 2\%$, it is not possible to differentiate or resolve subchrons of ~ 0.1 Ma duration prior to about 5 Ma ago. Therefore, the detailed interpretation of the time series may be restricted to certain parts of the time scale that are isolated by the suitability of the dating method. The omission of short subchrons could seriously undermine any statistical treatment and invalidate some hypothetised mechanisms.

A second aspect of some geological time series is that the measured variable is not completely independent, or free to vary entirely, due to some probability distribution. The effect is similar to that in the erroneous regression of $x-y$ data where x is not actually a completely independent variable (e.g. Chap. 7). The simplest and most readily comprehended examples in time-series analysis come from meteorology. For example, temperature-time graphs reveal inbuilt inertia or *persistence*; an anomalously hot day is likely to be followed by another until the hot spell abates. The measured variable at any time is not dictated by a probability process alone, but by some intransigent state of a pre-existing value, comparable to a memory. On the other hand, some time series may show a tendency to oscillate, one state setting in sequence a chain of events that may either accelerate or undo an earlier trend. Such processes may be likened to self-fulfilling or self-negating prophesies.

In the case of magnetic polarity, the inception of one attribute (normal polarity) cannot be independent of the other (reversed polarity) as both are caused by the same geodynamo mechanism. Moreover, a transition is involved which further questions the statistical independence of the two states. Thus, the polarity state cannot be an entirely free-willed probabilistic process, and a process memory of ~ 10 ka has been proposed.

More subtle reasons for interdependence of data can be expected in some other situations. For example, in the case of mine disasters discussed above, one might argue that an incident would provoke heightened awareness and safety measures, thereby delaying the recurrence of the next accident but mitigated by contemporary ore prices. Therefore, at any time, the recorded data are not strictly and equally controlled by some probability distribution. Instead, they arise from a complex feedback mechanism. In the mining example, this might include effects from politics, sociology, economics and mining engineering.

8.7 Moving Averages

The most common time-series plot observed values or some count of observations against time. Inevitably, there will be observational error, and in the case of earth science, much larger noise due to the intrinsic instability of the processes involved. The actual variation with time that is of importance to us is concealed beneath this. There are several ways to isolate the important component of the variation from the noise.

The crudest method is to generalise or smooth a curve by eye, drawing a pencil line through the estimated "centres of mass" of groups of data. This is hardly consistent, or objective. Different workers, or even the same worker at different times, would produce different results. The *moving average* is an arithmetic procedure closest to this intuitive and manual estimation of a trend. The technique of *moving averages* (sometimes also called running averages) achieves this, either with a simple calculation or automatically in spreadsheets like Microsoft Excel. Starting at one end of the data range (x_1), we select an odd number of values of the dependent variable, e.g. three values, y_1, y_2 and y_3. These are averaged and plotted at the position of the central value, x_2. Advancing one data point along to (x_3), we take the next three points $(y_2, y_3$ and $y_4)$ and repeat the process. This is repeated until the end of the data series is reached. In this simple version of the moving average, all data points are considered to be of equal significance or weight. The three-point moving average of the variable y at (x_i, y_i) is then given by:

$$ym_i = \frac{y_i + y_{i+1} + y_{i-1}}{3} \qquad (8.3)$$

and it is plotted as (x_i, ym_i). Larger windows may be taken using any odd number of points centred on the point at which the moving average is calculated. For example, the five-point moving average at point x_i, is given by:

$$ym_i = \frac{y_i + (y_{i+1} + y_{i-1}) + (y_{i+2} + y_{i-2})}{5}$$

and in general $ym_i = \dfrac{\sum\limits_{i-k}^{i+k} y_i}{2k+1}$ (8.4)

where the order $(2k+1)$ is odd.

Each calculated moving average is then plotted as (x_i, ym_i). The higher the order, k, the smoother the appearance of the *moving-average curve,* and the easier it may be to identify longer term trends at the expense of fine detail. The trends may also be enhanced by taking moving averages of previously determined moving averages.

A smoother curve results when the moving averages (x_i, ym_i) are joined. This method is analogous to a filter in electronic circuitry that suppresses high-frequency noise or spikes in the sequence of data. More specifically, it can be considered as a *low-pass filter* because low-frequency variations (long-period oscillations) are preserved. By increasing the number of data points in each averaging step, the curve becomes smoother, at the expense of losing some important detail in the data. Balancing a desirable degree of smoothing against loss of detail, which may not be spurious, is a matter of judgment and purpose. Peaks and troughs of the smoothed curve, produced by moving averages, are also displaced relative to their positions in the raw data. These features are illustrated in Fig. 8.5a, in which successively smoothed curves are displaced upwards successively by one unit, to improve clarity. Using an excessive number of points in each running-average window also causes the new curve to be a noticeably truncated version of the original data.

Examples of real data smoothed by simple moving averages are given in Fig. 8.5b–d for global mean annual temperature and for those of Nicosia, Cyprus. Despite their simplicity, a conventional five-point moving average effectively suppresses the spiky, noisy annual means and multi-annual temperature aberrations (Fig. 8.5b). The shift of the peaks and troughs to the right is slight and the truncation of the data sequence at the left end of the series is a minor sacrifice. Successively higher order moving averages for 9 and 19 years smooth global temperature variations considerably, making the increase between 1920 and 1950 very clear, as well as the plateau from 1950 to 1980, followed by a startling increase (Fig. 8.5c). Here, the choices of longer windows introduce some minor compromising

complications due to shifts of the peaks and troughs to the right and proportionally truncated data at the low end. The larger the moving window, the greater are these problems (Fig. 8.5d). For annual data with a ~100-year sample, a moving average of 5 (years) seems the optimum to compare temperature variations for Cyprus and the world (Figs. 8.5b, c and 7.9). However, this depends on the period of the variation, which we wish to preserve: all periods <5 years will be suppressed.

There are some obvious improvements to the simple moving-average method that may be appropriate under some circumstances. In its simple form, the moving average places an equal weight or importance to all the points in each step of the averaging process. Thus, we simply average the y-values of the points under consideration. We refer to this as a rectangular window because points outside this time interval are ignored and those within are equally considered. It may be more important to emphasise the role of values closer to the centre of the moving average. In practice, this is not done very commonly in the earth sciences, but in electrical engineering, *window carpentry* refers to modifications of the filter's shape to place different emphasis on data according to its distance from the position of the calculated moving average. A simple example is Sheppard's equation, using five data points, with different coefficients to determine the response of the filter. The central data point has a relative importance of 17, the adjacent points are rated relatively at 12, and the next-but-one data points are negatively weighted with a coefficient of −3. Thus, the weighted moving average at y_i is then given by:

$$ym_i = \frac{17y_i + 12(y_{i+1} + y_{i-1}) - 3(y_{i+2} + y_{i-2})}{35}$$ (8.5)

It is plotted by a point with coordinates (x, ym_i). Of course, considerations of specific data sets could justify different relative magnitudes, and signs, of coefficients to change the response of the filter. If the contribution of spurious neighbouring points is expected to behave somewhat like a nonsystematic observational error, a Gaussian filter may be selected. This assigns decreasing weights to adjacent points, each value being multiplied by a factor that decays with distance according to the Gaussian or Normal distribution (e.g. as used in studies of magnetic field variations with time, Védes 1970). The weighting coefficient for each value y_i could be tailored to suit the particular application but would have the general form:

$$e^{-k.\Delta x^2}   \text{where}   \Delta x = x_i - \bar{x}   k = \text{constant}   (8.6)$$

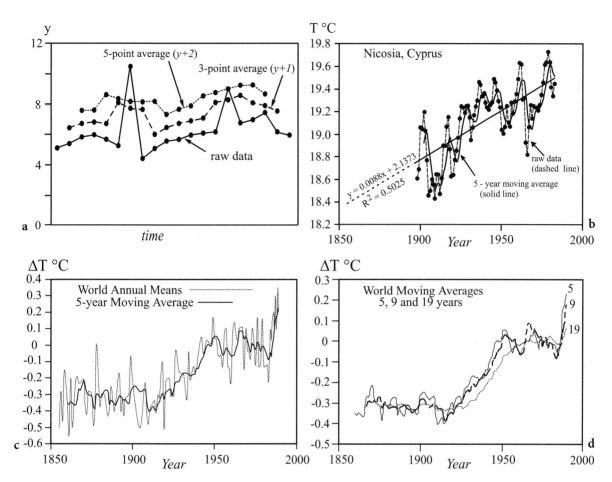

Fig. 8.5. Moving averages provide a simple, objective means of smoothing a curve. **a** Theoretical example of raw data smoothed by moving averages in three-point and five-point windows. For clarity, the three-point and five-point moving-average curves are displaced upwards by one and two units respectively. The series are progressively truncated and turning points are shifted to the right as larger windows are used. **b** Mean annual temperatures for Nicosia, Cyprus. The raw data are shown "smoothed" by a moving average in windows of five-points. However, the trend (shown by the regression line) is unaffected and the moving-average process may reveal periodicity close to the sunspot cycle (~ 11 years). **c, d** Global mean annual temperatures (1850–1990), smoothed in 5-, 9- and 19-year windows

Even more specialised filters may be applied via the moving average; they may be customised to the data set, using information derived from a preliminary analysis of the inherent frequencies present in the data.

8.8
Regression Lines and Curve Fitting

This subject has been treated mainly in regard to regression and correlation of bivariate situations (Chaps. 6, 7). However, examples of curve fitting to time-series data were introduced there incidentally, for example, where paleolatitude changes with time were investigated using linear regression and a variety of polyno-

mial fits (Fig. 6.6). The cautionary notes in Chapter 6 about the selection of an appropriate polynomial are evident here. Whereas the best fit to the data is provided by the polynomial of highest possible degree $(n-1)$, it will not suppress spurious observations and unimportant details. It should be noted that the example of Fig. 6.6 did not have the trend removed. Indeed, with such a short series, it would be very difficult to determine the trend with any confidence. Moreover, in that case, the trend was the item of foremost interest because it revealed the overall sense of plate movement. The first-degree polynomial, a straight line, may be the best estimate of a trend for a given purpose, although it does not describe the data adequately. In Chapter 7, the time series of spring flow (Fig. 7.7), and global and Mediterranean temperature changes (Figs. 7.9, 7.10a) were investigated by linear regression to identify their trends, which coincidentally were also time trends.

8.9
Data Stacking Where the Control Period Is Predictable

Under some circumstances, the period of the control variable may be known or reasonably suspected from theoretical reasoning or common sense. Seasonal or diurnal phenomena in meteorology provide the most obvious examples. Where this is the case, further investigation of the periodic phenomenon may proceed by stacking the data from separate cycles upon one another so that the spurious scatter from individual cycles may cancel out. For example, consider the discharge from a spring or the discharge of a river. Data could be collected monthly for many years and these would show a variety of variations due to short-term weather patterns as well as longer climatic trends. The noise in an individual annual record may be misleading. June may be exceptionally dry in one year, wetter another. However, common sense permits us to assume that periodicity is controlled by an annually recurrent mechanism. Therefore, in the absence of multi-year trends (i.e. climate change), it would be sensible to stack the data from successive years on a single graph, from 1 January through 31 December. The compilation should then reveal the shape of the annual cycle. Corrections may be necessary for the different lengths of the months on the time scale if detailed analysis is required. Hydrographs provide one of the simplest, successful examples of stacked time series. The discharge of a stream is expected to show a periodic variation which, e.g. at high latitudes or high altitudes, might peak with spring snowmelt or autumn rains. However, the time series for several years would not be expected to define the periodicity very well, due to weather variations masking the regular annual pattern. Thus, stacking the annual data may produce a clearer picture of the typical, expected annual discharge notwithstanding any long-term climate changes. The solar activity cycle represented by sunspot frequency probably represents the best-known cycle with the longest period, about 11 years. It is known to influence weather patterns and organic productivity as well as many geophysical phenomena. An example from the Nile basin (Hurst 1952) combines both sunspot data and hydrographic data and is used here to show the principle of data stacking in a known period to smooth out short-term random noise and attempt to identify any

patterns with a period of ~11 years (Fig. 8.6). Data were only available for 1900–1950, representing only four complete sunspot cycles. A conventional histogram for the period shows the periodicity of the solar activity, with some variation from one cycle to the next, but it is difficult to detect if there is an associated pattern in hydrographic data (Fig. 8.6a). The data are stacked, averaging the observations in the first year, second year, etc., of each solar cycle to produce a composite average cycle in which the random, idiosyncratic variations of individual years tend to cancel out (Fig. 8.6b). This certainly smoothes the sunspot time-series distribution, but it is now clear that no corresponding pattern can be discerned in the hydrographic data. Of course, the hydrographic data may show some cyclicity, but it just does not possess an 11-year period. In fact, any periodic structure in the hydrographic data is suppressed in this way. This reminds us that data stacking is counterproductive if we do not know the period. Of course, in most earth-science cycles, the controlling period is unknown but must be determined before other details of the cyclicity may be investigated. This is discussed below.

8.10
Autocorrelation to Reveal Periodicity

Many geological *renewal* processes yield observations with periodicity but without fixed-frequency components. The repetitiveness of such systems can be addressed in a simple manner by autocorrelation, which can confirm any repeating pattern and determine the wavelength of the repetition. *Autocorrelation*, sometimes known as *serial correlation,* is a technique by which a part of a series is compared with another part of the same series to detect any similarity in pattern. This is an objective means of recognising renewal processes without making any assumptions about their form. Any trend must be removed from the data series before attempting autocorrelation.

In essence, the procedure is to duplicate the sequence of y-values. The copy is slightly displaced in time with respect to the original sequence and the two are compared by conventional correlation. The copy of the sequence is displaced one unit relative to the original set to produce an out-of-phase copy. Thus, y_i becomes y_{i+L}, where L is the lag of one time unit. The correlation is determined and then the procedure is repeated for successively larger lags. The basis of autocorrelation is then to use a regression equation of the following form:

$$y_i = c + my_{i+L} \tag{8.7}$$

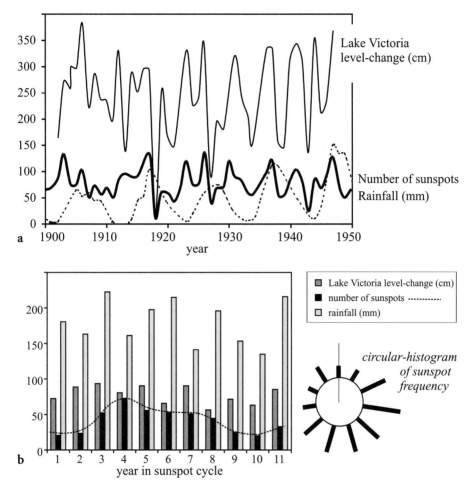

Fig. 8.6. Where the period of the repeating phenomenon is known, stacking and averaging the data from successive cycles may clarify the data pattern. Averaged in this way, random noise in individual cycles cancels out. This is illustrated with data taken from Hurst (1952). **a** Hydrographic data for the Nile basin compared with the number of sunspots, annually from 1900 to 1952. **b** Since the sunspot cycle has a period of 11 years, the data have been stacked in one 11-year window, to smooth the time series of solar activity in these four cycles. The hydrographic data do not share the 11-year periodicity. *Inset* Circular histograms are an attractive way of presenting cyclic data with a known period (see Chap. 9). Data stacking will obscure any pattern if the period assumed for the repeating phenomenon is inaccurate

As an example, consider oxygen-isotope oscillations with time. These are known to be a useful proxy for paleotemperature. The isotopic ratio varies sympathetically with other evidence for the glacial-interglacial cycles of the last 500 ka, but their variation with time is not sufficiently regular to apply any theoretical model in order to isolate the period of any renewal process. Therefore, the data set is suitable for analysis by autocorrelation. A crude illustration of the autocorrelation procedure shows the original data set displaced by successive lags of 50 and 100 ka (Fig. 8.7 a). Clearly, at a lag of 50 ka, the mismatch is nearly at its worst. On the other hand, at 100 ka, the match is better, leading us to conclude that the underlying renewal period is closer to 100 ka. Of course, in practice, many small incremental lags are considered and the definition of the renewal period is more precise. Correlating the set with its displaced copy for each chosen lag, gives one autocorrelation value:

$$\text{autocorrelation} \quad r_L = \frac{\text{covariance}\,(y_i, y_{i+L})}{s_y \cdot s_y}$$

$$\text{or} \quad r_L = \frac{\sum\limits_{i=1}^{n-L}(y_i - \bar{y})(y_{i+L} - \bar{y})}{s_y^2 \cdot n} \quad (8.8)$$

where s_y = standard deviation, $+1 \geq r_L \geq -1$

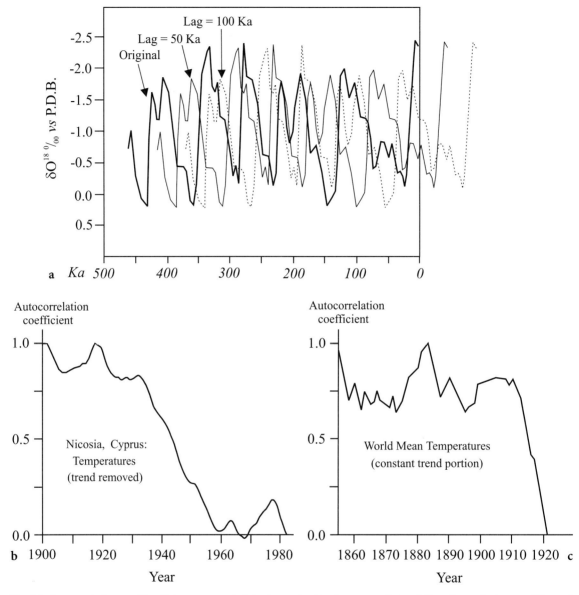

Fig. 8.7. Autocorrelation may identify important periods in time series. Simple autocorrelation shifts a time series by a small lag and then correlates the original sequence with its displaced version. This is repeated for different lags. The autocorrelation coefficient increases toward unity with improvement of the correspondence between peaks/troughs of the original and displaced series. The lag equals the period of the renewal process when the maximum value for the autocorrelation coefficient is found. **a** An example of the autocorrelation technique using the oxygen-isotope proxy for mean ocean temperatures during the Pleistocene glacial/interglacial fluctuations. Successive lags are displaced upwards for clarity. **b, c** Important periods in the oscillation of Cypriot and global temperatures are identified where the autocorrelation coefficient ≈ 1. The time scale may be considered in arbitrary decades. Possible periods (Cyprus ~ 15 a; world ~ 40 a) should be viewed very cautiously in view of the short data span and noisy data (data from Fig. 8.5)

(Covariance is also discussed in Chapter 7.) The auto-correlation r_L is calculated for many values of lag, L. A plot of r_L against lag then produces a *correlogram* in which peaks of perfect autocorrelation ($r_L \approx +1$) are separated by a number of lag (time) units that define the periodicity of the phenomenon. These peaks arise when correspondence is found between patterns in different parts of the time series. The great advantage of the autocorrelation method is that the observations, or their controlling influences, do not assume or require that the repeating pattern have any fixed waveform geometry. It is comforting to be able to relax this constraint in many earth-science applications. Unfortunately, the autocorrelation coefficient has a disadvantage in that it correlates variables that are clearly not independent. After all, it correlates a data set with a displaced version of itself. Therefore, it may not be used in statistical tests, such as hypothesis tests, like the ordinary correlation coefficient. A coefficient of determination, r_L^2, expresses the degree of variance explained by the autoregression just as with ordinary linear regression, although it will generally not be a very encouraging value.

The autocorrelation coefficient, literally representing the correlation of the series with a displaced version of itself, may vary in the range $+1 \geq r_L \geq -1$. Many sequences of geological, climatic and meteorological data have r_L values that are mostly positive. This is because rates of change are relatively slow, so the systems show *persistence*. For example, during the low temperatures of a glacial period, successive 100-year periods are likely to continue to be equally cold. By the same token, in interglacial periods, warm conditions tend to persist. Conditions dictated by major earth cycles tend to persist in their given state; they have a kind of inbuilt inertia because the scale and energies involved are so immense. Consider the following cycles and their fundamental, primary energy sources: the plate tectonic cycle (mantle radiogenic heat), the hydrological cycle (solar-biological), the CO_2 cycle (solar-biological-volcanic), and the geomorphological cycle (tectonic-solar). It is understandable that these immense complex systems are persistent, or reluctant to change. Thus, their autocorrelation will usually be positive, but only close to $+1$ when the period of the cycle equals the lag. Historical fluctuations in local and global mean annual temperatures often hint at periodicity. Unfortunately, industrial global warming disturbs the natural pattern that one would hope to read from recent meteorological records. Thus, it is important to identify and remove any trend. Because the sample interval is short (~ 100 a), we cannot make inferences about long periods, but one might suggest a periodicity of ~ 20 a

(Fig. 8.7 b, c). This is highly speculative and is not intended as a statement with any climatic consequences: it merely illustrates the data-processing technique. The simplicity and absence of assumptions for the autocorrelation procedure make it attractive for detecting the period of a renewal process, and thus perhaps for extrapolating past and future cycles. However, further evaluation of the significance of autocorrelation results is not possible because the correlated series are clearly dependent. In fact, it is the same series compared with itself at different lags.

Autocorrelation is affected by variance (s_y^2) along the whole data set. A modified procedure, cross-correlation, uses instead only the data from the overlapping portion of the original sequence and its displaced copy. In that case, the formula uses the covariance of the observations in the overlapped portion and, in the denominator, the product of the standard deviations in the overlapped section, (s_1, s_2), replaces (s_y^2). This technique may be preferable where only short data segments are considered suitable. In geological contexts, suitability might be dictated by selecting parts of the time scale dated by the same technique, or segments affected by specific geological events.

8.11
Isolating Fixed Waveforms by Fourier Analysis

Physics and engineering provide examples of quite complicated but precisely reproducing waveforms, e.g. concerning alternating electrical currents. Those phenomena are much more precisely repeatable or renewable than anything we find in earth science, so that the periodicity may be obvious by inspection and the phenomena lend themselves to precise mathematical analysis. It is tempting to apply similar techniques in natural systems. Well-defined waveforms, however complex or angular, may be represented by the sum of various sine-waves (Fig. 8.8 a–d). In many branches of science, there is an underlying implication that some of the component waveforms correspond to some controlling physical processes. A few earth-science situations, such as those related to astronomical activity (e.g. Figs. 8.6 and 8.10), or the meandering of stream channels (Leopold and Langbein 1966), appear to possess convincingly well-defined wavelengths that justify adopting sophisticated waveform analysis and permit association with some fundamental physical process.

Unfortunately, geological renewal processes do not occur with the precision of trigonometric functions and the time-base with which they are recorded may be

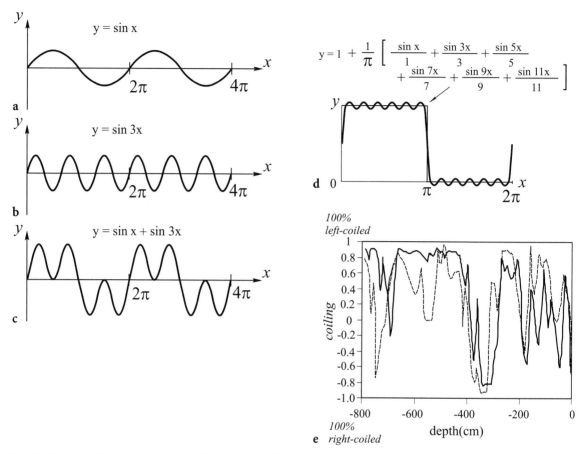

Fig. 8.8. Fourier analysis relies on the principle that any waveform may be synthesised from component sine-waves. Summing waveforms **a** and **b** yields **c**. Those shown differ in amplitude and wavelength but they may also differ in phase. **d** Suitably combined sine waves of appropriate amplitude and wavelength can reproduce a periodic form with any geometry, even square. Unfortunately, time series in the observational earth sciences rarely comprise component waveforms of precise geometry so that their analysis by Fourier transform may be of limited value. **e** The coiling sense of Foraminifera, which is influenced by sea temperature, shows a stratigraphic periodicity that relates to glacial cycles. Sediment cores from different parts of the South Atlantic show some correspondence (Eicher 1976). Such data do not provide regular waveforms, but there is clearly useful information in the periodic variation

variable in precision and accuracy. Undoubtedly important geological data tease us in this way. For example, the sense of coiling (left-handed or right-handed) of a certain Foraminifera, *Globotruncana truncatulinoides* is influenced by mean water temperature. Thus, the percentage of left- versus right-coiled individuals masquerades as a paleothermometer, and sediment depth is a crude clock (Fig. 8.8e; Eicher 1976). Intuitively, the correspondence of these two South Atlantic cores is credible because the variations in sedimentation rate (1–10 mm/ka) at and between the sites can easily account for the discrepancies. Non-geologists would probably show surprise that any visual match could be identified. The data set presented almost certainly contains valuable information on Pleistocene climatic variations (≤ 0.8 Ma from the minimum sedimentation rate), but the data quality (the variables are indirect proxies for time and temperature) defies simple quantitative analysis and even prevents direct correlation between sediment profiles from different regions. Such sets of data may be treated by Fourier analysis but warrant considerable caution in interpretation since it is unreasonable to suppose that the underlying processes possess well-defined frequencies, amplitudes and phase relations. Parts of the sequence might be better compared by cross-correlation which is more objective and does not yield so much potentially misleading and erudite information.

8.12
Deconvolution

Harmonic-component waveforms may be summed to simulate any arbitrary waveform using the Fourier approach. This leads us to the reverse procedure of *deconvolution*, where one may isolate competing or interfering renewable processes, at least qualitatively, provided that they are sufficiently different, especially with regard to period and phase. The hypothetical sequence of events shown in Fig. 8.9a indicates a complex process with significant peaks and troughs of the measured quantity, but also more frequent (shorter-period) variations. We will assume that errors of observation are trivial. A first step in deconvolution is to remove any vague background trend. This could be done with a moving average of very high order, or better with a curve fit, perhaps even linear regression. The values predicted by this curve (Fig. 8.9b) should then be subtracted from the initial data set. The simplified time series could then be treated by appropriate techniques, which go beyond the scope of this text, to identify the

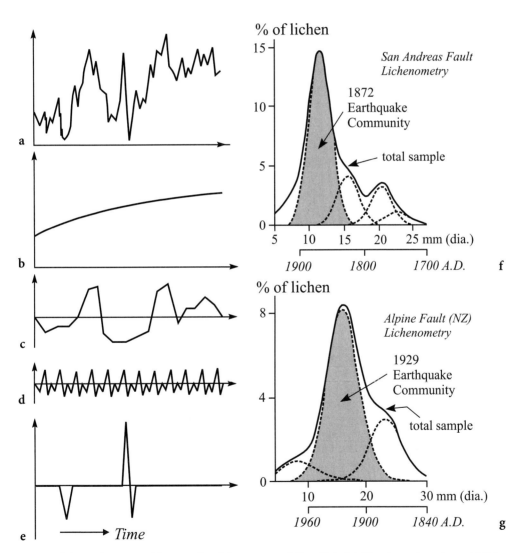

Fig. 8.9. a A natural time series may combine several useful components. **b** Trend is isolated by a low-order polynomial fit (or even by linear regression). **c, d** Subtraction of the trend from the original data reveals constituent waveforms of differing periodicity and amplitudes. **e** Some chaotic features may remain unexplained by any renewal process. **f, g** Lichen growth is proportional to the age of exposure of the substrate and may date rock falls, earthquakes, etc. Deconvolution of the frequency distribution of lichen diameter isolates populations that correspond to discrete historical seismic events. (Bull 1996a, b)

variation that occurs in different frequency windows. For example, this hypothetical data could comprise a low-frequency periodic variation (c), a high-frequency variation (d) and some random, catastrophic events. If the latter possessed any periodicity, the initial data set was simply too short to establish any regularity against the competition from spurious influences.

In reality, this graph could easily represent variations in the geomagnetic field vector; reversals, secular variation, diurnal variation, magnetic storms, or micropulsations. Sedimentology provides comparable examples with climatic change, seasonal trend, tidal effects and storms. Clearly, sedimentological time series, like many others in geology, may have predictable theoretical component frequencies, usually where there is some connection to a solar or astronomical cycle.

Deconvolution can also be used at a more intuitive level, even with series that have very poorly defined renewal behaviour. *Lichenometry* uses the size of lichen to determine the age of exposure of rock surfaces. Lichens grow at known rates so that their diameters provide a minimum-age estimate for the rock surface that they colonise. In geoarcheology, this may be used to estimate the age of stone monuments, pictographs, etc., and since different lichens grow at different rates, there is also potential for geomorphological and pre-historical seismic research. Varying lichen development on rock falls near the San Andreas Fault provides evidence of earthquakes prior to historical documentation and seismic records (Bull 1996a,b). New rock surfaces exposed by different earthquakes are colonised by lichen at different times. Since lichen diameter is almost a linear clock, a frequency distribution of lichen plotted against their diameter masquerades as a time series, whose modes correspond to optimum lichen-diameter achieved some time after the seismic event. Thus, in Fig. 8.9f, four constituent modes correspond to four lichen communities of different ages. Of course, there is always considerable variation in lichen behaviour due to local conditions, so that each new colonisation produces a unimodal frequency distribution of diameters. In turn, each of these modes corresponds to a seismic event whose age would otherwise be difficult to determine in the absence of historical documentation. A similar study is shown for the Alpine Fault of New Zealand in Fig. 8.9g. (Note that in lichenometry, as with other research related to life forms, the term "population" may be used in a nonstatistical sense to describe a group of organisms, and may require careful reading.)

An introduction to the Fourier approach is provided by an early, classic study comparing Pleistocene glaciation with suspected harmonic-component waveforms (Broecker and van Donk 1970). This pioneering study used $^{18}O/^{16}O$ ratios from pelagic sediments as a clue to global temperatures and thus ice volumes. The periodicity in oxygen-isotope ratio thus revealed the glacial and interglacial phases of the Pleistocene glacial period, with rapid temperature rises over ~10 ka followed by slow cooling over ≥50 ka (Fig. 8.10d). The

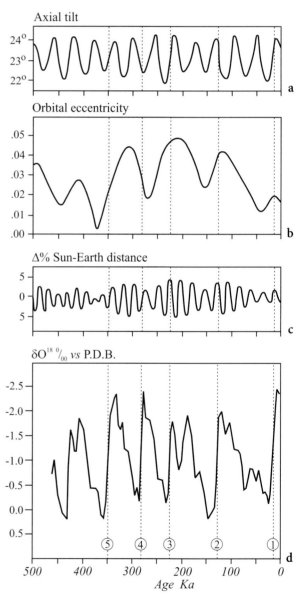

Fig. 8.10. Consideration of the earth's **a** axial tilt and **b** orbital eccentricity reveals **c** periodicity in earth-solar distance that was proposed as a cause of glaciations (Milankovitch 1938). **d** Greatest orbital eccentricity may correspond with the last three glacial-interglacial transitions where mean global temperatures are inferred from oxygen isotopes. (Burger 1988)

sudden increases in temperature that triggered deglaciation and sea-level rise are numbered one through five, with increasing relative age. It is remarkable that this curve represents the original, unsmoothed data set. It is possible that some of the high frequency, (minor) highs and lows are not significant. They are probably local geographic variations in conditions that affected that particular sample site rather than measurement errors. However, excuses are not necessary. The value of this classic set of raw data is almost self-evident. The original authors followed a line of investigation, which has intrigued geologists for many years. Could the glacial periodicity of the Pleistocene, now substantiated in terms of oxygen-isotope temperature inferences, be related to variations in solar energy received by the Earth?

This notion, first attributed in a quantitative sense to Milankovitch (1938), assumes that the solar energy received is dictated only by the variation in the Sun-Earth distance. In turn, this may be decomposed into the following known elements with fixed waveforms. The principal control is the periodicity of the globe's axial tilt, once per ~50 ka (Fig. 8.10a), and periodic variations in the orbital eccentricity (Fig. 8.10b) of the Earth about the Sun, which have periods ~100 ka. Their combination produces a synthetic time series predicting Sun–Earth distances (Fig. 8.10c), in which the effects of the component frequencies can be detected clearly. It was never suggested that this could solely account for the glacial maxima. Nevertheless, there is some correspondence between rapid deglaciation and extremes of orbital eccentricity (Fig. 8.10c) that would cause greatest seasonal variations and thus sea-level rise due to reduced ice volume (Chap. 7, Fig. 7.10b). There is good correspondence between deglaciations 1 and 2, but this is much less convincing for the preceding deglaciations (Burger 1988). This example is particularly instructive not only because of the clarity of the initial data set, but also because the hypothesised control, Sun–Earth distance, follows a well-known pattern that permits an intuitive example of deconvolution into constituent waveforms.

The Fourier analysis concept was introduced intuitively in Fig. 8.8. Consider that the signal is represented by Fig. 8.8c. The highest frequency component (longest wavelength, λ) identified from the data set is the *first harmonic* with a harmonic number of unity. The second harmonic has wavelength $\lambda/2$, of harmonic number two, etc. Note that each harmonic component may have a different amplitude and phase.

It can be shown that any precisely repeating waveform can be the sum of constituent sine and cosine functions of differing harmonic number n. Thus, however complex or unusual the shape of the curve, as long as it shows perfect repetition over many wavelengths, it may be described as the sum of several sine waves (e.g. Fig. 8.8d). At any point in the sequence, the amplitude of the variable is expressed by the sum of the component curves defined by:

$$y_n = \sum_{n=1}^{n} \left[a_n \cos \frac{2n\pi t_i}{\lambda} \right] + \left[b_n \sin \frac{2n\pi t_i}{\lambda} \right] \qquad (8.9)$$

where λ is the fundamental frequency and successive harmonics have different frequencies given by (n/λ). The wavelength of the first harmonic is λ, of the second harmonic is $\lambda/2$, and so on.

Computer analysis of some complex waveforms determines the component frequencies present in the preceding summation. The coefficients a and b control the importance of each harmonic (component waveform). Where a coefficient is zero, that particular harmonic is absent from the waveform. Any offset of a harmonic-component wave from the fundamental is determined by the relative magnitudes of the coefficients of the sine and cosine terms, a_n and b_n. This is summarised as a *phase angle* (ϕ_n) defined with:

$$\tan(\phi_n) = \frac{-b_n}{a_n} \qquad (8.10)$$

Even square waves can be reproduced or expressed by combining component harmonic trigonometric curves in this way that differ in period, phase and amplitude. A sequence of such harmonic waveforms is called a Fourier series. For data sets that form continuous output from electrical equipment, e.g. seismic recordings, the data sequence is uniformly reliable. In other situations, the appearance of the waveform may be an artefact of the sampling interval (e.g. Chap. 1, Fig. 1.2).

The solution by fast Fourier transform (FFT) yields the importance of each contributing harmonic, represented by its *power*, $(a^2 + b^2)$. The *power spectrum* is a convenient plot of power on the vertical axis and harmonic number on the horizontal axis. Moving averages may be used to improve legibility, and the raw-power axis may be normalised to unity for the convenient use and comparison of universal dimensionless units.

The paleotemperatures of the ocean are a sensitive measure of mean global temperatures that may be studied using oxygen isotopes: the heavier isotope ^{18}O is slightly more easily fixed or trapped (e.g. in fossils, ice, minerals, etc.) as the oceans cool (Fig. 8.10a). Thus, the relative abundances of the isotopes measured as a per-mil fraction, $\delta^{18}O$, determined from the same species of microfossil shell provide a temperature record. ^{18}O time series commonly provide intriguing historical records although the precision of the chronology may be questionable. Examples of power spectra for seawater temperature O^{18}-variations and Foraminifera abundance show dominant components with the most popular frequencies of approximately one cycle per Ka and one per 0.4 Ka (Fig. 8.11a; Briskin and Harrell 1980). Such high-frequency variations might be due to errors in ages, or to local environmental variations. Errors of measurement are probably negligible but it may be debatable how much the processed data actually tells us about paleoclimate. The slight mismatch of peaks could be an artefact of data collection and processing or, if real, they may indicate some lag between cause and effect (*hysteresis*).

On longer time scales, the abundance of dolomite in marine sediments is a paleoclimate indicator at a much more general level (Negi et al. 1996). Standardising the power axis to unit range facilitates comparison with other data sets. For example, could any of the prominent dolomite-production frequencies be compared to variation in Foraminifera abundance (cf. Fig. 8.11b with Fig. 8.11a)? The interpretation of such data adds an extra thought dimension to the interpretation of stratigraphic and paleoclimatic events, including oceanographic phenomena which in turn may be influenced by the configuration of tectonic plates. However, we must err on the side of caution; normally, only one or two of the lowest frequency peaks will be significant in geological time series. In particular, those of high harmonic number (high frequency) may be spurious due to sampling frequency, excessive noise and the imprecision of the time scale. Ultimately, the highest (Nyquist) frequency that could possibly be recognised is limited by the frequency of observations (Chap. 1, Fig. 1.2). Antarctic ice cores provide continuous, high-resolution and relatively easily dated stratigraphic records of atmospheric history from enclosed air bubbles (Fig. 8.11c). Furthermore, their oxygen isotopes also yield an estimate of mean global temperature. This data set from the Vostok station ice core provides a 150-ka record that closely corresponds to the recent part of the data set recorded earlier (Fig. 8.10d). The last two interglacial warming episodes are designated. Cursory inspection indicates that apart from the crude periodicity of interglacials (~100 ka) superimposed upon them, there are higher frequency variations in temperature and atmospheric composition with periods of ~10 ka.

8.13
Power Spectra of Geomagnetic Secular Variation

Previously, we noted that polarity switches of the geomagnetic field, on an Ma time scale provide an interesting example of a time series based on two states. In detail, the instantaneous or virtual geomagnetic pole is not axial, and its record with time at a given site yields the secular variation time series. The nonaxial geomagnetic field drifts westward quite rapidly (e.g. Figs. 8.2b, 8.12), completing approximately one revolution about the earth's axis in less than a millennium. Thus, at any point on the earth, the precessing field causes the inclination and declination to vary systematically, with a largely westward drift. Its angular velocity varies approximately in the range 0.2 to 0.8° longitude per annum, depending upon the latitude of the site. Generally, inclination is the more sensitive measure of secular variation. A comparison of secular variation of magnetic inclination for Japan and the eastern Mediterranean, sites separated by 110° longitude, are shown in Fig. 8.12. Their great angular separation is reflected in the lack of synchronisation of the prominent turning points in the archeo-inclination curve. The phase shifts are far from uniform, due to the complex, detailed geometry of the geomagnetic field as it sweeps by each location. For the locations indicated in the last two millennia, the cycles repeat at ~700-year intervals. Detailed Holocene curves document periodic secular variation (PSV) from many locations around the world. However, the periods and waveforms are complex and the chronological scales are highly variable in quality. Archeological dating is the most reliable, though restricted to the last few millennia and to regions with suitable cultural activity, e.g. Europe, southern Australia, the Far East and the Middle East. However, intra-continental lake sediments in post-glacial areas have steady sedimentation rates and relatively little disturbance in comparison with other young rocks. These may be dated to at least 20 ka B.P. using radiocarbon and there are complete PSV records for the last 20 ka from several recently glaciated regions, especially North America, northern Europe and China. Varve chronology and dendrochronology also provide age control in this range, but the uneven quality of the respective dating techniques risks over-interpretation of the time-series analyses.

As a consequence of westward drift, the dipole geomagnetic-field vector rotates clockwise with time (Runcorn 1959). A perfect dipole field and a magnetically homogeneous crust would cause the local field vector to process clockwise through a precisely pre-

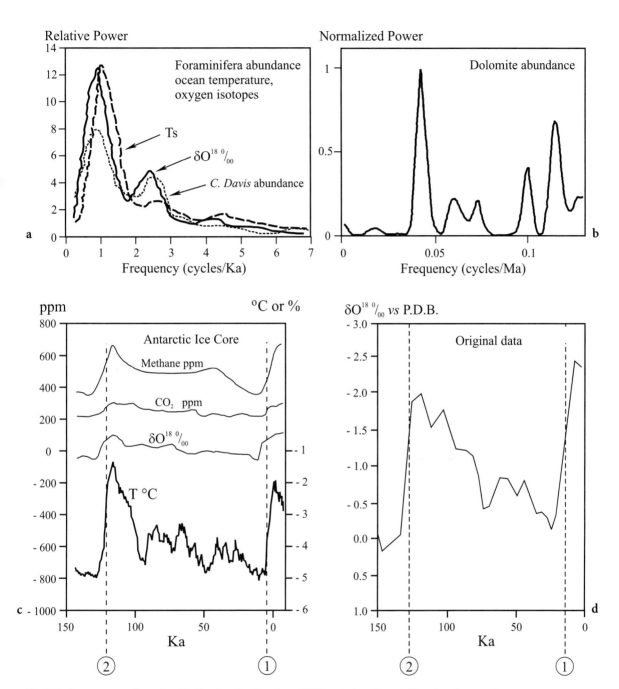

Fig. 8.11. Power spectra determined by Fast Fourier Transform (FFT) analysis. **a** Foraminifera abundance, and mean sea-water temperature inferred from oxygen isotopes. The characteristic frequencies correspond within the limits of precision of sampling and age determination. **b** Dolomite abundances in geological time. Three prominent periods may be ~40, ~100 and ~120 ka, perhaps corresponding to major oceanic and climatic conditions. **c, d** Compositions of atmospheric gases and stable oxygen isotopes in Antarctic ice bubbles provide information on temperature and atmospheric chemistry for the last 150 ka. This corresponds to the last two interglacial warming periods (numbered *1, 2*), see also Fig. 8.10

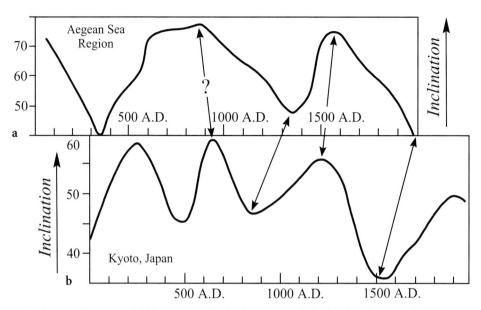

Fig. 8.12. During the generally westward drift of a non-axial, dipole-geomagnetic field, differently oriented field directions sweep by each point on earth. Thus, each site records different declinations and inclinations for the geomagnetic field with the passage of time as periodic secular variation (PSV). PSV is far from precisely regular, being poorly repeatable in space or time. Ideally, these two records from widely separated longitudinal locations should match when their time scales are adjusted to allow for the amount of westward drift, as shown

dictable cycle of inclinations and declinations. At any location, however, perturbations in the real geomagnetic field also drift beneath the site and can even cause local anticlockwise rotation of the field vector. Traditional "Bauer" plots of PSV show inclination plotted in *decreasing magnitude*, against declination, which preserves the clockwise sense of motion expected from westward drift. Points along the curve may be associated with ages to show the rate of secular variation. In this chapter, following graphic and mathematical convention, the inclination magnitudes are shown in ascending order, so that the characteristic Runcorn-loop here appears anticlockwise for a westward-drifting field (Fig. 8.13a). This well-accepted reference curve was established from several hundred archeological sites in southern Britain (Clarke et al. 1988), dating as far back as 2500 B.P. and provides a calibration curve against which enigmatic, otherwise undatable, sites are dated. Unfortunately, the traditional Bauer plot is difficult to read where several cycles of precession have elapsed. Separating declination and inclination as individual time series improves legibility. Also, as one author shows, it permits a better appreciation of the generous error bounds that must be tolerated with this data base (Fig. 8.13b, c; Batt 1997). For example, a non-dipole anomaly drifting beneath a site is expected to produce two oscillations of the declination for every oscillations of the inclination. This

may be observed more readily from separate graphs of declination and inclination with time in certain regions although much more complicated patterns may arise (Creer and Tucholka 1982). Certainly, deflections of declination appear more frequent than those of inclination (e.g. Fig. 8.3d). Lake sediments, dated by radiocarbon analysis of included organic material provide an extended PSV record for most of the Holocene. Their records date from 5,000 B.P. in Europe (Fig. 8.14; Turner and Thompson 1979) and from 20 ka for locations in China and North America (Fig. 8.15; Mothersill 1979, 1983; Lund and Banerjee 1985; Zhu et al. 1998). Bauer-type plots of archeo-inclination and archeo-declination are almost undecipherable where more than three secular-variation cycles are superimposed. Whereas they provide graphic evidence of the reality of Runcorn looping, they underscore the poor reproducibility at a site (Fig. 8.14b, c). The time-series graphic format is essential for any statistical characterisation or geological interpretation; it is sometimes shown in the geological literature with the time axis vertical. This corresponds with the orientation of the stratigraphic record obtained from lake-sediment cores (Fig. 8.14a) and is much more legible than the corresponding Bauer plot (Fig. 8.14c).

Lake sediments in the interiors of stable continental platforms provide useful sedimentary records for PSV; moreover, lacustrine records are normally amenable to

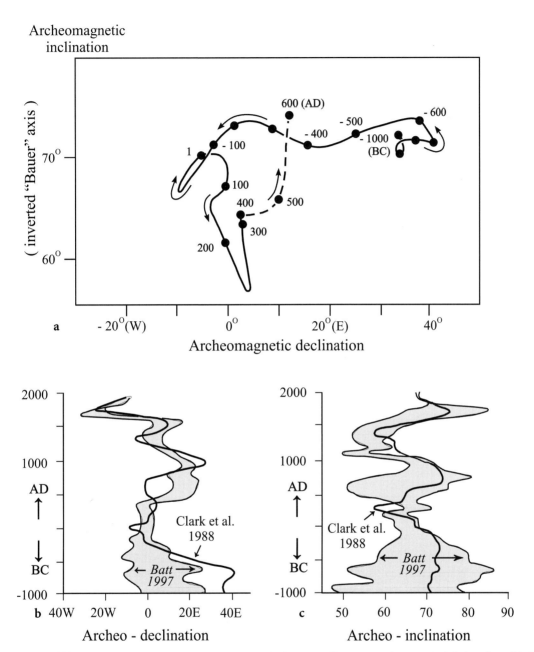

Fig. 8.13. a Traditional Bauer-type plot of secular variation data for the last two millennia in southern Britain (Clark et al. 1988) is difficult to read when fixing ages from archeomagnetic field orientations. **b, c** Separate time-series plots of inclination and declination improve legibility and permit an appreciation of the confidence limits on the declination and inclination through time (Batt 1997). Batt's data differ slightly because data from sediments suspected of inclination-shallowing were omitted. Her data were also smoothed by moving averages

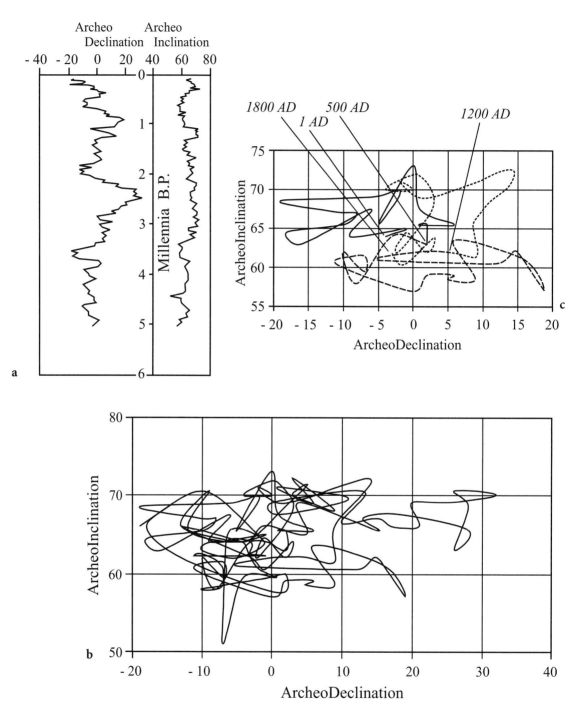

Fig. 8.14. Long secular-variation time series, extending to ages of 20 ka B.P. have been obtained from lake-sediment cores. This example is from Loch Lomond, Scotland (Turner and Thompson 1979, 1981). **a** Although the linearity of the time scale is open to question, it is clear that considerable order and pattern appear through the noise. **b** Traditional Bauer-style plots are illegible but underscore the irregular cyclicity of PSV. **c** Even two or three PSV cycles are difficult to decipher from such a plot

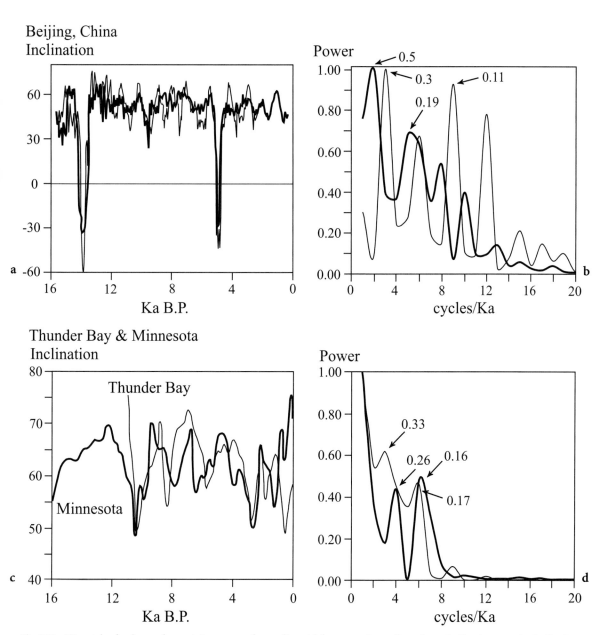

Fig. 8.15. Mismatch of paleosecular variation curves from adjacent lakes suggests nonlinearity and other inaccuracies in the time scales. **a** From Zhu et al. (1998). **b** From Mothersill (1979, 1983). **c, d** Notwithstanding difficulties of correlation and the lack of well-defined wavelengths, Fourier analysis may identify one or two characteristic component frequencies, whose power-spectra peaks are indicated with their frequencies

radiocarbon dating. Inclination data are shown for pairs of lakes from China (Zhu et al. 1998), Minnesota (Lund and Banerjee 1985) and northern Ontario (Mothersill 1979, 1983). In each region, the correspondence between the two close sites is disappointing apart from a few major features, such as the excursions (ephemeral near-reversals) of the geomagnetic field shown for China (Fig. 8.15a), and two minima in the Great Lakes region (Fig. 8.15c). Although the turning points have similar shapes in different locations, they mismatch along the time axis. This raises concerns about the accuracy of the time axis, but this is a common problem in the study of PSV. The raw data were smoothed by moving averages, and FFT then yielded the power spectra which are shown in Fig. 8.15b, d. The most prominent frequency in each location has a period in the range from 500 to 300 years in China, and from 260 to 330 years in the Great Lakes region. Some other frequency peaks occur at higher harmonic number and occasionally with higher power. Nevertheless, it is unlikely that more than two peaks are important in most geological power spectra because the time scale is of variable quality at different ages and there may be considerable uncertainty in some types of measurement.

Due to noise and imperfect repeatability in PSV records, as well as other natural renewal processes, it may be worth considering using autocorrelation instead of FFT. The two procedures are compared for the British PSV record in Fig. 8.16. In the light of information from the FFT, it is prudent only to consider the validity of the frequency at ~0.5 ka^{-1}. Autocorrelation indicates persistence in the frequency range up to ~0.5 ka^{-1} also. Little else can be concluded and the sophistication of FFT may have been unnecessary.

PSV records are limited in age range and best provided by archeological materials or ^{14}C-dated lake sediments. Global coverage is therefore inadequate and the precision of timing is variable and often poor. Nevertheless, a consensus seems to be developing that periodicities may be bracketed by several ranges including the following which appear related to precession of the dipole: 400–600 and 1000–3000 a. Longer periods have been claimed from disoriented pelagic cores obtained under the deep-sea drilling program, but these were not from direct measurements of paleo-declination and paleo-inclination. They were inferred from magnetic-intensity variations. The intensities had to be adjusted for the variation in ferromagnetic mineral content, using other magnetic properties that are now known to be influenced by paleoclimate (e.g. Thompson and Oldfield 1986). Therefore, some intensity-derived cycles may be attributable at least partly to paleoclimate changes rather than solely to geomagnetic variation. Very short period variations ~60 a may be caused by torsional oscillation in the core.

This superficial introduction to PSV illustrates some general cautions in the evaluation of geological time series with renewal components. First, time may proxy for several controls, some of which may interact. Second, time may be inaccurately determined with variable precision through the data set. Third, most

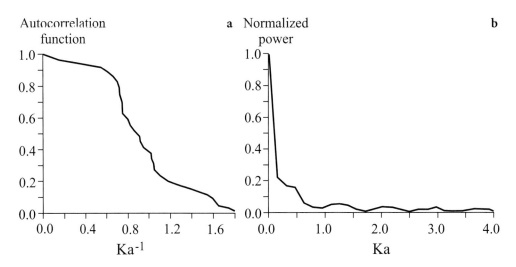

Fig. 8.16. The secular variation of geomagnetic inclination from Northern Britain (data of Fig. 8.13 a) treated by **a** autocorrelation and **b** FFT spectral analysis in an attempt to identify any important frequencies. Such vague and disappointing results are typical of many PSV sites and the two methods may not emphasise different frequencies equally. Autocorrelation does not assume regular waveforms whereas FFT does. A shoulder in **a** hints at a 500-a period, whereas the peaks in **b** suggest periods of approximately 500 and 1500 a

natural renewal processes do not possess a well-defined wave-geometry, which limits the transfer of techniques, such as FFT, from other branches of science.

8.14
Time-Series Analysis of Geomagnetic Polarity

More dramatic changes of the geomagnetic field appear in the form of reversals, over geological rather than archeological time intervals (e.g., Fig. 8.4c–e). Being discrete events, rather than measurements with a wide range of values, polarity-time series are more easily investigated in terms of the duration of each normal- or reversed-polarity interval. Throughout most of earth's history, these are measured in fractions or small multiples of 1 Ma. The more sophisticated studies analyse the time series or distribution of normal and reversed intervals separately, and have revealed statistical differences that belie an underlying asymmetry in the behaviour of the geodynamo. The geodynamo comprises a major dipole component that explains 80% of the earth's magnetic moment. It would be difficult to grasp how this could spontaneously reverse, but for the fact that core turbulence accounts for the remaining 20% in the form of dipole fluctuations (e.g. PSV, excursions) on a ~10 ka time scale and non-dipole fluctuations on a time scale <1 ka. In fortuitous combinations, these minor instabilities may trigger the more significant, but rarer polarity switches (Cox 1968). Since successive intervals are also statistically independent, a suitable model for their frequency distribution should therefore show some features of a Poisson process (Phillips et al. 1977; Laj et al. 1979).

From Chapter 3, Table 3.7 shows that the gamma distribution (Γ) provides a more general distribution covering the Poisson process but permitting more breadth in its statistical characterisation; probabilities of an event occurring between x_1 and x_2 are represented by the areas under the curve yielded by the following integral:

$$\text{Probability } (x) = \int_{x_1}^{x_2} x^{k-1} \exp(-x)\, dx \qquad (8.11)$$

Cox successfully used the Poisson end-member, for which the index $k = 1$, to establish that polarity intervals could be explained by such a probability distribution. Moreover, over the time period with sufficient resolution from ocean-floor magnetostratigraphy, the interval lengths appear to decrease from 0.96 to 0.18 Ma since 75 Ma (Late Cretaceous). Closer applica-

tion of the Γ-distribution shows that k varies with time and confirms the change in length of polarity intervals. These items favour core evolution (Naidu 1971). Different k-indices for normal and reversed polarity-interval distributions favour long-term asymmetry of the geodynamo.

We may fail to detect some short polarity intervals for parts of earth's magnetic history not represented by presently exposed ocean-floor magnetostratigraphy. Nevertheless, it may be still possible to investigate *polarity bias*. Fourier analysis is unsuitable for the spectral analysis, but maximum entropy analysis has corroborated fundamental polarity-bias periods of $\sim 80 \pm 10$ and $\sim 300 \pm 40$ Ma in several studies (McElhinny 1973; Merrill and McElhinny 1983). It has been suggested that the long periods may be due to the influence of galactic rotation and oscillation of the sun respectively, or due to core-mantle interaction. Again, we see that time proxies as a composite control for several interacting influences.

8.15
Precise, Time-Proxied Series

In conclusion, we should note that in instrumental analysis, very precise waveforms may occur such as those normally characteristic of engineering. This is a sequence of measurements either with time or with some accompanying variable changed at known times. However, a time axis is not necessary; both axes may be variables as in a bivariate x–y plot, but one acts as a progressive, sequential marker, like time. Since such measurements are now collected digitally, there is usually no shortage of data with which to document a waveform. Abundant digital data and precision timing make Fourier analysis a logical technique to apply to any laboratory-produced waveform. The example presented here is interesting because it shows that the instrumentation curve need not even be a readily recognisable periodic waveform, at first sight.

Our example comes from rock magnetism, in an experiment that measures the magnetisation (M) of a remanence-bearing mineral in the presence of an increasing field (H) which is then gradually decreased, reversed and gradually increased again. The curve for materials that may record a permanent magnetisation is a *hysteresis loop*. Hysteresis records the lack of synchronisation between cause and effect, between applied field (H) and magnetisation (M). Thus when the applied field is zero, some magnetisation remains (*remanent magnetisation*, permanent magnetisation or magnetic memory). This is typical of iron, nickel, mag-

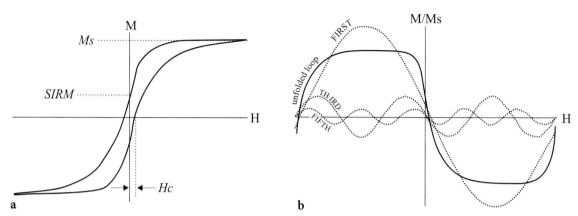

Fig. 8.17. a Hysteresis loop relates magnetisation in the presence of an applied field cycling between +H and –H; such data are useful in rock magnetism. **b** The unfolded loop may be Fourier analysed into odd harmonic components whose phase and amplitude precisely document the waveform. (After Jackson et al. 1990)

netite, and haematite, for example. The hysteresis loop and remanent magnetism is traditionally documented with three statistics read from the curve: saturation magnetisation (Ms), zero-field remanence after saturation or saturation-isothermal remanent magnetisation (SIRM), and coercivity (Hc) (Fig. 8.17a). Jackson et al. (1990) noted that considerable information is contained in such loops and that the specimen might be characterised more completely by using more than just three coordinates. Cursory examination would not suggest that the curve is suitable for Fourier transform analysis, but by unhinging the loop about one of its extremities and joining the two parts in mirror-image fashion, a regular waveform is synthesised (Fig. 8.17b). Despite its brevity, this precisely documented wave-

form, typically comprising more than 500 measurements for a single specimen, lends itself to Fourier analysis and the first, third and fifth harmonic components of the unfolded loop are illustrated in Fig. 8.17b. Jackson et al. used the first ten or so harmonics to characterise the waveform, successfully distinguishing between different hysteresis geometries that were previous barely documented by the few traditional statistics (Ms, Hc, SIRM). Thus, subtle distinctions between fat, pinched (or wasp-waisted as shown in Fig. 8.17a), and goose-necked loops could be quantified, and the differences associated with different mixtures of remanence-bearing minerals could be identified from the distribution of harmonics with respect to phase shift.

CHAPTER 9

Circular Orientation Data

Useful earth-science data occur in the form of orientations of lines that constitute an orientation distribution. Examples include fault trends, paleocurrent directions, and wind directions. Orientation data introduce some new challenges for presentation and for characterisation by statistics. Like constant-sum data, discussed in Chapter 7, they do not have an infinite range. Worse, the range is circular (0° to 360°, or 0 to 2π) so that the concept of an outlier, or a large variance, must be viewed cautiously. The most complete account of circular-orientation data is by Fisher (1993) although the concepts seem to have been first introduced formally into earth sciences by Cheeney (1983) and Till (1974), from the specialist monograph by Mardia (1972), subsequently expanded by Mardia and Jupp (2000).

Although our main concern will be with orientation data, many other interesting data sets may be considered in this way if they can be transformed into a circular, repetitive pattern. For example, the cyclic nature of data that relate to the annual calendar may be plotted on a circular diagram by *stacking data* from many years, averaging them in monthly intervals with due allowance for the unequal lengths of months. The first circular representation of a frequency distribution is attributed to the Chief Army Nurse of the British Crimea Campaign (1854–1855), the legendary Florence Nightingale. In an elegant scientific report, Nightingale (1858) carefully documented the mortality due to infection. In the first months of the campaign, infection mortality peaked at 73%; a rate that she noted was comparable to the Great Plague of 1655. If that had continued unabated, it would have extinguished the entire complement of British Troops in 10 months. Through the introduction of modern hygiene and dietary standards, she reduced the mortality rate to that of troops in training in Britain (Fig. 9.1). She presented

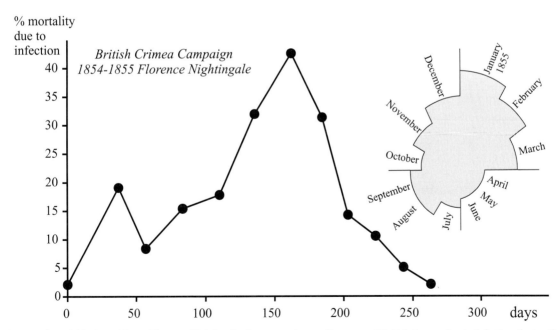

Fig. 9.1. The British Army Nurse, Florence Nightingale, documented mortality rates of British troops due to infection, during the Crimea Campaign (Nightingale 1858). She recognised the visual value of reporting calendar data on a cyclic diagram. That was a very early use of a rose diagram and possibly the first

her data in tables, but with great intuition appears to have invented the *rose diagram*, in effect a histogram wrapped around a circle. For Nightingale's report, this emphasised the time-related aspect of the data. However, the greatest benefit of Nightingale's rose diagram is realised when multiple cycles are superimposed, causing random fluctuations of individual cycles to cancel out.

Data-stacking in a cycle is routine in hydrology and meteorology where amongst other elements, river flow, precipitation and temperature vary in a similar cycle each year. Any given year shows its idiosyncrasies but if each January measurement is added for 10 years and averaged, unimportant short-term fluctuations are suppressed and a mean January record is established. If this is done for each month, a stacked data set is formed that usually shows a clear annual pattern or cycle. Such data stacking relies on the assumption that there is no long-term trend, and it is quite reasonable for weather- and climate-related phenomena over a few decades. However, in periods of climate change, data stacking over five decades may mix apples with oranges. Such a heterogeneous data set was shown in the example of flood-frequency curves (Chap. 6, Fig. 6.9). However, it is much easier to recognise such effects in circular plots. For example, consider the hydrographic data from the Nile basin and the contemporaneous sunspot frequency for the period 1900–1950 (Hurst 1952). From a conventional histogram, only the sunspot cyclicity is evident with an ~11-year cycle and any association with hydrographic changes is unclear (Fig. 9.2a). It is much easier to identify patterns when the data are stacked in an ~11-year sunspot cycle, plotting the annual averages (Fig. 9.2b). *Circular histograms* highlight the form of the average sunspot cycle but reveal that there is no simple relationship to the hydrographic data (Fig. 9.2c–e). Nightingale's rose diagram, the *circular-frequency line-graph* and the *circular-continuous-frequency distribution* convey the information still more effectively (Fig. 9.2f–h).

Many types of natural data are cyclical in character but if the period is inaccurately known or not constant, data stacking will be counterproductive, obscuring any pattern. One is reasonably safe stacking and circular-plotting data that are clearly related to annual cycles (weather, climate) or daily variations (magnetic storms). With geological data, the renewal process may not have a constant period so that data stacking may obscure the pattern rather than clarify it. For example, there are important similarities between physical changes associated with glacial advances and retreats (Chap. 8, Figs. 8.7a, 8.8e, 8.10, and 8.11). However, the cyclicity is

not sufficiently firmly tied to a fixed repeating period to warrant data stacking and circular plotting.

In the earth sciences, circular diagrams and circular statistics are mostly used for orientation distributions. The orientations are of linear features in a plane, such as paleocurrent directions, channel orientations and glacial striations. The plane need not be horizontal, of course. The nature of the oriented feature may fall into one of several categories that are now described.

9.1
Axes, Directions or Unit Vectors, and Vectors

Firstly, we consider *axial data* or lines without sense of direction (= *polarity*). For example, a fault line may crop out on a map running from east to west: its axial direction could be recorded *either* as 090° or 270°. A common practice is to record *both* possibilities for each observation, producing a symmetrical diagram (Fig. 9.3a, b). The repetition of orientations on diametrically opposite sides of the circle is unnecessary. Semicircular plots are adequate and remind us that the data are axial rather than directional (Fig. 9.3c, d). Nevertheless, many authors prefer to duplicate axial data on diametrically opposite sides of the full circle for completeness.

Secondly, another category of orientation data may possess a sense of polarity: these are *directions* or *unit vectors*. An example is the direction of transport of sediment, given by a sedimentary structure such as cross-bedding. This could only be reported with a unique azimuth, e.g. 090° if the paleocurrents flowed eastwards. Although we must discuss some further subtleties later, let us examine some examples of circular distributions. Examples of axial and vectorial (directional) observations will next be discussed with reference to Fig. 9.4.

A classic study of cross bedding from point bars along a channel yields directed lines, or unit vectors (Potter and Pettijohn 1977). These possess polarity and must be plotted around a full circle (Fig. 9.4a). True vectors possess both direction and magnitude, such as paleomagnetic vectors. The orientation aspect of a true vector may be plotted in a rose diagram, requiring a full circle in the same way as unit vectors. In contrast, most structural studies produce *axial* orientation data. The axes do not have a sense of direction; either of the two possible azimuths is equally valid. For example, petrofabric data concern the orientation of crystallographic axes, or of the long dimensions of grains as shown in Fig. 9.4b–d. The example shown gives three

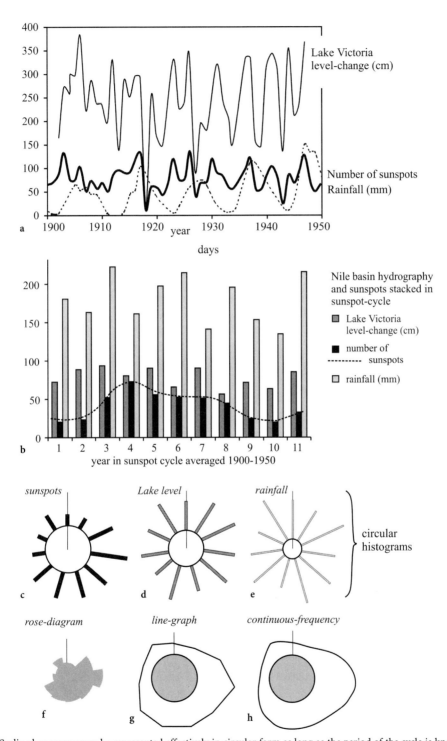

Fig. 9.2 a – h. Cyclic phenomena may be represented effectively in circular form as long as the period of the cycle is known. Multiple cycles are stacked and averaged, suppressing spurious fluctuations in individual cycles. This is illustrated with hydrographic data from the Nile and contemporaneous sunspot data (Hurst 1952). **a** A conventional histogram shows the cyclic nature of the sunspot data with an ~11-year period, but it is not clear that hydrographic data share this influence. **b** Stacking and averaging the data in the sunspot cycle using a conventional histogram show the shape of the average sunspot cycle. **c – e** The data in circular histograms emphasises the cyclic nature and indicate the peaks and troughs in their relative positions in a representative average cycle. **f – h** The rose diagram, line graph on the circle, and circular continuous-frequency distribution show the same data with greater visual impact

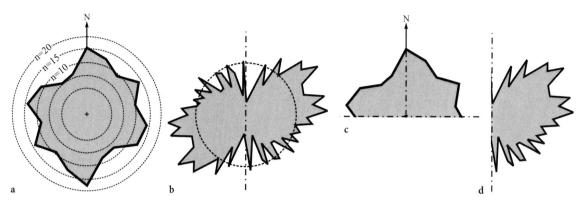

Fig. 9.3a–d. Axes are orientations without a polarity or sense of direction. Their orientation may be reported using the azimuth at either end of the line. Commonly, they are plotted on both sides of the circle, but a semicircular plot suffices to present all the necessary information. **a** An example of structural axes plots the midpoints of data bins at radial distances proportional to the number of counts (n) in each interval. **b** Orientations of metamorphic crystals (Schmidt 1917). **c, d** The above data may be plotted on a semicircle without loss of detail and serve to remind the reader that the data are axial and not directional

adjacent sections perpendicular to schistosity showing the improvement in crystal alignment caused by progressive strain (Borradaile 1976). It is redundant to use a full circle because each crystal long axis may be plotted as $x°$ or $(180 + x°)$. Moreover, a semicircular plot reminds the reader that we are dealing with axes. An example compares the trace of schistosity (S) with the mean-orientation ($\bar{\theta}$) of crystal long-axis orientations (Fig. 9.4b–d). There, orientations use an arbitrarily chosen reference. The radius of the rose diagram segments is proportional to the square-root of the number of observations it represents (Fig. 9.4). This reduces visual bias, a point that will be discussed further below.

One general lesson that we shall learn from this introduction is that there are several ways of presenting circular data; more than one method may be used to accentuate various features according to their importance in the study at hand. However, each method usually possesses some shortcomings and some advantages that deserve balanced consideration. All presentations of orientation data, both circular data discussed in this chapter and spherical data discussed in Chapter 10, suffer from the fact that it is difficult to show orientations on any diagram without losing some other attribute that may also be important. The most useful attribute is the observations' location but others include the age, rock type, metamorphic grade, measurement uncertainty or type of structure from which the orientation is acquired.

Orientations may be recorded directly in the field or taken from a map, mine plan, outcrop drawing or thin section, and transferred onto a circular diagram preserving the relation to some appropriate azimuth. Unfortunately, the geographical location of the data is not preserved in the circular diagram. There are usually

only two practical solutions that partly resolve this shortcoming. Firstly, we may create separate diagrams for data from different subareas. Each subarea should be homogeneous with respect to the orientations it provides. Spatial differences in the orientation distributions are then distinguished and clarified in separate directional plots, rather than merged and muddled on one. This subarea approach requires judicious examination of the data; it was used in the earliest petrofabric studies under the guise of *axial-distribution diagrams* ("AVA" of Sander 1930). Secondly, we may produce a single, stacked rose diagram within which different parts of the angular segments are coloured, or shaded differently to represent the percentages of data from different subareas. This is analogous to the familiar stacked bar graphs or histograms in which different parts of the columns are differently ornamented to indicate some attribute of part of the sample. These points are discussed with the following examples. It will become apparent that whatever approach is taken, the data must be examined to isolate homogeneous subgroups before the final plot is assembled.

An example of directions, or unit vectors, was provided by current data from sedimentary structures in Fig. 9.4a. The current directions from point bars along the channel scatter considerably and their geographic distribution is not of prime importance in the present discussion. It is more important to appreciate the angular variation of the current indicators with respect to the average channel orientation. This may lead us to determine whether it is reasonable to infer ancient channel directions from a small sample of paleocurrent directions in a stratigraphic sequence. Using this modern channel as a model, the current directions are plotted as unit vectors on a circle. Although the mean

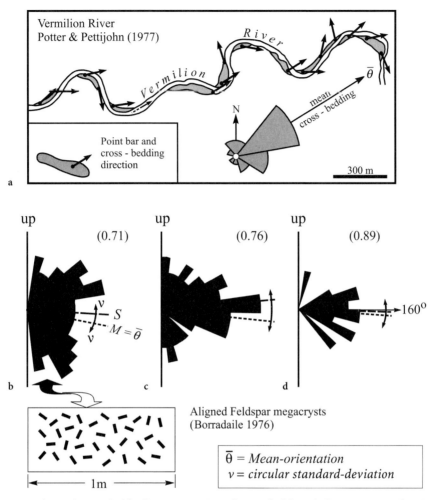

Fig. 9.4. **a** A unique azimuth must be specified for directions, requiring the use of a full-circle diagram. Current directions from cross-beds in channels yield vectorial data that must be plotted on a circle (Potter and Pettijohn 1977). **b–d** In contrast, orientations of megacrysts in a strained granite may be plotted on a semicircle (Borradaile 1976). Crystals are progressively aligned as the strain increases from **b** through **d** on a cliff-face exposure trending 160°. The mean-resultant values are $\bar{R}$ = 0.71, 0.76, 0.89, indicating the increasing preferred orientation. Mean orientation is indicated by a *dashed line* ($\bar{\theta}$) and the schistosity trace in the matrix by (S). The arc about the mean orientation in **b–d** represents the circular standard deviation (v)

direction corresponds to the average trend of the channel, there is considerable dispersion of current vectors due to the channel's meandering nature. This raises doubts about the accuracy with which ancient channel directions may be determined from a small sample of paleocurrent directions.

Another example is shown by the progressive alignment of feldspar megacrysts in a gneiss (Fig. 9.4b–d). The orientations were measured from photographs of a small vertical cliff, trending 160°. The orientations of the long axes of the megacrysts are plotted on a surface oriented with "up" and 160° (~left) as coordinates. Different parts of the cliff face show different degrees of

alignment due to different amounts of tectonic strain, and these are separated into homogeneous subareas (b, c and d), selected in a traverse perpendicular to schistosity. The long-axis orientations show a unimodal concentration in each case. As alignment and strain increase from (b) through (d), the mean of the crystals' long axes and the schistosity of the matrix are better aligned. In this example, it is clear that the selection of the subareas was critical. No variation in alignment would have been detected if the subareas were selected in a traverse parallel to schistosity.

Another example of the importance of spatial location is provided by the compilation of joint and frac-

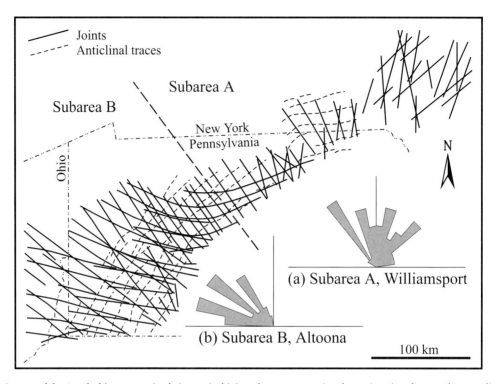

Fig. 9.5. In part of the Appalachian mountain chain, vertical joints show a progressive change in azimuth around an oroclinal bend (Nickelson and Hough 1967). Vertical joint trends are axial data, without sense of direction, and may be plotted on a semicircle without loss of information. **a, b** Two subareas with approximately constant fold-axial trend yield distinct orientation patterns for the joints. Using separate diagrams for homogeneous domains is the traditional way of isolating the geographical attributes of orientation data. Grouping all the data together would have produced a rose diagram that would be very difficult to interpret

ture orientations in the Northern Appalachians (Fig. 9.5), from Nickelson and Hough (1967). A well-developed bimodal pattern of fractures is identified. Bimodality is explained by the failure of rock on complementary or conjugate shear surfaces containing the principal compression direction in their acute angle. In this region the principal maximum compression would have been a subhorizontal axis, aligned approximately NW-SE. However, the pattern is slightly obscured by the oroclinal curvature of the belt from which the joints are sampled. Cursory visual inspection favours division of the region into homogeneous subareas, as with the previous example, yielding separate rose diagrams (Fig. 9.5a, b). These axes clearly show different mean orientations in the two halves of the belt.

Geographical location of the orientation data is an obviously interesting attribute. However, many orientations also possess other attributes of value to the scientist. It is difficult enough to recover some geographical control of our data as outlined above. It is more difficult, and much rarer that other attributes are acknowledged. It is unfortunate that this effort is not made because there is so much more to be learned

than just the orientations themselves. Logically, this leads us to the concept of a *vector* because each orientation is firmly associated with a measurable attribute, the vector's magnitude. Examples include paleomagnetic directions or a wind direction; these possess a polarized direction, and also magnitude. Clearly, the magnitudes are difficult to preserve when the data are merged into an orientation plot. An early successful attempt is found in meteorology (Panofsky and Brier 1965). Wind direction is easily presented in a rose diagram. Nevertheless, it is of paramount importance to know the wind speed. In Fig. 9.6a, the recorded wind speed has been simplified to a point represented by the vector's endpoint. Thus, each point identifies the direction, and its distance from the origin indicates wind speed. The endpoints can then be contoured according to their density on the diagram. Thus, both magnitudes and directions of the vectors are preserved in one clever diagram and described by contours. Unfortunately, this useful presentation seems to be used rarely in the earth sciences although one can realise beneficial applications in sedimentology, structural geology, petrofabrics and geophysics. The objectivity of the

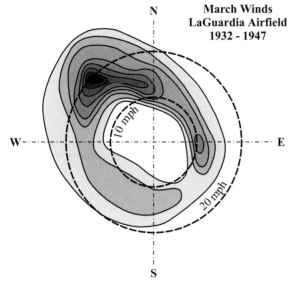

**March Winds
LaGuardia Airfield
1932 - 1947**

Fig. 9.6. Wind velocity is a true vector, possessing speed and direction, and it must be plotted on a full circle. It is normally difficult to show also the magnitude of the vector. However, here the wind-speed is indicated by the length of the line used to represent the vector's orientation. The wind-speed values at the end of those lines have then been contoured. This is a useful but little-used method by which to illustrate both orientation and magnitude of a vector. (Panofsky and Brier 1965)

density contours is strongly dependent on the contouring method, low wind speeds obviously cluster closely, despite big differences in direction. Vectors are obvious examples of orientations with an associated attribute (magnitude). However, a further development along these lines is to associate some other attribute with the orientations, although the observation is not of a traditional vector. For example, the orientation data could be of axes such as fault traces. There is often good reason to suspect that there is some special association between certain fault orientations and the displacements that occurred along them. Thus, if a fault had a displacement of two units, we would plot two counts in that angular bin for the *one* fault. If another fault had 5 units displacement we would plot 5 counts in the appropriately oriented bin. The resulting circular diagram would then represent the distribution of orientations, weighted according to the importance of the fractures, and could be contoured like wind speed (Fig. 9.6a). Another interesting and common attribute of geological fractures is their lengths. A small subarea of the Appalachian joint set study shown in Fig. 9.5 reveals joints with different orientation distributions according to lithology, and joints of different lengths in different orientations. It is understandable that this blend of lithological and fracture-length attributes produces a rose diagram that is not easily interpreted (Fig. 9.7a). However, separating the data according to the attribute of lithology provides some clarification

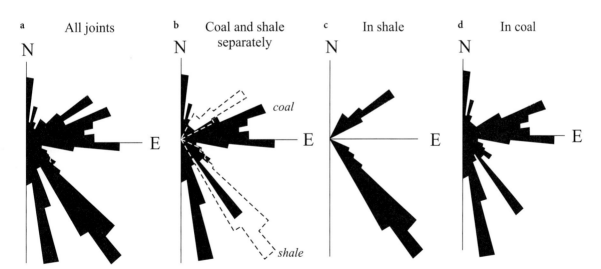

Fig. 9.7a–d. Trends of vertical joints in the Appalachians, from a small quadrangle in subarea B of Fig. 9.5 (Nickelson and Hough 1967), are plotted on a semicircle. The data are presented here in ways that draw attention to other attributes. **a** All data from coal and shale horizons produce no simple pattern that might be expected from a synchronous fracture set. **b** Data from coal and shale may be plotted with separate ornaments and superposed. **d, e** Separate histograms for coal and shale may also be used to compare and contrast orientations with different lithological attributes

(Fig. 9.7 c, d), and stacking the data on one rose facilitates comparison (Fig. 9.7 b). Nevertheless, these diagrams must be carefully evaluated if we wish them to convey something about the *fracture length*. The authors weighted the segment radii according to the average length of fractures with that orientation. Thus, the rose diagrams do not simply reveal an orientation distribution; instead they convey an important concept for the author's particular goals. It might be concluded that the rose diagram weighted for fracture length indicates the preferred *propagation direction* of fractures.

Although Fig. 9.7 conveys much information, showing the effect of rock type and the importance of certain directions for developing longer joints, it is very difficult to interpret. This is not a criticism of the authors' work; it is an illustration that diagrams designed to illustrate orientation inevitably sacrifice other attributes. We may retain some information about one attribute with difficulty but we must make a conscious decision about what other information we can afford to neglect. In this example, the radius of each sector blends the contribution of numbers of observations with fracture length. Thus, segments no longer simply represent preferred orientation. One may ask, "Does a long radius represent a few very long fractures or a large number of short ones?" The matter is further complicated because the number of observations per segment (n) may have been plotted directly, rather than ($\sqrt{n}$); this would diminish the importance of small numbers of data, whether or not they are weighted according to some attribute.

Examples of *non-geographical attributes* that could be associated with orientations of *axes* include:

1. plunge, curvature, or inter-limb angle of fold axes
2. length, width or displacement of fractures
3. mineral content of veins
4. size of aligned pebbles or other clasts in sediment
5. size of crystals

Examples of attributes that could be associated with orientations of *unit-vectors* include:

1. type of sedimentary structure associated with flow direction
2. inclination angle of cross bed, or height/width ratio of dune associated with current direction
3. tightness of slump fold associated with vergence-vector direction
4. ages of paleocurrent flow directions in a stratigraphic column
5. lithological control of paleocurrent flow indicators
6. lithological control on slickensides

Examples of *extra attributes* (beyond magnitude) that may be associated with *vectors* include:

1. blocking temperature of remanence for a given paleomagnetic direction
2. age of a paleomagnetic vector
3. Koenigsberger ratio of remanent magnetisation to induced magnetisation

9.2
Rose Diagrams

Most of the rose diagrams shown previously are essentially bar graphs wrapped around a circle. Just as with linear frequency distributions, the stepped appearance can be transformed into a smooth frequency distribution graph by joining the midpoints of the segments' ends. The visual appearance is agreeable and this was introduced into structural geology by Schmidt (1917; here, Fig. 9.3 b – d).

For conventional histograms on a line, the column height is proportional to the number of observations (n) it represents. However, when histogram columns are fanned around a point to form a rose diagram, a bin with *x* counts appears much less than half as important as a bin with *2x* counts. This is because the radius (r) of the sector is proportional to the counts in that sector but sector areas increase as r^2, overemphasising the contribution of preferred orientations. Since the bases of "circular-histogram" segments are pinched to zero width, low concentrations are reduced in importance and the slightest preferred orientations are exaggerated. *Equal-area rose diagrams* partly overcome this effect by plotting the height of the column, $r \propto \sqrt{(n)}$. Now the area of the histogram column is proportional to the counts it represents.

Cheeney (1983, p. 22) illustrated some of the important features of rose diagrams for geologists. One of his examples concerns the directions in which the openings of goniatite fossils are pointed, within a bedding plane. These orientations are unit vectors, possessing a polarity given by the direction towards which the goniatite opening faces. A full-circular presentation is required and alerts the reader to the fact that we are dealing with directions, not merely with axes. First, we should inspect the raw distribution of orientations around the circular diagram (Fig. 9.8 a). This is the simplest and most important method of avoiding pitfalls due to unfavourable bin width or bin-boundary positions (Fisher 1993). We count the number of observations (n) in a given angular interval; plotting the radius (r) of the segment proportional to n exaggerates

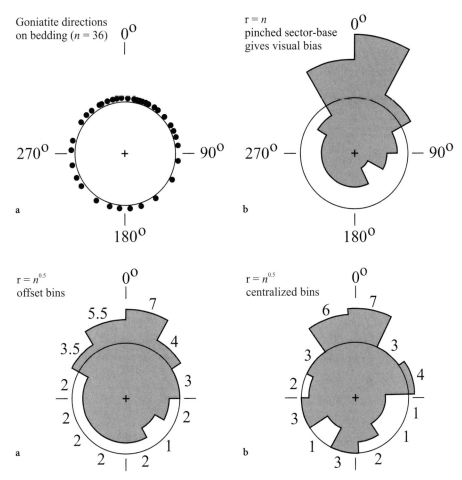

Fig. 9.8a–d. Cheeney's (1983) example of the preferred orientation of goniatites. **a** The direction in which their openings point yields unit vectors. **a, b** The tapering shape of angular sectors of a conventional rose diagram under-emphasises the contribution of angular bins with low counts, whereas the importance of sectors with many counts is exaggerated. **c** Plotting the sector-radius lengths proportional to $\sqrt{(n)}$ counterbalances the optical exaggeration of preferred orientations; this is called an equal-area rose diagram. **d** With the small sample (here $n = 36$), and degree of preferred orientation in this example, changing the positions of bin boundaries may influence the appearance of the orientation distribution

differences between maximum and minimum concentrations (Fig. 9.8b). Therefore, it is better to allow the *area of the segment* to represent the number of observations. Thus, plotting *an equal-area rose diagram* $(r \propto \sqrt{n})$ provides angular sectors whose areas more faithfully represent relative concentrations in different orientations (Fig. 9.8c), partly overcoming the problem of the pinched-base effect of sectors.

However, a further complication arises in that we are free to choose the positions of the *bin boundaries*. In Fig. 9.8c, the bin boundaries started with the left margin of the first bin at 0°. The orientation distribution of exactly the same data appears somewhat different when the first bin is centred on 0° instead. This reminds us of the importance of inspecting raw data

(Fig. 9.8a); changing the bin boundaries can affect the appearance of the rose diagram noticeably by shifting the observations from one bin to another (Fig. 9.8d). Such problems commonly arise even with seemingly large samples ($n \approx 100$ orientations) and 10° or 15° bin widths, although the actual pattern of data distribution prevents us from giving generally applicable rules. An unfortunate choice of bin boundaries changes the tallies in each segment and displaces the apparent location of the mode. Fortunately, modern programs like Spheristat allow us to experiment easily with different possibilities until the optimum conditions are found for characterising the data faithfully.

Unfortunately, the claim that larger samples and stronger preferred orientations are less susceptible to

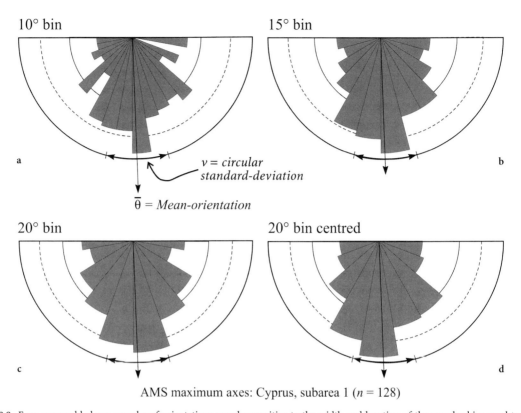

AMS maximum axes: Cyprus, subarea 1 ($n = 128$)

Fig. 9.9. Even reasonably large samples of orientations may be sensitive to the width and location of the angular bins used to construct the rose diagram (here $n = 128$). These orientations of the nearly horizontal, maximum magnetic susceptibilities represent tectonic-extension axes in the limestone sequence in Cyprus (Lagroix and Borradaile 2000a). Differently chosen bin widths change the appearance of the sample's orientation distribution. **a** 10° bins; **b** 15° bins; **c** 20° bins; **d** 20° bins centred with respect to north. *AMS* Anisotropy of magnetic susceptibility

the choices of bin width and bin boundary cannot be accepted as a universal rule. A study of magnetic susceptibility in Cyprus revealed neo-tectonic stress patterns where maximum susceptibility indicates the tectonic stretching direction due to alignment of clay minerals. A sample of 128 well-aligned north-south orientations is shown in Fig. 9.9. Bin widths of 10, 15 and 25° change the appearance of the orientation distribution considerably (Fig. 9.9a, b, c), progressively filling in gaps and smoothing out highs. However, these changes in appearance are just that. Essential statistics, like *mean orientation* $(\overline{\theta})$ and *circular standard-deviation* (v) are invariant, always being calculated from the raw data. Nevertheless, because the rose diagram features so prominently in reports, its appearance must be evaluated carefully, in the context of the problem that is to be researched. For example, in Fig. 9.9, changing bin-boundary positions and widths causes the concentrations to appear to occur either east of south, or west of south. Since we are dealing with the same sample in all cases, an adequate sample size ($n = 128$) and a well-defined orientation distribution, it seems unwise to attach any great importance to the differences among the four different representations. The only criterion for preferring one diagram over another might be the degree of simplification required in the report. The essential facts, mean orientation and circular standard deviation are fixed by calculation. Samples with $n < 100$ orientations may present difficulties in choosing boundaries and widths of bins. Although the dependence on sample size is important, it is not possible to generalise because the dispersion and the presence of multiple modes and outliers complicate matters. Samples are mostly small in structural geology, sedimentology or paleomagnetism. This is because each orientation must be taken from a different site so that the sampling strategy rapidly takes one into another subarea with a different orientation distribution. It is unproductive to combine heterogeneous samples so we are usually fortunate to compile a sample of more than 100 orientations for a homogeneous subarea.

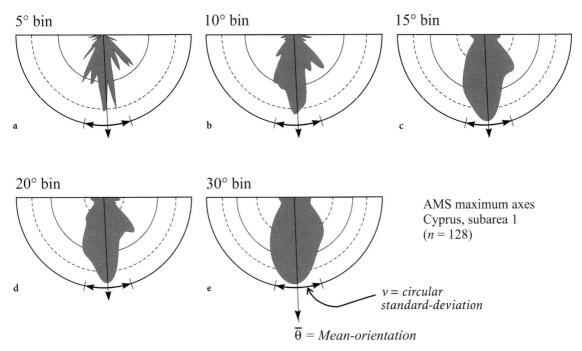

5° bin   10° bin   15° bin

a   b   c

20° bin   30° bin

d   e

AMS maximum axes
Cyprus, subarea 1
($n = 128$)

$v = circular$
$standard\text{-}deviation$

$\overline{\theta} = Mean\text{-}orientation$

Fig. 9.10. Computer programs such as Spheristat by Pangaea Scientific provide smoothing and contouring procedures that may enhance the appearance of circular-orientation diagrams. These may suppress spurious details and generalise trends. The data from Fig. 9.9 are here replotted using multiple overlapping bins to produce a continuous, smoothed rose diagram. The bin widths are 5°, 10°, 15°, 20° and 30° as shown in **a–e**. The orientations are of nearly horizontal, maximum magnetic susceptibilities, representing tectonic extension axes in the limestone sequence in Cyprus. (Lagroix and Borradaile 2000a)

A further enhancement of rose diagrams is made practicable by computer programs such as Spheristat (Pangaea Scientific Software) which can smooth the data to produce a meaningful continuous-orientation distribution function. Instead of merely joining the midpoints of discrete wedge-shaped bins (e.g. Fig. 9.9), the midpoints of multiple overlapping bins are contoured to give the continuous distribution. In this way, the data are replotted in bins of 5, 10, 15, 20 and 30° arc widths (Fig. 9.10). Clearly, an arc width of 5° produces many spurious maxima (Fig. 9.10a). On the other hand, bins with an arc width of 30° may over-generalise the data (Fig. 9.10e), suppressing an important and possibly significant minor mode trending NW-SE (Fig. 9.10b–d). The result is more than visually pleasing; it facilitates recognition of peaks and their relative importance. To expedite comparison of rose diagrams for different sample sizes, it is possible to standardise them so they show the proportion of orientations, rather than the actual number. The area shaded within each rose diagram is then equal.

Clearly, in each study the sample size, bin width, bin boundary and degree of smoothing must be tailored to suit the characteristics of the data and the nature of the orientation distribution. This requires some preliminary trial plots to determine how much detail is an artefact of the plotting technique. The best strategy should give reproducible results for different subsamples and different bin geometry. Natural samples are rarely large enough that we can ignore these issues. Generally, the inclusion of any attribute reduces the counts available and obfuscates the orientation interpretation. Usually, the prime goal is to understand the orientation distribution, which is sometimes challenging enough without the complications of attributes that weight the orientations differently.

An example of a study of magnetic fabrics in Cyprus revealed tectonic extension directions from oriented clay fabrics and magnetite (anisotropy of magnetic susceptibility, AMS) and younger fabrics due to alignment of magnetite alone (anisotropy of anhysteretic remanent magnetisation, AARM). Rose diagrams presenting the orientation distributions of maximum anisotropy values for AMS and AARM are shown for five subareas (Fig. 9.11). The data were contoured in the Spheristat program and a common bin-width of 15° was found satisfactory. Unfortunately, AARM measurements are difficult to acquire, and im-

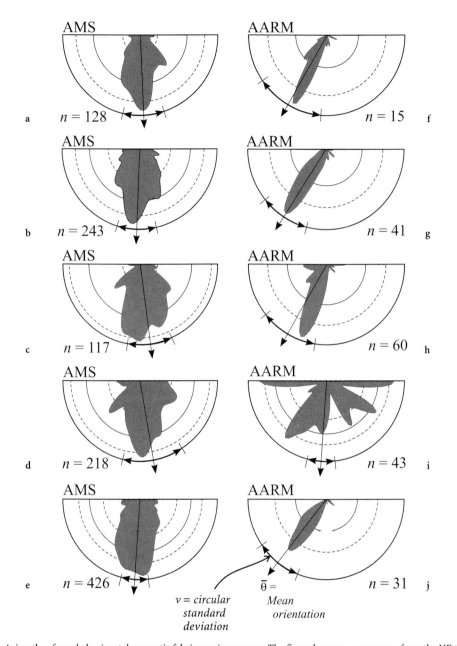

Fig. 9.11a–j. Azimuths of nearly horizontal magnetic fabric maximum axes. The five subareas **a–e** progress from the NE to SW of the limestone cover in southern Cyprus (Lagroix and Borradaile 2000). **a–e** The maximum susceptibility (*AMS*) due to clay alignment (*left-hand diagrams*) indicates fairly consistent north-south stretching with a minor NE–SW mode in subareas **c** and **d. f–j** The maximum imposed anhysteretic remanence (*AARM*) indicates NE-SW magnetite alignment, which is attributed to younger extension direction. *AMS* Anisotropy of magnetic susceptibility, due to alignments of all minerals; *AARM* anisotropy of anhysteretic remanent magnetisation, due to alignment of magnetite

possible for some sites so AARM samples are smaller and their histograms are consequently more scattered. Whereas a larger bin size could overcome this, it would complicate the comparison of AARM data with AMS data. AMS distributions show the presence of an N-S mineral alignment everywhere. One possibly significant minor NW-SE mode is recognised in subarea (a) and a NE-SW mode in subareas (c, d). The alignment of accessory magnetite is revealed by studies of the anisotropy of anhysteretic remanence (AARM, Fig. 9.11 f – j). Differences between subareas and between the different subfabrics represented by AMS and AARM respectively are due to the different ages of the subfabrics and the rotation of the stress field during their formation.

9.3
Idiosyncrasies of Orientation Data

The description of directional data poses some special problems. Since the data cannot represent an infinite distribution, they are comparable to constant-sum data, with the added complication that the upper and lower limits are indistinguishable due to the distribution wrapping around a circle. Some of these details are reviewed below with respect to the three main classes of orientation data.

9.3.1
Axes

Axes are non-directed lines – orientations without sense or polarity. Ambiguous azimuths are permissible, e.g. the east-west vertical fault may be correctly recorded as having a trend of either 090° or 270°. Unfortunately, axial azimuths are sometimes directly recorded in a 360° range rather than a 180° range, reducing the strength of any possible preferred alignment. In those circumstances, one should subtract 180° from any orientation >180° and plot the orientations in a semicircular plot. In the event that a concentration occurs close to 0° or 180° it may be preferable to present the orientations in a semicircular plot ranging from 90° to 270°, or to centre the distribution on the mean orientation.

The semicircular rose diagram suffices to present a faithful picture of the orientation distribution of axes. The semicircular form is a good reminder that *axial data* are shown. Although many authors replicate each orientation on the diametrically opposite side of a full circle, this adds no new information and has no intrinsic benefit. It is also important to preserve the semicircular data range when calculating statistics, e.g. circular mean and dispersion for axial data.

9.3.2
Unit Vectors

Unit vectors or "directions" are lines with a unique azimuth, such as the orientations of paleocurrents in sedimentary rocks. Even if the property does have an associated attribute or magnitude, we ignore it and assign equal weights to each azimuth. For example, the actual speed of the paleocurrent is unknown, and we are content to concern ourselves only with the orientation of the velocity vector. Each observation is plotted directly with the same unit weight, and the full 360° rose is obligatory, serving to remind the reader that the data are directed, or polarised.

9.3.3
Vectors

A vector possesses orientation, sense of direction, and magnitude. Its orientation, therefore, possesses a unique azimuth so that a full circular diagram is required. However, the orientations may be given a weight, proportional to the vector's magnitude. In many situations, the orientation aspect may be studied independently of the magnitudes. Such circular plots ignore magnitude and merely record the direction, just as with a unit vector. However, orientations may be weighted by their magnitudes. For example, if a vector has a direction 025° and a magnitude of 5 units, we could plot five unit vectors on the circular plot with direction 025°. In this way, a plot would be assembled whose orientation distribution reflected a combination of preferred direction and magnitudes. Of course, ambiguity then arises as to whether the five counts represent one vector of magnitude 5.0, or five vectors of magnitude 1.0, or some other combination that sums to 5.0. However, in some specialised applications, this may not be a concern (Fig. 9.7).

9.3.4
Special Characteristics of Orientation Data

Circular distributions of axes, unit vectors and vectors may all present challenges in interpretation, in characterisation, and in statistical estimations. Some of these idiosyncrasies merely deserve attention whereas others can represent insurmountable obstacles. The same problems arise in three-dimensional orientation distributions, as discussed in the next chapter.

The problems arise because orientation distribution is wrapped around a circle and loops back upon itself. The situation is worse than for constant-sum linear distributions (e.g. concentrations; Chap. 7), because the finite range of orientation data loops back upon itself. Let us consider the idiosyncrasies of orientation distributions on the circle in the following section, bearing in mind they are equally applicable to orientations in three dimensions (Chap. 10).

(a)
Mean Orientation and the Crossover Problem

If two orientations have azimuths of 355° and 005°, their arithmetic average of 180° is clearly in error. This is referred to as the *crossover problem*. Regardless of the class of orientation, treating all orientations like vectors overcomes this problem. Joining them from end to end in a series, in any order, results in a *resultant vector R*. The line joining this start and end of R is the mean orientation of the distribution of orientations (Fig. 9.12a). The mean orientation determined in this

way is independent of the nature of the distribution. A simple descriptor for the comparison of orientation distributions is R, normalised by the sample size to give $\bar{R} = (R/n)$. This is called the *mean resultant length*; for strong preferred orientations, $(R/n) \to 1.0$ and $R \to n$.

(b)
Unimodal and Multimodal Orientation Distributions and Their Dispersions

A distribution with more than one peak is *multimodal*. The mean orientation for the distribution may then have little meaning. In most cases, a homogeneous dataset contains only one mode. A multimodal orientation distribution may be characterised by sifting the data into separate homogeneous *unimodal* subgroups. These are often based on location in different homoge-

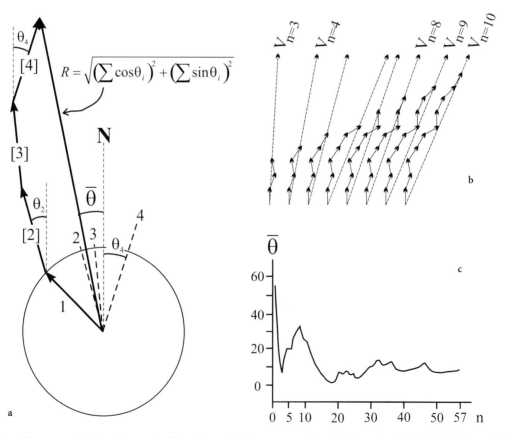

$$R = \sqrt{\left(\sum \cos\theta_i\right)^2 + \left(\sum \sin\theta_i\right)^2}$$

Fig. 9.12. a The mean orientation of a sample of lines (axes, directions or vectors) may be determined by joining their lengths in their correct azimuths, end-to-end. The order in which the lines are joined is immaterial, and the mean orientation is determined by the line joining the start and endpoints of the series. This example uses unit vectors also known as directions. **b** As the sample size increases, the estimate of the mean orientation stabilises. **c** Reiche (1938) showed that a cumulative vector sum plot (cf. CUSUM graphs of Chap. 8) establishes an optimum sample size for the determination of a reliable mean orientation. His data were from cross-bedding directions in sandstone

neous subareas (Fig. 9.5, 9.11), but homogeneous subgroups may be isolated on the basis of some other attribute such as stratigraphic age or rock type (Fig. 9.7). Since most natural orientation distributions can be analysed as separate homogeneous unimodal distributions, we will subsequently only consider statistical procedures for unimodal orientation distributions. However, the wrapped nature of circular distributions requires two special considerations.

First, in some large samples with weak preferred orientation, and in most theoretical models, as we will see later, there may be an *antimode*, some small concentration of directions antiparallel to the mode. The antimode makes the concept of range and the identification of anomalous outliers more difficult to appreciate than with infinite distributions on the line. Consequently, the concentrations of circular orientation distributions cannot be treated as neatly as the standard Normal distribution, for example. In the standard Normal distribution on the line, 99.99% of observations lie within $\pm 4\sigma$ of the mean. We shall see that there are circular equivalents to the Normal distribution, and circular equivalents to standard deviation. If our "circular standard deviation" is 45°, all of our data must lie within the circular equivalent of four standard deviations, not just 99.99%. Obviously, all ranges must lie within $\pm 180°$ of the mean.

Second, some natural distributions of vectors may have antiparallel members due to the special nature of the observation. Examples are provided by some marine currents and the paleocurrent directions produced by them, onshore and offshore wind-directions and normal-reversed polarities of the geomagnetic field. In those examples, bimodality is an intrinsic property of the observed phenomenon, requiring careful consideration. Depending on the subject and the goal, one may split the data into homogeneous single-mode sets, or one might reflect the data onto one semicircle so that the antiparallel modes are merged into one mode. Statistical characterisation could then be performed on the regrouped data.

(c)
Identification of Outliers

Since the data range is both finite and looped, it is difficult to determine what, if anything, may constitute an outlier. This is an irresolvable issue tackled only with some common sense and judgement. A true outlier is more obvious if the range or concentration of the sample is small. Such judgement depends on the nature of the data, the problem to be solved and the actual distribution of orientations. Fisher's linearised "unifor-

mity plot" (e.g. Fig. 9.15) may reveal outliers more clearly than the circular plot itself. A value suspiciously far from the main mode is readily distinguished in a linear distribution, if for example it is further than $\pm 3\sigma$ from the mean. The previous paragraph shows that such an approach is unlikely to guarantee success with orientation distributions. However, theoretical models are available for the circular distribution, most popular being von Mises' distribution. Fisher (1993) shows that the percentage points of von Mises' distribution could be used to reject unreasonably anomalous values on the basis of a suitable test statistic. He suggests the following test statistic, in which R is the *resultant length* (Fig. 9.12a, and explained below) for the entire sample and R' is the resultant length for the sample without the supposed outlier:

$$(R' - R + 1)/(n - R)$$

This may be compared with his tabled values for significance levels ($\alpha = 0.01$ or $\alpha = 0.05$). As we will see, sample size in earth science will usually be too small to distinguish between the possible theoretical models of circular distributions, but statistics and tests based on the von Mises or on the circular-Normal distributions are usually considered reliable.

(d)
Sampling Considerations

In Chapter 1, several aspects of spatial sampling were considered. However, that primarily concerned univariate data, distributed geographically. When we collect orientation data from different locations, we are confronted with a further complication in the implementation of a sampling strategy. To the dismay of the statistician, the locations are biassed, and the observations are spatially dependent; still worse, there may be a predisposition to sample certain orientations. Most rocks exhibit anisotropy, a variation in properties with orientation. Thus, at a given site, there may be a predisposition to the collection, appearance or availability of a certain orientation. Now we are confronted with two levels of sampling bias: spatial problems as usual (Chap. 1), confounded with orientation bias that may vary from site to site. Specialist knowledge of the subject may mollify the problem. For example, the trends of vertical joints in granite may be very difficult to observe if cliffs are parallel to the joints. However, cliff orientations could be weighted to counterpoint the bias given by their orientation. Alternatively, that bias could be sidestepped entirely by observing the orientation of vertical joints on drainage, from aerial photographs, giving a less detailed but less biassed view than on-site inspection.

There are no universal rules to define an adequate sample size. One simple test of a suitable, unbiassed sample, is to verify that further measurements produce no important changes in the mean orientation (Fig. 9.12b, c). In other words, once a representative sample size has been achieved, the mean orientation should stabilise. A more general test of the suitability of a sample is to resample the original material at a different time, perhaps using different workers or different methods. If the original distribution is reproduced in a second campaign of measurements, we may be more confident in the reality of the orientation distribution. Where a second sampling campaign is impracticable and sample size is small, computer resampling of the original may allow us to determine confidence limits. *Bootstrap resampling* generates >200 pseudo-samples from the original small sample ($n < 25$) (Chap. 4). The distribution of the pseudo-sample means establishes confidence limits for the mean orientation of the original sample.

9.4
Circular Statistics: No Theoretical Model Assumed

Statistical analysis of orientation data is complex and most geologists require only a few simple estimates of population parameters, chiefly the mean orientation and some measure of dispersion. However, advanced statistical procedures are available, just as with linear frequency distributions, and may be found in Fisher (1993), Mardia (1972), and Mardia and Jupp (2000). The latter two books cover circular and spherical distributions whereas Fisher's book focuses on circular distributions and contains more geological examples. The following subsections describe the statistics that may be estimated directly from the sample without assuming a theoretical model for the distribution.

9.4.1
The Mean Orientation

The first statistic that we require is the *circular mean orientation*, and in the case of circular-orientation statistics, it is fortunate that we may not need much more information because other descriptive measures are more complex and not simply interpreted (Fisher 1993). Fisher's historical examples are particularly valuable lessons in scholarship; surprisingly, it appears that in 1802, Playfair first estimated mean orientations in geology using the approach we describe below. The orientations are simply laid end-to-end, in any order.

The line joining the start and endpoint of the series then represents the mean orientation. It is not necessary to know the orientation distribution of the population or to assume any theoretical model; the method is universally valid (Fig. 9.12a, b). The illustration uses unit vectors shown by arrows of equal length. However, it is equally valid for axes, unit-vectors and vectors. Nevertheless, the mean orientation is not very helpful if the orientations are not clustered in one mode (e.g. Fig. 9.13b, c). Field studies that used this method to determine mean directions for unimodal clusters are shown for a sedimentological and structural studies in Figs. 9.4 and 9.14.

Although the looped nature of orientation distributions causes peculiar problems, the reader may appreciate that the restricted range could permit concentrations to be more readily recognisable with small samples. This begs the question, how can we be certain that a cluster is adequately identified? The *CUSUM* or *cumulative sum approach* (e.g. Chaps. 5 and 8) is of assistance here. Because orientations form a looped distribution, just like cumulative percentages, they lend themselves to the recognition of a stable value (*stationary state*, Chap. 8). As more orientations are progressively summed (in any order), the line connecting their start and endpoints develops a stable orientation (Fig. 9.12b), beyond which one quickly realises the benefits of the "law of diminishing returns". The optimum number of required observations depends on the nature of their distribution; the greater the dispersion, the larger must be the sample. Fig. 9.12c shows an example where a fairly stable mean orientation was recognised with $n = 30$: further sampling did not yield any noticeable improvement. The constant level of scatter in mean orientation $\bar{\theta}$ beyond $n = 30$ is an inherent property of that distribution's circular variance.

We can always determine the mean orientation for individual directions θ_i distributed between $0°$ and $360°$ or over any part of that range, using the procedure explained below. We do not need to make any assumptions about the sample's distribution, but the data should be inspected first to establish the usefulness of the mean orientation; e.g. consider the problems posed by nearly uniform distributions and equal antipodal clusters (Fig. 9.13a, b). Our knowledge of the subject should be used in conjunction with any calculation. For example, where the antipodal clusters are of axes, either cluster should be selected and the mean calculated for the distribution of orientations just in one semicircle. Either value may have physical significance, depending on the subject.

To determine a mean azimuth, we consider the orientations to be of equal weight. The method is thus

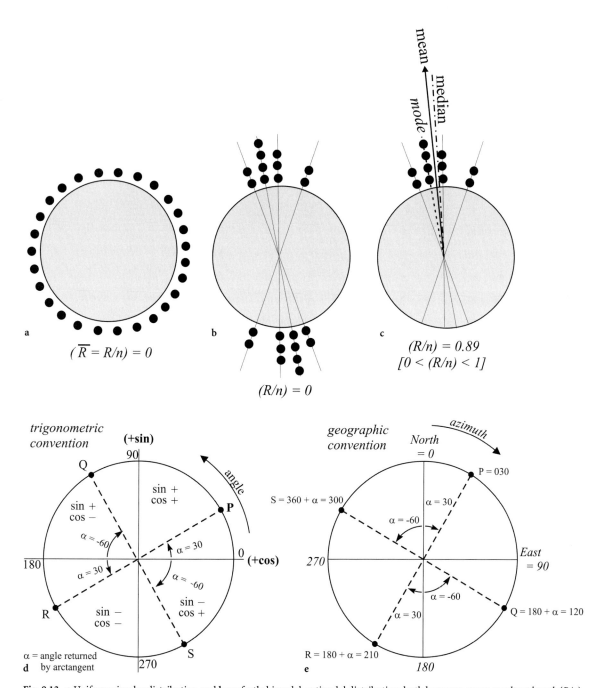

Fig. 9.13. a Uniform circular distribution and **b** perfectly bimodal-antipodal distribution, both have zero *mean resultant length* (R/n). **c** Modest preferred orientation gives high mean-resultant length, close to the maximum value of 1.0. Mean orientation, mode and median are shown. **d** Trigonometric conventions for calculation of mean orientation **e** The arctangent (α) returned in the trigonometric calculation requires a transformation to yield the mean orientation in geographic coordinates. This requires observance of the signs of the cosine and sine functions of the quadrant in question

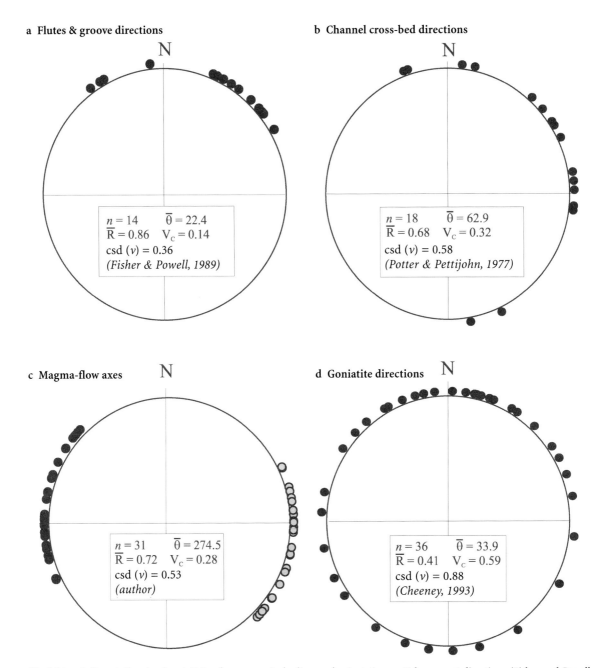

a Flutes & groove directions

N

$n = 14$ $\bar{\theta} = 22.4$
$\bar{R} = 0.86$ $V_c = 0.14$
csd $(v) = 0.36$
(Fisher & Powell, 1989)

b Channel cross-bed directions

N

$n = 18$ $\bar{\theta} = 62.9$
$\bar{R} = 0.68$ $V_c = 0.32$
csd $(v) = 0.58$
(Potter & Pettijohn, 1977)

c Magma-flow axes   N

$n = 31$ $\bar{\theta} = 274.5$
$\bar{R} = 0.72$ $V_c = 0.28$
csd $(v) = 0.53$
(author)

d Goniatite directions   N

$n = 36$ $\bar{\theta} = 33.9$
$\bar{R} = 0.41$ $V_c = 0.59$
csd $(v) = 0.88$
(Cheeney, 1993)

Fig. 9.14a–d. Descriptive circular statistics for progressively dispersed orientations. **a** Paleocurrent directions (Fisher and Powell 1989). **b** Cross-bed directions in a modern channel (Potter and Pettijohn 1977; see also Fig. 9.4). **c** Magma-flow axes in sills, N Ontario; note the azimuth of the sense of flow could be 270.5° or 104.5° as the flow indicators are axes, not vectors. Alternative possible senses for the flow are shown with grey dots. **d** Directions in which opening of goniatite fossils face on bedding. (Cheeney 1993)

directly applicable to unit vectors and to axes distributed over a range of 180°. Of course, the procedure is easily extended to true vectors of length x_i by substituting $x_i \cos \theta_i$ and $x_i \sin \theta_i$ for $\cos \theta_i$ and $\sin \theta_i$ in the following equations:

$$C = \sum_{i=1}^{n} (\cos \theta_i) \quad S = \sum_{i=1}^{n} (\sin \theta_i) \qquad (9.1)$$
$$R^2 = C^2 + S^2$$

$\bar{R}$ is the *mean resultant length* $(1 \geq (R/n) \geq 0)$. $\bar{R}$ expresses the strength of the alignment in a quantitative manner. An example of the mean orientation and resultant length for samples of progressively rotated feldspar megacrysts is shown in Fig. 9.4 b. The *mean resultant length* should not be taken at face value, without inspection of the samples' orientation distribution. For example, a uniform distribution (Fig. 9.13 a) and a perfectly bimodal–antipodal distribution in which the modes cancel (Fig. 9.13 b), both have $\bar{R} = 0$. However, $\bar{R} = 1$ does imply perfect preferred orientation, and quite modest concentrations have $\bar{R} \approx 1$ (Fig. 9.13 c).

The *mean orientation* $\bar{\theta}$ is now calculated as from:

$$\cos \bar{\theta} = \frac{C}{R} \quad \sin \bar{\theta} = \frac{S}{R}$$
$$\text{so we obtain} \quad \tan \bar{\theta} = \frac{\sin \bar{\theta}}{\cos \bar{\theta}} \qquad (9.2)$$

Unfortunately, the mean azimuth $\bar{\theta}$ is not given in meaningful geographical convention from the preceding equation. The following algorithm transfers the trigonometrically determined angle $(\alpha = arctan(S/C))$ into an azimuth:

$$\text{if } S > 0 \text{ and } C > 0 \text{ then } \bar{\theta} = \arctan\left[\frac{S}{C}\right]$$

$$\text{if } \qquad\quad C < 0 \text{ then } \bar{\theta} = \arctan\left[\frac{S}{C}\right] + \pi$$

$$\text{if } S < 0 \text{ and } C > 0 \text{ then } \bar{\theta} = \arctan\left[\frac{S}{C}\right] + 2\pi$$

$$\qquad (9.3)$$

The procedure may be followed from the illustration of trigonometric convention (Fig. 9.13 c) and geographic convention (Fig. 9.13 d) where, for convenience, α designates the angle returned trigonometrically from *arctan (S/C)*. Table 9.1 shows an example of the calculation for the four orientations P, Q, R and S from the figure.

Table 9.1. Deriving azimuth in geographic coordinates from trigonometric functions

Quadrant	P	Q	R	S
Azimuths, e.g.	30	120	210	300
Sine	0.500	0.866	−0.500	−0.866
Cosine	0.860	−0.500	−0.866	0.500
Tangent	0.577	−1.732	0.577	−1.732
α=Arctangent Transformation	30	−60 +180	30 +180	−60 +360
Azimuth	30	120	210	300

9.4.2
Mode and Median

We have noted already that in practice most simple earth-science circular distributions are unimodal. If they are not, it is usually recommended to divide the data set into unimodal subgroups, recognised on the basis of some other common geological or geographical attribute, most often location. The mode is simply the orientation of the peak of the unimodal angular distribution. If the data are grouped into bins, as previously noted, bin boundaries may influence the orientation of the mode. The median is simply the centre of the range of data where the range is designated as the shortest arc connecting all points around the circle. The median is only meaningful if the range is less than 360°. A simple example shows mode, median and mean orientation in Fig. 9.13 c.

9.4.3
Descriptors of Angular Dispersion

Statistical moments of orientation distributions are not as straightforward as those for frequency distributions on the line. Some useful guides for any unimodal concentration follow as described by Mardia (1972), and Mardia and Jupp (2000). Note that the similar terms *dispersion* (Δ) and *sample circular dispersion* (δ) embody different concepts; δ will be used later as a simple guide to compare theoretical distributions. The mean resultant is given as $\bar{R}$, using either expression, as suits clarity.

(a)
Dispersion (Δ)

The dispersion about a certain orientation (Δ) is defined by:

$$\Delta = \frac{1}{n} \sum_{i=1}^{n} [1 - \cos(\theta_i - \alpha)] \qquad (9.4)$$

where α is the mean orientation, Δ gives the circular variance (V_C), defined below (Mardia 1972; Mardia and Jupp 2000).

(b)
Sample Circular Dispersion

We have definitions of the first and second circular moments given as:

$$\text{1st moment } \overline{R} = \sqrt{(C^2 + S^2)}$$

$$\text{where } C = \frac{1}{n}\sum_{i=1}^{n}\cos\theta_i \quad \text{and} \quad S = \frac{1}{n}\sum_{i=1}^{n}\sin\theta_i$$

$$\text{2nd moment } \overline{R}_2 = \sqrt{(C_2^2 + S_2^2)} \tag{9.5}$$

$$\text{where } C_2 = \frac{1}{n}\sum_{i=1}^{n}\cos 2\theta_i \quad \text{and} \quad S_2 = \frac{1}{n}\sum_{i=1}^{n}\sin 2\theta_i$$

from which the sample circular dispersion is defined as:

$$\delta = \frac{1 - \overline{R}_2}{2R^2} \tag{9.6}$$

Note that the numerator contains the second moment, based on double angles. This definition of dispersion is useful for the comparison of theoretical distributions (Eq. 9.10 below).

(b)
Circular Variance (V_C)

In contrast to distributions on the line, circular variance has a finite upper limit and is defined as:

$$V_C = 1 - (R/n) \text{ and } 0 \leq V_C \leq 1 \tag{9.7}$$

(c)
Circular Standard Deviation (v or csd)

Unfortunately, circular standard deviation is not simply given by the square-root of circular variance but is derived by a logarithmic transformation to map the standard deviation onto a finite 2π range. It is shown that:

$$v = \sqrt{[-2\ln(1 - V_C)]} \tag{9.8}$$

For strong preferred orientations, V_C is small and (R/n) is large. Then we may use the approximations: $v \approx \sqrt{(2V_C)}$ and $v \approx \sqrt{(2(1 - R/n))}$ which are accurate to within 5% for $V_C < 0.18$ and $(R/n) > 0.82$ (Fisher 1993). These approximations are certainly satisfactory for most geological applications.

(d)
Chebyshev's Rule

In the statistics of frequency distributions on the line, we learned about a very simple, and entirely general, rule which predicts that a fraction of the observations given by $\geq (1 - k^{-2})$ must lie in the range $(\overline{x} \pm ks)$ where $\overline{x}$ and s are the sample's mean and standard-deviation respectively (Chebyshev's rule, Chap. 2).

This is true regardless of the nature of the frequency distribution. Similarly, for the circular case, the fraction of observations lying within $\pm \varepsilon$ of the sample's mean orientation is $\geq [V_C/2\varepsilon^2)]$, regardless of the form of the frequency distribution on the circle.

(e)
Typical Examples of Geological Circular Distributions

The nature of geological orientation distributions is that they vary considerably from site to site, so that a homogeneous subgroup commonly has quite a small sample size. What is perceived in the field as a preferred orientation usually has a circular variance, $V_C \leq 0.6$; circular standard deviation, $v \leq 0.8$; and a mean resultant $\overline{R} \geq 0.4$. The examples of Fig. 9.14 are typical. Figure 9.14c shows magma-flow axes, rather than vectors and, as such, there is inevitable ambiguity of the sense of flow if one uses only the information from orientation. Although other geological information may be available to discriminate between the two antipodal possibilities, the diagram shows the mean orientation calculated arbitrarily for the western end of the axes.

(f)
Is There a Preferred Orientation?

Before we discuss preferred orientations, it is logical to establish some simple criterion for their presence. Most distributions of interest reveal this by direct inspection. However, an objective graphic approach is available to assess whether the orientation distribution is significantly non-uniform. Each orientation (x_i) is plotted as $x_i/360$ for vectors, or $x_i/180$ for axes, against $(i/[n + 1])$ where n is the number of orientations (Fisher 1993). Uniformly distributed orientations should yield a straight line through the origin. Systematic departures from linearity will indicate preferred orientation in some manner. The trends of the maximum susceptibility directions from subarea 1 in Cyprus, shown earlier in Figs. 9.9 and 9.10, are presented here in a Fisher uniformity plot or *quantile-quantile plot* (Fig. 9.15). As an exploratory tool, its value in the

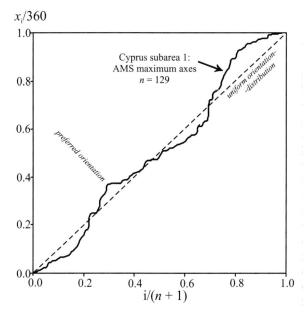

Fig. 9.15. The quantile–quantile linearised plot of circular orientation distributions provides a sensitive graphic test of the degree of uniformity or preferred orientation. This may be more objective than inspecting circular diagrams directly (Fisher 1993). Perfectly uniform orientation distributions would plot along the *dashed line* whereas the systematic departures indicate preferred orientation. The data shown are from the magnetic lineations of Fig. 9.9

identification of outliers was previously noted. However, systematic, periodic departures from the linear function clearly indicate the presence of a non-uniform distribution. Preferred orientations show values that oscillate above and below the line graph, corresponding to the quadrants in which the preferred orientations lie.

9.5
Theoretical Orientation Distributions

For linear frequency distributions, estimation and hypothesis testing require some theoretical model for the population from which the sample is drawn. That model provides a probability distribution whose low-probability tails have small areas (α) representing the probability of sampling extreme values of the variate beyond some critical value that is considered to define the ends of the representative range. Values of the variate lying in the rejection region of area α are then rejected as unrepresentative at a $100(1 - \alpha)\%$ confidence level (Chap. 4).

Although they are not yet in common use in geology, it is desirable to extend those applications to orientation distributions. This explains the need for a suitable theoretical model for the probability distribution of orientations. Unfortunately, orientation distributions cannot have extreme "rejection" regions defined simply and separately because the distributions loop back on themselves and have no tails. Instead, we have the concept of a *mode* as before, but also the concept of a single, contrasting antipodal low-density region called the *antimode* (Figs. 9.13, 9.16b, and 9.17a). We have also seen that the finite range prevents description of the spread in terms of constant multiples of standard deviation. Consequently, orientation distributions use instead the inverse concept of a *concentration parameter* (κ) about the mean orientation, using the same symbol for different formulations in different distributions. The densities of probability distributions for orientations are complicated functions of κ, or of $\bar{R}$ directly. The same symbol, κ, is used in different circular distributions and in the case of spherical distributions for three-dimensional orientations (Chap. 10). If a simple analogy is needed, the concentration parameter (κ) can be considered to be like precision. In linear statistics, *precision* is the inverse of variance ($1/\sigma^2$). Concentration ranges upwards from zero for a uniform distribution and has a range $0 \leq \kappa \leq \infty$. κ increases rapidly as the *preferred orientation*, or *precision*, increases, as is shown for the most useful theoretical model, due to von Mises (Fig. 9.17a–c). In natural orientation distributions, high values of κ may occur quite readily.

Four main theoretical distributions will be mentioned. For low concentrations, they converge towards the uniform distribution which has a probability density, $P(\theta) = 1/2\pi$. One distribution may converge towards another as extreme values of their parameters are considered. The distributions peak rapidly with small increases in κ. It will be apparent by visual inspection that the distributions are mostly very similar for low concentrations. For natural orientation distributions, quite large samples would be required to distinguish which of the four models most closely described the population. It is therefore almost of academic interest to practitioners whether the Circular-Normal or von Mises distribution is preferred; even the Cardioid or Wrapped-Cauchy distributions are not too different under some conditions (Fig. 9.16). The choice of model might not be very important for some natural samples of orientations, which are mostly rather small in geological orientation distributions with $n \leq 50$, as the typical examples of Fig. 9.14. However, the Wrapped-Normal, and especially the von Mises, distributions are preferred.

The four most popular circular distributions are all unimodal, symmetric and derived by transformations from distributions on the line. They show slight differences in dispersion and small departures from the uniform distributions at low concentrations ($\kappa < 1$). For simplicity of statistical evaluation, the antimode should be small in comparison with the mode, which is usually sufficiently true where $\kappa > 2$. We may recall that the inflexion points of the Normal distribution on the line identified the x-coordinates of the standard deviation (σ). Mardia (1972) shows that the inflexion points for a von Mises distribution are located at the following positions:

$$\bar{\theta} \pm \arccos\left[\left(1+\frac{1}{4\kappa^2}\right)^{\frac{1}{2}} - \frac{1}{2\kappa}\right] \qquad (9.9)$$

Although the inflexion points are not simply related to a statistic, they may be used in visual comparison of the circular distributions laid unwrapped onto a line, as in Fig. 9.16 a.

Each circular distribution varies its form with κ so that general visual comparisons are impossible. Therefore, Fig. 9.16 compares the four models for $\kappa = 1$, (R/n) = 0.45. Most natural orientation distributions are better concentrated than these.

Probability densities for the different models are given by complicated expressions that may be consulted in Fisher (1993) or Mardia and Jupp (2000), where percentage points are also tabled for the more important cases. The following list serves to compare them at an introductory level in terms of their *sample circular dispersion, δ.*

Uniform	$\delta = \infty$	$\bar{R} = 0$
Cardioid	$\delta = \dfrac{1}{2\bar{R}^2}$	$0 \le \bar{R} \le 0.5$
Wrapped Cauchy	$\delta = \dfrac{1-\bar{R}^2}{2\bar{R}^2}$	$0 \le \bar{R} \le 1$
Wrapped Normal	$\delta = \dfrac{1-\bar{R}^4}{2\bar{R}^2}$	$0 \le \bar{R} \le 1$
von Mises	$\delta = \dfrac{1}{\kappa[f(\kappa)]}$	
	$\bar{R} = f(\kappa); 0 \le \bar{R} \le 1$	

(9.10)

The preferred model for most situations is that of von Mises (1918), devised originally to examine the degree to which atomic weights approximated integral values. The Wrapped-Normal distribution and the von Mises distribution are very similar, and will be indistinguishable for most practical geological situations, but documentation and tests are more developed for von Mises distribution.

In the preceding list, the expression for the sample circular dispersion of a von Mises distribution included a function of κ in its denominator. That function, $f(\kappa)$, is the ratio of a first-order over a zero-order Bessel function of κ, written as $I_p(\kappa)$, in which P is the order. The Bessel functions are tabled in the cited books, but may also be calculated from expan-

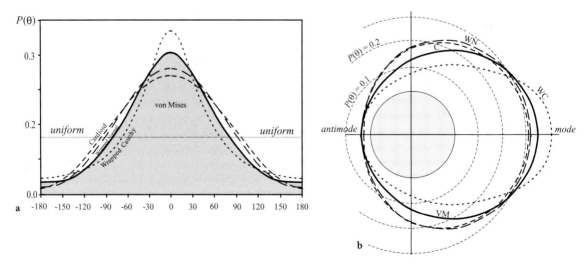

Fig. 9.16a, b. Comparison of the Cardioid, Wrapped-Normal, von Mises and Wrapped-Cauchy distributions for a concentration of $\kappa = 1$, $\bar{R} = 0.45$. These are shown **a** on the line and **b** on the circle. Note the antimode is quite large for the von Mises distribution and largest for the Wrapped-Cauchy distribution. The relative positions of the inflexion points are a visual clue to the different degrees of dispersion in the distributions. (We recall from Chapter 3 that the inflexion points of the linear Normal distribution are located $\pm\sigma$ from the mean)

sions. However, Fisher (1993) gives the following approximations for the function as shown in the above list:

$$f(\kappa) = \kappa^3 - 4\kappa^2 + 3\kappa \qquad \text{if } \kappa \geq 0.85$$

$$f(\kappa) = \left[\frac{0.43}{1-\kappa} + 1.39\kappa - 0.4\right]^{-1} \quad \text{if } 0.85 > \kappa \geq 0.53$$

$$f(\kappa) = \left[\frac{5\kappa^5}{6} + \kappa^3 + 2\kappa\right]^{-1} \qquad \text{if } 0.53 > \kappa$$

(9.11)

Most preferred orientations recognisable in earth-science studies have $\kappa > 2$, so the first approximation is probably all that we need. A Bessel function of zero order appears in the probability density of von Mises' distribution which is given for completeness below.

$$P(\theta) = \frac{1}{2\pi I_0(\kappa)} \exp[\kappa \cos(\theta - \bar{\theta})]$$

(9.12)

For strong concentrations, θ approximately follows a Normal distribution with standard deviation $(1/\kappa)$ wrapped around the circle. The mean resultant of the Wrapped-Normal distribution has a simple relation to the variance of the Normal distribution on the line, $\bar{R} = \exp(-1/2\sigma^2)$.

For low concentrations, the probability density of a von Mises distribution converges to that of the Cardioid distribution and has the following approximation:

$$P(\theta) \approx \frac{1}{2\pi}[1 + 2\bar{R}\cos(\theta - \bar{\theta})]$$

(9.13)

The appearance of a von Mises density distribution for different κ is shown in Fig. 9.17a. From our knowledge of typical orientation distributions (e.g. Fig. 9.14), the reader will appreciate that our applications mostly concern "strongly" clustered von Mises distributions with $\kappa > 1$. The ratio of von Mises' peak density at the mode, to the minimum value at the antimode, is given by $exp(2\kappa)$.

9.5.1
Estimation and Significance with von Mises' Model

Since the graph of density probabilities changes with concentration (κ), cumbersome tables are required to document the probabilities within a certain angle of the mean orientation. They are available in Mardia and Jupp (2000) as cumulative probabilities from the

antimode up to the mean orientation at the mode. Directly analogous to probability distributions on the line, they may be used to determine the angular range within which a certain percentage of orientations are expected to occur. For example, the angular range D_{95} that encloses 95% of orientations is useful (Fig. 9.17). This provides a useful percentage point of the probability distribution that helps us to quantify decisions and estimates, based on probabilities, just as with linear statistics. The reader is, however, referred to Fisher (1993), and Mardia and Jupp (2000) as statistical testing becomes quite complicated, depending on sample size and concentration. At the simplest level, it is reasonable to use von Mises' model and parametric estimations based upon it, if $\kappa > 2$, regardless of sample size. We have already noted that this is usually satisfied by most natural distributions that interest earth scientists. For weaker preferred orientations, the actual values of κ and n dictate whether it is better to use a non-parametric bootstrap approach, discussed later.

Fisher (1993) provides some useful approximations that are more than satisfactory for applied earth-science statistics. We have previously noted that the wrapped nature of circular distributions prevents us from working with constant multiples of standard deviations to define regions of certain probabilities, comparable for example to the $\pm\sigma$ and $\pm2\sigma$ limits for frequency distributions on the line.

For orientations, we may use multiples of von Mises' circular standard deviation (v_M) in a similar manner, but the factors are not integers and, like everything else in orientation distributions, they depend on κ. Von Mises' circular standard deviation is given by:

$$v_M = \sqrt{(-2\ln f(\kappa))}$$

(9.14)

and the probability (P) that a certain orientation lies within a certain range of the mean orientation $\bar{\theta}$ is given as:

$$P \approx \pm a v_M$$

(9.15)

where

$P = 0.90$	$a = 1.69$	If $\kappa \geq 0.65$, $\bar{R} \geq 0.31$
$P = 0.95$	$a = 2.06$	If $\kappa \geq 0.80$, $\bar{R} \geq 0.37$

The latter is a familiar range, containing 95% of observations, and it is valid for most geological concentrations ($\kappa \geq 0.80$). Nomograms are also provided in

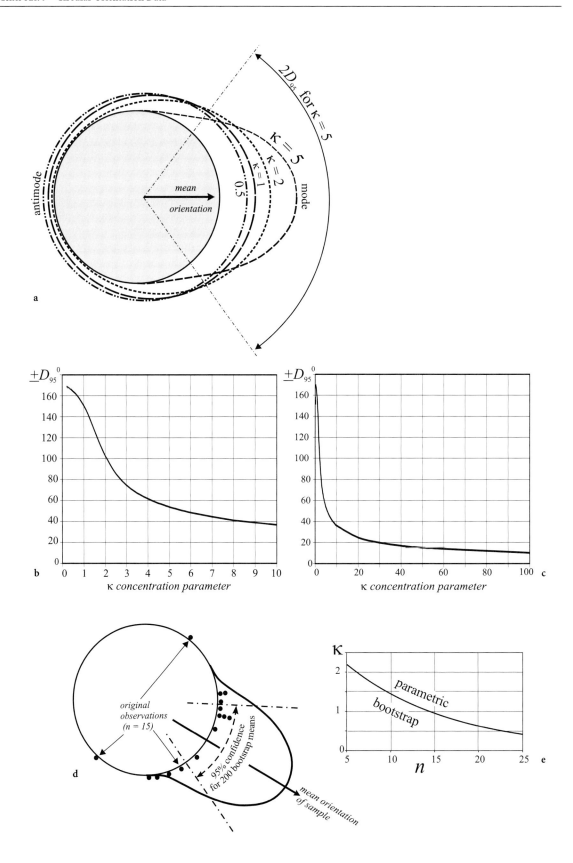

Fig. 9.17b, c for the angular range about the mean, D_{95}, which encloses 95% of the orientations in von Mises' distribution. It is given in two parts for ranges of κ that may be useful in different earth-science applications. Along the same lines, Fisher shows that since the standard error for the mean orientation is given by:

$$\frac{1}{\sqrt{n\overline{R}\,\kappa}} \qquad (9.16)$$

We may use the percentage points of the linear Normal distribution ($z = 1.96$, for a 95% two-tailed confidence region) to give the 95% confidence interval for the mean orientation, assuming a von Mises' population:

$$\overline{\theta} \pm \arcsin\left(\frac{1.96}{\sqrt{n\overline{R}\,\kappa}}\right) \qquad (9.17)$$

9.5.2
Estimation with Small or Dispersed Samples

It is feasible to estimate confidence regions using von Mises' model for $\kappa \geq 2$ and for most sample sizes. However, depending on their sizes and concentrations (n, κ), smaller samples may require a non-parametric approach. The bootstrap technique was described in Chapter 4 as a computer method for resampling a small sample to synthesise a large number of pseudo-samples. From an original sample of $n \leq 25$, at least 200 pseudo-samples, or bootstrap-samples, are generated. The frequency distribution of the means of each of these artificial samples is used to estimate the confidence range for the mean orientation $\overline{\theta}$, determined from the original sample in the usual manner (Fig. 9.12a).

An example is shown in Fig. 9.17d from Fisher (1993). From the original sample ($n = 15$), he generated the means of 200 bootstrap samples and plotted their smooth orientation frequency distribution (Fig. 9.17d). Each re-sample contains 15 orientations, but with replacement, so that each of the original observations may be omitted or repeated in the pseudo-sample. The principle captures the essence of the Central Limit Theorem of linear statistics (Chap. 4): the means of replicate samples are distributed about the mean of the population. An important difference is that the Central Limit Theorem involves real samples from the population whereas bootstrapping assumes that the sample is the population or, at least, that it embodies all of its properties. The large set of sub-sample means are then evaluated statistically just as if it were a large set of well-behaved observations, according to the procedure outlined in Chapter 4. We should not lose sight of the fact that the results are based on the clever manipulation of a small sample. The addition of even one extra observation to such a small sample could change the outcome of the experiment, and we must be confident that the initial small sample is a truthful representation of the uncertainty contained in the population, obtained by ideal statistical sampling procedures.

Fisher (1993) gives guidelines to assist in the choice between a parametric approach, as described in the previous section, with a bootstrap technique. Most importantly, regardless of sample size, he recommends the parametric approach whenever $\kappa \geq 2.0$, and a bootstrapping method whenever $\kappa < 0.4$. For intermediate concentrations, the sample size must be taken into consideration, and his suggestions are presented in the form of a nomogram in Fig. 9.17e.

Fig. 9.17. a Von Mises' distributions for different concentrations (κ). For $\kappa < 2$ the antimode (data concentration antiparallel to mode) is quite noticeable. For $\kappa = 5$, 95% of orientations lie within the range $\pm D_{95}$ of the mean orientation. **b, c** The ranges $\pm D_{95}^{\circ}$ on either side of the mean in which 95% of orientations are found in von Mises' distribution, for various concentrations κ. **d** For small samples ($n < 25$), it is possible to estimate confidence limits by bootstrap resampling of the sample. In this example, Fisher (1993, p. 91) drew 200 resamples with replacement from the original small sample. The distribution of their means is shown and its 95% percentage points give the 95% confidence limits ($2D_{95}$ range) for the mean orientation of the original sample. The mean orientation was calculated in the usual way, according to Fig. 9.12a. Bootstrapping is popular also with small samples in spherical orientation distributions (Chaps. 10 and 11)

CHAPTER 10

Spherical-Orientation Data　　　　　　　　　　　**10**

The previous chapter introduced the study of orientations that were constrained to lie in a plane, hence their shorthand name *circular* data. In nature, all orientations exist in three-dimensional space and the circular-orientation data that we collect in earth science are simply special cases whose angular distribution is controlled by the orientation of some planar feature in which they lie. For example, current directions on a bedding plane or the trends of vertical joints on a horizontal map projection may be represented without any important loss of information as lines on a plane. However, many geological and geophysical orientation data require specification in three dimensions. Therefore, each orientation requires three pieces of information to define its orientation. These may be the direction cosines of an axis or the *x, y, z* coordinates of the endpoint of a unit vector, or for that matter of a true vector. Statistical characterisation therefore requires some special graphic and statistical techniques. The principal graphic construction is the stereogram which permits the projection of orientations in three dimensions onto a plane so that they may be visualised on paper. Whereas computer techniques are not just faster, but technically superior, some old-fashioned manual techniques devised for plotting and contouring orientations are described. This is justified because extensive experience has shown that students who have not learned to plot and contour manually do not grasp the principles of sterographic projection as readily as students who proceed directly to the software manipulation of orientation data. For example, without some paper-and-pencil experience, newcomers to orientation distributions may experience difficulty with low-inclinations axes, especially with their density contouring and with vectors spread over two hemispheres.

10.1
Visualising Orientations in Three Dimensions: The Stereogram

Earth scientists use a simple two-dimensional representation of a sphere called a *stereogram* on which to record the orientations of axes or the directions of vectors. The manual use of such a diagram is briefly described in Appendix II. If any attribute must be associated with the orientation, such as its location, age or rock composition, it may be extremely difficult to illustrate that information with the orientations in one diagram. Similarly, it may be very difficult to show the magnitude of vectors on an orientation diagram, even though this is an intrinsic piece of information. In structural geology and petrofabrics, different parts of an orientation diagram were sometimes shaded differently (the axial-distribution diagrams or "AVA" of Sander 1930), or the orientation symbols were directly labelled. To say the least, such diagrams become very cluttered with any reasonably large sample or even with a small sample if the orientations are tightly clustered.

There are few studies in which both orientations and magnitudes of vectors are simultaneously required. Rock magnetism provides a familiar example and in Fig. 10.1a the vector components of remanent magnetism are shown for a specimen in three dimensions, to give some idea of their orientation and magnitude simultaneously. In paleomagnetism, the magnitude is required to isolate a direction of characteristic magnetisation (ChRM). In geophysical surveying, the magnitude and direction of remanence must be known in order to model its contribution to magnetic anomalies. For the treatment of individual paleomagnetic samples, where both orientation and magnitude are simultaneously considered, a vector-plot is used (Zijderveld 1976; Tarling 1983; Fig. 10.1b). Commonly, this superimposes plan and cross-sectional projections of the vectors on a single graph, but separate graphs may also be drawn. Since the horizontal and vertical projections do not present the true lengths of vectors, an additional graph of intensity values may be provided (Fig. 10.1c). The lengths of the vectors are scaled proportional to their magnitude. Practitioners of paleomagnetism become skilled at reading the vector plot although it may confuse novices. Alternatively, using a computer monitor, rotatable three-axis (*x, y, z*) graphs are more readily visualised (Fig. 10.1a). Although pre-

3-D view of paleomagnetic
vector components

n = 16

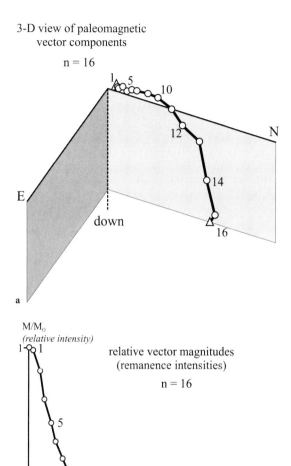

a

M/M₀
(relative intensity)

relative vector magnitudes
(remanence intensities)

n = 16

c   A.F. demagnetization steps

traditional "Zijderveld" vector plot: superimposed
plan and vertical cross-section

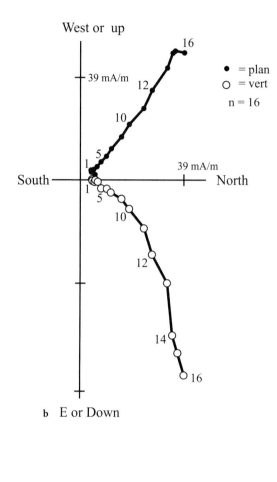

b   E or Down

Fig. 10.1. In earth science, orientations rarely need to be associated with some attribute. However, paleomagnetic vectors are an exception. To isolate characteristic magnetisation components from individual samples, one needs to simultaneously associate orientations of vector components with their magnitudes. Where successive components have similar orientations, they may be taken to define a significant paleofield direction. The length of the vector components is essential to this construction. **a** This is relatively easy to visualise in three dimensions, if the diagram is rotatable on a computer monitor. **b** However, traditionally, a plan view and a vertical cross section are superimposed on the same graph. Occasionally, stable directions may be obscured by this method. This may be overcome by examining the intensity plot or simultaneously plotting a differently oriented vertical section, or a stereogram (see later)

cise readings of values cannot be made from the image, this is rarely a disadvantage as the software provides that digitally, more precisely than any graph reading.

Commonly, the vector's magnitude may not be required, either because its value has been realised in the preliminary stages of analysis, as in classical paleomagnetism, or because the direction alone is of interest. In these cases, the orientation distribution of directions in three dimensions may be envisioned as radiating from the centre of a sphere. The orientations strike the sphere at points whose distribution on the sphere,

like stars in the sky, may form clusters and other patterns whose frequency distribution may be examined statistically. For convenience of presentation and communication, we project an image of the sphere and the endpoint distributed on it, downwards onto a plane. The sphere's planar projection is a *stereogram*; the orientations are represented by points in the projection circle to form an *orientation distribution*.

The stereogram is almost exclusively used to represent orientations in three dimensions. For axes, the diagram is the projection of a hemisphere for which

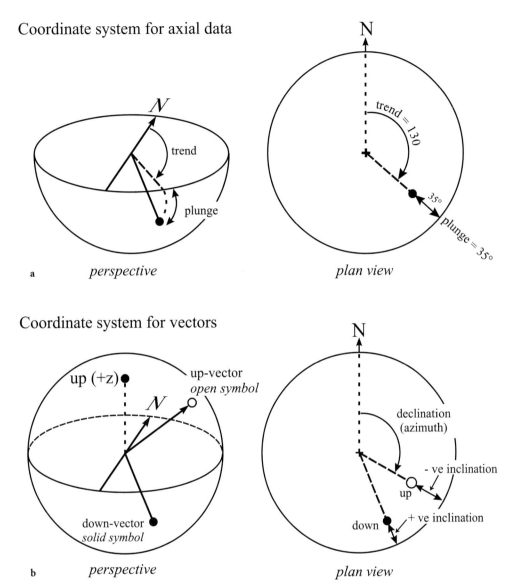

Fig. 10.2. Coordinate systems for orientation data in hemispherical and spherical projections. **a** Axial data or undirected lines are specified uniquely by a trend (also known as azimuth or declination) and a plunge (also known as inclination). Two numbers, e.g. 130/35 would specify the trend/plunge for the axis shown. **b** Vectors and unit vectors possess a polarity so that their orientation requires the use of both upper and lower hemispheres. The convention in geology and geophysics is that vectors directed upwards have negative inclinations whereas vectors directed downwards have positive inclinations. The up-vector shown would be specified as 110/–30 and the down-vector would be specified as 165/+20

the trend and plunge of an axis suffice for its specification (Fig. 10.2a). For the example shown in Fig. 10.2a, the trend and plunge of the linear element would be recorded in degrees as (130/35). On the plan-view projection, the axis is represented by a point where the axis touches the lower hemisphere. In structural geology and geophysics, most observed orientations are of axes

for which the lower hemisphere is used by tradition. Mineralogists and crystallographers also work with axial data but project it onto the upper hemisphere since they prefer to consider their crystal axes radiating upward, a logical approach in the historical, manual investigation of crystals in the laboratory. Where vectors are recorded, it is necessary to use both upper and lower hemispheres since the line is polarised, i.e. it matters which way it points.

Vectorial data are mostly met in paleomagnetism, the remanence acquired by rocks being parallel to the paleomagnetic field direction. Thus, the orientation of the magnetism has polarity because the magnetic vector is north seeking. For this reason, normally both hemispheres are superimposed on one circular diagram. Upper-hemisphere orientations are usually shown as open symbols whereas lower-hemisphere symbols are shown solid. Since we are not using a convention to constrain us to one hemisphere, as in structural geology, three pieces of information are now needed to specify the orientation of the vector. These are azimuth and inclination but also the sense of inclination; up (−ve) or down (+ve). For the two examples shown in Fig. 10.2b, these would be recorded as (110/−30) for the up-example and (165/+20) for the down-example.

For convenience, the orientations must be recorded on a plane projection or *stereogram* using one of several possible geometrical projection methods, two of which are considered here (Fig. 10.3).

Placing, locating and manipulating the orientations of planes and lines on the stereogram may be performed manually, quite quickly and simply, with an accuracy of ±2° in most cases, using a template known as a *stereonet*. The orientations are normally pencilled onto a sheet of tracing paper that is centred above the stereonet. Rotating the tracing paper about the centre of the net facilitates fast plotting with the use of a drawing pin (thumbtack) to act as an axis of rotation. Techniques are described in most structural geology textbooks and briefly in Appendix II. Computer programs plot the orientations with greater accuracy, and permit data manipulation and analysis that would be prohibitive by manual techniques. However, novice students may experience difficulty with the underlying concepts if they have not also plotted some data manually. Apart from consistency and accuracy, computer programs do not provide any distinct advantages for plotting individual orientations. The problems of

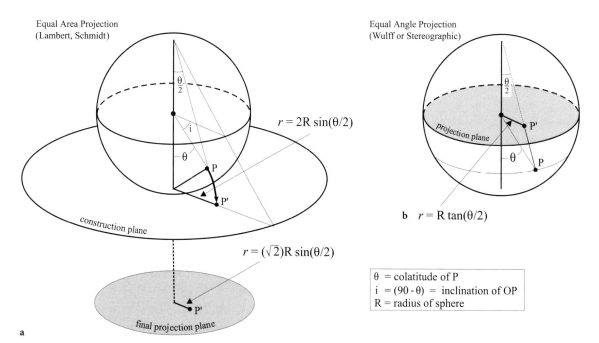

Fig. 10.3. This shows methods by which orientations on the sphere or hemisphere are projected onto a plan view, as shown in Fig. 10.2. In each case, the point representing the endpoint of the line is plotted at a distance "r" from the centre of the stereogram where θ is the co-latitude, or complement of the inclination, ($\theta = 90–$inclination). The trend/azimuth/declination of the line is recorded directly as the angle from north, in geological and geophysical conventions, i.e. east would have a trend of 090°. However, there are different methods of projecting the endpoint of the line on the sphere down onto the plane. The two common choices in the geosciences are shown. **a** The equal area projection ensures that any particular solid angle always subtends the same area on the projection albeit at the cost of distorting its shape (e.g. see Appendix II, Fig. AII.5a). The plane of construction is a stretched version of the equatorial section: to reduce it to the same radius as the sphere, the factor "2" in the equation should be replaced by $\sqrt{(2)}$. **b** The equal-angle projection attempts to minimise the distortion of angular differences between lines but results in considerable distortion of areas; the same angular cone projects as a smaller area near the centre of the stereogram than at the edge (e.g. see Appendix II, Fig. AII.5b)

spherical distortion due to projection onto a plane, described below, persist with sterograms produced by any method. However, computer programs have immense advantages when statistics must be calculated and density contours of orientation distributions must be produced, particularly in the freedom to view the orientation distribution from different perspectives, thereby reducing the effects of projection distortion. The software *Spheristat* by Pangaea Scientific is particularly recommended, and the essential elements required by programmers may be found in Diggle and Fisher (1985).

The manner by which the sphere or hemisphere is projected onto a plane may be achieved in several ways. This is true whether we are creating a physical template (as shown in Fig. 10.4) or using a computer program. All methods involve some degree of distortion in the plane projection. Two projections are common because they minimise certain types of distortion (Figs. 10.3, 10.4). We should remember that the angular relationships between different points in the projection can always be determined reliably, but the manner of projection causes visual distortion, which may condense or disperse points that would otherwise seem to

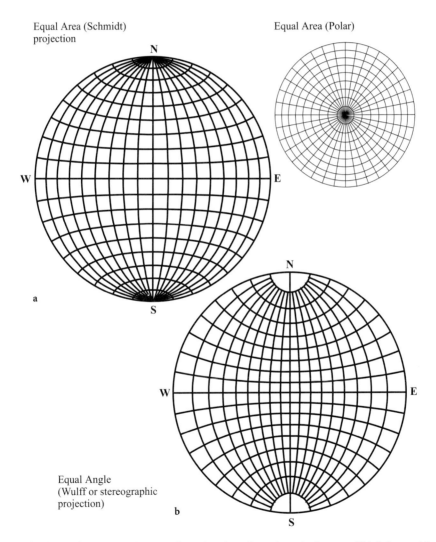

Equal Area (Schmidt) projection

Equal Area (Polar)

Equal Angle (Wulff or stereographic projection)

Fig. 10.4. a Template for the equal area projection; note that 10°×10° quadrants have similar areas. This is favoured for statistical uses where clustering of orientations as an artefact of projection would be undesirable. However, it is misleading where interangular relationships are important. **b** Template for the equal angle projection. Note 10°×10° quadrants have similar areas so that the distortion of inter-line angles is minimised and not too dependent on their orientation. However, the 10°×10° quadrants differ in area by a factor of 4 between the centre and margins of the net. Such distortion of orientation distributions is unacceptable in statistical applications

be differently distributed. It should be noted that the equatorial projection shown for equal-area and equal-angle stereograms in Fig. 10.4 provides sufficient flexibility to record the orientation of any plane or line and to determine angles between them. The polar view of the projection may be used for rapidly plotting lines (Fig. 10.4a, inset). Computer programs exist for plotting any arbitrary view of the spherical grid onto a plane surface (e.g. Starkey 1966b).

In branches of earth sciences such as structural geology and paleomagnetism, we are less interested in the preservation of equal inter-angular relationships in different parts of the spherical projection. Instead, the statistical curiosity of the investigator is directed toward relative densities or concentrations of orientations in different parts of the spherical projection. For example, if two parts of the projection subtend equal solid angles and each contains 10% of the orientations, then this equivalence should be apparent visually. An *equal-area projection* such as those attributed to Lambert or Schmidt accommodates this. The manner in which an orientation on the lower hemisphere is dropped to the plane of the equal-area projection is shown in Fig. 10.3a. The corresponding template is shown in Fig. 10.4a and we may see that the graticules are similar in area over the stereogram, for example by comparing $10° \times 10°$ segments at the perimeter and centre. Nevertheless, there is distortion since the $10° \times 10°$ segments have different aspect ratios in different parts of the projection. Distortion is an inherent property of any two-dimensional projection of a sphere and is not overcome by any computer-generated diagram. This is an important consideration when we compare density distributions by contouring them as Appendix II shows (especially Fig. AII.5a, b). The equal-area projection is used in most studies concerned with orientation distributions, orientation statistics and density contours.

Where one prefers to preserve relative angular relationships in different parts of the stereogram, a Wulff or stereographic projection is preferred. Equal distances between points on the stereogram correspond to equal interdirectional angles in three dimensions (Figs. 10.3b, 10.4b). Consequently, this is also known as an *equal-angle projection*. It finds most use in mineralogy and crystallography where the angular relationships between directions are of paramount importance. The template reveals that there is a considerable distortion of areas in this case, as the plane-projection area of segments (e.g. $10° \times 10°$) is four times larger at the edge of the stereonet than at the centre (Fig. 10.4b). This would be quite unacceptable in the evaluation of orientation distributions where the density of orienta-

tions must not be diluted or concentrated as an artefact of their position on the plane projection. No projection of a spherical distribution onto a plane can preserve a truthful visual representation in all parts of the diagram. For example, we prefer and will only use the equal-area projection in this text, as our interest is in frequency distributions of orientations whose densities must be faithfully preserved in a plane projection. However, the equal-area projection will distort the shapes of the density distributions; for example, two axes separated by 10° appear in equal-area projection as points separated by different distances depending on their mean orientation. Thus, a concentration of axes will have the same density but will change in shape depending on its mean orientation.

Although our data are invariably plotted as axes or vectors, many important geological data elements are not lines but planar features such as beds, schistosity, joints or faults. These are normally described by an orthogonal axis, the normal or perpendicular line to the plane, which is substituted for the planar element. Thus, actual bedding planes are only plotted as great circles in special circumstances; more commonly one would plot their normals. These would be axes in the cases of most beds and other planar features. However, if the sense of younging or way-up of the beds was important, the normal would be plotted as a vector.

10.2
Recording and Reporting Three-Dimensional Orientations

Although we are primarily concerned here with the manipulation, plotting and interpretation of orientation distributions of axes and vectors, we must understand something about the way in which geologists record planar data. This is then transformed into axial data, or sometimes vectorial data, to be plotted as normals to the plane, using a stereographic projection. The traditional record of a planar element involves recording of two items.

1. The *dip* of the surface, being the maximum inclination within the plane.
2. Either the *dip direction* or the *strike*. Strike is a horizontal line on the plane, perpendicular to the dip direction.

For the example of an eastwards-dipping plane shown in Fig. 10.5a, the angle of dip or inclination is 20°. The strike of the plane is north–south. Any system that requires the strike to be recorded involves an inherent

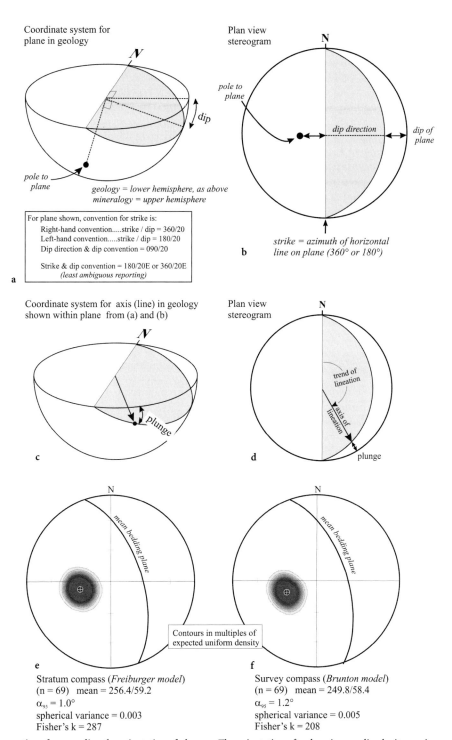

Fig. 10.5. Conventions for recording the orientation of planes. **a** The orientation of a plane is recording by its maximum slope (= dip), and the direction of dip. Instead of the dip direction, the trend of a line perpendicular to dip may be recorded. That horizontal line on the pane is known as the strike. **b** When plotted on a stereogram, the plane is not represented by its cyclographic trace but by its normal. This permits the orientations of many planes to be recorded and their orientations densities to be contoured, see **e, f. c, d** A line's orientation is recorded simply as trend and plunge. **e, f** Concentrations of bedding normals from sedimentary rocks. The sediments were tectonically flattened causing bedding-plane roughness to be reduced. Using different styles of compass, for **e** the dip and dip direction were measured simultaneously, whereas for **f** strike and dip were measured separately

problem; for shallow dip angles, the azimuth of a horizontal line on the surface becomes quite inaccurate. Consider the extreme case of a horizontal plane; the strike is indeterminate and could be reported anywhere in the range 0–360°.

There are several conventions for recording the orientation of a plane. The best method should be least ambiguous and least susceptible to error in the field. To some extent, this depends on the type of compass used. Traditional surveying or *transit compasses* always require the strike to be recorded using the compass dial. The inclination of the feature is then measured in a separate manipulation using a clinometer, which is usually built into the compass. On the other hand, *stratum compasses* permit the measurement of strike to be bypassed and yield dip direction and dip automatically in one reading because the lid of the compass is used to imitate directly the orientation of a plane. The slope of the compass lid gives the amount of dip and its trend records the direction of dip. The edge of the stratum compass's lid may be used to record the trend and plunge of a lineation.

Table 10.1 lists possibilities for recording the orientation of a plane. All are valid, but some records would not be self-explanatory. A notebook record would have to be prefaced with some explanation in those cases. The least ambiguous method specifies three pieces of information; after all, the plane is a structure that exists in three dimensions. Thus, a record that includes strike, dip and the direction of dip with respect to the strike line (e.g. NW, NE, etc.) is recommended. Thus, for the plane shown in Fig. 10.5b, a record "360/20E" would be understood by all geologists in any language and, moreover, it could only refer to a plane. The notation for a lineation would be trend and plunge, e.g. 160/15 for the lineation in Fig. 10.5c,d; this notation could be confused with the designation of a plane under other conventions for reporting a plane (Table 10.1). For simplified and consistent manual plotting with protractors or mapping compasses, some workers avoid using a reference to south, thus all azimuths would appear as N25E, N60W. Whatever scheme one adopts, it must be consistent and should be explicit. Unfortunately, as discussed in the next paragraph, shorthand notations commonly assume that the reader of the notebook is familiar with some convention.

Inevitably, repetitious measurements in extensive field surveys encourage the use of shorthand. The *right-hand rule* is very common, especially in North America, as it requires only two numbers to be reported; strike and dip (e.g. 360/20 for the plane in Fig. 10.5a,b). The ambiguity is resolved because the proponents of this system adopt the convention that their recorded dip-direction must be located clockwise (*right-hand*) with respect to the azimuth chosen for strike. Thus, the geologist must note the dip direction of the plane and then consciously report the end of the strike line such that the dip direction lies on its right-hand side. It is less ambiguous to record the strike and the dip direction fully (Table 10.1), so that a convention need not be remembered and notebook records are universally intelligible. Unfortunately, some geologists use a left-hand-rule.

In the field, the choice of convention for recording planes' orientations is biassed to some degree by the type of compass used. Most geological compasses are most conveniently used to measure strike and dip separately; these are based on traditional surveying or *transit compasses* like the familiar Brunton model. Thus, the right-hand rule or strike-and-dip methods are convenient. However, for repetitive structural measurements, the *stratum compass* is superior as it reduces fatigue and operator error; such compasses are manufactured in Germany by Freiburger and by Breithaupt and in Japan by Showa-Sokki. The Japanese model is dual-purpose and retains all the surveying functions of a traditional transit compass. The lid of the stratum compass may be laid against the plane to be measured (or the lid edge along a lineation to be measured) and the lid's inclination and orientation are simultaneously recorded. That would favour a recording method using dip direction and dip. In this way, it is possible to make measurements by one hand and on a greater range of awkward surfaces, e.g. overhanging

Table 10.1. Conventions for recording orientations of a plane, example of Fig. 10.5

Dip direction	Strike (always three digits)	Dip (always two digits)	Convention	Report
090°		20°	Dip and dip direction	090/20 (S)[a]
	000° or 360°	20°E	**Strike, dip and dip sense**	**000/20 E** or **360/20 E**[b]
	000° or 360°	20°	Right-hand rule	000/20 (S)[a]
	180°	20°	Left-hand rule	180/20 (S)[a]

[a] For security, such records should be noted as surfaces (S) to avoid confusion with a lineation whose trend and plunge would also simply be denoted by two numbers, e.g. 170/20 (L).
[b] Least ambiguous convention.

or with small areas. A comparison of measurements for the two types of compass is shown for a well-bedded intertidal sequence is shown in Fig. 10.5 c, d. For the stratum compass (c), dip direction and dip were measured more rapidly than the separate strike and dip measurements with the surveying compass (d). Those measurements are from an outcrop in which bedding-plane roughness was subdued by tectonic flattening. The consistency of orientation is high for both types of compass but perhaps marginally superior for the stratum-type compass; the statistics shown are described later. However, it may be appreciated that the mean directions differ at the 95% confidence level.

No method of recording the orientation of a plane is fundamentally superior to another, provided that the convention is used consistently and is explicitly stated for the benefit of those who may later need to use the field measurements or field-oriented specimens. It is generally preferable to choose a method that requires the least assumptions by subsequent users of the information, and the author prefers full specification of strike, dip and approximate azimuth (e.g. NE or SW) of the dip direction to avoid ambiguity of dip on either side of the strike line. Such records are immediately understood from a notebook without any assumed conventions. However, the greater ease and precision of stratum compasses favour the "dip direction and dip" convention for planes. Its facility for averaging out the roughness of a surface can be enhanced, with any compass, by placing a clipboard or other easily manipulated flat object on the geological surface and then measuring its orientation.

10.3
Field Measurements, Oriented Samples and Associated Errors

We must not lose sight of the fact that almost all geological orientation data still come from field measurements using hand-held compasses. It is difficult to read an azimuth from the compass, or the inclination from the inclinometer (dip needle) to yield observations reproducible within ±2°, although that is the normal fine graduation of their scales. Moreover, and more importantly, the structures we measure have intrinsic roughness that prevents reproducibility of individual measurements to better than ±4°. These are conservative and generous concessions to the precision that most geologists would claim. This is compounded by the following insurmountable problem. Consider a steep linear structure, inclined nearly 90°; it will be clear that its trend or azimuth will be difficult to ascertain. In-

deed, when the lineation is vertical, any value may be assigned to the trend. Conversely, consider a plane that has a very shallow dip: estimating strike is difficult. In the extreme case of a horizontal plane, any value could be assigned to its strike. Thus, the intrinsic errors of observation due to compass manipulation are compounded by the errors of trend for high or low inclinations. Nevertheless, with suitably concentrated orientations and a sufficient number of measurements, mean orientations may be quoted with a meaningful sensitivity of ±0.1°, despite much greater inherent variance of the orientations. However, some branches of geoscience claim such high precision for *mean* orientations routinely with few observations, e.g. in paleomagnetism. Errors in dip associated with the incorrect choice of strike, and errors in inclination associated with the wrong choice of sectional plane can be inferred from Fig. 10.6. The intrinsic variability of the structural element that we are measuring may also contribute to measurement error. Depending on the depositional environment, beds will show varying degrees of roughness. Even for a uniformly well-bedded, intertidal sediment, in which the bedding planarity has been improved by tectonic flattening, there is still recognisable variation from field measurements (Fig. 10.5 e, f; $n = 69$).

Many geological and geophysical orientation data are not collected directly in the field. Rather, they must be extracted from specimens that were oriented in the field with respect to geographic coordinates using a magnetic compass, or using a sun compass in the proximity of rocks with strong remanent magnetism. Immediately, we recognise that we are at least doubling the possible occasions in which orientation errors can be accumulated by measurement and also by recording. Petrofabrics, petrophysics, magnetic fabrics and paleomagnetism are the principal studies in which specimens must first be collected with a known field orientation, and then subsampled in the laboratory to provide the final preparation (microscope thin-section, X-ray texture goniometer slide, or paleomagnetic rock cylinder). The confounding of errors of orientation at each stage may be appreciated after reviewing Appendix I.

The more common methods of retrieving oriented samples from rocks are:

1. *to remove oriented hand specimens that may then be prepared (perhaps core-drilled) in the laboratory under controlled conditions (structural geology, petrofabrics, magnetic fabrics, paleomagnetism).*
2. *drilling in the field using a portable, gasoline- or electric-powered drill and then orienting the core as*

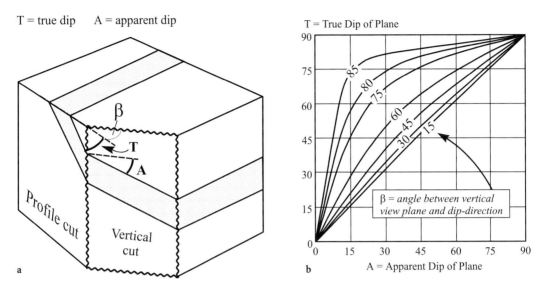

Fig. 10.6. a The true dip angle of a plane appears to be less when viewed on a plane not parallel to the dip direction. **b** Nomogram relates true and apparent dip

it is retrieved from a shallow hole usually ~10 cm deep, producing a core ~25 mm in diameter. This is most common in paleomagnetism.

3. *sampling an exploration drill-core of great length. The inclination of such cores is usually well-documented at each depth, but it is quite expensive and time-consuming to preserve the azimuth of such core as it is retrieved in sections from a long hole. In some cases, a young component of viscous magnetisation in the core may indicate the north azimuth (e.g. Tarling 1983), but that is a risky procedure since drilling, using a large commercial rig, commonly magnetises core before it is retrieved.*

Here, the discussion is restricted to the first two approaches since these most concern the students who may use this book. The first approach requires a large hand specimen (~2 kg) with some unique flat surface on which an orientation is recorded with a permanent marker or diamond scribe. It is preferable to select a surface on the specimen with some structural or geological significance, such as a joint or bedding plane. This is useful because it provides a meaningful reference, permits resampling and makes subsequent verification of data easier. The orientation convention should be completely unambiguous: the specimen will be handled many times, stored perhaps for years and may become separated from the associated field notes, so it is sensible to use the *strike/dip/sense-of-dip* convention, as shown in Fig. 10.7a. In the unlikely event that one must mark an inverted surface, this should be indicated as

shown in Fig. 10.7b. The sample may then be returned to the laboratory and, when clamped in a rotatable, tilting vice, it is possible to saw thin-section slices or drill cylindrical cores in any desired orientation. The procedure outlined is preferable because it permits the final sample preparation to be placed back into geographic coordinates. For example, the core shown in Fig. 10.7a is vertical and the top arrow points north. Such a core can be archived without the fear that it may become useless many years in the future if the associated field notes are misplaced. Large specimens for laboratory coring may be collected by a single worker, which reduces cost. However, the weight of samples and shipping costs may be prohibitive. The author prefers this method since his laboratory uses specimens for rock magnetism, petrofabrics, magnetic fabrics and paleomagnetism. A few hundred grams of drill core would not suffice. Moreover, the laboratory-prepared cores are of very high quality, permitting use of a high-sensitivity, high-speed spinner magnetometer and also the determination of magnetic anisotropy. These are not particularly strong advantages for paleomagnetists who require just a few hundred grams of drill core, especially if they use a cryogenic SQUID magnetometer in which specimen shape need not be so perfect.

The first approach, selecting oriented specimens directly from outcrop, has a refined variation which is used for rare material or where high orientation precision is required. This is used particularly in archeomagnetism and geoarcheology where precise paleofield orientations must be determined. The sample-

Fig. 10.7. Orientation of specimens required for laboratory work. **a** Hand-specimen orientation noted by strike, dip and sense of dip. (Laboratory core may be prepared directly in geographic coordinates.) **b** Convention for recording specimen using an overhanging surface. **c** Technique for orienting drill hole in field, after core has been extracted. The orientation must then be marked on the core

orientation error between outcrop and measurement machine may be $\geq 1°$ (Henry and LeGoff, IPGP, Paris, pers. comm.). A cap of resin or plaster is placed on top of the part of the material to be sampled at the outcrop. A horizontal-level gauge ($=$ spirit-level) is placed on the surface of the setting material to guide a horizontal surface. When the synthetic cap is solid, its horizontal surface is marked with a north arrow and the specimen is then removed (e.g. Tarling 1983). Unfortunately, this procedure is not too common, probably because it is relatively time-consuming. A less precise procedure presses a plastic sample cube into sediment or soil. The cube has an orientation arrow and a venting hole to permit air to escape as the specimen invades the container. A lid is then placed on the container. Both cubic and cylindrical containers are available, with a volume of ~ 8 cm³ and they may have internal ridges to prevent movement of the specimen in transport. Their use is restricted mainly to archeomagnetism of soft materials, and they carry the risk of disturbance of the speci-

men as it intrudes the container and the risk of dehydration–disaggregation during transport.

The second approach retrieves short drill core in the field with a portable gas-powered, or more rarely with an electrical, battery-powered drill. This is generally the practice in paleomagnetism and has the advantage that small features may be targeted, such as chilled margins of igneous rocks. However, the logistical difficulties (gasoline, supply of cooling water), necessity of an assistant and low-durability drill bits make this procedure quite time-consuming and expensive. The resulting specimens are inferior to those produced in the laboratory, where drilling is better controlled. Moreover, field cores are in some arbitrary orientation, usually at a high angle to some outcrop surface requiring careful documentation and archiving of field notes. Nevertheless, this procedure serves paleomagnetism very well for its high productivity and low shipping costs of samples from remote regions. The core is best oriented with the aid of an orienting tube in the core hole, after the core has been carefully extracted. The tube has a swivel-base plate on which a compass may be placed to determine the trend and plunge of the drill hole. One of the greatest sources of error in this technique is extracting the core without rotation, thus losing its correct azimuth.

10.4
Stereogram Density Contours

Plotting the plane's trace on a stereogram as a great-circle, or *cyclographic trace* (e.g. Fig. 10.5a) is useful for the manipulation of a few data and constructions or illustrations concerning the intersections of planes. For example, one may readily recognise the direction of the intersection of a fault with a bedding plane from such *cyclographic* plots. For stereograms with larger numbers of planes, it is more effective to plot the *normal* to the plane (Fig. 10.5a, b). Thus, the orientation of each plane is reduced to a point projected on the stereogram, representing the orientation of an axis that is perpendicular to the plane. In this way the orientations of many planes are recorded, compared, and treated statistically without producing a cluttered diagram. When the number of points, representing orientations, is reasonably large and somewhat concentrated, it is helpful to replace the points with shaded areas that represent the frequency of observations in any given angular segment of the stereogram. This is referred to as the orientation density and introduces the concept of density contours of orientations on the sphere. An example of bedding poles for a well-bedded sequence is shown in Fig. 10.5e, f.

The greatest value of any orientation-contouring procedure is that it does not assume any mathematical form or model for the data. It may be compared to a non-parametric technique in formal statistical analysis. Although stereogram density contours do not directly provide statistics as such, the objectivity they bring to the summary description of axes and vectors cannot be underestimated. Mode, pattern and symmetry of the orientation distribution may be revealed painlessly by contoured stereograms. They were introduced to contour axes in structural geology and extended to more and more complex situations, each time facing new complications. The principal concerns are as follows:

1. *Each axis plotted carries equal weight, and density contours mask any attribute such as geographical location or measurement uncertainty of individual orientations.*

2. *Where orientations of vectors are plotted, both upper and lower hemispheres are required causing difficulty in the visual comparison of vectors on different hemispheres, especially antiparallel vectors with low inclinations or high inclinations.*

3. *For some vectors, the magnitude is not merely a subsidiary attribute but is essential to the interpretation. Contours suppress this information.*

4. *When applied to tensors, contours may mislead. In Chapter 11, we will see that a single second-rank tensor is a set of three principal axes, with different magnitudes. It describes the directional variation in some physical property, often visualisable as an ellipsoid, at a single location. The three principal axes of the tensor must be orthogonal. Stereograms are commonly plotted to show the orientations of the three axes of many tensors, especially in magnetic fabric work. Unless the tensors are very well aligned, the constraint of mutual orthogonality of the axes from individual sites greatly complicates the interpretation of the density contours. Furthermore, if tensors from different sites have different mean magnitudes, the interpretation may be confounded further, as with vectors (2) above. The density peaks for the three principal axes, contoured separately, will usually be non-orthogonal, whereas this is a requirement for each individual tensor.*

Our main concern with density contours is that they force equal weighting of all orientations in the sample. However sophisticated the analysis of the contoured orientation distribution, it may be unreliable if the uncertainties of individual observations are large or uneven. Unusual orientations, outliers, may be inaccurate, yet they are included in the density contours and influence the shape of those contours to the same extent as more accurate orientations. In the case of vectors, orientation uncertainty may be confounded by magnitude uncertainty.

Structural geologists very rarely consider the orientation uncertainty of individual axes, their stereograms commonly assume all orientations are equally precise. However, paleomagnetists are especially sensitive to the orientation uncertainty of their paleomagnetic vectors, and, in the initial stages of data reduction, the vector magnitudes are also evaluated. They are less inclined to use density contours on their stereograms, but when they do, the uncertainties of the orientations have been carefully evaluated and are very similar. Paleomagnetism also evaluates orientation uncertainty in different ways. For example, one may plot all observations assuming uniform orientation uncertainty and investigate their mean orientations and dispersion.

Alternatively, since different sites produce different data quality, one may plot site means with comparable uncertainties, from which to estimate the overall mean and dispersion.

Assume that it is valid to assign equal weights to all observations. For the benefit of those who do not have a structural-geology background, the following is a simple guide to the traditional crude procedure of manual density plotting. The orientations are plotted in the usual manner as axes, unit vectors or normals-to-planes. Then, using a counting cell of constant area, one records the number of points that fall within the cell. One may record the number of observations directly, or the number as a fraction of the total number of points on the stereogram. The principle is intuitive with contours of any map data (Chap. 1) and only requires minor modification for stereograms; the main difference is the closed nature of the system, giving rise to a stereogram edge effect (Appendix II). In structural geology and petrofabric work, a circular cell cut into a plastic sheet would be moved across the diagram. The number of observations lying within the cell would be recorded on the stereogram at the centre of the cell. The cell would typically have an area equal to 1% of the stereonet projection circle. The choice of its diameter may be varied according to the number of data and required detail of contours but that is difficult in manual operations, computer-generated stereograms are much more flexible in that regard. Successive positions of the cell are chosen at regular intervals and the spacing may be adjusted to produce more smooth generalisations, within reason. When it is necessary to place the counting cell so that it partly goes beyond the edge of the stereonet, counts must be included from the diametrically opposite margin of the net (Appendix II, Fig. AII. 5).

Manual density contouring of the stereogram is still a useful initial training exercise. The leap to the use of computer programs may leave the uninitiated somewhat mystified as to the complexity of the underlying pitfalls and assumptions. Clearly, computer algorithms are superior in their speed and consistency. At a fundamental level, they still involve the same difficult choices as to cell diameter, the frequency of counting sites and the interpolation of contour levels. The technical superiority of computer algorithms for density contours was exploited by several structural geologists including Robin and Jowett (1986), Starkey (1996a,b) and perhaps most fully documented with source code by the orientation statisticians, Diggle and Fisher (1985). Fisher et al. (1987) summarise advantages and choices offered by computer algorithms and their success in achieving desirable compromises; some but not all options are available in commercial software although Spheristat by Pangaea Scientific has proven very satisfactory for the author's team. Even where the options are not all explicitly available, their appreciation helps one to understand what the program may or may not be doing with our data.

The first step in assessment of orientation distributions, including density contouring, is the appreciation of the true nature of the distribution unimpeded by the artificial constraints of the net's edge effect and different degrees of density distortion for different orientations. This is achieved by "centering" the data so that it is more easily visualised, for example, rotating the data so that a cluster is located at the centre, or a girdle along a N–S, E–W or peripheral great-circle. This is very useful for reconnoitering weak concentrations and is only practicable by computer.

The second step is to establish the degree of detail that one wishes to preserve in the density plot. In other words, how much smoothing of the contours is desirable? This depends on the goal of the application, the strength of the concentration and the number of data. In general, fewer data may require more smoothing. Large numbers of data may not need any smoothing or contouring because the density of points achieves the same effect visually and directly (e.g. Fig. 10.11 d). The degree of smoothing is sensitively proportional to the term $1/(\kappa n^{1/3})$ where κ is the Fisher-concentration parameter for the sample.

The third step involves choosing contour levels so that contours are not too crowded at clusters, so that low-significance contours are not constructed through sparse data and so that the contours of different values have some internal consistency. N. I. Fisher lists some possibilities:

1. *contour levels equally spaced between the peak and trough frequency in the distribution.*
2. *contour levels dividing the sample into groups of equal numbers of observations. These contours then separate orientation space into groups representing equal probabilities of the orientation distribution.*
3. *contour levels whose value increases with the concentration so that each contour interval encloses more data than the next lower interval.*

In this book, density contours follow the method provided by Pangaea Scientific's Spheristat software. This is essentially a combination of (3) above, with a constant reference, like (2). The contours are at multiples of the density expected if the sample were distributed uniformly over the sphere. The contour levels are then in multiples of the expected uniform density. This appears to produce satisfactory contours in a wide range of situations and the software permits their smoothing at three different levels.

A paleomagnetic study of the Kapuskasing Structural Zone of northern Ontario provides examples of contoured vectors. These are characteristic remanences (= ChRM), believed to be of primary, or other, significant origin, acquired when lower crustal rocks were thrust up to the surface. They form part of an unpublished study by Borradaile, Werner and Lagroix. Individual vectors plot as points, and since there are both normally magnetised, downward-seeking magnetisations and upward ones, we require separate upper and lower hemisphere projections for clarity (Fig. 10.8a,b). In some studies, both up-vectors and down-vectors are plotted on one circular projection that combines upper and lower hemispheres, as the down- and up-vectors are conventionally distinguished by closed and open symbols anyway. However, for more than ten observations, this normally becomes confusing if there are antipodal vectors at the margin or centre of the stereogram. Separate plots of the upper and lower hemispheres overcome this problem. Still, the orientations themselves may not present a clear picture. Simple statistics to describe the dispersion may involve assumptions about the nature of the distribution. A mean direction is shown in Fig. 10.8, together with its *confidence cones*. The confidence limit is usually at the 95% level so that there is a 5% probability that the mean direction of the population lies outside the sample's confidence cone; this will be discussed later.

For more than 20 orientations in a hemisphere, clarification is normally achieved by indicating the approximate frequency of orientations at any point. Thus, the density of observations is contoured. Many schemes are available, but we shall present all contoured stereograms with contour levels in multiples of the expected uniform density, if there were no preferred orientation for the same number of observations. Examples of paleomagnetic vectors are contoured in Fig. 10.8c,d. Density contours characterise the distributions empirically; there are no preconceptions of the nature of the orientation distribution of the population from which the sample is drawn. However, the contouring procedure must be carefully considered, and the number of observations must be sufficient to characterise the distribution. It is reassuring if the pattern of contours may be replicated from subsamples of the main sample. The actual number of required observations depends on the idiosyncrasies of the distribution: strong preferred orientations are characterised by fewer orientations than weakly concentrated distributions.

For the paleomagnetic example, the downward-seeking remanences scatter in a WSW–ENE girdle (Fig. 10.8c) whereas the upward-seeking remanences form a more nearly isotropic cluster in the NE quadrant (Fig. 10.8d). Stupavsky and Symons (1982) showed the usefulness of density contours in the initial evaluation of paleomagnetic data; contouring is generally underused in that field. Separate stereograms are required for the upper and lower hemispheres if there are both upwards- and downwards-directed vectors. The side-by-side comparison of a lower hemisphere and an upper hemisphere projection of density contours may be somewhat confusing, especially for the novice. Fisher et al. (1987) overcame this disadvantage with a simple technique. Instead of viewing upper and lower hemispheres in geographical coordinates (Fig. 10.8c,d), with the required mental adjustment for lines radiating either upward or downward from the origin, they presented the two hemispheres as though they were halves of a horizontally sliced grapefruit, both facing cut-side upwards. Their construction was trigonometrically logical; they unhinged the sphere about the north coordinate in order to invert the upper hemisphere. However, for geologists, it is convenient to retain north at the top of both hemispheres. Thus, in Fig. 10.8e,f, the sphere has been opened and hinged about the east (090°) coordinate so that the *right half* of the diagram is the *inverted* upper hemisphere. This is probably less confusing for most earth scientists. Comparison of the form of the contours in the two versions of the upper hemisphere shown in Fig. 10.8f,d will reveal the effect of the reflection caused by hinging the hemispheres. Most workers, especially newcomers, may find this presentation less confusing when dealing with orientations distributed on both upper and lower hemispheres, as is commonly the case in paleomagnetism, or with any other vectors. The traditional, combined upper and lower hemisphere projection may be preferred for small numbers of orientations as in paleomagnetism, but it is very difficult to use them with density-contoured data. The flipped hemisphere projections recommended by Fisher et al. (1987) provide one way in which to appreciate density contours of vectors on both upper and lower hemispheres, simultaneously. Alternatively, some people may find it easier to not invert the upper hemisphere at all but simply rotate it by 180° (Mike Jackson, pers. comm.).

Of course, most structural data and petrofabric data are axial in character and are presented on the lower hemisphere alone. A one-hemisphere projection is also possible for vectors if they are fortuitously directed into one hemisphere, like the downward-seeking remanences of Fig. 10.8a,c.

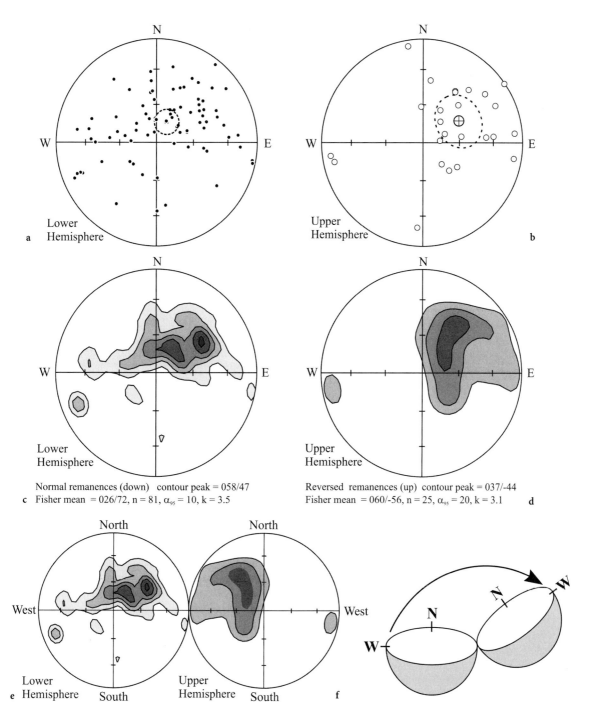

Fig. 10.8. Presentation of vectorial data and the use of density contours using an example of paleomagnetic remanence vectors from the Kapuskasing Structural Zone of Canada (Borradaile, Werner and Lagroix, unpubl. work). **a** Lower hemisphere projection of downward-seeking paleomagnetic vectors with 95% confidence cone of mean shown *dashed*. **b** Upper hemisphere projection of upward-seeking remanences. The Fisher means refer only to the orientations that plot in that hemisphere. **c, d** The same data, contoured in multiples of the orientation density, expected if the same orientations were uniformly distributed in each hemisphere. Density contours are difficult to visualise using traditional upper and lower hemispheres in this way. **e, f** The same data are shown as if the hemispheres were unhinged, as *inset* (*right*) following the convention of Fisher et al. (1987). This facilitates visual comparison of the contours across and between the hemispheres

10.5
The Mean Orientation of an Axis, Direction or Vector

Orientation distributions in three dimensions are among the most complex types of data described in this primer, and our treatment will be largely qualitative. An early clear introduction to the most commonly met situations is available in Watson and Irving (1957). Mardia (1972) provided the first textbook coverage, which was largely superseded by Fisher et al. (1987), a comprehensive, thorough and clear analysis of theoretical models and appropriate statistical tests. Clear, concise explanations of the fundamental statistics and other treatments related to orientations in paleomagnetism may be found in McElhinny and McFadden (2000) and in Butler (1992). Paradoxically and fortunately, this involved aspect of earth-science statistics can be handled to the satisfaction of some routine problems in a largely qualitative manner. That is achieved with density contours, the calculation of the mean orientation and some clever approximations that permit estimates of confidence limits around the mean orientation. We shall proceed initially with some robust techniques, independent of any assumed model for the population from which the sample of orientations is drawn. We have introduced the concept of contouring the data to represent concentrations of observations in different orientations, which is largely self-explanatory. This leads to the first estimate for principal data concentration. The centre or peak of the highest contour loop is an indication of the strongest preferred orientation direction. It is directly analogous to the mode of a linear frequency distribution, and is usually defined with sufficient precision for most geological problems by visual estimation from a stereogram. Examples are given by the contour-peak values of Fig. 10.8c,d. The next procedure for estimating the preferred orientation is trigonometric and uses the vector-sum approach (cf. Fig. 9.12), extended to three dimensions. The orientations are considered as unit vectors or axes. In the case of axes, the sense of direction of the summed orientation is immaterial and the appropriate summed direction is plotted on the lower hemisphere as an estimate of the mean direction. Geophysics and structural geology normally use the following convention for its coordinate system:

$+x$ = North $+y$ = East $+z$ = down

Of course, declinations are measured as positive, clockwise from zero at North. This contrasts with trigonometric convention where angles are positive anticlockwise with zero at east, and $-z$ = down. Downward inclinations are positive, whereas upward inclinations are negative (Fig. 10.9a).

When projected onto the coordinate axes, individual orientations or unit-vectors have lengths (x, y, z). Where the orientations are specified in a geological coordinate system as declinations (D) and inclinations (I), components of each unit vector on the coordinate axes are obtained in the following steps:

direction cosines of each vector:

$$L_i = \cos I_i \cos D_i \quad M_i = \cos I_i \sin D_i \quad N_i = \sin I_i$$

resultant vector R is given by:

$$R^2 = \left(\sum_{i=1}^{n} L_i \right)^2 + \left(\sum_{i=1}^{n} M_i \right)^2 + \left(\sum_{i=1}^{n} N_i \right)^2$$

giving coordinates for end of mean vector:

$$\bar{x} = \frac{1}{R} \sum_{i=1}^{n} L_i \qquad \bar{y} = \frac{1}{R} \sum_{i=1}^{n} M_i \qquad \bar{z} = \frac{1}{R} \sum_{i=1}^{n} N_i$$

from which the mean vector has declination & inclination:

$$Dec = \alpha = \arctan\left(\frac{\bar{y}}{\bar{x}} \right) \qquad Inc = \arcsin(\bar{z}) \qquad (10.1)$$

The inclinations (Inc) have a negative sign for the upper hemisphere, in the geological coordinate system. The angle α returned for the declination must be placed in the correct quadrant of the complete circle of possible azimuths. To determine the quadrant in which the declination lies requires observation of the signs of the mean vector's endpoint coordinates $(\bar{x}, \bar{y}, \bar{z})$, as shown in Table 10.2 and Fig. 10.9b. Orientations in three dimensions may be referred to a polar coordinate scheme (Fig. 10.9c) which observes strict trigonometric conventions for the signs of longitude ($= 90° -$ declination) and co-latitude ($90° +$ inclination). The concept of co-latitude is used frequently in paleomagnetic calculations.

In paleomagnetism, it is often necessary to make an approximate determination of an angular variance, and from it a 95% confidence cone about the mean

Table 10.2. Declination from direction cosines (see Fig. 10.9)

Sign of $\bar{x} = (\sum L)/R$	Sign of $\bar{y} = (\sum M)/R$	Where $\alpha = \arctan$ $(\bar{x}/\bar{y})$ declination (Dec) is given by
+ ve	+ ve	α
– ve	+ ve	$180 - \alpha$
– ve	– ve	$180 + \alpha$
+ ve	– ve	$360 - \alpha$

Geological convention

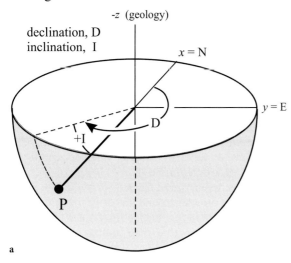

a

Declination from $\alpha = \arctan(y/x)$

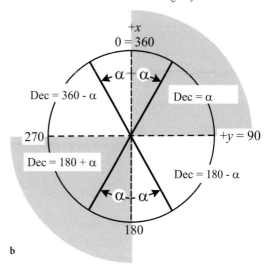

b

Polar coordinate convention

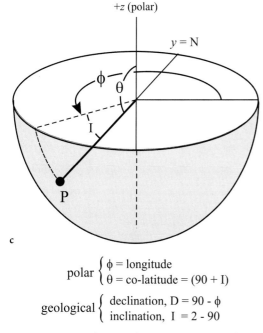

c

$$\text{polar} \begin{cases} \phi = \text{longitude} \\ \theta = \text{co-latitude} = (90 + I) \end{cases}$$

$$\text{geological} \begin{cases} \text{declination, } D = 90 - \phi \\ \text{inclination, } I = 2 - 90 \end{cases}$$

Stereonet: change of viewpoint (rotation of coordinate system)

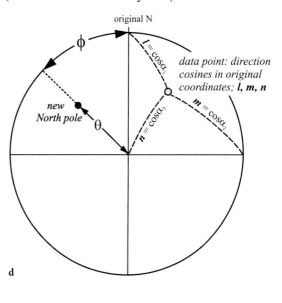

d

Fig. 10.9. a Most software and trigonometric manipulations commence with a polar convention initially which uses the longitude (ϕ, positive anticlockwise) and co-latitude (θ) to specify the orientation. Geological convention uses the declination (D) and the inclination (I). Geological convention assigns a negative sign to the inclination of up-vectors. **b** The calculation of a mean declination from trigonometric functions may cause some ambiguity which can only be overcome by considering the signs of the cosine and sine functions of the returned angle α. The transformation of α required to give the appropriate longitude (ϕ) is shown

vector. One could imagine the mean vector defined by all or part of the sequence of vector components of Fig. 10.1 a, and the cone of confidence that would encircle it. Such information is useful in the preliminary analysis of the paleomagnetic component vectors determined in a demagnetisation experiment. For that purpose, Stupavsky and Symons (1982) suggested estimates of the angular variance (σ_A) and 95% confidence cone-radius (θ_{95}) by analogy with the Normal distribution on a line. However, the more usual expression of angular variance is given as follows, where the angle each orientation makes with the mean orientation is Δ_i:

$$\sigma_A^2 = \frac{\sum_{i=1}^{n}(\Delta_i)^2}{(n-1)} \qquad (10.2)$$

However, this does not simply lead to a definition of spherical standard deviation or confidence limits. That requires some preconceived statistical model for the distribution and will be postponed for now.

10.6
Theoretical Orientation Distributions in Three Dimensions

Orientation statistics fall outside the curriculum of traditional statistics courses and are mentioned only briefly in a few of the earth-science applied-statistics texts, and then mainly for the circular case (e.g. Cheeney 1982; Till 1974; Davis 2002). The most relevant explanations for earth scientists creep into disciplines relying heavily on orientation statistics such as paleomagnetism and magnetic-fabrics (Butler 1992; Tarling 1983; Tauxe 1998; van der Voo 1993). There are a few rigourous statistical treatments on orientation distributions in three dimensions, heavily theoretical but with interesting examples from astronomy, earth, life and natural sciences by Fisher et al. (1987), Mardia (1972), and Mardia and Jupp (2000). The text by Fisher et al. is most useful for geologists and geophysicists due to its applied approach and familiar examples, but unfortunately few of its recommendations and techniques have yet found their way into routine applications. Throughout, we shall deal with orientation distributions that show a single mode, or which may be decomposed into single modes for analysis.

An important part of the advanced texts just cited, concerns theoretical models of orientation distributions. These are essential for advanced analysis and testing samples as in linear-statistical testing (Chap. 4). Earth scientists mostly need a theoretical model for the

population of orientations in order to construct confidence limits around the mean orientation. A statistical model is not required for the calculation of central tendency or angular variance. For example, in most unimodal-orientation samples, central tendency is characterised by a mean orientation calculated like a vector mean (Fig. 9.12), or we may easily determine the median or the mode. Alternatively, the eigenvectors of the orientation-distribution matrix may be calculated objectively from the data, locating the principal axes of the concentration; one associated with the peak concentration and two orthogonal axes that determine the shape of the modal cluster.

However, we may need to estimate confidence limits for the mean orientation, and this demands some theoretical model for the population from which the sample is drawn. The same is true when we study frequency distributions of scalars (e.g. Chap. 2.4). If we possess a theoretical model for the population, its percentage points and critical values may be used in estimations and in statistical tests. N.I. Fisher (1993) and Mardia and Jupp (2000) provide tables of the appropriate values for the most popular model, due to R.A. Fisher (1953). However, the expressions for the test statistics and critical values are usually much more complex than we meet in linear statistics and should be consulted from the cited texts. Fortunately, much can be achieved with the mean orientation, and its 95% confidence limits. However, tests for the non-uniformity of the sample, or to compare two sample means, are not too difficult to apply.

We shall only consider models and theories for orientation distributions that have a single cluster or some form of girdle, since these describe most natural geological distributions, either directly, or by decomposing the sample into homogeneous domains that do posses such simple organisation.

For completeness, comments and descriptions of some of the more important and useful theoretical distributions are given below. Each model describes an orientation distribution that is unimodal, but some are general enough to include the case of an elliptical cluster or girdle. Orientation concentrations on the sphere that fall in the cluster-ellipse-girdle spectrum suffice for almost all homogeneous, single-event, co-axial orientation distributions in geology, as we shall see later in Fig. 10.23. The arguments are transferable equally among axes, unit vectors and vectors. For the purpose of thought experiments, it is best to consider the peak of the orientation distribution rotated, for convenience, to some position close to the centre of the stereogram. *Centering* overcomes the visual distraction of a concentration that is arbitrarily split on diametrically opposed sides of the stereonet (e.g. Fig. 10.10 a, c).

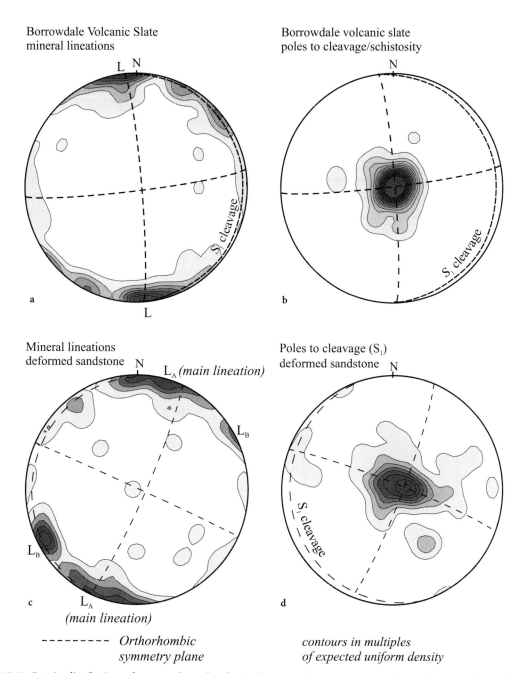

Borrowdale Volcanic Slate
mineral lineations

Borrowdale volcanic slate
poles to cleavage/schistosity

Mineral lineations
deformed sandstone

Poles to cleavage (S₁)
deformed sandstone

- - - - - - - -   *Orthorhombic*
symmetry plane

contours in multiples
of expected uniform density

Fig. 10.10. Density distributions of axes may be analysed using the orientation-tensor concept. This defines three planes of symmetry for the dispersion pattern which intersect in the eigenvector orientations. Each vector has an associated eigenvalue (maximum, intermediate and minimum), indicating the strength of the concentration in that direction. The maximum should be nearly parallel to the peak density, and the peak density may be elongated toward the intermediate value, depending on the shape of the orientation-distribution ellipsoid (prolate versus oblate). **a, b** Mineral lineations and normals to schistosity in Borrowdale Volcanic Group slates, northern England. **c, d** Mineral lineations and normals to cleavage in sandstones

The software package Spheristat (Pangaea Scientific) readily centres orientation distributions, facilitating recognition of the shape of the distribution on one hemisphere. It involves examining the spherical projection from a different viewpoint by re-expressing the data with respect to a new set of coordinates; it may be performed manually (Fig. AII.4) according to traditional structural geology practices, but in practice it is more conveniently and effectively performed by software. We must note the new reference direction in terms of its polar longitude (ϕ) and co-latitude (θ) in the old coordinate system. Consider just one data point and note its direction cosines (l, m, n) in the old coordinate system (Fig. 10.9 d). The orientation of the data point in the new coordinate system is then given by l', m' and n' where:

$$\begin{bmatrix} l' \\ m' \\ n' \end{bmatrix} = \begin{bmatrix} \cos\theta\cos\phi & \cos\theta\sin\phi & -\sin\theta \\ -\sin\phi & \cos\phi & 0 \\ \sin\theta\cos\phi & \sin\theta\sin\phi & \cos\theta \end{bmatrix} \begin{bmatrix} l \\ m \\ n \end{bmatrix} \quad (10.3)$$

(e.g. Mardia 1972). A more complicated rotation matrix allows for the simultaneous rotation about the polar axis by introducing an extra trigonometric function in each element of the rotation matrix so that there are no non-zero terms (Fisher et al. 1987). However, in most practical applications, any desired view may be obtained by two successive rotations using the simpler rotation matrix given above.

The simplest orientation distribution is a cluster, scattered closely and evenly about the mean orientation. The deviations should be non-systematic and isotropic, i.e. equally scattered in any direction away from the mean orientation. Such circular-symmetric clusters of paleomagnetic vectors are commonly encountered in paleomagnetism where a lava flow is rapidly magnetised in a uniquely oriented geomagnetic field as it cools quickly. However, the lava flow must be free from any magnetic anisotropy that could bias the magnetic remanences from the paleofield direction. Nor should the rock possess some strong fabric that could deflect the paleofield and scatter the remanence directions. Fine-grained lava is an ideal candidate to trap accurately the paleofield orientation due to its isotropic character and rapid cooling. Such distributions have density contours that define a small-circle cluster as circular symmetric about the mean orientation (Fig. 10.11 a). They show a similar simple circular-symmetrical cluster, like many observations in structural geology such as normals to planar schistosity (Fig. 10.10 b,d), or normals to bedding (Fig. 10.5 e,f). However, paleomagnetic vectors probably come closest to providing a circular-symmetric distri-

bution of orientations since they owe their origin to lines of magnetic flux that point to the ancient magnetic pole (Fig. 10.12 a). Paleomagnetic applications triggered the development of the first theoretical model for orientation distributions by R.A. Fisher (1953). Subsequently, several more elaborate theoretical models have been proposed to describe more complicated orientation distributions. These treat complexities such as asymmetry, multiple clusters and dispersions along small-circles, though not all within any single model. They are explained in detail by N.I. Fisher et al. (1987), and a few important models and their main features are listed in Table 10.3. The most common use of these models is to provide equations for the calculation of spherical variance, or confidence limits around the mean value. This may be quite complex: an elliptical confidence cone requires several parameters to describe its orientation and dimensions, more complex distributions need further descriptors. Commonly, tables must be consulted, each for different parameters that describe the distribution. Some theoretical models require up to eight parameters for their full specification and discourage routine use. However, in almost all cases, a Bingham-type model for an elliptical dispersion of points on the sphere provides the most sophistication required, and the Fisher model for a circular distribution of points on the sphere may suffice in many studies. The simple circular and elliptical concentration patterns described by these two models are shown in Fig. 10.11. Fisher statistics suffice for the circular clusters but dispersions toward a partial or a great-circle girdle indicate an anisotropic distribution of points on the sphere. These usually require elliptical rather than circular confidence limits for their description. For that, Bingham statistics or the Le Goff approximations are appropriate.

Why are most orientation distributions described satisfactorily by circular and elliptical dispersions on the sphere? It is a natural consequence of the symmetry between cause and effect in nature. Given a single event or process, acting homogeneously on the material that will constitute a sample, the axes, directions or vectors invariably fall into the cluster-girdle spectrum. Their *orthorhombic symmetry* is characteristic. More complex situations, for example small-circle dispersions and multi-modal distributions, almost invariably arise from multiple or heterogeneous processes. Long before statistics were applied in earth science, structural geologists showed how heterogeneous tectonic strain was expressed as folded strata that dispersed enclosed linear features along girdles and into multiple clusters. Fisher et al. (1987) discuss the complicated statistics of non-orthorhombic orientation distribu-

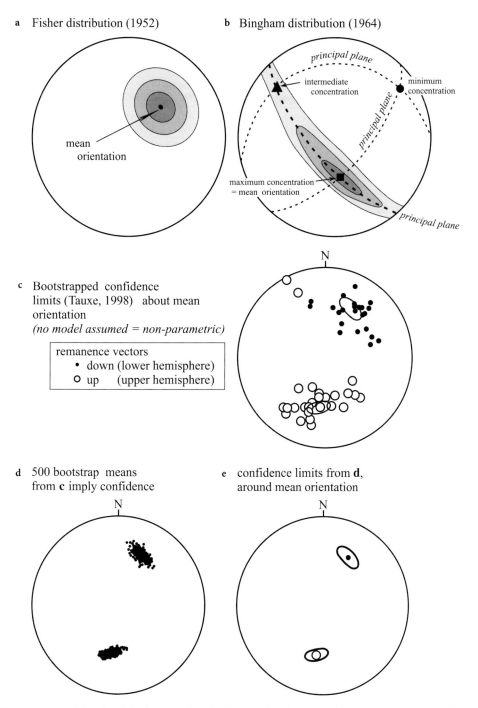

a Fisher distribution (1952)

mean orientation

b Bingham distribution (1964)

principal plane

intermediate concentration

minimum concentration

principal plane

maximum concentration = mean orientation

principal plane

c Bootstrapped confidence limits (Tauxe, 1998) about mean orientation
(no model assumed = non-parametric)

remanence vectors
- • down (lower hemisphere)
- ○ up (upper hemisphere)

N

d 500 bootstrap means from **c** imply confidence

N

e confidence limits from **d**, around mean orientation

N

Fig. 10.11. Simple, symmetrical unimodal orientation distributions may be characterised by a mean orientation and a confidence cone around the mean. On an equal-area projection, the cone will appear as an oval distorted ellipse, unless it occurs at the centre of the net for a vertical mean orientation. **a** Fisher distribution: circular-symmetrical cluster of orientations. Density contours are represented by cylindrical cones subtended from the centre of the sphere. Its confidence cone is circular on the sphere with a radius α_{95}. **b** Bingham distribution has a cluster of orientations that does not possess circular symmetry but rather an orthorhombic symmetry. Density contours range in shape from ellipses to complete great-circle girdles. The orthorhombic symmetry planes intersect at principal values, locating the eigenvectors of the distribution. **c–e** Tauxe's non-parametric bootstrap approach determines the distribution of mean orientations of many subsamples (**d**), drawn with replacement from the original sample (**c**). In turn, their mean and confidence limits characterise the original distribution (**e**) by analogy with the Central Limit Theorem of linear statistics (Chap. 4)

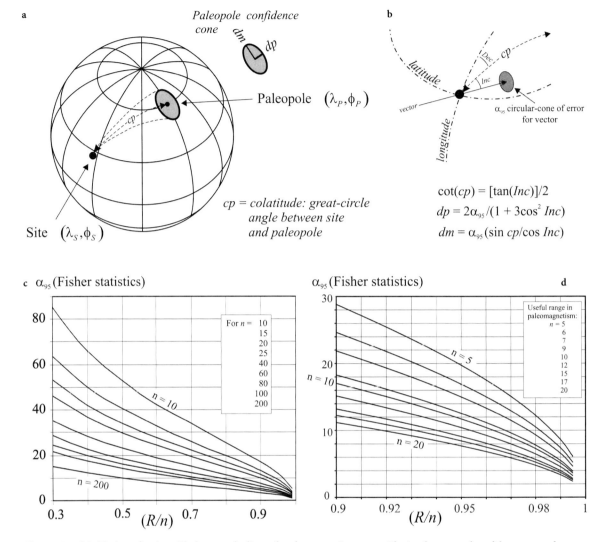

Fig. 10.12. a World view of a site with downwards-directed, paleomagnetic vector with circular cone of confidence α_{95} and corresponding uncertainty ellipse for paleopole. **b** Details showing inclination and declination of vector. **c** Radius of the 95% confidence cone on the sphere for a circular-symmetric distribution of orientations that follows Fisher's (1953) model. α_{95} is the angular radius of the 95% confidence cone about the mean orientation for the sample size n and mean resultant R. **d** Useful range of values for routine paleomagnetic applications

tions and multimodal orientation distributions, but we shall avoid them. Usually, it is possible and simpler to decompose such data sets on geological criteria into more homogeneous sub-samples that do possess orthorhombic symmetry.

Descriptions of theoretical spherical distributions are complex but described by *spherical probability-density functions (PDF)* which give the probability distribution of points over the sphere, each of which corresponds to an orientation in space. Just as the Normal distribution reports the most common *frequency* distribution on the line, so the von Mises distribution describes a circular equivalent (Chap. 9), and we shall see that, on the sphere, the Fisher distribution is a close analogy to the linear Normal distribution. The reader will find reference to the *wrapped Normal distribution* on the circle and the *bivariate-Normal distribution on the sphere*. Although these are directly transferred and expanded from the Normal distribution on the line, they are unsatisfactory models for most work with orientations.

Just as formal equations are available for important theoretical frequency distributions on the line (Chap. 3, Table 3.7), so we find more intimidating equa-

Table 10.3. Selected theoretical distributions of orientations in three dimensions (see Fisher et al. 1987 for full details)

Theoretical statistical model (references)	Stereogram appearance	Probability density element $PDE = Ce^T dA$	T
Uniform	Random scatter	$dA/4\pi$	
Fisher (1953)	Circular-symmetric cluster around mean orientation	$\dfrac{\kappa}{4\pi \sinh \kappa} e^T dA$	$\kappa[\sin\theta \sin\overline{\theta} \cos(\phi-\overline{\phi})] + \kappa \cos\theta \cos\overline{\theta}$
Watson (1956)	Bipolar axes or girdle	$\dfrac{1}{4\pi \int\limits_0^1 e^{\kappa u^2} du} e^T dA$	$\kappa[\sin\theta \sin\overline{\theta} \cos(\phi-\overline{\phi}) + \kappa \cos\theta \cos\overline{\theta}]^2$
Bingham (1964), Onstatt (1980), etc. [*Bivariate Fisher approximation by Le Goff 1990; Henry and Le Goff 1995*]	Elliptical clusters about mean orientation, may tend to girdle	$\dfrac{1}{4\pi\kappa'} e^T dA$	$\kappa_1 (\mathbf{x}', \xi_1)^2 - \kappa_2 (\mathbf{x}', \xi_2)^2$
Kent or FB$_5$ *Fisher-Bingham five-parameter* (Kent 1982; Wood 1982)	Girdles and clusters	$\dfrac{e^{-\kappa}(\kappa^2 - 4\beta^2)^{\frac{1}{2}}}{2\pi} e^T dA$	$T = \kappa \mathbf{x}' \xi_1 + \beta \mathbf{x}' \xi_3 - \beta \mathbf{x}' \xi_3$
von Mises on a circle (*for comparison*)	Symmetric hump wrapped around circle	$\dfrac{1}{2\pi [I_0(\kappa)]} e^{\kappa\cos\theta}$	

κ, Concentration parameter, β, shape parameter $= k_{INT}/k_{MIN}$; $\overline{\theta},\overline{\phi}$ = mean co-latitude, mean longitude, $\mathbf{x}' = (\sin\theta \cos\phi, \sin\theta \sin\phi, \cos\theta)$

tions for theoretical distributions of points on the sphere where each point corresponds to an orientation in space. Although we shall refer to the functions as *spherical PDF*, their comparison is simplified by using an expression to represent the density of points in a unit area on the surface of the sphere using a *spherical probability-density element (PDE)*. The element of area that comprises a unit may be expressed in the form

$$dA = \sin\theta.d\theta.d\phi$$

where co-latitude is θ and longitude is ϕ. For the different spherical distributions, the probability density elements have the general form, $Ce^T dA$, and they are compared for the most common distributions in Table 10.3 where the von Mises distribution on the circle is included also for comparison.

10.6.1
Circular Concentration: Fisher Model

The Fisher distribution, being of rotational symmetry, is appropriate where we recognise or expect that orientations cluster symmetrically about a central axis. It may be considered as the three-dimensional equivalent of a Normal distribution wrapped over the surface of the sphere. The variance of the orientations is equal in

all directions from the mean orientation and a cone that intersects the sphere as a small-circle describes the confidence region. The radius of the 95% confidence circle around the mean orientation on the sphere is α_{95}.

Fisher's model is summarised in its simplest form by a probability density function (PDF) that defines the frequency (density) of orientations in a small area on the sphere as:

$$PDF = \frac{\kappa}{4\pi \sinh \kappa} \cdot \exp(\kappa \cdot \cos\delta) \tag{10.4}$$

Here, δ is the angle between the mean orientation and the direction in which the density is defined by PDF (Fisher 1953). κ is a concentration parameter ranging from zero for a uniform distribution of orientations, rapidly increasing without limit as the preferred orientation improves. The probability density over the whole sphere is unity, and the density decreases rapidly away from the mean orientation, especially for large κ. The decrease in density from the mean is illustrated in Fig. 10.13a. A simple approximation to κ is given by Fisher's k:

$$k \approx \kappa = \frac{n-1}{n-R} \tag{10.5}$$

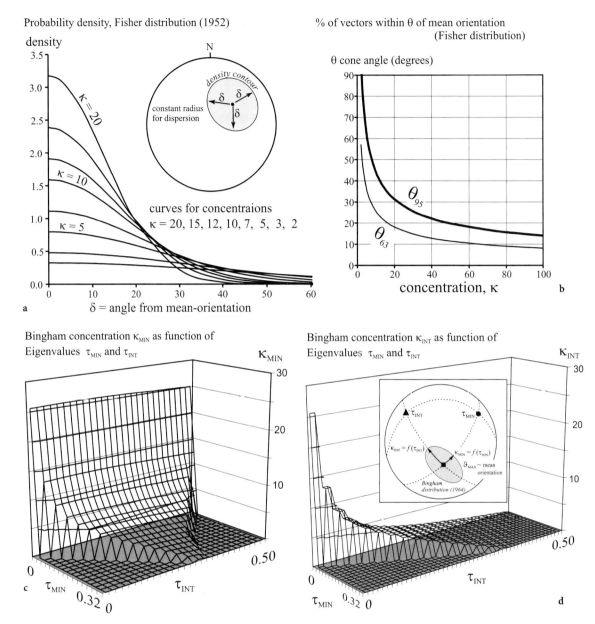

Probability density, Fisher distribution (1952)

% of vectors within θ of mean orientation
(Fisher distribution)

curves for concentraions
κ = 20, 15, 12, 10, 7, 5, 3, 2

δ = angle from mean-orientation

Bingham concentration κ_{MIN} as function of
Eigenvalues τ_{MIN} and τ_{INT}

Bingham concentration κ_{INT} as function of
Eigenvalues τ_{MIN} and τ_{INT}

Fig. 10.13. a Probability density of the Fisher distribution for different concentration parameters (κ) at different angles from the mean orientation. The total density over the whole sphere is unity, as with all probability distributions. **b** Percentage of orientations lying within $\theta°$ of the mean orientation according to the Fisher model. 63% of data lie within one spherical standard deviation of the mean (θ_{63}). The values for θ_{63} and θ_{95} may be divided by the sample size (n) to give standard error estimates for the mean orientation. **c, d** The confidence cone for a Bingham distribution of orientations is elliptical (*inset*, **d**). A limited, reasonable range of values of the ellipse's radii may be determined from the concentration parameters κ_{INT} and κ_{MIN} which are deduced from the intermediate and minimum eigenvalues (τ_{INT} and τ_{MIN}) of the orientation-distribution matrix as illustrated here and in Table 10.4, from the paper of Mardia and Zemroch (1977)

Here, n is the number of orientations sampled and R is the resultant that was defined earlier, during the calculation of mean orientation. Fisher's k-value is simply calculated and the approximation is reasonable for $n \geq 10$ and $\kappa \geq 3$. Most paleomagnetic studies and all structural geology or petrofabric research projects can potentially satisfy these constraints.

The angular radius of the confidence cone about the mean direction is traditionally designated by α. Unfortunately, this is the symbol for the significance level in conventional statistics and hypothesis-testing (Chap. 4). Therefore, we substitute the symbol p for significance level in this context. For spherical distributions in earth sciences one rarely sees any other choice than $p = 0.05$, corresponding to a $(1-p) = 95\%$ confidence level. At the confidence level $(1 - p)$, the angular radius of the confidence cone is given by:

$$\cos \alpha_{(1-p)} = 1 - \left(\frac{n-R}{R} \right) \cdot \left(\left(\frac{1}{p} \right)^{\frac{1}{(n-1)}} - 1 \right) \qquad (10.6)$$

The angle α is the generator of a cone which intersects the sphere as a circle, although it appears as a distorted ellipse, in most spherical-projections. However, a useful shortcut to estimating confidence of the mean orientations comes from considering the angular standard deviation, θ_{63} which encloses 63% of the orientations surrounding the mean, and θ_{95} which encloses 95% of the data. Note that one standard deviation of the mean encompasses only 63% of the observations because the Fisher distribution is tighter than the Normal distribution on a line. The relationship of these critical angles to Fisher's concentration parameter (κ) is shown in Fig. 10.13 d. Thus,

$$\theta_{63} = \frac{81^0}{\sqrt{(\kappa)}} \quad \text{and} \quad \theta_{95} = \frac{140^0}{\sqrt{(\kappa)}}$$

and since $\kappa \cong kn$ the confidence limits for the mean are:

$$\alpha_{63} = \frac{81^0}{\sqrt{(kn)}} \quad \text{and} \quad \alpha_{95} = \frac{140^0}{\sqrt{(kn)}} \qquad (10.7)$$

The approximations are reasonable for $k \geq 10$ and $n \geq 10$. McElhinny and McFadden (2000) show that the approximations may be improved by substituting the more fundamental approximation (kR) in place of (kn). This is a useful improvement for small samples $(n < 50)$, but for strongly concentrated, large samples, R is very similar to n anyway. However, knowing both R and n, the cone radius at the 95% confidence level (α_{95}

in degrees), may be inferred from graphs drawn in two ranges. One is presented for weak concentrations found in structural studies (Fig. 10.12 c), and the other for the stronger concentrations useful in paleomagnetism (Fig. 10.12 d).

In paleomagnetic studies, samples oriented in the field must be manipulated in the laboratory, then components of characteristic remanence (ChRM) must be isolated from them by a complicated incremental process of measurement and demagnetisation. During this campaign, there are many possible sources of error. However, all being well, the determined characteristic remanence fixes a paleopole responsible for the magnetisation, or at least the paleolatitude of the site (Fig. 10.12 a, b). Major tectonic conclusions concerning plate movements may hinge upon this information. In view of the sensitivity of the important goals to which the results lead, and of their potential for compounded errors, the 95% confidence limit is a minimum and conservative level at which to accept results.

The simple circular form of the confidence limits of the mean paleomagnetic vector is usually put to some further purpose which may require further processing. A common goal is establishing the position of the paleomagnetic north pole associated with that paleofield. The confidence cone from a sample of n orientations has a radius designated as α_{95}. In most cases, samples from N sites are merged into a larger sample in order to determine a collective confidence cone for the pole-position as A_{95}.

Continental masses change position in terms of longitude, as well as latitude but only the latter can be detected paleomagnetically. Thus, records of *absolute* longitudinal motions elude us. (Relative longitudinal motions of adjacent plates during separation or collision may be detectable.) On average, the latitudinal component of movement is a fraction of a degree north or south, per million years. Well-defined paleopole positions are obtained from paleomagnetic vector distributions with $\alpha_{95} \leq 5^\circ$. The necessity of such precision becomes apparent when we research applications of paleomagnetic data in tectonics. Not only are precise paleogeographic reconstructions attempted without knowledge of absolute paleolongitude, but still more demanding interpretations use the relative position of the paleopole along an apparent polar wander path, to infer age of magnetisation. With poorly calibrated paths, a precision of ± 10 Ma is sometimes claimed, as discussed by Tarling (1983). Of course, individual situations can vary dramatically in complexity, depending on the rate of plate movement, which was faster in early earth history, and on the latitudinal component of the motion, which depends on the unknown direction of plate-motion. Moreover, most tectonic models assume that the rotation axis for the plate is fixed. However, if such applications have any value, they must be based on precisely documented paleopole positions. This has become possible since the development of sensitive equipment and techniques (e.g. Symons 1975 a,b) and with favourable material *paleopole confidence limits* $(= A_{95})$ have now been reduced to ~3°, (e.g. Henry et al. 2001; Le Goff 1990; Le Goff et al. 1992; Derder et al. 2001). This is comparable to the precision with which an individual specimen may be oriented in the field, and it is difficult to see how further improvements are possible. Nevertheless, plate tectonic reconstructions require the highest possible precision in paleopole definition (Butler 1992; van der Voo 1993). For Paleozoic and older plate reconstructions and those using paleomagnetic data from any deformed rocks, further limitations are imposed by the extremely difficult orientation corrections required to remove the effects of tilting and penetrative strain (Borradaile 1997). Consequently, apparent polar wander paths and paleogeographic reconstructions are less well-defined, and one may have to remain content with precision of the order of $A_{95} \leq 10°$ for most paleopoles.

The fundamental formulae used by paleomagnetists determine the following: the paleolatitude (λ), the *present* latitude (λ_p) and longitude (ϕ_p) of the paleopole. These are given by the following formulae where I is the inclination of the remanence, D is its declination and λ_s, and ϕ_s are the present-day site latitude and longitude respectively. An important reference angle in all calculations is the angular distance between the site and the paleopole, termed the paleo-colatitude $(cp = 90 - \lambda)$. These quantities are illustrated in Fig. 10.12 a,b. The following important relationships are then used to determine the paleopole position in terms of latitude (λ_p) and longitude (ϕ_p) with respect to the present North Pole, and the paleolatitude of the site (λ):

(a) $\tan(I) = 2 \tan(\lambda)$

(b) $\sin(\lambda_p) = \sin \lambda_s \cdot \sin \lambda + \cos \lambda_s \cdot \cos \lambda \cdot \cos D$

(c) $\sin(\phi_p - \phi_s) = \dfrac{\cos \lambda \cdot \sin D}{\cos \lambda_p}$ (10.8)

For the 95% Fisher cone of confidence about a mean remanence vector, these then yield corresponding uncertainties in the paleolatitude (dp) and in the orthogonal meridional direction (dm) shown for example by Butler (1992) and Tarling (1983) as:

$$dp = \Delta \lambda_p = \frac{2\alpha_{95}}{1 + 3\cos^2 I} \quad dm = \Delta m = \alpha_{95} \left(\frac{\cos \lambda}{\cos I} \right)$$

(10.9)

Thus, even a circular-symmetric orientation distribution of paleomagnetic vectors yields, in general, an elliptical confidence cone for the paleopole (Fig. 10.12 a,b). As an approximation, the Fisher-equivalent circular confidence cone would have a radius given by

$$A_{95} \approx \sqrt{(dm.dp)}.$$

Paleomagnetic vectors provide an interesting case in which the uncertainties of the mean orientation may be evaluated with equal statistical validity in different ways. In general, in many complex earth-science studies it may be possible to treat data in incremental stages, chosen so that errors, uncertainties and variations tend to cancel out in that stage of the analysis.

For example, there is considerable data processing involved in any paleomagnetic study. Multiple specimens are collected and oriented, either drilled in the field or laboratory. Multiple sites are sampled in which the acquisition of magnetic remanence may have been by different processes. The specimens are then incrementally demagnetised and the orientations of vector components removed in each stage are used to determine a characteristic remanence believed to be parallel to the instantaneous contemporaneous ancient geomagnetic field for that site. Unfortunately, a single record of the geomagnetic field at one point in geological time will be an unreliable indicator of the paleopole due to secular variation of the geomagnetic field (Chap. 8), and perhaps also due to excursions and reversals. Thus, data from many sites must be combined to remove those ambiguities.

The preceding formulae offer a simple approach in which we could average all the characteristic remanence vectors from all the specimens and all the sites as a single bundle. But is there a way of treating the data hierarchically so that uncertainties at certain stages may cancel out? Paleomagnetic experts show that there are good subject-specific reasons for following the following scheme (e.g. Butler 1992; van der Voo 1993; McElhinny and McFadden 2000), apart from the simple statistical reasoning used here. First, consider the data grouped at the site level. These are presumably affected by a constant paleofield orientation, and any variation in paleomagnetic orientation errors between specimens and cores are largely non-systematic and therefore tend to cancel out. Thus, one recognises a preference for calculating a site-mean orientation and, from that, calculating the position of a *virtual geomagnetic pole* (VGP). The VGP is thus effectively insulated from orientation errors introduced in sampling, core-preparation and measurement. Of course, the site VGP will be a poor approximation to a true paleopole as it usually represents a single record of the paleofield without any time-averaging of secular variation. Therefore, secondly, one combines the numerous site VGPs; their variation is chiefly attributable to secular-variation of the geomagnetic field and by averaging the VGPs, those uncertainties cancel out, giving a better approximation of the true paleopole. VGPs are known to follow a Fisher distribution much better than remanence vectors (Butler 1992).

The paleopole is determined from the VGPs as follows. For each VGP we substitute its latitude for I and its longitude for D in $x = \cos I \cos D$, $y = \cos I \sin D$ and $z = \sin I$ and proceed with the vector summation as earlier (Eq. 10.1). The mean paleopole then has latitude $\lambda_p = \arcsin(N)$ and longitude $\phi_p = \arctan(M/L)$. As VGPs conform rather well to a Fisher distribution, the uncertainty of the mean may be calculated by Fisher statistics yielding an A_{95} circular confidence pole about the mean paleopole. A_{95} is calculated using the approximation in Eq. (10.5) where k and n refer to the dispersion of VGPs.

10.6.2
Elliptical Concentration: Bingham Model

The Bingham (1964) statistical model describes populations of points on a sphere that may range from clusters with circular symmetry, through partial girdles to full girdles with axial symmetry. Structural geologists refer to girdles with axial symmetry as great-circle girdles. In general, the confidence cone about the mean orientation of a Bingham distribution will be elliptical. In the case of the Fisher model, a single concentration parameter (κ) suffices to describe the concentration of itsaxially-symmetric, circular concentration of points (Fig. 10.13a,b). However, for the Bingham case, the anisotropy of the orientation distribution distorts the confidence region about the mean orientation. The concentrations ($\kappa_{INT}, \kappa_{MIN}$) in directions perpendicular to the peak concentration determine the shape of the elliptical confidence cone (Fig. 10.13c,d).

The most useful development of the Bingham distribution is a five-parameter hybrid Fisher-Bingham distribution ("FB$_5$") according to Kent (1982). The probability-distribution element (PDE) defines the orientation concentration in a small area of the sphere (dA) and is given by a complicated equation using the following:

$\kappa_{INT}, \kappa_{MIN}$ = concentration of the points in the directions of intermediate and minimum density (the directions perpendicular to the maximum concentration)

β = ratio of κ_{INT} to κ_{MIN}

ζ_{MAX} = mean direction of distribution

ζ_{INT} = direction of density (perpendicular to ζ_{MAX} and ζ_{MIN})

ζ_{MIN} = direction of minimum density.

Large values of κ occur for highly concentrated distributions and the larger the value of β, the more elliptical and less circular is that concentration. The concentration of orientations striking the surface of the sphere per unit area (dA) is the probability-density element (Fisher et al. 1987) with the approximation C_K being valid for strong concentrations:

$$PDE = C_K e^T dA \quad \text{where} \quad C_K \approx \frac{e^{-\kappa}(\kappa^2 - 4\beta^2)^{\frac{1}{2}}}{2\pi}$$

$$\text{where} \quad T = \kappa \mathbf{x}'\xi_1 + \beta \mathbf{x}'\xi_2 - \beta \mathbf{x}'\xi_3$$

$$\text{and} \quad \mathbf{x}' = (\sin\theta\cos\phi, \ \sin\theta\sin\phi, \ \cos\varphi) \tag{10.10}$$

Dispersion estimates for a circular-symmetric cluster from the Fisher model and from the Bingham model should agree where $\beta = 1$. However, they do not (Onstott 1980). Calculations and programming involving Bingham-type models are quite taxing, but κ_{INT} and κ_{MIN} for the original Bingham distribution may be estimated using published tables (Mardia and Zemroch 1977), knowing the eigenvalues (τ) of the orientation-distribution matrix described below. Their relationship to the eigenvalues is illustrated in Fig. 10.13.

Orientation concentrations ranging from clusters through ellipses may be characterised partly by the use of an orientation-distribution (OD) matrix introduced to geology by Scheidegger (1965), Woodcock (1977), and by Fisher et al. (1987) in more general terms. The OD matrix provides principal directions of concentration and an indication of symmetry and of relative concentration in the three principal directions. The convention for the orientation of axes must exclude the possible ambiguity of their two possible azimuths (= trends). Usually, the confinement of consideration to one hemisphere automatically preempts that problem. Coordinates of the endpoints of the unit-length orientation lines are (x_i, y_i, z_i) and the orientation-distribution matrix is:

$$\mathbf{T} = \begin{bmatrix} \sum x_i^2 & \sum x_i\,y_i & \sum x_i\,z_i \\ \sum x_i\,y_i & \sum y_i^2 & \sum y_i\,z_i \\ \sum x_i\,z_i & \sum y_i\,z_i & \sum z_i^2 \end{bmatrix} \tag{10.11}$$

The matrix is in geographical coordinates (x, y, z) but it may be manipulated by standard matrix algebra into a form such that it is transformed into fabric coordinates (see Fig. 11.5), in which the reference axes are the orientations of mutually perpendicular maximum, intermediate and minimum concentrations of the ob-

Table 10.4. Concentration parameters, κ_{INT}, κ_{MIN} for the Bingham distribution from the eigenvalues τ_{INT} and τ_{MIN} of the orientation distribution. (Mardia and Zemroch 1977; see Fig. 10.13c,d)

τ_{INT} / τ_{MIN}	0.02	0.04	0.06	0.08	0.10	0.12	0.14	0.16	0.18	0.20	0.22	0.24	0.26	0.28	0.30	0.32
0.02	25.55[a]															
	25.55[b]															
0.04	25.56	13.11														
	13.09	*13.11*														
0.06	25.58	13.14	9.04													
	9.00	*9.02*	*9.04*													
0.08	25.60	13.16	9.07	7.04												
	6.98	*7.00*	*7.02*	*7.04*												
0.10	25.62	13.18	9.08	7.04	5.80											
	5.76	*5.78*	*5.79*	*5.80*	*5.80*											
0.12	25.63	13.19	9.09	7.04	5.79	4.92										
	4.92	*4.93*	*4.94*	*4.94*	*4.93*	*4.92*										
0.14	25.64	13.20	9.09	7.03	5.77	4.90	4.23									
	4.30	*4.30*	*4.30*	*4.29*	*4.28*	*4.26*	*4.23*									
0.16	25.65	13.20	9.08	7.02	5.75	4.87	4.20	3.66								
	3.80	*3.80*	*3.79*	*3.78*	*3.76*	*3.73*	*3.70*	*3.66*								
0.18	25.65	13.19	9.07	7.00	5.73	4.84	4.16	3.62	3.16							
	3.38	*3.38*	*3.36*	*3.35*	*3.32*	*3.29*	*3.25*	*3.21*	*3.16*							
0.20	25.64	13.18	9.05	6.97	5.69	4.80	4.12	3.57	3.11	2.71						
	3.03	*3.01*	*3.00*	*2.97*	*2.94*	*2.91*	*2.86*	*2.82*	*2.77*	*2.71*						
0.22	25.63	13.17	9.03	6.94	5.66	4.76	4.07	3.52	3.05	2.65	2.29					
	2.71	*2.70*	*2.67*	*2.64*	*2.61*	*2.57*	*2.52*	*2.47*	*2.41*	*2.35*	*2.29*					
0.24	25.61	13.14	9.00	6.91	5.62	4.71	4.02	3.46	2.99	2.58	2.22	1.89				
	2.43	*2.41*	*2.38*	*2.35*	*2.31*	*2.26*	*2.21*	*2.16*	*2.10*	*2.03*	*1.96*	*1.89*				
0.26	25.59	13.12	8.97	6.87	5.57	4.66	3.97	3.40	2.93	2.52	2.15	1.81	1.50			
	2.18	*2.15*	*2.12*	*2.08*	*2.03*	*1.98*	*1.93*	*1.87*	*1.81*	*1.74*	*1.66*	*1.58*	*1.50*			
0.28	25.57	13.09	8.93	6.83	5.52	4.61	3.91	3.34	2.86	2.44	2.07	1.72	1.41	1.11		
	1.94	*1.91*	*1.87*	*1.83*	*1.78*	*1.73*	*1.67*	*1.60*	*1.53*	*1.46*	*1.38*	*1.29*	*1.20*	*1.11*		
0.30	25.54	13.05	8.89	6.78	5.47	4.55	3.84	3.27	2.79	2.36	1.98	1.63	1.31	1.00	0.71	
	1.72	*1.68*	*1.64*	*1.59*	*1.54*	*1.48*	*1.42*	*1.35*	*1.27*	*1.20*	*1.11*	*1.02*	*0.92*	*0.82*	*0.71*	
0.32	25.50	13.01	8.84	6.73	5.41	4.48	3.77	3.20	2.71	2.28	1.89	1.54	1.21	0.89	0.59	0.29
	1.51	*1.47*	*1.42*	*1.37*	*1.31*	*1.25*	*1.18*	*1.11*	*1.03*	*0.94*	*0.85*	*0.76*	*0.65*	*0.54*	*0.42*	*0.29*
0.34	25.46	12.96	8.79	6.67	5.35	4.42	3.70	3.12	2.62	2.19	1.79	1.43	1.09	0.77	0.46	0.15
	1.31	*1.27*	*1.22*	*1.16*	*1.10*	*1.03*	*0.96*	*0.88*	*0.79*	*0.70*	*0.60*	*0.50*	*0.39*	*0.27*	*0.14*	*0.00*
0.36	25.42	12.91	8.73	6.61	5.28	4.34	3.62	3.03	2.53	2.09	1.69	1.32	0.97	0.64		
	1.12	*1.07*	*1.02*	*0.96*	*0.89*	*0.81*	*0.74*	*0.65*	*0.56*	*0.46*	*0.36*	*0.25*	*0.13*	*0.00*		
0.38	25.37	12.86	8.67	6.54	5.21	4.26	3.54	2.94	2.43	1.99	1.58	1.20				
	0.94	*0.89*	*0.82*	*0.76*	*0.68*	*0.61*	*0.52*	*0.43*	*0.34*	*0.23*	*0.12*	*0.00*				
0.40	25.31	12.80	8.60	6.47	5.13	4.18	3.45	2.85	2.33	1.87						
	0.76	*0.70*	*0.64*	*0.56*	*0.49*	*0.40*	*0.31*	*0.22*	*0.11*	*0.00*						
0.42	25.25	12.73	8.53	6.39	5.05	4.09	3.35	2.74								
	0.59	*0.52*	*0.45*	*0.37*	*0.29*	*0.20*	*0.10*	*0.00*								
0.44	25.19	12.66	8.45	6.31	4.96	3.99										
	0.42	*0.35*	*0.27*	*0.19*	*0.10*	*0.00*										
0.46	25.12	12.58	8.37	6.22												
	0.25	*0.17*	*0.09*	*0.00*												
0.48	25.04	12.50														
	0.08	*0.00*														

[a] Upper cell entry = κ_{INT}.   [b] Lower cell entry (in italics) = κ_{MIN}.

served orientations. Intuitively, we appreciate that there must be some intrinsic, internal reference system which defines axes of maximum and minimum concentration, as well as an intermediate value. Specifically, the maximum concentration axis is referred to as *the* principal axis, but looser phrasing refers to maximum, intermediate and minimum concentration axes as *principal axes*.

To determine the orientation and intensities of the principal concentrations, conceptually replace the dots on the sphere with points of equal mass. The axis about which these masses have a minimum moment-of-inertia is the maximum concentration axis. For a circular cluster of points, it would pass through the centre. Its eigenvector describes the orientation of the maximum concentration; it has an associated *eigenvalue*, τ_{MAX}, describing its strength. The axis for which the moment of inertia is maximised is perpendicular to any cluster and perpendicular to any girdle or partial girdle. Its orientation is given by the minimum eigenvector with an associated magnitude τ_{MIN}. Thus, a set of three orthogonal *eigenvectors* describe the orientation concentration in three dimensions. Their associated eigenvalues ($\tau_{MAX} \geq \tau_{INT} \geq \tau_{MIN}$) record the relative concentration intensities in those orientations.

The mean orientation for a Bingham distribution may be given by the orientation of the maximum Eigenvalue, τ_{MAX}. The 95% confidence limit about the maximum concentration is clearly going to be drawn out towards τ_{INT} and τ_{MIN} according to their relative magnitudes (normally the eigenvectors' magnitudes are normalised to their mean, so that their sum is unity). The determination of the concentration parameters and the angular radii of the elliptical confidence cones for τ_{MAX} towards τ_{INT} and τ_{MIN} is complex and can best be accomplished using the results of Mardia and Zemroch (1977), simplified here in Table 10.4.

10.6.3
Elliptical Concentration: Henry-Le Goff Bivariate-Normal Approximation

A computationally simpler approach to the treatment of non-circular distributions was developed by Le Goff (1990) for paleomagnetic vectors and adapted to certain tensors by Henry and Le Goff (1995) that will be discussed in Chapter 11. The Le Goff model involves some approximations that were tested and shown to be quite sufficient for the most demanding needs of paleomagnetism. The paleomagnetic vectors are replaced with n unit vectors oriented with respect to the mean direction at co-latitude θ and at longitude ϕ with mass m_i located at their ends. The hypothetical mass may represent the intensity of the magnetisation of a sample, thereby integrating measurement uncertainty from the *individual observations* into the sample statistic. This feature is absent from other approaches, but for simplicity, we will assume unit mass for the moment ($m_i = 1$) for compatibility with the other statistical approaches determining mean direction and confidence limits. The polar-coordinate system with respect to the mean orientation may be replaced by Cartesian coordinates: $x = sin\theta cos\phi$, $y = sin\theta sin\phi$ and $z = cos\theta$. From the sample, an *inertia tensor* is then formulated as:

$$\mathbf{D} = \begin{bmatrix} n - \sum x_i^2 & -\sum x_i y_i & -\sum x_i z_i \\ -\sum x_i y_i & n - \sum y_i^2 & -\sum y_i z_i \\ -\sum x_i z_i & -\sum x_i z_i & n - \sum z_i^2 \end{bmatrix} \quad (10.12)$$

For a circular-symmetric distribution, amenable to Fisher statistics, the tensor would be symmetric about the z-axis in the form

$$\mathbf{D} = \begin{bmatrix} A & 0 & 0 \\ 0 & B & 0 \\ 0 & 0 & C \end{bmatrix} \quad (10.13)$$

Here, $A = B$ and $A + B + C = 2n$. For strong concentrations of directions, one may use small angle approximations $\theta \approx sin\theta$ and $\theta \approx \sqrt{(2 - 2cos\theta)}$ and, using Fisher's concentration parameter:

$$k = \frac{n}{n - \sum cos\theta_i} \quad (10.14)$$

Le Goff (1990) and Le Goff et al. (1992) show satisfactory approximations for the Fisher case are expressed by the diagonal terms of the inertia tensor as:

$$A = B = \frac{kn}{1+k} \quad \text{and} \quad C = \frac{2n}{1+k} \quad (10.15)$$

He extended this to the case of elliptical distributions and showed that the concentration parameters along the maximum and minimum axes of the elliptical confidence cone were given by:

$$k_{MAX} \cong \frac{A+B}{C+B-A} \quad k_{MIN} \cong \frac{A+B}{C-B+A} \quad (10.16)$$

From these the radii of the 95% confidence cone were obtained as:

$$\alpha_{95}(max) = \frac{140}{\sqrt{k_{MAX}n}} \quad \alpha_{95}(min) = \frac{140}{\sqrt{k_{MIN}n}} \quad (10.17)$$

where $\sum w_j n_j$ replaces n if observations have different weights.

The weighting factors (w) may be qualitative, for example, related to the precision of the age of the rocks, their composition, geographical location or of some feature of a laboratory-measurement technique. Alternatively, it may be intrinsically quantitative, related to the vectors' intensities or orientation concentration.

The azimuth (Ω) of the maximum radius with respect to the meridian passing through the mean direction is derived from the eigen-directions of the inertia tensor after rotation into geographic polar coordinates of latitude and longitude. This is given by:

$$\Omega = cos^{-1}\left(\frac{sin I_{INT}}{sin I_{MAX}}\right) \quad (10.18)$$

where I_{MAX} represents the mean orientation.

The chief advantages of the Le Goff model are that via some reasonable and satisfactory approximations it permits the definition of elliptical confidence cones for orientations that do not show circular symmetry about their mean attitude. Secondly, it permits weighting of individual orientations within samples, or of grouped means from different sites, which is necessary in paleomagnetism and magnetic fabrics (Chap. 11). Confidence limits determined by other statistical models only include uncertainty of individual orientations, whereas Le Goff's method integrates other aspects of uncertainty of individual observations to be included in the expression of overall confidence. Moreover, individual orientations may be weighted according to a wide range of quantitative or qualitative factors of subject-specific importance such as formation age, error in geochronological age, or some quality aspect of the remanence (TRM, CRM, DRM, etc.).

10.7
Statistical Models: Circular or Elliptical Clusters?

Examples of some linear elements in geology closely follow a Fisher distribution. For example, the axes or hinge lines of cylindrical folds (Fig. 10.14a, b) form axial-symmetric clusters that follow Fisher's model, and α_{95} is a meaningful representation of the confidence of their mean orientation. However, the most common use of Fisher statistics is in paleomagnetism because the geomagnetic field vector is more or less constantly oriented at any instant. Trapped in a rock as a paleomagnetic vector, any variation in orientation between samples may be considered, to a first approximation, as random noise. There is usually no strong reason to suspect that this should not produce a circular distribution of density contours (e.g. Fig. 10.11a), and most paleomagnetic studies present their uncertainties in this way, the diameter of the confidence region being of constant radius (α_{95}) on the sphere. (Of course, for most mean orientations and most projection methods, the shapes of circular confidence cones are distorted to subelliptical oval shapes.) Fisher statistics describe a geometrical end member, rather than a general case, recognized in orientation distributions. Nevertheless, the Fisher model finds application to many geometrical elements found in geology. For example, consider the fold hinges of Fig. 10.14a. For such a cylindrical fold, the minor fold axes cluster symmetrically about their mean orientation giving circular density contours on the sphere (Fig. 10.14b), which are amenable to Fisher statistics. However, for most non-cylindrical

folds, the fold axes may not form a circular-symmetric distribution on the sphere; here, the Bingham model is appropriate (Fig. 10.14c, d). In petrofabrics, structural geology and magnetic fabrics we shall see that orthorhombic distributions are the general case and the Bingham model is almost always preferred. Even in paleomagnetism, there are situations where an orthorhombic distribution of paleomagnetic vectors could be expected, and since it is the *general case*, there is no harm in considering its use anyway. The Bingham model and Le Goff's bivariate model include and exceed the information inherent in the circular-symmetrical distributions required by the simple Fisher model.

Consider the paleomagnetic study in which, at a given site, there is considerable age difference between the paleomagnetic vectors, and that the site has been subject to plate tectonic motion. The remanence directions can no longer form an axial-symmetric cluster: instead, they will be smeared to an elongate cluster with orthorhombic symmetry, to a first approximation. Many such cases arise in global tectonic-paleomagnetic studies (e. g. Le Goff et al. 1992). In another much less common exception, consider paleomagnetic vectors determined from a site in which the rocks have anisotropic magnetic properties. These may have refracted the paleofield towards a certain direction in which the remanence-bearing minerals are aligned, such as in a bedding plane. (The subject of magnetic anisotropy was developed to alert us to this possibility; more in Chap. 11.) Consequently, the axial-symmetric paleofield may have been dispersed through the rock to impose an elliptical-cone distribution of paleomagnetic vectors. In archeomagnetism the same effect may be caused by the geometrical configuration of walls and other architectural features or landscape topography. Confidence regions about the mean orientation of a Bingham distribution require two angles to describe the maximum and minimum radii of the elliptical confidence cone, which will be similar in shape to the orientation distribution's density contours (Fig. 10.11b).

There are two important justifications for continued use of the Fisher model to estimate precision in paleomagnetism. First, our results are then compatible with the large existing database of Fisher statistics. Second, paleopole distributions tend to approximate the Fisher model better than the remanence orientations (Butler 1992). Otherwise, more general Bingham-type models that include the circular-symmetrical Fisher distribution might take precedence. A better case for Bingham-type models may be made in structural geology and petrofabrics where distributions on the sphere are much more likely to be orthorhombic rather than circular-symmetrical.

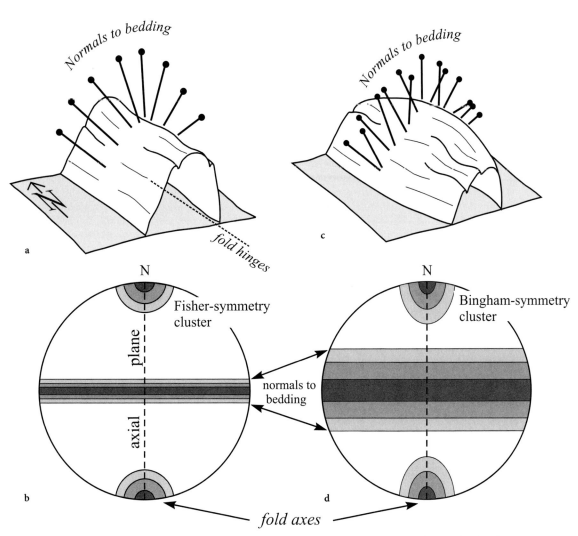

Fig. 10.14. The expected symmetry of single-generation structures and fabrics is invariably at least orthorhombic in symmetry. This concept may be extended to heterogeneous structures such as folds. **a** Normals to a cylindrical-folded layer form a great-circle-girdle density distribution. The minor fold hinges would form a point concentration or cluster, corresponding to the Fisher model. **b** Stereogram of the structures in **a**. **c** The concept may be extended to most first-generation structures even if they have more complex geometry, as with this canoe-shaped fold for which the orientation distribution of structural elements still retains at least orthorhombic symmetry. **d** Stereogram of structural elements in **c**, showing that fold axes follow a Bingham-type distribution

Precision may be estimated by a quite different, non-parametric approach that sidesteps the thorny issue of selecting any theoretical model (Fisher et al. 1987; Tauxe 1998). The *bootstrap* technique is recommended only where a parametric approach is impossible and is most applicable to small samples with $n \leq 25$ (Fisher et al. 1987). Otherwise, the large-sample approaches may be preferred, taking advantages of the known probability distribution for the population. The bootstrap approach assumes that all the uncertainty of the population is adequately represented in the sample.

This may be reasonable for high-quality paleomagnetic data, but not for more dispersed distributions, as in some structural geology problems.

In Chapter 4, the principle of bootstrapping was detailed. The original sample of say 20 orientations is resampled, repeatedly drawing 20 values at random but with replacement, so that a "re-sample" may duplicate or omit any of the original observations. The mean orientation of each re-sample is calculated. After at least 200 resampling experiments, obviously with a computer, the mean orientations are plotted to produce an orientation distribution of pseudo-sample means (Fig. 10.11 c). Inevitably, this beautifully defines a clear circular or elliptical distribution of pseudo-sample means. This procedure must be applied with caution as noted by Fisher et al. (1987) and Tauxe (1998). With less than ten original orientations, one may generate a deceptively agreeable, high-symmetry distribution of pseudo-sample means. 95 % of the bootstrap means lie within a cone that defines the 95 % confidence limit about the mean orientation. The mean orientation is calculated directly from the original observations, in the usual way whereas its confidence region (Fig. 10.11 e) is determined from the distribution pseudo-sample means (Fig. 10.11 d).

Clearly, one or two aberrant orientations in the original sample, or the subsequent inclusion of a meaningful outlier in an extended campaign of data-collection might produce a different distribution of bootstrap-means. Bootstrapping is only sound if the original sample was unbiassed, and representative of the population. The example from Tauxe (1998) shows two groups of paleomagnetic vectors (Fig. 10.11 c). One might propose that each of these was drawn from a Bingham-type distribution. However, the distribution of 500 bootstrap means is more reassuring (Fig. 10.11 b) and *their* mean and confidence cone (Fig. 10.11 c) compares favourably with the original data when superimposed upon it (Fig. 10.11 a). The bootstrap approach will probably be used more often in the future although the original data should always be presented with the confidence-cone for the bootstrap means. However, where the sample size is adequate ($n \geq 25$), a parametric approach should take precedence.

Statistical tests exist to determine whether the Fisher or the Bingham model is appropriate (e.g. Fisher et al. 1987). However, in routine applications visual inspection of the distribution is usually considered adequate. Even where that indicates an absence of circular symmetry, in some paleomagnetic studies the Fisher model will be used anyway as an approximation to permit simple reporting of a single α_{95} radius. In deciding which of these two models is appropriate, we are really assessing the importance of the degree or lack of circular symmetry of the points on the sphere. In a purely theoretical approach, we may apply the models and examine the results. However, in our applied approach we should take advantage of practical subject knowledge and be aware that our samples may be too small or the sampling may be biassed. We may well question whether a feebly elliptical distribution is actually an unfavourable sample from a circular-symmetric population. Therefore, let us consider three possible sources of orthorhombic (elliptical) symmetry

that would genuinely warrant treatment as a Bingham-type distribution. Keep in mind that a Bingham-type elliptical distribution should be the general case for most geological orientation distributions, with more dispersion of orientations from the mean orientation (τ_{MAX}) towards τ_{INT} than towards τ_{MIN}.

The first source of elliptical symmetry may be due to errors of measurement, sampling or laboratory treatment that introduce an anisotropic bias. One may broadly group these as errors of measurement, for the sake of argument. For example, if a field geologist measures mineral lineations (fabric lineation) from the alignment of minerals in outcrops of a plutonic rock, the range and accuracy of orientations recorded will be influenced by many factors. Even though the true orientation distribution of minerals' long axes may be circular symmetric about their mean orientation, the availability of suitable outcrop surfaces and the biassing cut-effect of certain outcrop surfaces may provide an unrepresentative sample that has an elliptical dispersion about the sample's mean orientation. Moreover, measurement errors may introduce bias, steep orientations are more accurately measured from steep surfaces, and shallow orientations are better estimated from shallow surfaces. Such factors may readily conspire to cause a circular-symmetric Fisher-type distribution to falsely imitate one of elliptical or orthorhombic symmetry.

The second main source of Bingham-type dispersions occurs where (1) the natural process imparts anisotropy initially or (2) where the process anisotropically disturbs an originally circular-symmetric orientation distribution. In both cases, in most situations, the symmetry of the dispersion will appear to be drawn from a Bingham-type distribution (Fig. 10.11 b). Tectonic flow and rock magnetism furnish excellent examples in both categories. Primary elliptical-orientation distributions of axes are produced by flow within a plane, but directed preferentially in one direction. The flow of magma within an igneous sill will produce a concentration of mineral-long axes parallel to the flow direction, but the constraints imposed by wall friction as well as the freedom of movement within the sill plane usually disperse the flow lineations to form an elliptical-orientation distribution, stretched in the plane of the sill (Fig. 10.15). An example shows how the regional orientation of magma-flow axes determined in this way may be related to regional rifting (Fig. 10.15 c–e). We previously mentioned that paleomagnetic vectors may also be imparted with primary elliptical symmetry about the mean orientation, due to the anisotropy of the magnetised rock caused by bedding or schistose mineral align-

Poles to flow-foliation in sills from
AMS minimum axes (n = 819)

Flow-lineations in sills from
AMS maximum axes (n = 819)

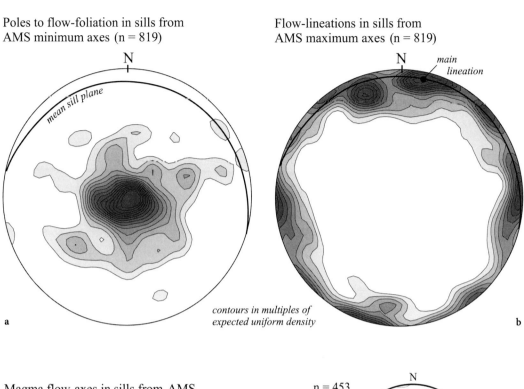

*contours in multiples of
expected uniform density*

Magma flow-axes in sills from AMS
maximum axes, northern Ontario

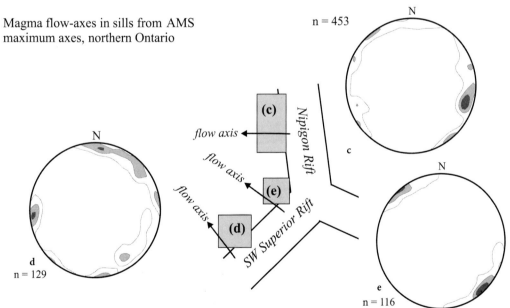

Fig. 10.15. Flow fabrics defined by mineral orientations in sills (Proterozoic, northern Ontario) deduced from anisotropy of magnetic susceptibility (AMS). **a** Slightly elongated cluster of normals to magnetic foliation which represents the magmatic flow plane. **b** Girdle of mineral lineations define a pronounced orthorhombic symmetry with symmetry planes almost horizontal, NS and EW. **c–e** AMS maximum axes in horizontal sills reveal magma-flow axes adjacent to the Proterozoic rift system, in northern Ontario

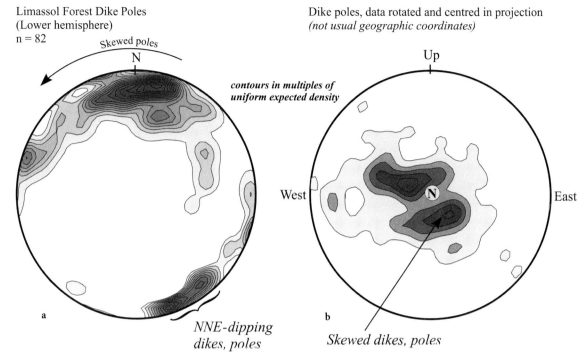

Fig. 10.16. Ophiolite dykes of Cyprus are sheared by a transform zone in the Limassol Forest (Borradaile 2001). **a** Normals to dykes show two modes; the normals to NNE-dipping dykes (*top* of stereogram) do not show a symmetric cluster, but are skewed. This is an example of a mode that does not possess orthorhombic symmetry. **b** When the same data are rotated and centred in the stereogram (note geographic coordinates on stereogram), the two modes are more readily interpreted, but the asymmetry of the skewed mode is not so exaggerated by the edge effect of the equal area projection as it was in **a**. The skewed dispersion of dyke normals is due to tectonic shearing against a transform fault

ments (Fig. 10.17 c, d). The second type of process-dispersion is not an original feature. Some process disperses an originally circular-symmetric distribution to an elliptical form. The elliptical orientation distribution of axes may not preserve symmetry about the mean orientation even though that was a feature of the original orientation distribution. In structural geology, the shearing of orientations may also produce an asymmetric, elliptical dispersion of orientations. An example is shown by the deflected normals to dykes, sheared against a transform fault (Fig. 10.16). Tectonics and paleomagnetism provide further comprehensible illustrations. For example, remagnetisation scatters remanences along a great circle between the least and most remagnetised directions, connecting the orientation of the primary magnetising field with that of the remagnetising field (Fig. 10.17 a). The distribution of remanences will normally be asymmetrically smeared along the great circle, producing imperfect elliptical symmetry although sample sizes may obscure this. Tectonic deformation may also scatter paleomagnetic vectors and where the deformation is reasonably coax-

ial and homogeneous the resulting distribution should possess orthorhombic symmetry (Fig. 10.17 b). The skewness of the distribution will depend on the completeness of the smearing mechanism as well as the success of sampling.

Unfortunately, some natural orientation distributions conform neither to the Fisher model nor to the Bingham model although they may be decomposed into subgroups or subsamples amenable to those simple models. For example, consider the primary paleomagnetic vectors from the lower crustal gneisses at Kapuskasing (Fig. 10.18). The contours show that the distribution is fairly complicated. Most of the data is on the lower hemisphere, giving one major downward-directed group of vectors. However, that mode is not circular and there is at least one other important mode. It is not surprising that the confidence cones differ in shape for the two assumed distributions. However, even their mean orientations are different. The Fisher mean is calculated as a vector mean whereas the Bingham mean is centred on the maximum Eigenvector. For a well-defined unimodal distribution, corresponding

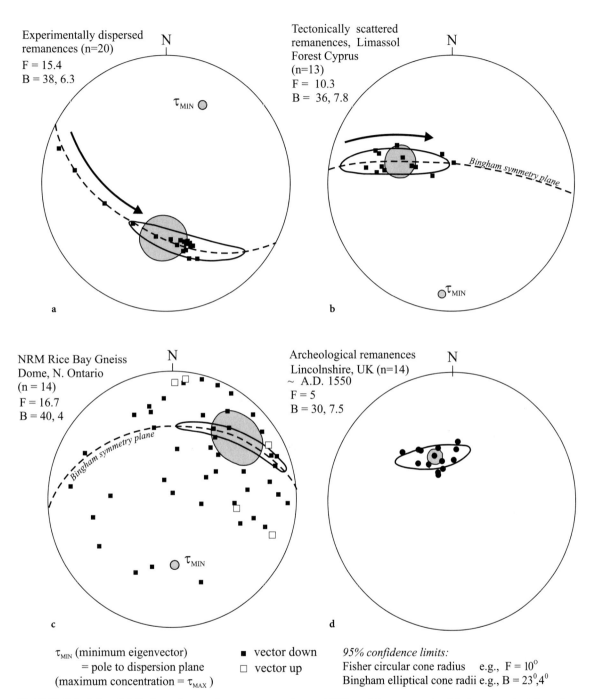

Fig. 10.17. Orientation distributions of magnetic-remanence vectors and 95% confidence cones about their mean orientation calculated according to Fisher and Bingham statistics. The circular Fisher cones of confidence appear slightly elliptical due to the inherent distortion of the spherical projection. **a** Remanences dispersed due to syn-stress remagnetisation in an experiment (Borradaile 1996a). **b** Paleomagnetic vectors dispersed by shearing of ophiolite dykes adjacent to a transform fault, Cyprus (Borradaile 2001). **c** Non-circular distribution of paleomagnetic vectors due to paleofield deflection by magnetic anisotropy in gneisses. **d** Archeological remanences with some orthorhombic dispersion indicated by Bingham confidence cone (Borradaile et al. 2001)

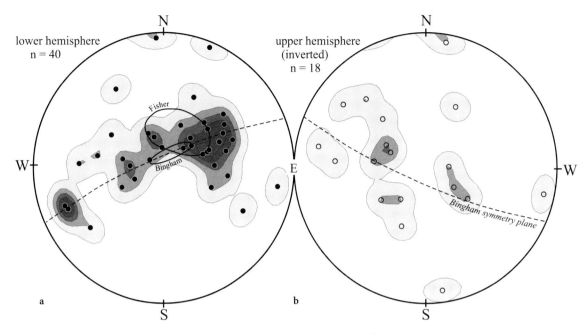

Fisher vector mean   037/61   95% cone radius   27.3° (k = 1.5)
Max. Eigen vector   020/78   Bingham 95% cone radii   14.5°; 3.7°

contours in multiples of 2U where U = expected uniform density

Fig. 10.18. Primary remanence components from the Kapuskasing Gneisses, northern Ontario. **a** Lower and **b** upper hemisphere projections of the vectors with contours in multiples of the expected (E) concentration for an equivalent number of uniformly distributed orientations. (The upper hemisphere is unhinged about the eastern azimuth and inverted, as shown in Fig. 10.8.) In this example, the orientation distribution satisfies neither the Bingham model nor the Fisher model, being girdle-like and multimodal. Different mean orientations and confidence limits are obtained from Fisher and from Bingham statistics, probably because the distribution is heterogeneous or inadequately defined

to either the Fisher model or Bingham model, the two confidence cones would be coaxial. However, the orientation distribution may be so heterogeneous that the maximum eigenvector is differently oriented from the vector mean. The plane of symmetry according to the Bingham model (the plane containing the maximum-intermediate Eigenvectors) does connect the modes so that we might suspect some tectonic smearing of an original unimodal distribution. However, these metamorphic rocks were also uplifted about 40 km so that remanences will have been superimposed by differently oriented paleofields at different times as the rocks rose through the Curie isotherm. Thus, one may accept and justify various reasons for subdividing the distribution into separate modes, which may permit the calculation of statistics according to a simple model.

10.8
Tests for Preferred Orientation and Confidence Limits for the Mean Orientation

Most geologists use stereograms qualitatively and, fortunately, the degree of order imposed by geological processes and structures commonly produces quite strong preferred orientations which forgive a qualitative approach. For example, there is little doubt that the flow foliation is horizontal in the example of Proterozoic diabase sills illustrated in Fig. 10.15. Foliation poles cluster in a single subcircular clump centred on the vertical direction. It would seem superfluous to test whether the preferred orientation was significant or whether the mean pole was significantly different from the vertical. However, a unique orientation distribution may not be so clear in other cases, such as the ophiolite dyke normals shown in Fig. 10.16a. In other examples, one might even doubt the presence of a convincing preferred orientation, e.g. Fig. 11.12. However, quantifying the significance of a preferred orientation and the confidence in the mean orientation is more important if that information is required for further calculations. Nowhere is this more evident than in paleomagnetism, where after much processing a small number of characteristic remanence orientations are used to define a paleofield which, in turn, will be used to calculate an ancient paleolatitude to reconstruct plate movements. The possibilities for confounding errors and uncertainty are serious.

The treatment that follows is elementary and intended only as an introduction to the underlying concepts; now we may take advantage of procedures that were formerly impossible without specialised professionally developed software. The reader is referred to Fisher et al. (1987) for a definitive account.

10.8.1
Tests for Preferred Orientation

The development and application of most orientation-statistics procedures in earth science concern paleomagnetism because the characteristic directions are used in further calculations upon which depend major tectonic conclusions. However, the methods may be applied to axes as well as vectors. Most structural and petrofabric orientations are of axes whereas some may be unit vectors (= directions). Even in the case of paleomagnetism, the magnitude of the remanence vector is not important in the final stages of orientation-studies, so that it too is treated as a unit-vector or direction. Whether we deal with an axis or a direction, the following methods are equally applicable, but appropriate attention must be given to handling the data. Distributions of axes are best "centred" to avoid splitting a maximum artificially on either side of the hemispherical projection, and in general, there will not be any major problem if there is a single concentration in the form of a cluster or partial girdle. The mathematical procedures are available for dealing with multimodal distributions, significant outliers, and girdles, but these become complex and require special programs. It is now possible to provide definitive answers to many useful questions beyond merely testing for the presence of "any" preferred orientation. For example, it is possible to test whether a distribution follows one theoretical model more closely than another, whether a specified preferred orientation agrees with some other orientation, how close a distribution is to a cluster or to a girdle, etc. (Fisher et al. 1987). However, this discussion remains at an introductory level.

The simple and logical way to address the question of the presence of any preferred orientation and its sig-nificance is to pose the null hypothesis (H_0) that the distribution is uniform. Some expression is then formulated so that if it exceeds a certain limiting value, we reject the hypothesis that our sample of orientations was drawn from a population of uniformly distributed orientations. The principle was explained with conventional variables distributed along a line in Chapter 4.

For a cluster of orientations with almost circular symmetry, a simple statistical procedure follows from the Fisher model (1953). Treating the orientations as unit vectors, their mean resultant, R, can be large for significantly clustered orientations. If, for a given sample size (n), R exceeds a certain threshold value, we must reject the hypothesis that the sample is drawn from a uniform distribution. Thus, we may infer that the sample does show a preferred orientation. A nomogram gives the minimum value of R for a sample of n orientations for their distribution to be significantly clustered at the 95% confidence level (Fig. 10.19a; Watson 1956; Irving 1964). The relationship is approximated by:

$$R_{\min} = \left[\sqrt{(2.605n)} \right] - 0.04 \qquad (10.19)$$

This simple test suffices for approximately circular-symmetric cluster distributions such as are commonly found in sets of paleomagnetic remanences. However, often we need a more general approach that does not assume any underlying theoretical model for the distribution and which is applicable to distributions that might be ordered in some other way, for example as a girdle, partial-girdle or perhaps even as a bimodal distribution. The non-parametric, or distribution-free χ^2-test, introduced in Chapter 5, has also been used for this purpose (Watson and Irving 1957; Vistelius 1966). As we may recall, it can be applied to compare any two distributions. We may apply it to determine whether or not the distribution is significantly different from a uniform distribution. The stereogram must be divided into a number of cells of equal solid angle. These provide the m classes required for the χ^2-test, but their choice is arbitrary with regard to shape and size which permits them to be arranged so that there are ≥ 5 observations expected in each class. In the example shown, the classes were 24 spherical triangles (Fig. 10.19b). However, the classes of the χ^2-distribution may be more conveniently chosen using latitude and longitude. The number of orientations falling in each of the classes is tallied (f_O) and compared with the frequency expected if the distribution had been uniform (f_E) for the whole distribution using the χ^2-statistic:

$$\chi^2 = \sum_{i=1}^{m} \frac{(f_O - f_E)^2}{f_E} \qquad (10.20)$$

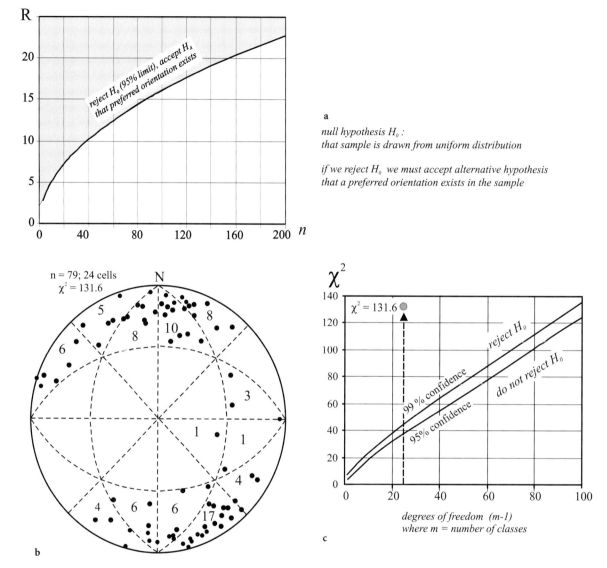

a

null hypothesis H_0 :
that sample is drawn from uniform distribution

if we reject H_0 we must accept alternative hypothesis
that a preferred orientation exists in the sample

Fig. 10.19. Elementary tests illustrate how one may establish the presence of a preferred orientation at a certain confidence level. Following the reasoning of Chapter 4, it is better to establish a null hypothesis that may be firmly rejected. Therefore, we establish the null hypothesis H_0 that the sample is drawn from a uniform distribution. **a** If a sample of n orientations is drawn from a uniform distribution, its vector resultant (R) should be less than shown by the curve. If that is not the case, we reject H_0 and conclude that a preferred orientation exists. **b** Vistelius's (1966) χ^2-test for preferred orientation (see Chap. 5 for the χ^2-test). The χ^2 cells are 24 spherical triangles of equal area. If the χ^2-test statistic determined from data exceeds the value graphed for the appropriate number of classes (k), we reject the hypothesis that the sample is drawn from a uniform distribution and therefore conclude that a preferred orientation exists at the 95% level, in this case also at the 99% level

For the 79 directions shown, $\chi^2 = 131.6$ with 23 degrees of freedom. We must reject the null hypothesis (H_0) that the sample is drawn from a uniform distribution because χ^2 lies above the critical limit at both the 95 and 99% levels (Fig. 10.19c). Therefore, we conclude that a preferred orientation exists.

Fisher et al. (1987) should be consulted for serious attempts at testing hypotheses; the preceding examples merely provide an introduction to the field. Fisher et al. provide tabled values of percentage points and the appropriate test statistics for model distributions. These may be used to compare mean orientations of samples and sample distributions with different theoretical models similar to the procedures in hypothesis-testing for distributions on the line. Unfortunately, the test-statistics are mostly much more complicated than in linear statistics. Moreover, for small samples ($n \leq 25$), the large-sample, probability-distribution approach must be replaced by a non-parametric approach that does not assume a model. That requires the bootstrap-computer technique, which is difficult to implement.

10.8.2
Approximate Comparison of Mean Orientations with Confidence Cones

A simple, informal check of the similarity of two mean orientations is achieved by examining the degree of overlap of their confidence cones. One may choose to define the confidence region either as a circular-symmetric or elliptical area on the sphere according to the approximations given earlier:

circular-symmetric distribution:

$$\alpha_{95} \cong \frac{140}{\sqrt{(kn)}}$$

elliptical distribution:

$$\alpha_{95}(\max) \cong \frac{140}{\sqrt{k_{MAX}n}} \qquad \alpha_{95}(\min) \cong \frac{140}{\sqrt{k_{MIN}n}} \qquad (10.21)$$

These are valid for a significance level of 0.05. Should the confidence regions for two samples not overlap, we may reject the null hypothesis that the mean directions are similar at the 95% confidence level. Examples of confidence limits for circular-symmetric (Fisher) and elliptical (Bingham-Le Goff) distributions on the sphere will be shown below. As long as one mean is outside the confidence cone of another mean, we may reject the hypothesis that the samples are drawn from the same population (i.e., the sample means are the same) at the chosen confidence level. For Fisher-distributed samples whose confidence regions overlap but do not enclose another mean, the F-test (Chap. 4) is applied. The F-statistic is defined from the sample sizes (n_1, n_2) and the resultants for the individual (R_1, R_2) and combined samples (R) using:

$$F = (n_1 + n_2 - 2) \left[\frac{R_1 + R_2 - R}{n_1 + n_2 - R_1 - R_2} \right] \qquad (10.22)$$

One rejects the null hypothesis that the samples are drawn from the same Fisher population if the calculated F-value exceeds the Tabled value for degrees of freedom ($v_1 = 2$, $v_2 = 2[n_1 + n_2 - 2]$), using Table 4.6.

Some paleomagnetic orientation distributions and almost all those seen in single-event structural geological processes fail to achieve circular symmetry. Instead, they form orthorhombic-symmetry distribu-tions that may fall anywhere in the spectrum from a cluster through a partial girdle to a full great-circle girdle. Such orientation distributions are a natural consequence of most single-event, homogeneous coaxial processes. For example, undisturbed normals to bedding planes commonly form circular clusters. Paleo-current lineations on the bedding plane disperse along a great-circle girdle or partial girdle of the great circle. In structural geology, cylindrical folding produces a great-circle girdle of bedding normals. Its fold axes form a circular-symmetric cluster as do the normals to its axial-planar cleavage. Structural orientation distributions that do not follow the cluster/great-circle spectrum are less common and usually result from some heterogeneity of the process (e.g. conical folding) or multiple causes or events (multiple fabrics, multiple deformations).

Certain combinations of structural elements place mutual constraints that fix one or the other to some part of the girdle/partial-girdle/cluster spectrum as a consequence of a geological process (Lisle 1989). For example, striations caused by faulting (slickensides or slickenfibres) are constrained to lie in the plane of the fault. However, as a consequence of laws of solid-state physics, many physical properties show an orthogonal relationship of three principal values at a point in a material. This is well known to geologists who use *tensors* to describe the variation with orientation of the dielectric constant, magnetic susceptibility or thermal conductivity (Chap. 11). However, the tensor concept may also be applied to certain physical states, such as stress and strain, familiar to structural geologists. As an extension, the orientation distributions of each of the three orthogonal principal values from a homogeneous sample of specimens should also show orthorhombic symmetry. The three clusters of principal axes are also mutually constrained to be orthogonal clusters or girdles. Girdle-cluster distributions for principal orientations of tensor properties are discussed in Chapter 11.

10.9
Symmetry Compatibility of Orientation Distributions and their Causative Processes

Single-event, homogeneous, coaxial processes usually impart an orientation distribution with orthorhombic symmetry to the material in question. The orthorhombic symmetry will be better defined if any pre-existing orientation distribution is completely overprinted. The generality of these remarks may seem vague but it is hopefully clarified with a few examples. Consider the vertical shortening of a rock, producing an alignment of minerals into a plane (= schistosity) and some degree of lineation within that schistosity (e.g. Fig. 10.20). The stress-field is invariably orthorhombic, possessing, by definition, orthogonal maximum, intermediate and minimum stresses. The effect is to produce an alignment of grains compatible with the symmetry of either the stress ellipsoid or finite-strain ellipsoid (a debatable point in structural geology depending on the coaxiality of the strain history). Maximum and intermediate axes of grains lie parallel to schistosity and minimum axes are perpendicular to schistosity (Fig. 10.20 a). Similar reasoning allows us to intuitively appreciate the shape of density contours and confidence ellipses for orientation distributions in which the linear component of the fabric is increased through L-S and L-fabrics (Fig. 10.20 b, c). This argument could also be applied to sedimentary compaction, aligning grains in the plane of bedding: in that case the symmetry would probably be very close to that of Fig. 10.20 a.

Consider also magmatic flow. The pressures (stresses) controlling flow have orthorhombic symmetry: mineral long axes align parallel to the maximum flow (minimum pressure) and mineral short axes align perpendicular to the flow plane. Intermediate stress is perpendicular to the other two. The expected symmetry would be similar to Fig. 10.20 b. An actual example of magmatic flow fabrics from large sub-horizontal sills is quite similar to this pattern (Fig. 10.15).

The generality of orthorhombic symmetry for orientation distributions arising from single-event, coaxial processes is attributed to the structural work of Flinn (1962, 1965). The mutually perpendicular stresses ($\sigma_{max} > \sigma_{int} > \sigma_{min}$) are invariably all compressive and may therefore be conceptualised as an ellipsoid describing the three principal magnitudes of stress and their orientation. The stresses acting over time produce corresponding finite compressions and extensions in a rock. These are denoted $Z \leq Y \leq X$ in modern convention (e.g. Ramsay 1967; Ramsay and Huber 1983) although the original definitions were in reverse order (Flinn 1962, 1965). Even if the stress changes orientation with time, as it generally does, the body will change shape to have three mutually orthogonal finite-strain axes of lengths, $X \geq Y \geq Z$. The hypothetical shape adopted by an initially spherical portion of the material is referred to as the *finite-strain ellipsoid*. Generally, a penetrative single-generation event that runs to completion (i.e., overprinting any earlier texture) must therefore produce a fabric or texture of orthorhombic symmetry. Consequently, it follows that the Bingham-type distribution and statistics based on

it are generally most appropriate. The symmetry arguments may be applied to fabrics due to magmatic flow, paleocurrent flow in sedimentary environments and solid-state flow in tectonic deformation and metamorphic crystallisation. Where sufficient sampling defines convincing non-orthorhombic orientation distributions of structural, petrofabric or paleomagnetic data, they are normally attributable to multiple events or heterogeneity. In those cases, sub-sampling may yield more easily analysed homogeneous data groups. This is discussed in the following paragraphs.

The first is rather simple: the sample is too small so that the sample distribution is unrepresentative of the population. You may recall that orientation data are more sensitive to sampling problems than scalar data (see Chap. 1). It is difficult to recommend a minimum sample size for a reproducible orientation distribution because this varies with its preferred orientation. As the concentration of orientations increases in a particular direction, fewer measurements are needed to characterise their distribution. Two possible precautions spring to mind. One may increase the sample size. If successive increases do not change the pattern considerably, then the lack of orthorhombic symmetry is due to some other cause, such as discussed below. Conversely, if the sample size is adequate, subsamples drawn from it randomly should retain the asymmetric characteristics of the original larger sample.

The second situation in which non-orthorhombic orientation distributions arise is due to the mixing of heterogeneous fabric elements. In other words, the sample orientation distribution mixes observations from different populations. The different populations may simply be different sub-areas or regions with different orientation distributions. This may be resolved by regrouping the observations according to external geological criteria. The most troublesome kind of mixed distribution arises where the observations are of different structures or orientations that are otherwise incompatible. For example, a paleomagnetic orientation distribution may be sub-orthorhombic because primary and secondary magnetic vectors have been mixed together. Similarly, in structural geology, an orientation-distribution of mineral lineations may be sub-orthorhombic because lineations of two different ages have been mixed together, or perhaps an intersection lineation was erroneously recorded as a mineral lineation and included with the mineral lineations. These problems cannot be resolved from the orientation data alone, careful re-sampling is necessary. In structural geology, the combination of different fabric elements was formerly used to define tectonites of monoclinic and triclinic symmetry, an extension of the

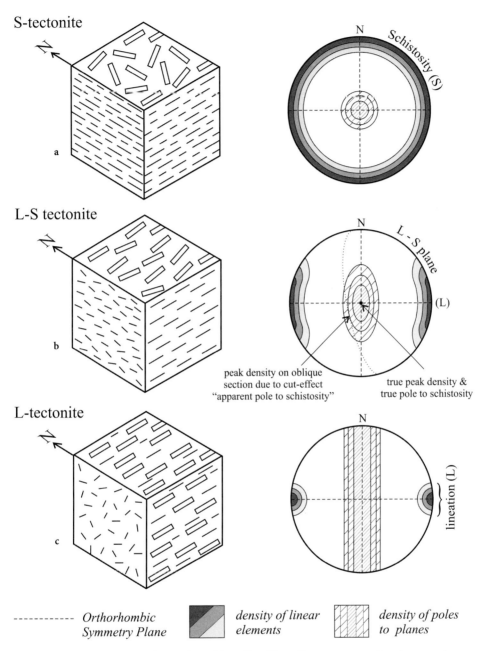

S-tectonite

L-S tectonite

peak density on oblique
section due to cut-effect
"apparent pole to schistosity"

true peak density &
true pole to schistosity

L-tectonite

- - - - - - - - -   *Orthorhombic
Symmetry Plane*

*density of linear
elements*

*density of poles
to planes*

Fig. 10.20. Simple orientation distributions of minerals or objects aligned by sedimentary, magmatic and structural processes possess at least orthorhombic symmetry. This is generally true if the fabrics are penetrative, pervasive, homogeneous and developed in one coaxial event. The orientations of normals to micas and of their long axes define orthorhombic symmetry for the full spectrum of fabric types (Flinn 1965). **a** S-fabrics or planar fabrics. **b** L-S fabrics. **c** L or linear fabrics. Note that where axes (lines) are clustered, the associated *normals to planar elements* would define girdles, and vice versa. In the text, discussion concerns linear elements but the arguments are directly transferable to normals to planes. In the stereogram of **b**, reference is made to the view of a preferred orientation on a general, oblique surface. The preferred orientation will appear skewed and its attitude misrepresented on that plane (see Fig. 1.17 f for block diagram). This is a special *cut-effect* problem

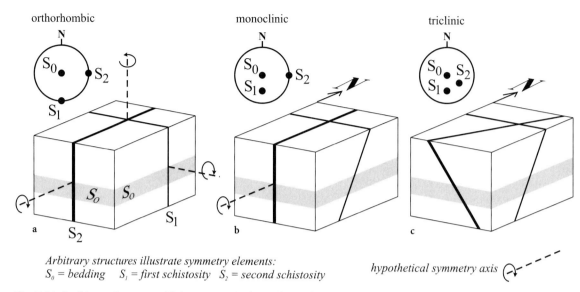

Fig. 10.21. In older works on petrofabrics, symmetries lower than orthorhombic were cited for fabric descriptions. Invariably, these were of orientation distributions that were incompletely sampled or as shown here, of asynchronous or heterogeneous fabric components. The assignment of monoclinic and triclinic symmetries had little productive value

concepts of crystal symmetry to less regular fabric elements of different age, such as schistosity, mineral lineation and bedding. Thus, in the past tectonically deformed rocks were routinely described as possessing triclinic or monoclinic symmetry (Turner and Weiss 1963). This pigeonhole classification provides few profitable avenues for further analysis. Successive, quite different fabric elements like bedding (S_0) and schistosities (e.g. S_1, S_2) have largely fortuitous geometrical relationships so that assigning orthorhombic, monoclinic or triclinic symmetry has little further use (Fig. 10.21). Normal sample sizes rarely provide the precision to convince us that the orientation distribution is truly non-orthorhombic.

Flinn's (1965) *L-S* fabric scheme introduced a new way of looking at fabrics, which recognised the fundamental orthorhombic symmetry of the orientation distribution of *each individual fabric element* (Fig. 10.20). This not only has direct interpretative value: it is justified in practice because as sampling approaches the size of the population, the perfection of the symmetry is usually corroborated. Furthermore, from first principles, most single-event, single-process geological orientation distributions are intrinsically orthorhombic in symmetry, and a Bingham-type model is generally appropriate (Fig. 10.11 b). This refers to homogeneous, coevally generated, paleomagnetic vectors; magmatic-flow directions; mineral lineations in metamorphic rocks; fold axes; intersection lineations; poles to schistosity, etc.

There are rare processes where one may argue for a non-orthorhombic fabric generated as a homogeneous entity, during a single episode by a single mechanism. One way in which this can occur is where the fabric is controlled by the non-coaxial application of stress, or the non-coaxial accumulation of strain during flow. The flow may be magmatic, or due to solid-state crystal plasticity during metamorphism, or due to strain during tectonic deformation caused by any mechanism, or due to depositional flow in many kinds of sediment. Most of those flows produce non-coaxial histories since the controlling stresses vary in orientation with time so that an asymmetric or skewed density distribution of fabric elements arises. An example is provided by ophiolite dikes, sheared against a transform fault in Cyprus (Fig. 10.16). One might argue that a mechanism encompassing change in geometry or mechanism is not a single process. However, non-coaxial stress or strain sequences are continuous sequences and in such cases, the departure from orthorhombic symmetry carries some meaning for the geological history. This pattern of asymmetry may be expected in any orientation-producing mechanism that includes a component of flow or shear. These have been shown to include magmatic flow, sedimentary current alignments, slumping, mylonite and cataclastic fault rocks, and more subtle situations due to non-coaxial strain histories in many metamorphic rocks.

For metamorphic rocks, the non-coaxial accumulations of strain may be recorded by different minerals growing at different stages. It is now possible to quickly and precisely isolate the petrofabric orientation distributions of paramagnetic silicates, and of several different ages of magnetite by magnetic-anisotropy measurements from the same sample using non-destructive techniques. These magnetic-anisotropy methods show beyond doubt that individual fabric-forming events have intrinsic orthorhombic symmetry and that successive events may be progressively inclined to one another during non-coaxial flow or strain processes. Magnetic anisotropy is a tensor and the orientation-distribution of tensors provides much more information than that from other petrofabric elements. These matters are discussed in Chapter 11.

10.10
The Fabric Ellipsoid

Since it is usually only the relative magnitudes that are of interest, we divide each of τ_{MAX}, τ_{INT} and τ_{MIN} by their sum, yielding normalised Eigenvectors (τ') so that $\tau'_{MAX} + \tau'_{INT} + \tau'_{MIN} = 1$. For a uniform distribution, the normalised Eigenvectors have equal magnitudes of 1/3

and the orientation-distribution ellipsoid is spherical, in other words, the fabric is isotropic. Where $\tau'_{INT} = 1/3$ and the other two differ, the ellipsoid is described as neutral (the *plane-strain ellipsoid* of structural geology). The plane containing the maximum and minimum principal directions may be termed the foliation if $\tau'_{INT} > 1/3$; the orientation distribution would correspond to *S*-tectonite symmetry. Where $\tau'_{INT} < 1/3$, the orientation distribution would be described by a rod-shaped ellipsoid, compatible with the symmetry of an *L*-tectonite. As the ratio τ'_{MAX}/τ'_{MIN} increases, the ellipsoid becomes more elongate but it may be shaped like a disc or a rod, depending on the value of τ'_{INT}. The shape of the orientation-distribution ellipsoid and its relation to the Eigenvalues and to the associated fabric terminology of field geologists is given in Table 10.5 and Fig. 10.22.

The orientation-distributions provide a description of frequency-distributions of directions over the sphere. We have presented this in terms of mineral-orientations within individual rock-samples which was the original application of the *L-S* fabric scheme (Fig. 10.20, Fig. 10.22). However, the concepts may be applied with equal success and validity to structural

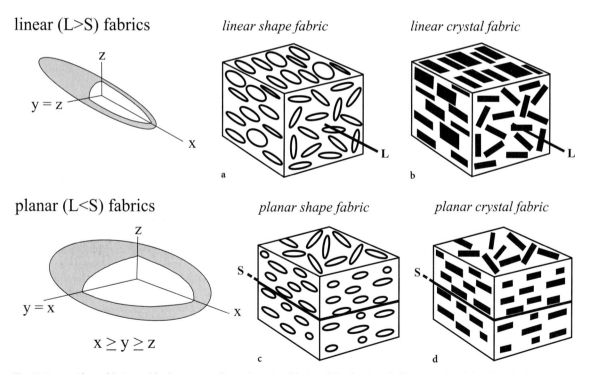

Fig. 10.22. a, c Shape fabrics and **b, d** corresponding orientation fabrics of the dominantly linear L > S and dominantly planar L < S types, with lineation (*L*) and foliation (*S*) indicated. For clarity, the associated orientation-distribution ellipsoids are illustrated as perfectly prolate and perfectly oblate (after Flinn 1962, 1965). The orientation-distribution ellipsoid may be coaxial to the finite-strain ellipsoid, and of the same general shape

Table 10.5. Fabric types from orientation-distributions of linear elements and their normalised eigenvalues (τ'): $\tau'_{MAX} + \tau'_{INT} + \tau'_{MIN} = 1$

Fabric type	Structural feature (Fig. 10.22)	Orientation distribution (Fig. 10).20	Relative τ'
L	Lineation alone, "pencil cleavage"	Circular cluster//max	$\tau'_{MAX} > 1/3 > \tau'_{INT} = \tau'_{MIN}$
$L > S$		Elliptical cluster//max stretched toward min	$\tau'_{MAX} > 1/3 > \tau'_{INT} > \tau'_{MIN}$
$L = S$	Neutral fabric, schistosity and integral lineation equally developed	Girdle in max-int plane, pinched width at int axis, bulges at max	$\tau'_{MAX} > (\tau'_{INT} = 1/3) > \tau'_{MIN}$
$L < S$		Great-circle girdle in max-int plane, minor concentration at max	$\tau'_{MAX} > \tau'_{INT} > 1/3 > \tau'_{MIN}$
S	Perfect schistosity	Perfect great-circle girdle // max-int plane	$\tau'_{MAX} = \tau'_{INT} > 1/3 > \tau'_{MIN}$

features sampled from many outcrops. For example, in a homogeneously strained subarea of metamorphic rocks, the poles to schistosity or the outcrops' mineral lineations (e.g. Fig. 10.20) may be described easily in the *L-S* scheme and their orientation-distributions treated statistically with a Bingham-type model. Even in heterogeneously strained domains such as folds, the fold-hinges and the dispersion-girdles of bedding-normals (Fig. 10.14) produce Bingham-type distributions.

10.11
Shape of the Orientation-Distribution Ellipsoid

Flinn (1965) introduced a diagram for plotting the shapes of ellipsoids which is still popular in structural geology. He introduced it to plot mean shapes of deformed objects, as well as preferred orientations of crystals that arose due to tectonic strain. The Flinn plot has as its vertical axis the ratio *maximum/intermediate* and as its horizontal axis *intermediate/minimum*. Originally, from the optical indicatrix convention, he used $z \geq y \geq x$ to designate the principal axes. Subsequently, following Ramsay (e.g. 1967) most structural geologists use $x \geq y \geq z$. For the purposes of computation and matrix operations, some subscript notation is usually preferred, e.g. $x_3 \geq x_2 \geq x_1$. The axes of the Flinn plot are normally a = maximum/intermediate, and b = intermediate/minimum. Commonly, the shape ratios are more readily determined than the relative lengths of individual axes $(x{:}y{:}z)$, for example from orientations or shapes on selected two-dimensional cuts through specimens. If we arbitrarily set the ellipsoid volume to unity, i.e., $x.y.z = 1$, each axis may be then standardised (x', y', z') to the radius of a sphere of equivalent volume:

$$x' = \frac{x}{(xyz)^{1/3}} \quad y' = \frac{y}{(xyz)^{1/3}} \quad z' = \frac{z}{(xyz)^{1/3}} \quad (10.22)$$

using $a = x/y$ and $b = y/z$:

$$x' = a^{2/3}b^{1/3} \quad y' = a^{-1/3}b^{1/3} \quad z' = a^{-1/3}b^{-2/3}$$

These equations are rather useful since the standardised ellipsoid is readily compared with the unit sphere which is of great value in strain and fabric analysis. For example, suppose the fabric-anisotropy measurements were directly determined as:

20.48 : 19.20 : 10.40

These numbers are rather difficult to evaluate. Standardising with the ratios a and b as shown gives us:

1.28 : 1.20 : 0.65

This is of "unit volume" and directly comparable with an isotropic state *1.00 : 1.00 : 1.00*. Immediately, we appreciate the disk-shaped nature of the anisotropy. In contrast, some authors simply standardise the axes to the magnitude of the minimum axis, which is not quite as useful. For the anisotropy we have just discussed, the values would be:

1.96 : 1.85 : 1.00

Some anisotropies are extremely feeble, with standardised axes such as *1.02 : 1.01 : 0.97*. These commonly arise with ellipsoids describing the concentrations of orientations (orientation-distribution ellipsoid) or magnetic anisotropies that are used for the same purpose. In these cases, plotting *ln(a)* and *ln(b)* clarifies weakly anisotropic data, near the origin (1,1) where a perfectly isotropic fabric is represented by a sphere. Woodcock (1977) applied the Flinn fabric diagram to the description of orientation distributions, greatly clarifying the integration of orientation distributions of minerals or structural elements with strain, petrofabrics and also of anisotropic physical properties described by tensors (e.g. see Chap. 11). The Flinn-Woodcock plot is shown in Fig. 10.23a, relating ellipsoid shape to the type of orientation distribution, ranging from a circular cluster, through neutral to a great-circle girdle. It should be noted that this is taken as understood that such fabrics are due to a single-coaxial structural process; more complex distributions, multimodal, or small-circle distributions do arise rarely as a result of special structural processes that are the focus of structural geology and beyond the scope of our present interests (see, e.g. Ramsay and Huber 1987).

Jelinek's (1981) equations describe the shape (*Tj*) and eccentricity (*Pj*) of an ellipsoid more efficiently than previous parameters. His shape of a parameter, *Tj*, is symmetrical, ranging from – 1 (prolate) to + 1 oblate with *Tj* = 0 for neutral ellipsoids. This is a distinct advantage over the Flinn-Woodcock plot, where shape must be described by (*a/b*) which ranges from 0 to 1 for disc-shaped ellipsoids, but from 1 to ∞ for rod-shaped ones. On that plot, ellipsoids of equivalent shape lie along lines radiating from the origin.

On Jelinek's plot, the eccentricity of the ellipsoid or the intensity of the fabric which it describes is represented by *Pj*. For a sphere, *Pj* = 1, ranging upwards, logarithmically, with ellipticity. This aids in the discriminations of weakly eccentric ellipsoids. Moreover, some properties described by ellipsoids develop progressively and combine as products (e.g. strain) so that a

logarithmic parameter is advantageous. The *Pj*–*Tj* parameters may be used to describe any ellipsoid such as a strain ellipsoid in structural geology, a shape ellipsoid for some objects, or the spatially varying concentrations of orientations in an orientation distribution. In geophysical studies, the directional variation of seismic velocity, magnetic anisotropy and many other anisotropic physical properties may be recorded similarly. Although the parameters were originally introduced for the description of magnetic anisotropy (Chap. 11), it is transferable to the anisotropy of orientation concentrations that concern us here. If the ellipsoid describing the concentration of orientations has axes $x_1 \geq x_2 \geq x_3$, Jelinek's anisotropy parameters are defined as:

$$Tj = \left[\frac{2(\ln x_2 - \ln x_3)}{(\ln x_1 - \ln x_3)} \right] - 1$$

$$Pj = \exp\left(\sqrt{[2 \, (a_1^2 + a_2^2 + a_3^2)]} \right) \tag{10.24}$$

where $\quad \overline{x} = \dfrac{x_1 + x_2 + x_3}{3} \quad$ and $\quad a_i = \ln\left(\dfrac{x_i}{\overline{x}} \right)$

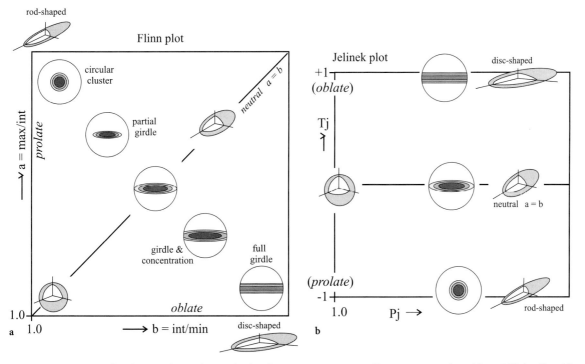

Fig. 10.23. Orientation distributions show at least orthorhombic symmetry corresponding to some member of the L-S fabric-ellipsoid spectrum (Fig. 10.22). Distributions of axes range from circular-symmetric clusters (L-type) to full great-circle girdles (S), with elliptical clusters in between. Typical stereograms of axes and their associated fabric ellipsoids are shown on the **a** Flinn (1965) – Woodcock (1978) diagram as well as on the **b** Jelinek (1981) plot. The orientation distributions of normals to planes are inverse to those for axes. For example, a great-circle girdle of normals would be perpendicular to the associated circular-symmetric cluster of axes

The Jelinek plot with Tj as the vertical axis provides a rectilinear distribution of fabric ellipsoids and associated orientation distributions that are readily comprehended, especially in view of the linear, symmetrical shape scale (Tj). The Jelinek plot applied to fabric orientations is shown in Fig. 10.23b.

Note that the fabric plots illustrated here present the orientations of lines, i.e. axes or vectors, so that a point cluster of such lines is associated with the maximum axis of an orientation-distribution ellipsoid or the fabric ellipsoid. Great-circle girdles correspond to disc-shaped ellipsoids. In other words, the concentration of orientation of lines is proportional to the radius of the ellipsoid in that attitude. However, structural geology mostly uses planes to represent bedding, schistosity, cleavage, faults, joints, etc., and for large numbers of observations it is better to plot their normals. Similarly, in petrofabrics and X-ray goniometer studies, data are collected in terms of normals to basal planes of minerals. In these cases, *the concentration of normals is inversely proportional to the radius* of the fabric ellipsoid in that orientation. *Clusters of poles correspond to the short axis of disc-shaped ellipsoids,* and *girdles of normals are perpendicular to the longest axis* of the orientation-distribution ellipsoid. Thus, we must appropriately exchange girdle and cluster distributions if we wish to consider the distributions of poles to planar elements using Woodcocks' plot, or the Jelinek version (Fig. 10.23).

CHAPTER 11

Spherical Orientation Data: Tensors   **11**

This text first introduced us to the study of *univariate* observations, those documented by a single magnitude value. Fortunately, in their initial training, geologists mostly deal with measurements that are univariate and scalar, such as rock density or a chemical abundance. Our discussions progressed to *bivariate* observations, described by two scalar values ($x-y$), and then to *multivariate* observations, each of which required three or more values for its description. From Chapter 9 onward, complexity was incremented with a new dimension, quite literally. Orientation became an issue in sampling. Although it is disregarded in almost all introductory statistics courses, even the novice earth scientist must become adept at managing orientation data, usually without formal study. Our introduction to orientation data concerned *axes*, *directions* and *unit vectors*. *Orientation distribution* replaced our introductory obsession with *frequency distributions along a line* (Chaps. 2–8). Although true vectors possess an associated magnitude, the arguments mostly concern the orientation distribution. The lack of attention to magnitudes was excused because they are little used in most aspects of earth science. Usually, only the orientation aspect of the vector is of interest, e.g. direction or axis of paleocurrents, paleowinds and magma flow. For other vectors such as paleomagnetic remanences, the direction is of prime interest but the associated magnitudes are mainly used during the investigation. There are a few branches of geoscience such as plate kinematics where both the vector's direction and magnitude are equally important in interpretation and generally equally measurable, especially during the last 180 Ma or in modern geodetic studies.

This concluding chapter is concerned with a rather involved kind of quantity, a *second-rank tensor*, that combines multiple values and orientations to define a single observation at one point. Unfortunately, all aspects of the tensor are needed for any useful interpretation. In least-technical terms, a tensor describes the variation of magnitude of some physical property with orientation, at a single point in a material. Materials that vary in physical properties in this way are termed *anisotropic*. For example, permeability of a sandstone

to fluid flow may be higher parallel to bedding and much reduced perpendicular to bedding. In contrast, the permeability of granite may be equal in all directions, its permeability would be *isotropic*.

In some branches of earth science, the aficionados need very little knowledge of the detailed mechanics of second-rank tensors, yet may manage and interpret the data competently with stereograms, magnitude graphs and in some cases a *magnitude ellipsoid*. For example, structural geologists are familiar with second-rank tensors for strain and stress, the tensor's magnitude ellipsoids are familiar to them as the *strain ellipsoid* and the *stress ellipsoid* respectively. Finite strain and stress are second-rank tensors. Most geologists first encounter ellipsoidal representations of anisotropy with the *optical indicatrix,* which illustrates the variation with orientation of refractive index of light through crystals. However, refractive index is not itself a second-rank tensor but it is related to the dielectric constant which is a tensor (Nye 1957). Magnitude ellipsoids are directly related to the principal values of the second-rank tensor, but an ellipsoidal representation is not possible for second-rank tensors in which all the principal values are not of the same sign. (The most common example in earth science is provided by some stress tensors.) For those making measurements of anisotropic physical properties, the following mathematical summary is probably an essential introduction. However, those primarily concerned with applications could postpone the mathematical details. Applications and interpretations mostly use stereograms that reveal the orientations of maximum and minimum values of the property (*principal values*) and certain diagrams that show their magnitudes or ratios of their magnitudes. Applications of second-rank tensors in structural geology (for stress and strain) are covered in an excellently focused text by Means (1976). In geophysical studies, second-rank tensors are discussed frequently in connection with the anisotropy of magnetic susceptibility and of magnetic remanence (Tarling and Hrouda 1993). The ease of measurement and wide application of magnetic anisotropies make them the most widely studied group, but others such as thermal conductivity, electrical resistivity, and dielectric constant are commonly required and studied, though sometimes with great experimental difficulty. We shall see that second-rank tensors require two subscripts to describe the property (e.g., k_{ij} where k is magnetic susceptibility) and they fall into three categories:

1. those that describe a *physical property* and relate *cause-and-effect vectors*. Most anisotropic properties that interest geophysicists and structural geologists fall into this category (e.g. magnetic susceptibility, thermal conductivity).
2. those that describe an *anisotropic state* at a *point in a material*, i.e. stress, finite strain: these are of paramount importance in structural geology, rock mechanics, tectonics and geophysics.
3. those that describe an *orientation distribution*, the measurement in any direction being the density of orientations recorded along that axis. This is very common in structural geology, petrofabrics and in the interpretation of magnetic fabrics where the orientation distribution represents the degree of preferred orientation of mineral grains or crystals. This leads to the wondrous world of interpreting mineral alignments due to magmatic, tectonic and paleocurrent flow. The latter and larger part of this chapter is concerned with the interpretation and documentation of orientation-distribution tensors, and those readers interested in geological applications may proceed there directly.

Finally, although it will certainly not concern us here, some physical responses involve more interaction between cause and effect. Third-rank tensors are required to document the piezoelectric anisotropy, of interest in mineralogy and seismicity. These require three subscripts in their notation, e.g. e_{ijk}. Anisotropy of elasticity, of interest in engineering geology, requires a fourth-rank tensor, involving four subscripts.

11.1
Second-Rank Tensors in Geology

A "tensor" is in fact an umbrella term for all of the above types of quantity, based on the required extent of the association of their magnitude components with a reference orientation. Consider three orthogonal axes to describe orientation, x_1, x_2, x_3, and some magnitude that may be associated with an axis as denoted by the subscript i or $j = 1, 2, 3$.

1. *Scalars* are described by a single value, not associated with any axis. For example, temperature (T) is

obviously not associated with direction so i is irrelevant, there is no subscript needed and it is thus a *zero rank tensor*,

2. *Vectors* (bold symbols here) require three values, e.g. components projected on the axes x_1, x_2, x_3. Temperature-gradient, $\mathbf{T}'$ is fully specified by three components formed by its projection on the axes T_1', T_2', T_3': thus *one subscript* is required T_i' where $i = 1$, 2 or 3; thus we have a *first-rank tensor*.

3. *Second-rank tensors* may relate cause and effect vectors for an anisotropic material. They describe an anisotropic property on which some cause is imposed in a three-axis frame of reference ($i = 1, 2, 3$) and measured in a three-axis reference frame ($j = 1$, 2, 3). For example, thermal conductivity $[\kappa]$ is anisotropic in many natural materials, especially rocks and minerals. It relates the vector heat flow h_i to the vector thermal-gradient T_j', intuitively, we suspect that *two subscripts* will be needed to describe the linking beast, which we call a *second-rank tensor of thermal conductivity* $[\kappa_{ij}]$. Each of the three directional components of heat flow (h_1, h_2, h_3) is influenced by each of the three components of temperature gradient (T_1', T_2', T_3').

$$h_1 = \kappa_{11}T_1' + \kappa_{12}T_2' + \kappa_{13}T_3' \quad \text{or} \quad h_1 = \sum_{i=1}^{3} \kappa_{1j}T_j'$$

$$h_2 = \kappa_{21}T_1' + \kappa_{22}T_2' + \kappa_{23}T_3' \quad \text{or} \quad h_2 = \sum_{i=1}^{3} \kappa_{2j}T_j'$$

$$h_3 = \kappa_{31}T_1' + \kappa_{22}T_2' + \kappa_{33}T_3' \quad \text{or} \quad h_3 = \sum_{i=1}^{3} \kappa_{3j}T_j' \tag{11.1}$$

Thus there are nine items of information required to specify the directional control of thermal conductivity on heat flow ($\mathbf{h}$) due to the thermal gradient ($\mathbf{T}'$), giving a second-rank tensor of thermal conductivity:

$$\kappa_{ij} = \begin{bmatrix} \kappa_{11} & \kappa_{12} & \kappa_{13} \\ \kappa_{21} & \kappa_{22} & \kappa_{23} \\ \kappa_{31} & \kappa_{32} & \kappa_{33} \end{bmatrix} \tag{11.2}$$

From Eq. (11.1), the relationship between the vectors using the second-rank tensor is usually abbreviated:

$$h_i = \sum_{j=1}^{3} \kappa_{ij}T_j' \quad \text{or just as} \quad h_i = \kappa_{ij}T' \tag{11.3}$$

Note that i and $j = 1, 2$ or 3.

Our second-rank tensor describes a variation with orientation that is intrinsic to the material at a point and does not change when the reference axes change. Thus some sample may be measured in some arbitrary coordinate-reference frame and the definition of the tensor will be the same. A systematic matrix transformation exists for transferring a tensor described in one coordinate system to another. The common denominator of second-rank tensors for geoscience is stress (σ); it concerns so many branches of the subject. Earth scientists usually use the convention that compressive stress σ is positive. Measured in an arbitrary coordinate system (x_1, x_2, x_3), the tensor would be represented by nine components. It may be considered that the state of stress acts upon an infinitesimally small cube, shown in Fig. 11.1 with the appropriate stress-tensor components. A component σ_{ij} acts along the j-axis and on the plane perpendicular to i:

$$[\sigma_{ij}] = \begin{bmatrix} \sigma_{11} & \sigma_{12} & \sigma_{13} \\ \sigma_{21} & \sigma_{22} & \sigma_{23} \\ \sigma_{31} & \sigma_{32} & \sigma_{33} \end{bmatrix} \quad (11.4)$$

Recall that the stress tensor, like the finite-strain tensor, does not relate two vectors but rather a state of being. The stress tensor is an ephemeral state that may only be truly measured at the instant in time at which it acts. The finite-strain tensor records a finite state

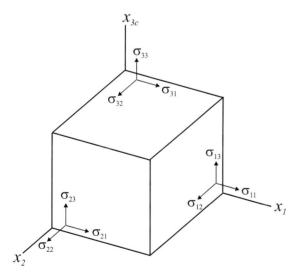

Fig. 11.1. The nine components of a second-rank tensor $[\sigma_{ij}]$ in an arbitrary Cartesian reference frame (x_1, x_2, x_3). The generally accepted notation and geometrical convention is that a component σ_{ij} acts along the j-axis and on the plane perpendicular to i. The example is used to discuss the stress tensor, which describes the ephemeral state of stress at a point, here represented by an infinitesimally small cube. In the case of stress, the eigenvectors $\sigma_{11}, \sigma_{22}, \sigma_{33}$ (σ_{ij}) are termed principal stresses (i = j) and the off-diagonal tensor components (i ≠ j) are called shear-stress components

available for measurement at any time after its accumulation.

For a cubical element in a stable state, components off the leading diagonal (top left to lower right), i.e. $i \neq j$ must balance so that $\sigma_{ij} = \sigma_{ji}$. Otherwise, the cubical element would rotate and it would not represent a stable state (Fig. 11.1). This analogy extends to other second-rank tensors and the off-diagonal terms are seen as an expression of orientation. Stress components that act parallel to a plane with $i \neq j$ are called shear stresses.

We noted that tensors may be re-expressed in a differently oriented coordinate system, but its description of the physical state (in this case, of stress) will not change. Thus, suitable matrix procedures can find a coordinate system such that the tensor is transformed to the following form (e.g. Nye 1957; Means 1986).

$$\begin{bmatrix} \sigma_{11} & 0 & 0 \\ 0 & \sigma_{22} & 0 \\ 0 & 0 & \sigma_{33} \end{bmatrix} = \begin{bmatrix} \sigma_1 & 0 & 0 \\ 0 & \sigma_2 & 0 \\ 0 & 0 & \sigma_3 \end{bmatrix} \quad (11.5)$$

where $\sigma_1 \geq \sigma_2 \geq \sigma_3$ are the *principal stresses*.

The coordinate-reference frame that achieves this is that of the *principal axes*; the leading diagonal now contains the *principal magnitudes*. The off-diagonal components are zero, indicating no inclination of the principal values to these unique reference axes. In the case of stress, structural geologists refer to these as *principal-stress orientations* (= *trajectories* = *axes*) and the *principal stresses*. The axes which present this essential information of the tensor are known as *eigenvectors* as they are intrinsic to the material properties (in German, eigen = characteristic, ownership). The principal values may also be called the *eigenvalues*. The terms eigenvector and eigenvalues are popular for tensors used to describe the anisotropic state of an *orientation distribution*, because no recognisable units of measurement are involved as with stress, magnetic susceptibility and most other second-rank tensors. They are frequency distributions in space. The *mean value* or *bulk value* of an *anisotropic property* is given by $\sigma_{\mathrm{MEAN}} = (\sigma_{11} + \sigma_{22} + \sigma_{33})/3$ for most purposes. However, in special applications, such as strain analysis where the differences between the principal magnitudes $Z \leq Y \leq X$ may be quite large, the geometric mean $(Z.Y.X)^{1/3}$ may be more appropriate. These means are only sensible where all principal values have the same sign, and the geometric mean is only meaningful if the values are > 0. This caution is necessary for magnetic fabrics where bulk susceptibilities may be negative or zero.

In the principal-axes frame any second-rank tensor may be represented by a *quadric surface* which may be considered as a continuum of possible topologies from a hyperbolic tube through a pair of hyperbolic cones to an ellipsoid, as two, one or none of the principal magnitudes achieve values of the same sign (Fig. 11.2a–c). The stress ellipsoid was chosen for discussion here because stresses may indeed be positive or negative. Similarly, in some circumstances, weakly magnetic materials with an important diamagnetic component have an anisotropy of (low-field) magnetic susceptibility (AMS) so that principal values may be of either sign. However, most common anisotropic properties represented by second-order tensors have principal values that may only be positive, which leads to a visual simplification, as we shall see.

For example, finite strain is recorded most simply as stretches (new length/original length) that can only be positive. Similarly, properties such as thermal conductivity or electrical conductivity cannot be negative and their tensors are all representable by ellipsoids.

However, the *magnitude of the property* is not illustrated directly as the tensor's quadric. In the reference frame of principal axes, the quadric is defined by:

$$\sigma_1 x_1^2 + \sigma_2 x_2^2 + \sigma_3 x_3^2 = 1 \tag{11.6}$$

whereas the generalised formulae for the quadric-ellipsoid family of surfaces is:

$$\frac{x_1^2}{r_1^2} + \frac{x_2^2}{r_2^2} + \frac{x_3^2}{r_3^2} = 1 \tag{11.7}$$

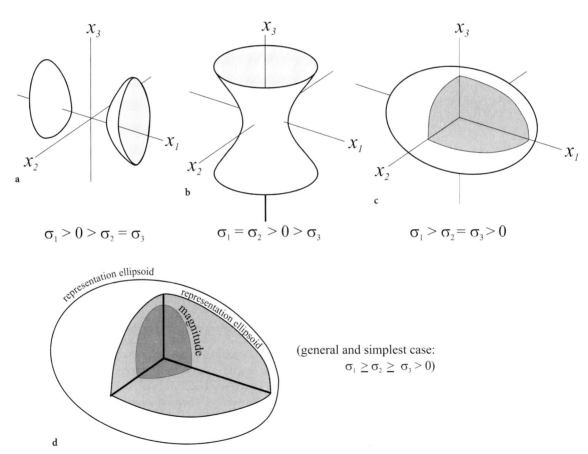

$$\sigma_1 > 0 > \sigma_2 = \sigma_3 \qquad \sigma_1 = \sigma_2 > 0 > \sigma_3 \qquad \sigma_1 > \sigma_2 = \sigma_3 > 0$$

(general and simplest case:
$$\sigma_1 \geq \sigma_2 \geq \sigma_3 > 0)$$

Fig. 11.2a–d. The quadric representation surfaces for second-rank tensors in the coordinate system of principal axes. The form of the quadric changes according to whether the principal values have **a** two negative principal values, **b** one negative principal value or **c** no negative principal values. Note: for simplicity of illustration the quadrics shown are axial symmetric, i.e. two principal values are equal. **d** Only where all principal values have the same sign, is it possible to draw a *magnitude ellipsoid* (*dark-shaded inset* ellipsoid) that illustrates directional variation of the actual magnitudes of the property. This is the ellipsoid of interest to geologists (e.g. stress, strain or AMS ellipsoid). Its radii are given by the inverse-square-root of the principal values of the second-rank tensor

Comparing these equations shows that the magnitude of the property σ, which earth scientists really want to visualise, is given by the inverse-square radius of the quadric surface, $1/r^2$. When earth scientists speak of strain ellipsoid, stress ellipsoid or magnetic-susceptibility ellipsoid, they refer to a *magnitude ellipsoid* not the quadric-ellipsoid representation of the tensor. The quadric is the first choice for the mathematician and most physicists, being more general and capable of representing tensors with mixed-sign principal values.

Thus, the magnitude ellipsoid is represented by the dark-shaded ellipsoid in Fig. 11.2d. It has semi-axes that we might label simply σ_{MAX}, σ_{INT}, σ_{MIN} or σ_1, σ_2, σ_3, in structural geology. Now we may realise why the magnitude ellipsoid is not universally useful. In the case of stress, magnitude may be positive or negative, so that one, two or all magnitudes may be of the same sign. Principal stresses of the tensor may differ in sign at a point in mines or in near-surface bedrock, but for most geological environments all three stresses are *compressive* and thus have the same sign, by convention positive in geology. However, a representation quadric may be drawn for any combination of signs (Fig. 11.2a–c). This is of little comfort for the geology student as the hyperbolic tube and hyperbolic cone are difficult to visualise, especially if they lack rotational symmetry. For simplicity, in Fig. 11.2a–c, the three quadrics possessed rotational symmetry about one-axis (x_3 in a, b; x_1 in c). The quadric ellipsoid in Fig. 11.2d is more general having three unequal axes, and it is shown in relation to the magnitude ellipsoid which we use for routine descriptive purposes. Geologists would refer to this as the strain ellipsoid, stress ellipsoid, magnetic-susceptibility ellipsoid, etc., according to the application. Fortunately, in geological applications, almost all representation quadrics are ellipsoids.

11.2
Measurement of Second-rank Tensor Anisotropy

Earth scientists are concerned with the way in which direction through the earth, or rocks, or minerals reveals different values for a physical property (e.g. electrical conductivity) or a state such as stress, strain or preferred orientation. Intuitively, we could take a specimen of any size, even the whole earth, and measure that property in numerous directions across the specimen. Somehow, the array of different values recorded, notwithstanding errors of measurement and orientation, could define a magnitude ellipsoid (Fig. 11.3a). If the number of measurement directions available were unlimited, the distribution of values in different orientations would reveal the magnitude ellipsoid automatically. This is the case in petrofabrics where the orientations of thousands of crystals are measured in each of hundreds of orientations through a specimen with the assistance of an automated X-ray texture goniometer. In each orientation, the intensity of reflections corresponds to the frequency with which a certain rational crystallographic surface is oriented. Thus, the *orientation distribution* is revealed directly by the frequency of counts in each direction. (Generally, fine-grained rocks are required to maximise the number of counts.) Such information is presented on stereograms directly without an explicit use of tensors. In the same way, an older, simpler version of the same approach is performed with a petrographic microscope on medium and coarse-grained rocks. The microscope is fitted with a *Universal stage* that rotates and tilts about four axes so that a transparent thin section of the specimen may be viewed in a wide range of orientations. With the use of optical tests, the crystallographic orientation of several hundred minerals may be determined in less than a couple of hours, depending on the mineral in question and the texture of the rock. Again, the frequency of counts of grains in a given orientation is presented as an orientation distribution on contoured stereograms. Both techniques work with plane-sectioned specimens so that it may be necessary to combine information from two or more different cuts through the specimen. This ensures that the full range of orientations has been sampled. Also, it minimises the problems of cut-effect that may inadvertently present a biased view of the preferred orientation, as shown in Fig. 10.20b and discussed more generally in Chapter 1. These approaches produce a stereogram visualisation of the orientation-distribution tensor directly.

The simple direct approach avoids the use of any data manipulation and does not require any knowledge of tensor manipulation. However, three main branches of anisotropy measurement do require mathematical intervention. We shall consider only examples involving second-rank tensors. The three categories of anisotropy measurement that use tensors directly are:

1. *Single-crystal specimens*: measurement of a fundamental anisotropic physical property such as thermal conductivity, magnetic susceptibility, velocity of light or electrical conductivity. These may be of basic use in physics and petrophysics showing lattice control of the anisotropy. Symmetrical association of the anisotropy with crystal symmetry may be useful in applications (Neumann's law: *the sym-*

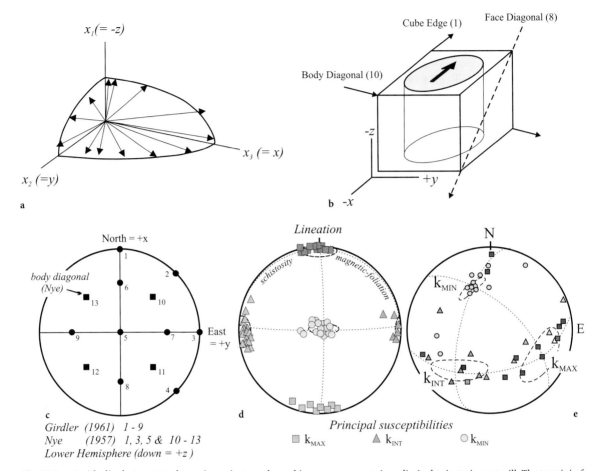

Girdler (1961) 1 - 9
Nye   (1957) 1, 3, 5 & 10 - 13
Lower Hemisphere (down = +z)

Principal susceptibilities
☐ k_{MAX} △ k_{INT} ○ k_{MIN}

Fig. 11.3. **a** An idealised attempt to determine anisotropy by making measurements in unlimited orientations at will. The termini of the vectors would define a magnitude ellipsoid: the principal axes are shown. **b** Geophysical measurements, for example magnetic anisotropy may use a right-cylindrical core specimen of the material, placed in a cubical holder. The measurements are then conveniently made in orientations determined by the holder's edges, face diagonals and body diagonals. The anisotropy would then be determined in specimen-holder coordinates. **c** Stereographic projection of potential measurement orientations for a specimen-holder arrangement as in **b**; of course, measurements may be made in antiparallel directions also since tensor components $k_{ij} = k_{ji}$. **d** Anisotropy of magnetic susceptibility (AMS) in specimen coordinates; for convenience the specimens were cut so that mineral lineation and pole to schistosity corresponded to x and z (author's data). **e** Magnetic anisotropies of a group of specimens in geographic coordinates, showing individual as well as mean orientations for their principal axes (data in Henry 1997); the principal axes of anisotropy were measured in specimen coordinates and then rotated into geographical coordinates

metry elements of any physical property must include the symmetry elements of the crystal's point group.) Many of those of interest to geologists may be represented by a magnitude ellipsoid.

2. *Polygranular specimens* (= rocks): measurement of an extrinsic anisotropic property such as permeability of fluid flow, or heat conduction have practical value but are not necessarily directly related to a fundamental crystal anisotropy. Nevertheless, the anisotropies mentioned can be represented by the magnitude ellipsoid of a second-rank tensor.

3. *Polycrystalline specimens*: measurement of a fundamental property controlled by the anisotropy of single crystals. The anisotropy of a physical property is controlled precisely by crystal structure but the crystal orientations are inevitably varied in the rock specimen. Therefore, the polycrystalline anisotropy must be a subdued version of the single-crystal anisotropy. This may have some direct geophysical value. For example, the geophysicist surveying for ore deposits needs to know the anisotropy magnetic susceptibility, electrical conductivity etc., of the rocks, not of their constituent crystals. However, there is another useful application, which occupies most of this chapter. Polycrystalline anisotropy is a meaningful blend of single-crystal anisotropy with

the anisotropy of crystal orientation distribution. For example, if all the crystal-lattices were perfectly aligned, the polycrystalline specimen should have the same anisotropy as an individual crystal. The anisotropy of the polycrystalline specimen decreases as the alignment of crystals becomes less perfect. This may seem convoluted but polycrystalline anisotropies provide a very sensitive measure of crystal-orientation distribution. In particular, since the mid-1970s, geophysicists and structural geologists have made great progress in understanding problems concerning tectonic, magmatic and other forms of flow in rocks from the measurement of anisotropies of magnetic susceptibility in low field (AMS), of isothermal remanent magnetisation (AIRM), of anhysteretic remanence (AARM) and of conductivity (complex susceptibility, ACMS). The applications in this chapter concern studies of polycrystalline anisotropy. The following paragraphs describe the generalised approach of laboratory measurement for those interested in the acquisition of such data. Although this requires specialised laboratory facilities, it may be useful for geologists who use and interpret such data to appreciate what is involved.

11.3
Measurement Procedures
for Anisotropy

For most specimens it is unreasonable to suppose that we may measure the required property in any direction (Fig. 11.3a). Rather, some suitably prepared specimen should be measured along predetermined axes of experimental convenience. This permits us to treat all specimens with the same routine experimental configuration and the measurements will correspond to the same orientations each time so that a consistent algorithm may be invoked to perform the necessary calculations. Here, we illustrate the approach used in the measurement of magnetic anisotropies. A right-cylinder is prepared, usually 22 mm high and 25 mm in diameter, as traditional in paleomagnetism. The specimen's size may be varied according to requirements of the study but the height:diameter ratio is kept constant at ~0.88 to minimise spurious specimen-shaped contributions to the measurement. The specimen should be precisely machined, and fit in a machined cubical holder manufactured from some material that should not interfere with the measurements. Plexiglass is a suitable inexpensive polymer; Vespel is more durable.

A holder correction may be applied, subtracting previous measurements of the empty holder to ensure that only specimen contributions are considered.

We now have a specimen that may be manipulated in a coordinate system controlled by the holder (Fig. 11.3b) and one may choose suitable axes along which to measure susceptibility. The cube edges (orientations 1-3-5 in Fig. 11.3b, c) and face diagonals (orientations 2-6-7) are simple choices and these form the basis of the most commonly used systems (Girdler 1961). However, they do not provide very complete coverage in three dimensions. There is a practical advantage to fill in the stereogram with body-diagonal measurements, which is the procedure used by Borradaile and Stupavsky (1995) for the Sapphire Instruments AMS system. Numerous schemes exist, depending on the nature of the equipment used and the idiosyncrasies of specimen preparation. However, at least six differently orientated measurements are made and they should be in suitable directions to provide the fullest directional coverage. Moreover, it is usually convenient to arrange them in some symmetrical pattern with respect to the measurement-coordinate axes for simplicity of the algorithm. The chosen array of orientations constitutes a *tensor design,* of which several are discussed by Hext (1963) and Jelinek (1976, 1978). In the author's laboratory the 13 possible orientations in specimen coordinates of Fig. 11.3c are used, with some replicate measurements in the reverse direction along axes. However, making many measurements may not be a practical advantage if the measuring equipment is sensitive to drift. In the author's laboratory, using the Sapphire Instruments anisotropy system, measurements along seven axes suffice to determine the anisotropy of almost all rock samples, even those with bulk susceptibilities as low as 50×10^{-6} SI. For technical reasons, it becomes doubtful whether anisotropies of rocks of lower susceptibility are representative since their AMS may be dictated by a very small number of high-susceptibility grains whose preferred orientation may not be representative of the overall preferred orientation in the rock (Borradaile and Henry 1997).

The fundamental concept of measurement of low-field susceptibility anisotropy (AMS) is to make susceptibility measurements in the presence of an applied field (H) that is sufficiently weak (~0.1 mT) so that it does not noticeably permanently magnetise any "ferro" magnetic minerals. Thus, for low-field magnetisation there is a linear relationship, $M = kH$, with the magnetisation (M) related to the field by a dimensionless constant, the susceptibility (k). More precisely, in a given direction ($i = 1$) there is a linear relationship between the magnetisation vector and the components of the magnetic field along the three principal axes ($j = 1, 2, 3$) of the coordinate system:

$$M_i = k_{ij}H_j + k_{ij}H_j + k_{ij}H_j \qquad (11.8)$$

The three possible linear equations between the two vectors (M, H) may be assembled into a second-rank tensor:

$$\begin{bmatrix} M_1 \\ M_2 \\ M_3 \end{bmatrix} = \begin{bmatrix} k_{11} & k_{12} & k_{13} \\ k_{21} & k_{22} & k_{23} \\ k_{31} & k_{32} & k_{33} \end{bmatrix} \begin{bmatrix} H_1 \\ H_2 \\ H_3 \end{bmatrix} \qquad (11.9)$$

Unfortunately, magnetic anisotropy of *permanent* magnetisations, which may give useful information for paleomagnetism or the petrofabrics of ferromagnetic minerals, causes a complication. To determine the anisotropy of remanence, the simplest approach is to give successive permanent magnetisations in different directions, as with AMS. Unfortunately, this involves large fields (e.g., isothermal remanent magnetisation, IRM), and the relationship between magnetisation and field is not linear. By keeping the fields reasonably small (e.g. <30 mT) the relationship may be sufficiently linear to permit a second-rank-tensor relationship to be established, however errors may be introduced because the resulting magnetisation may be rather small (Daly and Zinsser 1973). A more complex tensor with nonlinear terms was introduced to avoid the approximations incurred by assuming low-field magnetisation linearity (Jelinek 1996). However, the traditional anisotropy tensor with linear terms is appropriate in the case of anhysteretic remanence, which has technical advantages in rock magnetism and magnetic petrofabrics. *Anhysteretic remanence* uses small *direct fields* yet imparts measurable remanences because they are acquired during simultaneous exposure to a large, decaying alternating field that leaves no trace (Jackson 1991; Jackson and Tauxe 1991; Stephenson et al. 1986). It may also be applied selectively to subfabrics of different mineralogy, grain size or domain structure within the same specimen.

Having made measurements along suitable axes (e.g. Fig. 11.3c), we now wish to determine the anisotropy in some meaningful coordinate frame. Our initial measurements are in specimen coordinates or geographic coordinates, and it is most unlikely that these correspond to the principal axes of the anisotropy. Thus, from values obtained by measurements along axes as shown in Fig. 11.3c one may determine the orientations of the principal axes for the tensor of each specimen. In the case of Fig. 11.3d, the cylindrical specimens were prepared perpendicular to schistosity

with their mineral lineation along the specimen's "North" reference line. More usually, the axes of the specimens' tensors will be oriented elsewhere (e.g., Fig. 11.3e). Some matrix algebra and trigonometry achieve the transition from the experimental measurements (Fig. 11.3c) to the useful principal orientations (Fig. 11.3d, e; Nye 1987; Tauxe 1998). The algorithm must be designed around the particular experimental design for the choice of measurement directions; those used by Sapphire Instruments use the convention of Fig. 11.4 (Borradaile and Stupavsky 1995). Most other systems measure along some combination of body edge and face diagonals (Fig. 11.3b), such as the Girdler system (1961) or that of Jelinek (1976) for the Kappabridge instrument, which uses 15 measurements (Tauxe 1998). Common parts of all manipulations from the measurement matrix to the anisotropy tensor involve the following procedures:

1. *Inclusion of the experimental-design matrix*: using boldface notation for matrices representing vectors and tensors we have
   a) experimental design-matrix of direction cosines (e.g. from Fig. 11.4), **D**
   b) matrix of measurement vectors (**M**)
   c) **K** matrix of second-rank tensor describing the anisotropy
   Matrix algebra then determines our anisotropy from the measurements according to the experimentally designed system of orientations:

$$\mathbf{K} = [\mathbf{D}^T \mathbf{D}]^{-1} \mathbf{D}^T \mathbf{M} \qquad (11.10)$$

Here, T denotes transpose. Just as the anisotropy is determined from the measurements, so is the uncertainty $\Delta\mathbf{K}$ related to the measurement errors $\Delta\mathbf{M}$ in the same way.

2. *Transformation of orientation-reference frame*: measurements in one coordinate system x_1, x_2, x_3, must be expressed as new orientations in another coordinate system x_1', x_2', x_3', the angles between the two reference axes being known. In our case, these would be the experimental-measurement coordinates and the specimen or geographical coordinates. For example, measurements made according to the design of Fig. 11.3c are in specimen coordinates. However, there is usually no knowledge of the AMS orientations, and the specimens may have been collected in various orientations in the field. Therefore, the results of the anisotropy measurements must then be restored to geographical coordinates. The procedure sounds involved but it is a routine trigonometric bookkeeping task performed by software. However, operator errors are easily in-

Borradaile-Stupavsky
7-orientation scheme

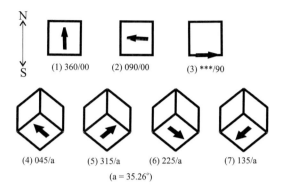

Borradaile-Stupavsky
13-orientation scheme

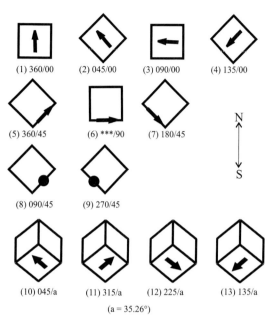

Cubical specimens, or specimen-holders shown in plan view for measurement of anisotropic property (e.g., magnetic susceptibility) along N-S axis.

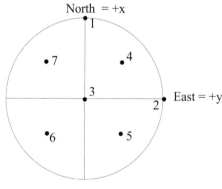

a

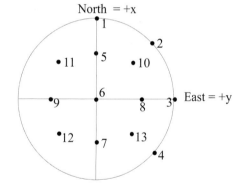

b

Fig. 11.4. Measurement-scheme for anisotropy of magnetic susceptibility (AMS) and anisotropy of anhysteretic remanent magnetisation (AARM) used with Sapphire Instruments equipment. The cubic specimen holder is viewed from above in **a** 7 and **b** 13 orientations

curred due to misunderstandings about conventions for recording and inputting orientations from field records and specimens. Recall that the tensor comprises three vector components, and the following treatment is applied to each of them. Consider just one vector V in its old coordinate system (x_1, x_2, x_3, Fig. 11.5). The procedure simply requires finding its projection on each of the axes of the new coordinate system (x_1', x_2', x_3'). We first take its component on the x_1-axis of the old system and in turn, deter-

mine its projection on x_1' of the new system, which is $V_1 C_{11}$, using the appropriate direction cosine (C_{11}). Similarly, we find V_1 components on the other two axes (x_2', x_3') of the new system. Their vector sum defines V_1' in the new frame of reference from:

$$V_1' = C_{11}V_1 + C_{12}V_2 + C_{13}V_3$$

and similarly for the other two components of V

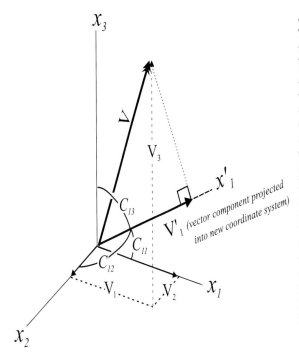

Fig. 11.5. Determination of anisotropy and the re-expression of anisotropies from specimen coordinates to geographic coordinates requires the ability to transform the orientation of the vector components of tensors from one coordinate system to another. Consider one of the three vectors V, with components V_1, V_2 and V_3 in the original x_1-x_2-x_3 system. Using the direction cosine C_{1i} between x_i and the x_1' axis of the new coordinate system, we obtain the component V_1' parallel to x_1' in the new system. The same procedure is applied for each component of V_i' parallel to each new axis, thus defining its components in the new coordinate system

$$V_2' = C_{21}V_1 + C_{22}V_2 + C_{23}V_3$$
$$V_3' = C_{31}V_1 + C_{32}V_2 + C_{33}V_3$$

which are summarized as

$$V_i' = \sum_{j=1}^{3} C_{ij}V_i \qquad\qquad (11.11)$$

11.4
Interpretations of Anisotropy Using Second-Rank Tensors in Rocks and Minerals

The remainder of this chapter is concerned with applications, chiefly using anisotropy of magnetic susceptibility (*AMS*), anisotropy of anhysteretic remanence (*AARM*) and the consequent interpretations required of mineral orientation distributions in rocks. In the large literature on magnetic anisotropy of rocks, almost ex-

clusive attention is paid to the alignments of minerals. The degree of their crystallographic alignment means that the magnetic anisotropy of the rock must therefore be related to the preferred orientation of the minerals. Although we shall see that crystals do not show a simple one-to-one mapping of "shape" with their magnetic-anisotropy magnitude ellipsoid, it is remarkable that the orientation distribution of magnitude ellipsoids for specimens does reflect accurately and reproducibly even weak preferred orientations of minerals. It is sensible to consider magnetic anisotropy at the crystallographic level and then proceed to the interpretation of orientation distributions of magnitude ellipsoids shown by the alignments of their maximum, intermediate and minimum axes, as in Fig. 11.3d. We shall not digress into the interesting field of magnetic anisotropy too deeply. Reviews are available from Hrouda (1982), Stephenson et al. (1986), Jackson and Tauxe (1991), Rochette et al. (1992), Tarling and Hrouda (1993) and Borradaile and Henry (1997). With apologies to the experts, a few key items and exceptions should be noted:

1. AMS is shape controlled for magnetite and a few other minerals of very high susceptibility. For multidomain magnetite, k_{MAX} is parallel to the long dimension and k_{MIN} is parallel to the short dimension of the grain. For single-domain magnetite, the magnitude ellipsoid is *inverse*; k_{MAX} is parallel to the *short* dimension and k_{MIN} is parallel to the *long* dimension (*inverse fabrics*).

2. AMS of almost all other minerals is crystallographically controlled with a strict relationship between the orientation of the magnitude ellipsoid and the crystallographic axes, which arc only mutually perpendicular in the case of high symmetry minerals (e.g. tetragonal, orthorhombic). Depending on crystallographic symmetry, all three, one or none of the magnitude-ellipsoid axes may be parallel to the crystal axes. In general, k_{MAX} will be parallel or close to the longer crystal dimension and k_{MIN} can be close or parallel to the short-dimension of the crystal. However, in some cases inverse-crystallographic symmetry exists where k_{MIN} is closest to the long-crystal axis (e.g. calcite, quartz, tourmaline, goethite).

3. Notwithstanding the preceding note, high-susceptibility inclusions (e.g. iron oxides) may completely mask the intrinsic crystallographically controlled AMS of their hosts.

4. Susceptibility varies greatly in magnitude (from $\sim +2.5 \times 10^0$ SI to -13×10^{-6} SI), and also in anisotropy. The two factors may compete so that a highly aligned, strongly anisotropic silicate orientation distribution may have its AMS masked by a trace

content of differently oriented magnetic grains that are less anisotropic but of vastly higher susceptibility (Borradaile and Henry 1997).

11.5
Crystallographic Control of AMS

The key elements of crystal-symmetry control on AMS are outlined in Table 11.1. One must remember that our interest is usually in alignments of *rock-forming minerals*, such as silicates and calcite, because they are most instructive in the interpretation of tectonic deformation, strain, magmatic flow and paleocurrent alignment. However, their mean susceptibilities are low. That does not present a measurement problem with modern equipment like the AGICO Kappabridge or the Sapphire Instruments SI2B, which can measure even diamagnetic-susceptibility anisotropy routinely, as for example in quartz, calcite or feldspars (mean susceptibilities $\sim 13 \times 10^{-6}$ SI). The problem is that those minerals inevitably contain high-susceptibility inclusions, like chlorite, amphibole, or magnetite. Those trace amounts may invalidate any attempt to determine the AMS of the orientation distribution of matrix minerals (Borradaile and Henry 1997). Work on single-crystal AMS is in fact hindered by those inclusions as it may be difficult to isolate the AMS of the

host lattice (Borradaile 1994; Borradaile and Werner 1994; Lagroix and Borradaile 2000b). Nevertheless, it is reassuring that the measurements confirm symmetry compatibility between the AMS tensor and crystallographic axes. A visual overview of some relationships is presented in Fig. 11.6. It was mentioned that anisotropic properties must include the symmetry elements of the point group (*Neumann's principle*). Thus the symmetry of the representation quadric, mostly visualised as an ellipsoid in geological applications, must correspond with but exceed the symmetry of the crystal-point group. For example, a prolate ellipsoid represents the anisotropy of a trigonal crystal like calcite or quartz. The crystal has only a three-fold symmetry axis whereas the ellipsoid has perfect revolutional symmetry. One could express this concept naively that a crystal symmetry is limited by regular faces (rational atomic planes) whereas the physical property must show a continuous variation with direction through the lattice. Moreover, some crystals, like tourmaline, lack a centre of symmetry, but all magnitude ellipsoids are symmetrical about their centres because the physical property is equally measured in either sense along an axis ($\sigma_{ij} = \sigma_{ji}$).

Most rock-forming minerals have low symmetry, most commonly monoclinic, so the best we may expect is one principal susceptibility parallel to the crystal

Table 11.1. Anisotropy and its second-rank tensor in relation to crystal-symmetry (simplified from Nye 1957), using anisotropy of low-field magnetic susceptibility (AMS) as an example

Analogy from optical crystallography	Crystallographic Symmetry class and some relevant minerals	Rotational symmetry crystal axes, a, b, c	Magnitude ellipsoid	AMS tensor referred to crystal coordinates
Isotropic	**Cubic** (e.g. garnet: spinel group including magnetite*, maghemite, chromite)	Four three-fold axes about body – diagonal	Sphere (* usually shape rather than crystal control for magnetite)	$\begin{bmatrix} k & 0 & 0 \\ 0 & k & 0 \\ 0 & 0 & k \end{bmatrix}$
Uniaxial (one circular section of optical indicatrix)	**Tetragonal** **Trigonal** (e.g. calcite, quartz, tourmaline, haematite, ilmenite) **Hexagonal**	One axis of rotational symmetry about crystal axes *a, b* and *c*	Prolate or oblate ellipsoid (rotational symmetry about one axis)	
Biaxial (two circular sections of optical indicatrix)	**Orthorhombic** (e.g. olivine, orthoamphiboles, orthopyroxenes)		General ellipsoid coaxial with *a, b, c*	$\begin{bmatrix} k_1 & 0 & 0 \\ 0 & k_2 & 0 \\ 0 & 0 & k_3 \end{bmatrix}$
	Monoclinic (e.g. epidote, micas, chlorites)	One axis of rotational symmetry about *b*	General ellipsoid, one axis//*b*	$\begin{bmatrix} k_{11} & 0 & k_{13} \\ 0 & k_{22} & 0 \\ k_{31} & 0 & k_{33} \end{bmatrix}$
	Triclinic (e.g., plagioclase, clinoamphiboles, clino pyroxenes)	No axes of rotational symmetry	General ellipsoid, no axes// to crystal axes	$\begin{bmatrix} k_{11} & k_{12} & k_{13} \\ k_{21} & k_{22} & k_{23} \\ k_{31} & k_{32} & k_{33} \end{bmatrix}$

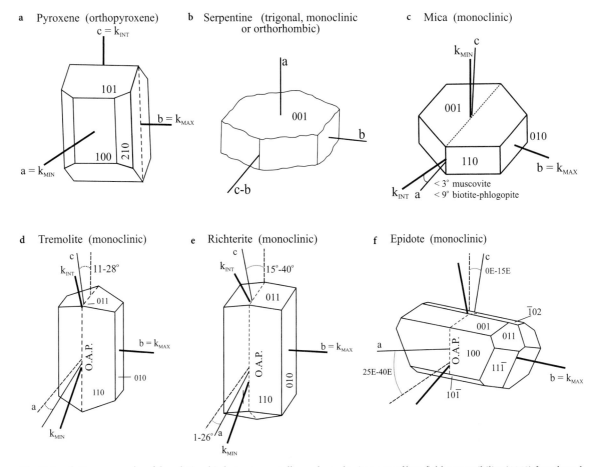

Fig. 11.6a–f. Some examples of the relationship between crystallography and anisotropy of low-field susceptibility (AMS) for selected rock-forming silicates (Borradaile 1994; Borradaile and Werner 1994; Lagroix and Borradaile 2000a). Note that whereas AMS axes, as with all tensors, are mutually perpendicular. However, many common rock-forming silicates are monoclinic; this means that AMS axes cannot map one-to-one on the orientation distribution of crystallographic axes. The situation is worse with triclinic crystals in which no crystal axes are perpendicular. The most common rock-forming silicates that provide a simple coaxial relationship between AMS and crystal axes belong to the orthorhombic system

b-axis. Unfortunately, k_{MAX} is not always defined precisely by the long axis of the mineral. Fortunately, the angular discrepancies are small in many cases. AMS, averaged over hundreds of grains, can still faithfully record even a weak preferred orientation. Nevertheless, a unique or even definitive statement about preferred crystallographic orientation may not be inferred from AMS; there will always be fundamental ambiguity. The high-symmetry AMS tensor simplifies and conceals the underlying complexity of the polycrystalline aggregate of low-symmetry crystals. The shape of the magnitude ellipsoid is best recorded by Pj and Tj of Jelinek (1978) that were defined at the end of the Chapter 10. Tj solely describes the shape of the ellipsoid, ranging from prolate ellipsoids with rotational symmetry about the maximum axis, to oblate ellip-

soids with rotational symmetry about the short axis. The departure of the ellipsoid from sphericity is denoted by Pj, also known as the anisotropy degree, or corrected anisotropy degree. These descriptors are more symmetrical and separate the intensity and shape aspects nicely, unlike less informative, simple ratios used by structural geologists ($a = max/int$; $b = int/min$). Determinations of mineral AMS are shown in the Jelinek plot and compared with the anisotropies of polycrystalline metamorphic rocks with strong preferred mineral orientations (Fig. 11.7b–d). The anisotropies of magnetic susceptibility of individual silicate minerals may be quite high (Fig. 11.7b), compared with the rocks comprising such minerals. One may argue that we do not know what role is played by high-susceptibility inclusions, but the examples shown are

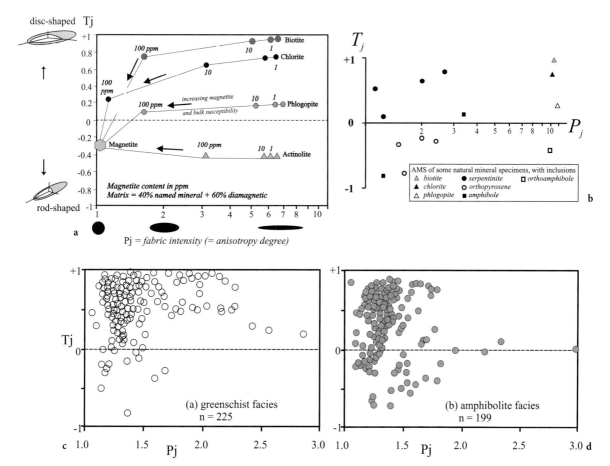

Fig. 11.7a–d. Anisotropy of magnetic susceptibility (AMS) of rocks shows a competition between the contributions of low-susceptibility, high-anisotropy minerals like some silicates, with those of high-susceptibility, low-anisotropy minerals like magnetite. **a** The susceptibility-versus-anisotropy balance principle explains how the anisotropy contribution of matrix silicates is weakened by lower anisotropy but very high susceptibility of magnetite, even though it is present only in traces (Borradaile 1987). **b** Jelinek (1981) plot shows examples of AMS for selected minerals (Borradaile 1994; Borradaile and Werner 1994; Lagroix and Borradaile 2000a). Note in the experimental determination AMS combines contributions from the silicate plus its inclusions. **c, d** Since rock anisotropy is a subdued version of mineral anisotropy, the AMS of metamorphic rocks is less intense (lower Pj) than the constituent, well-aligned, highly anisotropic silicates like mica and chlorite (Borradaile and Henry 1997). Similarly, the shape of the rock's AMS ellipsoid is usually more neutral than that of the constituent minerals (Tj closer to zero)

of natural crystals and there is no reason to suppose they are atypical. However, the AMS of rocks comprising such minerals is inevitably much lower in degree, Pj (Fig. 11.7c, d). There are two main reasons for this.

First, the *alignment of minerals* will *not* be *perfect*. Their orientation scatter reduces the AMS of the rock from that of the constituent minerals. Very strong alignments of minerals, or saturation alignments, occur for example where crystal orientation was controlled by stress during nucleation. However, that is quite rare. Normally, rock AMS is a subdued version of the mineral AMS.

Second, *the rock* will be *polymineralic* in most cases. For simplicity, let us assume that the minerals are per-

fectly aligned. There is a large variation in mean susceptibility of minerals, and a large variation in their anisotropy. Therefore, one appreciates intuitively that a mineral of low anisotropy and high susceptibility will counteract the expression of anisotropy of lower susceptibility, high-anisotropy minerals. In fact, a *susceptibility-anisotropy balance principle* operates. It is important to note that high-susceptibility accessory minerals like magnetite, ilmenite, pyrrhotite and haematite may have susceptibilities thousands of times larger than the matrix-forming minerals. The illustration presents the *rock AMS* defined by (Pj, Tj) for a rock comprising high-anisotropic but low-susceptibility silicate (e.g. biotite) and other matrix minerals of negligible susceptibility

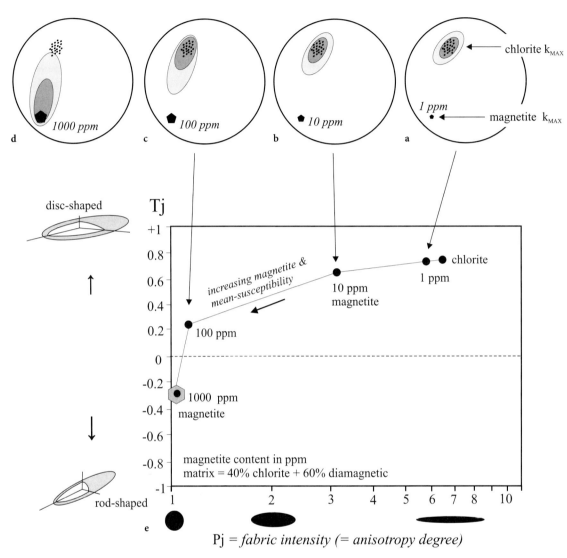

Fig. 11.8. The mean susceptibility versus anisotropy balance principle also influences the orientation distribution of the AMS tensors. Make the reasonable assumption that the chlorite matrix of a greenschist is well aligned and that the later magnetite traces are aligned in a different orientation. **a** In the absence of magnetite, the maximum susceptibilities (k_{MAX}) of chlorite grains control the AMS orientation distribution shown by their density contours. **b–d** Imagine that we could increase the abundance of magnetite. For higher concentrations of magnetite, the orientation distribution of k_{MAX} becomes progressively more controlled by the orientation distribution of magnetite. **e** The AMS ellipsoid's shape and intensity are related to the orientation-distribution ellipsoid change and change sympathetically with subtle variations in magnetite abundance ranging from 1 to 1000 ppm

(quartz, feldspar) with varying proportions of magnetite (Fig. 11.7a). Magnetite has very low anisotropy but its susceptibility is so high that the inclusion of a few parts per million (ppm) reduces the anisotropy (and increases the mean susceptibility) of the rock from that of the higher anisotropy mineral (biotite) towards that of magnetite. This explains why rocks have much lower anisotropies, but higher mean susceptibilities than the rock-forming minerals that comprise their bulk.

Our concern in this chapter is with the orientation distributions of tensors, in short, what controls the preferred orientation of the susceptibility ellipsoids for minerals within a specimen and for the mean AMS ellipsoid of a sample of many specimens. Before we commence that statistical avenue of investigation, it is necessary to realise that the susceptibility-anisotropy balance principle does not just affect ellipsoid shape, as in Fig. 11.7a, but also its orientation. In the preceding discussion, we assumed that the minerals were all similarly aligned, for the sake of argument as a saturation alignment. However, in general, they are not and, in particular, high-susceptibility accessory minerals such as magnetite, ilmenite, and pyrrhotite commonly form later and therefore have a strong possibility of corresponding to a different part of the kinematic history. Simply put, they will be differently aligned. Therefore, the balance principle extends to orientation distributions.

Consider the balance principle operating in a rock comprising a well-oriented fabric of chlorite that has high anisotropy but rather low susceptibility $\sim500 \times 10^{-6}$ SI, comprising 40% of the rock-matrix (Fig. 11.8e). The remainder of this model rock's matrix comprises diamagnetic material such as feldspar, whose contribution to susceptibility and to AMS is comparably negligible. Magnetite is a common accessory mineral in such rocks; it has very low anisotropy but high mean susceptibility, $>2.5 \times 10^0$ SI. The relative positions on the Jelinek plot reflect the differences in their anisotropy, connected by tie lines. If 1 ppm magnetite is added to the pure chlorite-feldspar rock, the mean susceptibility increases a little and the diagram shows a slight displacement of the anisotropy. As the magnetite content increases, the rock's anisotropy rapidly moves away from that of chlorite and towards that of magnetite. However, we have noted that magnetite would probably have a different orientation from the chlorite. With a concentration of one part per million of magnetite, the orientation distribution of chlorite would control the AMS. For example, the orientation of k_{MAX} would be almost entirely due to the chlorite alignment (Fig. 11.8a). However, as the magnetite content increases through 10 and 100 ppm, the orientation-distribution of AMS's k_{MAX} moves progressively further toward the alignment of magnetite and away from the chlorite alignment. With a magnetite content of 1000 ppm, the contribution of chlorite is comparatively negligible, compared to the rock's AMS and k_{MAX}-defined magnetite-grain-shape orientation (Fig. 11.8d).

11.6
Property Anisotropy
of a Polycrystalline Aggregate

One readily appreciates the fundamental control of crystal lattices on physical properties. In earth science however, our concern is with rocks that are polycrystalline and usually polymineralic aggregates. For the moment, let us postpone the consideration of multiple minerals and concentrate on the orientation distribution of a single-mineral species. We may know the anisotropy of magnetic susceptibility (AMS) for the mineral, and the orientations of its principal axes would be indicated on a stereogram with traditional symbols:

maximum   = *square*
intermediate = *triangle*
minimum   = *circle*

If the minerals were perfectly aligned as a *saturation alignment* the polycrystalline aggregate would in fact behave as a monocrystal and the specimen's AMS magnitudes and orientations would correspond directly to that of the mineral. If the orientation distribution were perfectly uniform, the anisotropy would disappear and the physical property would be isotropic, the AMS ellipsoid being the special spherical case. In general, the orientation distribution of the rock lies between these two end members: the eccentricity of the ellipsoid (Pj) is less than that of the mineral and the shape will be closer to spherical ($|Tj|$ is less than that of the mineral). It is therefore optimistic to assume that one may readily quantify the alignment of minerals: in reality the problem becomes almost insurmountable with multiple minerals and imperfect alignments (Borradaile and Henry 1997; Hrouda and Jelinek 1999; Hrouda and Schulmann 1990). Nevertheless, theoretical models using somewhat unrealistic and restrictive assumptions provide boundary cases and benchmark models for comparison. Our most reliable summary is that the AMS orientation distribution for the rock is a subdued version of mineral anisotropy. Thus, its shape is closer to the origin ($Pj = 1$, $Tj = 0$) of the Jelinek plot (Fig. 11.7c, d) than that of the constituent minerals, and its principal axes will be controlled by the balance between abundances and bulk susceptibilities of the constituent minerals (Fig. 11.8a–d). Furthermore, we should be attentive to the looseness of the correspondence between crystal shape and rock AMS; crystals with large maximum susceptibilities easily give rise to a linear magnetic fabric with a prolate AMS magnitude ellipsoid even if the minerals are platy and somewhat planar-oriented (Henry 1997).

Our concern in this book is largely with the treatment of large numbers of measurements and their characterisation. This becomes quite challenging dealing with stereograms that illustrate the AMS of polymineralic assemblages. Initially we shall concentrate on assemblages where all the mineral phases contribute similarly to AMS, with coaxial principal-susceptibility axes. That is optimistic, given our understanding of crystal symmetry and AMS. What follows is a simplified treatment of certain aspects of the special features of tensors that affect our appreciation of their orientation statistics and their errors of measurement (Jelinek 1978; Henry and Le Goff 1995; Werner 1997; Owens 2000a,b; Borradaile 2001).

11.7
Mean Tensor and its Uncertainty for Multiple Specimens

The treatment of axes, unit vectors and vectors normally proceeds simply since the dispersion ranges from a cluster with circular symmetry, through a partial girdle, to a full great-circle girdle (Fig. 10.20). No

other symmetry constraints are placed on the distribution if it satisfies the model of belonging to a homogeneous, single-event orientation distribution. The mean orientation, calculated by the previously described vector-summation procedure (Fig. 9.12), and a concentration parameter then suffice to describe the central tendency and dispersion.

On the other hand, any anisotropic physical property such as magnetic susceptibility that identifies the orientation distribution of minerals is much more complicated. *Each observation* yields three mutually perpendicular orientations (maximum, intermediate and minimum values) that are interdependent and define a magnitude ellipsoid whose orientation and shape characterise the orientation distribution of min-

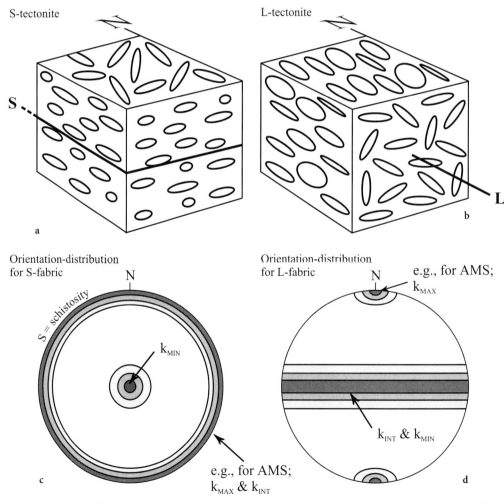

Fig. 11.9. Note the arrangement of AMS specimen tensors with orientation distributions that correspond to **a** L-fabric and **b** S-fabric. The orientation distributions of those specimen tensors can be considered in the same way as the orientation distribution of minerals or deformed objects (e.g. Figs. 10.20, 10.22). **c** and **d** show stereograms with density contours of the orientation distributions of principal susceptibilities for AMS fabrics corresponding to the drawings above them

erals within that specimen. If the fabric is homogeneous and shows little variation in orientation, a stereogram of *many observations* will show three concentrations, one for each of the maximum, intermediate and minimum axes of the measured specimen tensors. Any of the three concentrations may range in form from a circular cluster, through partial girdle, to a full girdle. However, every combination of distribution freedom does not exist for each principal value in the same fabric. Reference may be made to Fig. 10.20, where the gamut of possible L-S fabrics was illustrated for minerals with a platy habit, such as micas and chlorite. In that case, only minima (k_{MIN} ~//poles to mica *001*-cleavage) and maxima (k_{MAX} ~//long axes of mica) were illustrated. Nevertheless, it is obvious that certain combinations of axial distribution cannot exist, for example great-circle girdles of minima and maxima. Similarly, girdles of minima may only lie in the plane of the minimum-intermediate axes. We have seen that the orientation distribution of individual platy minerals may be described by an L-S fabric scheme on a stereogram. In the same way, measurements of anisotropic specimens can be plotted to define an orientation distribution of specimen tensors. Conventionally, the anisotropy of the specimens is represented by AMS-magnitude ellipsoids. This is shown simplistically, in Fig. 11.9, that may be interpreted as with Flinn's conceptual models of L-S fabrics from mineral orientations or aligned objects (Fig. 10.22). The orientation distributions of AMS ellipsoids presented in the stereogram define the fabrics in the L-S range by summarising the combined contributions of many observations of individual tensors, represented by the ellipsoids. The concept may be applied equally to the combination of AMS ellipsoids of minerals within a specimen or to AMS ellipsoids of multiple specimens at a site. On the larger scale, the individual magnitude ellipsoids may be averaged to yield a mean tensor also with axes designated k_{MAX}, k_{INT} and k_{MIN}. The sample comprises many individual specimen ellipsoids whose individual axial orientations define certain possible combinations of clusters, partial girdles and full gir-

dles. However, just as individual anisotropies are described by a tensor with orthogonal principal axes, so the tensor mean's axes are constrained to be mutually orthogonal. Possible dispersal patterns, clusters, partial girdles and girdles for the principal axes are summarised in Table 11.2. The mean-tensor sums contributions from imperfectly aligned individual specimen tensors and is therefore less anisotropic than the average, individual specimen tensor. Thus, the mean tensor is a subdued version of the specimen tensors, less eccentric (*Pj* closer to unity) and more neutral in shape (*Tj* closer to zero). The mean tensor would only have the same shape as the individual specimen tensors if the latter were perfectly aligned. This is not always a major concern; the main use of the mean tensor is to understand the orientation distribution and to determine the mean orientations of k_{MAX}, k_{INT} and k_{MIN} and their confidence cones.

11.8
The Mean Tensor: Hext-Jelinek Statistics

Calculating a mean tensor from a sample of specimen tensors requires that the varying magnitudes (shapes) and orientations of individual specimens are somehow incorporated into a single mean tensor. Its axes must be orthogonal and its shape will be a subdued version of the individual tensors. The mean tensor of several specimens is found in the same way as the best tensor from a set of measurement vectors for a single specimen (Eq. 11.10), where the measurements matrix is replaced by an average of individual-specimen tensor components referred to the same coordinate system. The orientation of the mean tensor is particularly sensitive to the magnitudes of specimen tensors. This effect is particularly troublesome in AMS because minerals have average or bulk susceptibilities that may range over five orders of magnitude. For this reason, individual specimen tensors are sometimes automatically standardised so that all have an average magni-

Table 11.2. Simplified classification for orientation distributions described by second-rank tensors

Fabric ellipsoid shape	Dispersion pattern of principal axes		
	Maximum (e.g. k_{MAX} for AMS)	Intermediate (e.g. k_{INT} for AMS)	Minimum (e.g. k_{MIN} for AMS)
L	Circular cluster centred on L	Great-circle ⊥ to L	Great-circle ⊥ to L
L-S	Partial girdle || S, centred on L	Partial girdle || S, ⊥ to L	Partial girdle ⊥ to S, ⊥ to L
S	Great-circle girdle || S	Great-circle girdle || S	Circular cluster ⊥ to S

Note: The orientation-distribution ellipsoid shape is a subdued version of the specimen-tensor ellipsoid, and its eccentricity (Pj, |Tj|) may not exceed that of the individual tensors (specimens).

tude of unity. This prevents an unusually oriented specimen tensor with high bulk susceptibility unduly influencing the mean tensor's orientation. However, standardisation of specimen-tensor magnitude is not obligatory and under some circumstances it may conceal some useful information, as we shall see below.

The confidence limits of the mean tensor's orientation are critical to understanding and interpreting the orientation distribution of a sample of specimens. The oval confidence regions around the orthogonal axes of the mean tensor require involved procedures introduced by Hext (1963) and Jelinek (1978), and summarised by Tauxe (1998). The uncertainty cone around each principal axis is elliptical in cross-section and intersects the surface of the sphere as an ellipse. Its projection onto the stereogram produces an oval confidence region that is normally defined at a significance level of $\alpha = 0.05$ corresponding to a 95% confidence level. (Other confidence levels are unlikely to find much application in geological orientation distributions.) Each stereogram shows three oval confidence regions centred on the mean orientation for the appropriate principal axis of the tensor (k_{MAX}, k_{INT}, k_{MIN}). Each confidence region may be described by the dimensions of its major and minor axes recorded in degrees of arc, as with Fisher and Bingham confidence cones for vectors (Chap. 10).

The three principal axes of the mean tensor are constrained to be orthogonal and they are connected by orthogonal planes of symmetry, the principal planes. However, the three confidence ellipses around the mean tensor's principal axes need not be parallel to principal planes. That is to say, they need not possess orthorhombic symmetry. In our introduction to the measurement of tensors in rock magnetism, we saw that the dispersion of the principal axes should be approximately symmetrical with respect to the principal planes of the mean tensor (Fig. 11.3 d). Although it seems initially counterintuitive, the confidence cones are in some cases inclined to the principal symmetry planes of the tensor. An example from the literature is shown in Fig. 11.3 e; this is of considerable interpretive value, a matter which will be discussed and illustrated later.

Werner (1997) compared various methods of computing mean-tensor principal orientations and their 95% confidence cones. The Hext and Jelinek methods usually agree closely, although Hext's confidence cones may be somewhat more conservative (Fig. 11.10). Differences in the actual mean orientations are normally negligible, but most software uses the Jelinek algorithm (e.g. AGICO Kappabridge system, Sapphire Instruments SI2B system), but see Tauxe (1998) for fur-

ther comparisons. However, as we have seen in other statistical procedures, a parametric approach that makes no assumptions about the distribution of the data may be beneficial in some cases. Thus, Tauxe has applied the bootstrap method of generating synthetic datasets by repeated sampling with replacement of the original sample. This permits huge numbers of synthetic mean orientations to be calculated for each principal axis. Using this procedure alone, the mean axial orientations would not be constrained to be mutually orthogonal since each group of axes, k_{MAX}, k_{INT} and k_{MIN} are treated separately. However, this may not present a problem for a well-defined anisotropy using the large number of pseudo means available by Monte-Carlo simulation. To ensure their orthogonality, Tauxe determines the mean orientations by Hext-Jelinek tensor statistics. She uses the Monte-Carlo simulation only to determine the dispersion of pseudo-means and hence the confidence limits for the mean orientations (pers. comm., 31 January 2002). Whatever approach we use to find the mean orientations for the tensor, the bootstrapped confidence regions are liberated from any symmetry constraints imposed by tensor-symmetry planes. This might provide useful petrofabric information, as a later discussion will reveal. Werner compared bootstrapped mean orientations with Hext and Jelinek means for a small dataset of 11 specimens. His study of magnetic fabrics used anisotropy of anhysteretic remanence (AARM) rather than AMS, but the principle is the same. He confirms Tauxe's comparisons of Hext and Jelinek confidence regions. However, for small datasets which benefit most from statistical characterisation, the bootstrapped confidence regions may possess unusual shapes and orientations although they are smaller than Hext-Jelinek confidence regions and the mean orientations agree well with Hext-Jelinek statistics. In particular, the small bootstrapped confidence regions are much more likely to lose orthorhombic symmetry as they are derived from independent averaging of each principal orientation (Fig. 11.10b).

11.9
Examples of Tensor Orientation Distributions and the Confidence Cones of Their Principal Axes

An introductory example of a petrofabric-magnetic fabric orientation distribution is shown in Fig. 11.11. This illustrates magnetic anisotropies of 74 specimens of the Borrowdale Group volcanic slate, located near the author's ancestral home in Northern England. The slate is a greenschist-facies, slaty tuff with well-de-

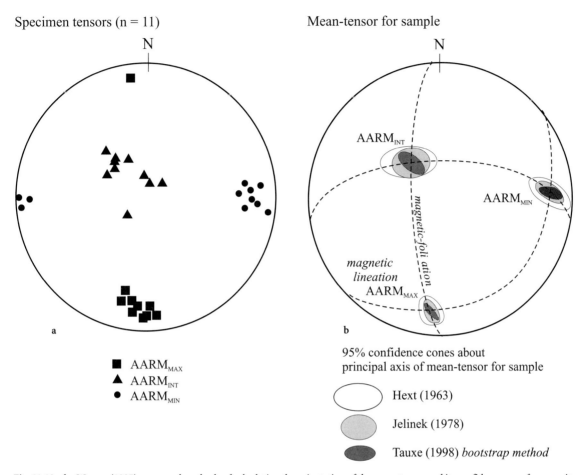

Specimen tensors (n = 11)

Mean-tensor for sample

■ AARM$_{MAX}$
▲ AARM$_{INT}$
● AARM$_{MIN}$

AARM$_{INT}$

AARM$_{MIN}$

magnetic-foliation

magnetic lineation
AARM$_{MAX}$

95% confidence cones about principal axis of mean-tensor for sample

Hext (1963)

Jelinek (1978)

Tauxe (1998) *bootstrap method*

Fig. 11.10a,b. Werner (1997) compared methods of calculating the orientation of the mean tensor and its confidence cone from a suite of specimens. He illustrated this with a study of anisotropy of anhysteretic remanence (AARM) in peridotites from Poland. The AARM tensor may be considered exactly in the same way as the AMS tensor, represented by a magnitude ellipsoid for each specimen. **a** Specimen AARM tensors. **b** Orientation of axes of *mean tensor* are almost identical by three methods. However, the confidence cones differ in size and in their degree of parallelism to the principal planes. (In the simplest and ideal fabric confidence cones would be parallel to the principal planes and thus show orthorhombic symmetry; Fig. 11.15)

fined cleavage (S) and consistently oriented mineral lineation (L), visible on the cleavage plane. The observations are presented in specimen coordinates with the cleavage horizontal and mineral lineation North-South. Usually, orientation data are given in geographic coordinates, i.e., with respect to North, South, East, West etc. However, inspection and interpretation in petrofabrics are sometimes facilitated by rotating the orientation data into sample coordinates, as here. This is called *centering the orientation distribution*. Clearly, the fabric is of the S-type with a weak L-component, i.e. L ≪ S, by analogy with previous hypothetical summaries (Fig. 11.9, Table 11.2). Quantification within the L-S range is possible and will be mentioned later. From the point of view of simple descriptive statistics, the first requirement, and for most studies the only re-

quirement, is to determine mean orientations of the principal axes. These may be interpreted in terms of syn-crystallisation stress orientations for metamorphic rocks; in other examples, the mean orientations may isolate flow directions or strain axes, with due caution for interpretation (Borradaile and Henry 1997). Naively, one might be tempted to calculate the mean orientation for each type of axis in turn, first for the minimum axes, then for intermediate axes etc., as if they were independent vectors. In that manner, mean orientations of individual groups of axes plot as the "+" symbols in Fig. 11.11. Indeed, in this fortunate case, these correspond quite closely with the peak concentrations of the density contours because this sample has particularly well-defined fabrics. However, closer inspection reveals that the individually estimated

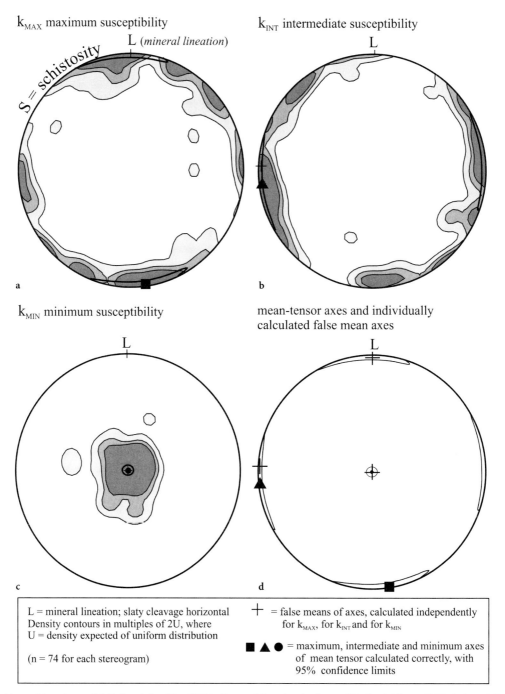

Fig. 11.11 a–c. Lineation and foliation defined by AMS in Borrowdale volcanic slates of England. The AMS fabric is mainly due to chlorite girdles of k_{MAX} and k_{INT}, and the circular cluster of k_{MIN} indicates an almost oblate symmetry (S-fabric), but there is a subtle preferred alignment (L) in the schistosity (S) plane shown by the k_{MAX} concentration in the girdle. Nevertheless, the mineral lineation is evident in the field from strain shadows and aligned chlorite. **d** The mean tensor, determined from individual specimen AMS tensors by Jelinek statistics, yields small, well-defined confidence cones around the mean tensor's principal axes. These show almost perfect orthorhombic symmetry. For comparison, the mean orientations of k_{MAX}, k_{INT} and k_{MIN} were calculated independently for each group of principal axes, treating each group (e.g. k_{MAX}) as if they were vectors. Although this does not constrain the mean axes to be orthogonal, they are reasonably perpendicular in this fortunate case because the orientation distribution of AMS specimen tensors is so well defined and clearly orthorhombic. In fact, only tensor means calculated using tensor statistics guarantee the required orthogonality of the mean tensor's principal axes

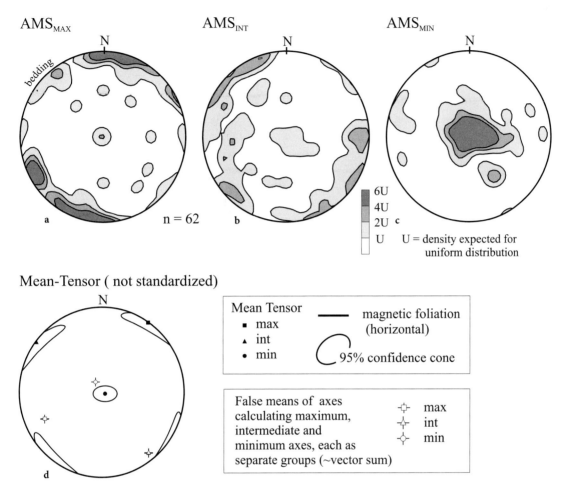

Fig. 11.12. a–c Contoured density distributions of principal susceptibility axes for a suite of sandstones determined by anisotropy of magnetic susceptibility (AMS). Contour levels in multiples of equivalent expected uniform density. **d** Mean orientations of AMS axes calculated using Jelinek's tensor statistics. The 95% confidence cones about the axes show orthorhombic symmetry compatible with the bedding-planar fabric. Mean orientations of principal directions calculated independently, for the maximum, for the intermediate and for the minimum values, as if they were separate groups of axes, yield false non-orthogonal mean orientations

mean principal orientations are not perfectly orthogonal (Fig. 11.11 d). For weak preferred orientations, independently calculated means for the three principal axes may be grossly non-orthogonal (Fig. 11.12 d). These have little meaning or use since the mean tensor for the sample must possess orthogonal axes (Figs. 11.3 d, e, 11.10 b), like individual specimen tensors (Fig. 11.2).

We need a way in which to calculate the mean orientations of each type of axis (minimum, intermediate, maximum) so that their orthogonality is guaranteed. Jelinek (1978) investigated this subject thoroughly and developed a system of complex calculations of tensor statistics that provided for orthogonal principal mean orientations with some reasonable

approximations and limitations. Moreover, his formulae provide confidence limits about the mean principal orientations. The confidence cones possess the orthorhombic symmetry of the orientation distribution so that each axis has elliptical confidence limits mimicking a "Bingham type" distribution of its principal directions and parallel to the principal planes that connect principal axes (Fig. 10.11). Projections of 95% confidence cones calculated from the tensor-statistical approach of Jelinek (1978) are shown for several distributions of field structures and magnetic susceptibility (AMS) fabrics (Figs. 11.11 d, 11.12 d, 11.13, 11.14, 11.16, 11.7, and 11.18). The 95% confidence cones have the appropriate principal value at their centre. Following the same basic concepts explained in Chapter 4, we

(**a** to **c**) Kapuskasing granulite gneiss; contours in multiples of 2U where U = density of uniform distribution

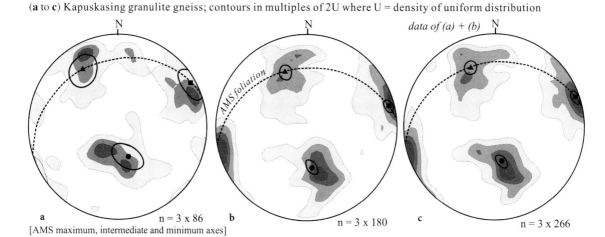

Fig. 11.13 a–e. Orientation distributions of AMS principal axes, Kapuskasing gneiss. **a** For 86 specimens collected by the author's students. **b** For 180 specimens collected by the author at the same localities. **c** The two sets of data combined (*n* = 266). Well-defined clusters correspond approximately to maximum, intermediate and minimum magnetic susceptibility and the elliptical, 95% confidence limits about the orientations of the axes of the mean-tensor are shown. **d, e** Welsh slate; mean orientations of the tensors of principal magnetic susceptibilities are shown for specimens of purple slate and green-spotted slate. **d** The purple slate's AMS is controlled largely by the preferred orientation of chlorite as haematite's abundance and susceptibility are too low. **e** Green-spotted slates in which haematite is locally replaced by high-susceptibility magnetite. The magnetite in the green spots records a later, differently oriented stress field. Consequently, the AMS mean tensor differs in orientation between purple and green-spotted slate

must accept one in twenty chances that the population's mean orientation could lie outside that limit. We should also note that large confidence cones raise suspicion on technical grounds because Jelinek's statistics involve calculations that are strictly only valid for mean-tensor confidence cones with radii $\leq 25°$.

The necessity of tensor statistics is not a matter of seeking unrealistic perfection in the interpretation of normally poorly defined petrofabric data. Almost any normal petrofabric responds unfavourably to inappropriate treatment of its principal tensor axes as inde-

pendent features. Consider the tectonically deformed sandstones of Fig. 11.12. Most structural geologists work routinely with preferred orientations no better defined than the density contours of the principal axes illustrated. However, it is difficult to use peak densities to determine tectonically significant strain axes, because the contours are multimodal and their peak values are rarely orthogonal. Using separate calculations for each type of axis (i.e. first for the maximum axes, then for the intermediate ones, etc.) and then incorrectly averaging their orientations as with unit vectors,

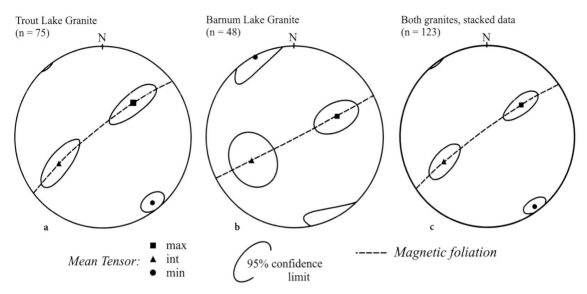

Fig. 11.14a–c. When samples from congruous subareas are combined, their mean tensor should be better defined than that of either subsample. In this case, the principle is illustrated with two adjacent granite plutons that experienced the same mild, post-emplacement, tectonic stress field. **a** Trout Lake Granite. **b** Barnum Lake Granite. **c** Combined samples yield better defined mean orientations with smaller confidence ellipses and clearer orthorhombic symmetry

produces grossly non-orthogonal, *false mean orientations* that are incompatible with peak densities, in part because they are divided between small separate modes (Fig. 11.12d). However, the mean orientations calculated by Jelinek statistics are orthogonal and the 95% confidence limits for the mean axes provide useful information not just about the significance of the mean principal orientations, but also on the *best-fit symmetry* of the entire orientation distribution of the measured tensors.

It is also instructive to compare density distributions of tensor axes with the confidence cones for the mean tensor's axes. The magnetic susceptibility anisotropy (AMS) of lower crustal gneisses from the Kapuskasing structural zone in northern Ontario is well defined with a NE, shallow-plunging k_{MAX} (lineation) and steeply southerly inclined k_{MIN} over areas exceeding tens of km². This is clear from the three maxima of the orientation distribution (specialist programs used by the author distinguish the clusters by the type of axis, but it is not necessary here). Because individual samples over the area depart from the mean orientation in complex patterns, the three maxima are not precisely orthogonal and they do not show perfect circular or orthorhombic symmetry. However, individual *specimen tensors* are much more likely to show confidence cones with orthorhombic symmetry (and obviously including orthogonal axes). Consequently, the *mean tensor* has orthogonal axes distinguished in

Fig. 11.13a–c by the *traditional symbols: square (maximum), triangle (intermediate) and circle (minimum)*. Note that the tensor-mean axes cannot precisely correspond to the peak concentrations of the density distribution; that will only rarely occur with well-defined fabrics. Density plots require absolutely no assumptions about the population from which the sample was drawn. Depending on the method of preparation, 95% confidence limits for tensor-mean axes may involve some assumptions. However, both density contours and mean tensors should be used for a full evaluation of the fabrics. The degree to which the mean tensor's confidence cones satisfy orthorhombic symmetry may provide very valuable information. In particular, the shape of the confidence cones for the mean tensor's axes may afford further interpretation as will be discussed later. Figure 11.13a shows data from 86 specimens collected by two of the author's students. Figure 11.13b shows data for 180 specimens collected by the author *from the same area*. From the density distribution, but especially from the confidence cones one may compare the greatly improved precision of 180 specimens as opposed to 86. [Of course, the greater precision of (b) might owe something to the fact that the instructor collected the data!] It is interesting to note that the stacking of (a) and (b) to give 266 samples in (c) produces relatively little improvement over (b). For the definition of sample mean and dispersion, excessive sampling often stumbles into the law of diminishing returns. In Chapter 2 we saw that a sample size of $n \sim 30$ usually sufficed *for the definition of mean and variance,* in the simple case of scalars that follow some hump-shaped distribution along the infinite line range. However, in three-dimensional orientation distributions, the data have more degrees of freedom yet are limited to the spherical 4π solid-angle that may obscure the recognition of outliers. Therefore, it is not possible to make a universal recommendation about the sample-size needed to characterise the mean orientation and dispersion of an orientation distribution. Unfortunately, each distribution must be evaluated individually, not just on common sense statistical intuition but also in the light of specific subject knowledge involving mineralogy, petrology and structural geology. Inspection of the traditional density contours is a good first step to evaluating the suitability of the sample distribution.

11.10
Orientation Distributions Involving Multiple Fabrics and Fabric Heterogeneity

It may be possible to isolate subfabrics due to subordinate orientation distributions of minerals with different age, composition or kinematic history. The same mineral may be present in two or more different orientation distributions if it formed at different ages in a metamorphic history; biotite and chlorite are good examples. Tensor statistics may aid in the isolation or inference of multiple orientation distributions, or in the common parlance of structural geology, multiple fabrics. The Lakehead Rock Magnetism group has found several examples of this in Archean rocks ranging from greenschists to amphibolites, and this led to the determination of major transpressive motion along Archean terrain boundaries (Borradaile and Spark 1991; Borradaile and Dehls 1993; Borradaile and Kukkee 1996; Werner and Borradaile 1996; Borradaile et al. 1998, 1999a).

The first metamorphic rock was investigated petrofabrically by Sorby (1860; see Ramsay 1967), from which grew the interest in strain and preferred mineral orientations. Sorby made the first petrographic observations on slates, coined the term *slaty cleavage* and thus it is fitting to use the example of his material, the Welsh Slate of Penrhynn, North Wales. It provides excellent material for study by magnetic petrofabrics due to the fortunate oxidation transitions shown by its accessory iron oxides (Borradaile et al. 1991; Jackson and Borradaile 1991; Nakamura and Borradaile 2001). The matrix comprises very well-aligned green chlorite with

high intrinsic *crystal AMS* but the colour is masked by traces of hematite, rendering the slate purple. Locally, reduction of hematite to magnetite increases bulk susceptibility and changes the *rock AMS*. Moreover, these specimens are readily identified as green patches, spots and beds due to the reduction of the hematite pigment. Due to its high susceptibility, traces of magnetite can dominate the rock AMS, according to the susceptibility-anisotropy balance principle explained earlier (Figs. 11.7a, 11.8e). In the purple slate, AMS is controlled by chlorite so that the maximum susceptibility is parallel to the mineral lineation and the magnetic foliation is parallel to the slaty cleavage. However, in green varieties, the presence of a differently oriented magnetite subfabric deflects magnetic lineation ($= k_{MAX}$) and foliation ($= k_{MAX} - k_{INT}$ plane) with respect to the mineral lineation within the cleavage (Fig. 11.13d, e). Nakamura and Borradaile (2001) have suggested that the spots postdate cleavage formation. In that case, they are not deformed objects from which strain may be estimated from the reduction spots at that locality. However, in other regions, ellipsoidal reduction spots in slates are convincing strain markers, showing the variation in orientation and shape that are necessary consequences of finite strain acting upon an initial dispersed orientation distribution of spots (e.g. Ramsay 1967).

Jelinek's confidence cones also provide a means of detecting and comparing fabric heterogeneity. Anisotropic petrofabric, structural and geophysical properties invariably show spatial heterogeneity. This means that anisotropy of suites of samples must be interpreted carefully with some knowledge of the geology to guide sampling and grouping of specimens into compatible, approximately homogeneous groups. In this way, each subarea should produce a stereogram that is relatively simple to interpret, with well-defined peak concentrations and better-defined orthorhombic symmetry. However, it is not usually possible to define homogeneous subareas in advance. One should inspect, sift and regroup the data until satisfactory homogeneous stereograms are obtained. For this purpose, computer programs that plot stereograms with density contours and also Jelinek statistics provide a rapid method of determining how congruous and compatible are the data within a subarea. If a subarea is homogeneous, the Jelinek confidence cones will be small and have good orthorhombic symmetry. Conversely, where heterogeneous data have been merged from unlike terrains or heterogeneous fabrics, confidence cones for the mean tensor may be large and non-orthorhombic. Tensor statistics for orientation distributions from two compatible subareas are shown in Fig. 11.14. In fact, the two subareas represent two adjacent, essentially synchronous granite intrusions that subsequently experienced mild tectonic deformation. The tectonic overprint imposed anisotropic magnetic fabrics in the magnetite grains. Each granite intrusion thus represents a separate subarea with its own magnetic fabric. The two differ only slightly, and each provides a satisfactory stereogram with quite small confidence cones, clear orthorhombic symmetry of the cones, similar orientations for the principal axes of the orientation distribution and similarly oriented magnetic foliation. (The magnetic foliation, like any other foliation, is the plane containing the maximum and intermediate principal directions.) Stacking, or combining the datasets, and recalculating the Jelinek statistics confirm the congruity of the tectonic magnetic petrofabrics of the two granites. The stacked data show that the confidence cones become much tighter, their orthorhombic symmetry is improved and a sensible "mean" orientation is obtained for the combined principal directions and magnetic foliations of the two granites (Fig. 11.14c).

The preceding discussion has referred primarily to the usefulness of the confidence cones to infer the significance of the principal directions and of the foliation plane, and of the the homogeneity of a sample. However, it also raised the question of the orthorhombic symmetry revealed by the arrangement of the elliptical confidence cones. The confidence cones should possess orthorhombic symmetry where the following conditions are satisfied.

1. *the sample size is adequate*
2. *the subarea is homogeneous*
3. *the fabric was formed by a single event or process with coaxial symmetry*

The ideal confidence-cone arrangements for orientation distributions of tensors are given for $S \gg L$, $S \ll L$ and neutral ($L = S$) tectonites in Fig. 11.15a–c. The foliation plane containing the maximum and intermediate principal values of the orientation distribution need not necessarily correspond with any visible foliation or planar structure in the rock. Firstly, this is because for $L = S$ and $S < L$ fabrics, the arrangement may not be robust enough to express itself as a planar feature. Secondly, if these orientation distributions are determined from the anisotropy of some physical property (e.g. magnetic susceptibility, electrical conductivity) this need not correspond to minerals of a suitable habit or grain shape that can form a visible planar structure of the corresponding "fabric shape".

How then may non-orthorhombic symmetry arise in the orientation distribution of tensors? The mathematical procedures, beyond the scope of this book and

usually available in professionally developed software, can only yield mutually orthogonal principal axes. Clearly, nothing else would be a logical conclusion of any successful tensor-averaging process. Individual tensors have orthogonal axes, and this must be true also for the mean tensor. Why then may the *confidence cones* show sub-orthorhombic symmetry in some instances (e.g. Figs. 11.3e, 11.10b)? It must be understood that we are not free to ignore the magnitudes of the property associated with the principal directions. The stereogram shows only the tensors' orientation distribution, and that is the prime concern and preoccupies most of the interpretation for geologists. However, magnitudes of the individual tensors can greatly influence the appearance of the mean-tensor orientation for the set of observations, especially where the mean magnitude (bulk susceptibility) of the individual specimens varies greatly. This is discussed below.

11.11
Heterogeneous or Multiple Fabrics and the Mean Orientation of Sample Tensors

In dealing with orientation of axes or unit vectors, every observation had the same weight or importance. Subsequently, we saw that when dealing with orientation distributions of vectors, the magnitude of the vector may be taken into consideration. For example, if we measure the orientations and lengths of fractures, we may choose simply to plot the orientations as axes (all observations of equal weight) or we may decide to consider longer fractures "more representative" and weight them accordingly. In paleomagnetism, attention is almost exclusively devoted to the orientation of the characteristic component of the vector of magnetic remanence. Its magnitude is an arbitrary response due to rock type and to magnetisation process, but its orientation is of fundamental and paramount importance, e.g. in plate tectonics.

Unfortunately, with tensors, the magnitudes of the property measured in the three principal directions cannot be ignored. They are an intrinsic part of the definition of the tensors' mean orientation, and the magnitudes must be considered in the calculation of the mean orientation and its confidence limits. We shall see that some choice exists as to how we proceed, with vastly different outcomes, depending on the questions we wish to address. There are two situations that arise commonly and lead to our appreciation of the importance of magnitude in understanding tensor orientation distributions.

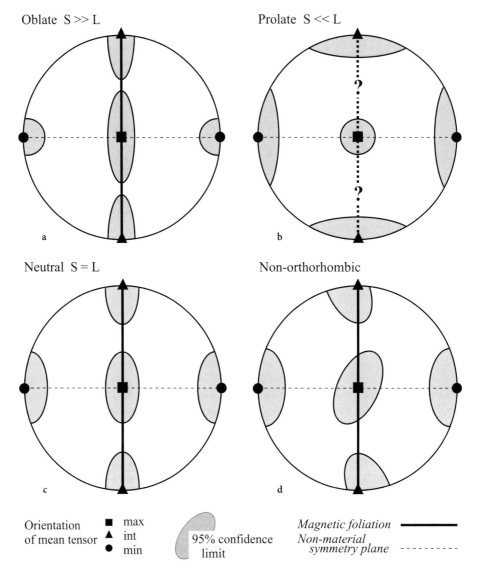

Fig. 11.15a–d. This shows the idealised symmetry of confidence cones about the mean orientations of the principal axes of the mean tensor for a petrofabrically homogeneous group of specimens. The cones' shapes and symmetry should conform to the density contours of principal axes. **a–c** Compatibility of confidence cones with the spectrum of orientation distributions in the L-S fabric scheme. These all show orthorhombic symmetry since the confidence ellipses are parallel to principal planes. **d** Non-orthorhombic confidence cones may arise due to the presence of multiple fabrics or heterogeneities associated with anomalous specimens of high intensity (= large magnitude)

First, consider the situation in which most specimens have similar mean susceptibilities, that is to say, their tensors may be represented by magnitude ellipsoids of similar volumes, as shown in Fig. 11.16a. The average fabric is visually obvious, but measurements of an anisotropic property (e.g. magnetic susceptibility) to detect this would be biased by a specimen with unusually large magnitude. The presence of heterogeneous mean magnitudes may be detected by standardising each specimen, dividing its principal magnitudes (maximum, intermediate and minimum) by their arithmetic mean. In this way, all samples are assigned equal weight, and become represented by magnitude ellipsoids of equal "size" in Fig. 11.16b. The mean orientation of the susceptibility-normalised fabric will reflect the most commonly recognised orientation. Of course, it is neither more nor less correct than the non-normalised data, but by comparing their tensor means

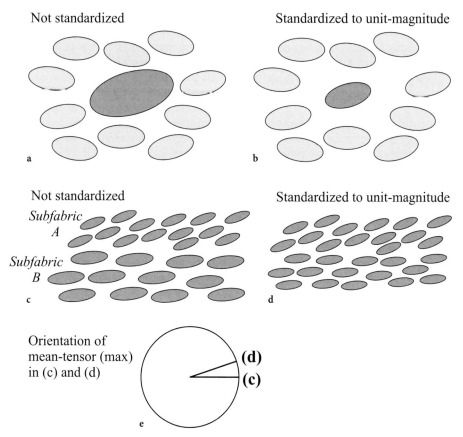

Fig. 11.16a–e. The orientation distributions of tensors for a suite of specimens is not just affected by the orientation of individual tensors but also by their individual magnitude. The latter is represented by the volume of the ellipsoid. **a** A few specimens with high magnitude may mask the overall orientation distribution; the mean-tensor orientation for the specimens is deflected toward the orientation of the few large-magnitude specimens. **b** Standardisation of specimens to a common magnitude forces them to contribute equally to the orientation distribution. This is achieved by dividing the principal susceptibilities of each specimen by its average susceptibility. **c** Multiple subfabrics also complicate the interpretation: the mean tensor's orientation is biased toward that of the higher magnitude subfabric. **d** Contribution of a higher magnitude subfabric is neutralised and detected by standardisation

and confidence cones, we may be alerted to the role of high-intensity specimen outliers that may cause anomalous tensor-mean orientations and confidence-cone symmetry. One must caution the use of normalised tensors with very weakly susceptible material. For example many carbonates and quartzites have mean susceptibility close to zero, either positive or negative, due to the mixture of a few paramagnetic impurities (large positive susceptibility) in a matrix that is diamagnetic (-14×10^{-6} SI). Clearly, dividing by very small values of bulk susceptibility (~ 0) would not produce meaningful standardised anisotropies.

In general, however, bulk susceptibilities are large positive values and standardising specimen magnitudes before calculating orientations and confidence cones for the mean tensor is particularly valuable. For simplicity, we shall assume similar magnitudes for all

samples, but include the possibility of two subfabrics (A, B), each with a characteristic orientation distribution. In Fig. 11.16c they are shown in spatially distinct locations but, of course, the specimens of different mean susceptibility are often penetratively interspersed. Subfabric (B) has higher average magnitudes and thus dominates the orientation distribution. When each specimen's tensor is normalised, all specimens have equal weight and the orientation distribution blends the two subfabrics. Again, it is not suggested that normalised and non-normalised sample tensors differ in "correctness". However, by inspecting both, we may detect the presence of multiple, competing subfabrics or sample heterogeneity.

The mean principal directions for the orientation distributions may be affected greatly by normalising sample tensors where either small numbers of anomalously high-magnitude samples are present or where subfabrics conflict, e.g. a bedding fabric and a tectonic one. Moreover, the orthorhombic symmetry of the confidence cones may be improved or degraded by normalisation. Comparison of normalised and non-normalised Jelinek statistics will probably benefit interpretation, whatever the outcome.

Some examples show the dramatic effects of normalisation. Magnetic fabrics provide a common example where a matrix of relatively low-susceptibility minerals contains a subfabric of high-susceptibility accessory minerals, such as magnetite. The latter, even where present in traces, can mask the contribution of the susceptibility tensor of the matrix, which is normally of greater interest in tectonic studies. As an extreme example, consider a suite of quartzites that combine a matrix of negligible susceptibility and low-anisotropy quartz with a small concentration of more susceptible and higher anisotropy mica (Fig. 11.17 a, b). A few specimens of the quartzite, contaminated with magnetite, control the orientation distribution of magnetic anisotropy for the sample suite (Fig. 11.17a). Thus, without normalising the specimen tensors, we are deceived into accepting an apparently well-defined fabric (Fig. 11.17a). However, normalisation de-emphasises the anomalously high-susceptibility specimens, changing not only the apparent principal orientations but also revealing the true scatter of the mineral orientation fabric from the enlarged confidence cones (Fig. 11.17b). Even worse, the 95% confidence ellipses do not show orthorhombic symmetry. This indicates that setting aside a few influential, magnetite-rich specimens, the bulk of the sample suite actually has a very poorly defined preferred orientation of matrix minerals. The interpretation might be that the non-normalised mean tensor defines the bedding fabric through the alignment of heavy minerals whereas the normalised tensor detects an incipient metamorphic fabric (cf. Fig. 11.17a, b).

A further example using relatively high-susceptibility schists (k_{MEAN} = 6181 µSI) shows that the mean tensor for the non-normalised sample suite is not really compatible with the well-developed schistosity fabric. However, standardisation of specimens to unit-value mean susceptibility resolves this (cf. Fig. 11.16c, d). We can prove that this is due to the presence of anomalous outliers, i.e., specimens of high susceptibility. When we omit four samples with extraordinarily high susceptibility (k_{MEAN} = 41,973 µSI), the oblate schistosity-parallel S-fabric is very well defined (Fig. 11.17e, f).

Two further examples serve to illustrate the power of normalisation in the interpretation fabrics represented by anisotropy. Again, the example uses the anisotropy of magnetic susceptibility (AMS), which is effective in determining preferred crystallographic orientations in rocks that have granular textures, e.g. granites and high-grade gneisses. The first example is of a weakly strained granite intrusion at Trout Lake, northern Ontario (Fig. 11.18a). In this case, normalisation changes the positions of the mean orientations of the orientation distribution slightly, but it does improve the orthorhombic nature of the confidence ellipses, elongating those for the maximum and intermediate directions along the magnetic foliation plane. Thus, it isolates a slightly more planar fabric than is recognised from the non-normalised, raw data. This suggests that some specimens with high susceptibility and spurious orientations mask the preferred orientation of the silicate matrix. The second example concerns specimens of lower crustal, high-grade gneisses that are thrust to the surface in the Kapuskasing Structural Zone of northern Ontario (Fig. 11.18b). Comprising pyroxenes, feldspars and garnet, their textures are granoblastic, which causes extreme difficulty in recognising well-defined or consistent fabric orientations in the field. Nevertheless, preferred crystallographic orientations are present and rapidly detected by AMS. Normalisation of these samples preserves the well-defined maximum orientation that indicates the mineral lineation. However, it dramatically shrinks the confidence cones for the intermediate and minimum values: this improves definition of the intermediate and minimum directions represented by the silicate matrix, but makes the L-symmetry of the orientation-distribution less convincing (cf. Figs. 11.18b, 11.15). One interpretation may be that high-susceptibility specimens, for example dominated by the younger magnetite fabric, emphasise the planar component to the fabric, giving the impression that the gneisses are S-tectonites. Normalisation reduces the influence of high-susceptibility specimens, suggesting that the majority of the specimens show instead an L or L > S fabric (cf. Fig. 11.15).

11.12
Specimen-Level Uncertainty and Its Contribution to Sample-Mean-Tensor Confidence Limits

11.12.1
Jelinek Statistics

The calculation of the mean-tensor orientation and its elliptical confidence cones is a great benefit of tensor statistics (Jelinek 1978). The confidence cone for the orientation of each principal axis of the mean tensor

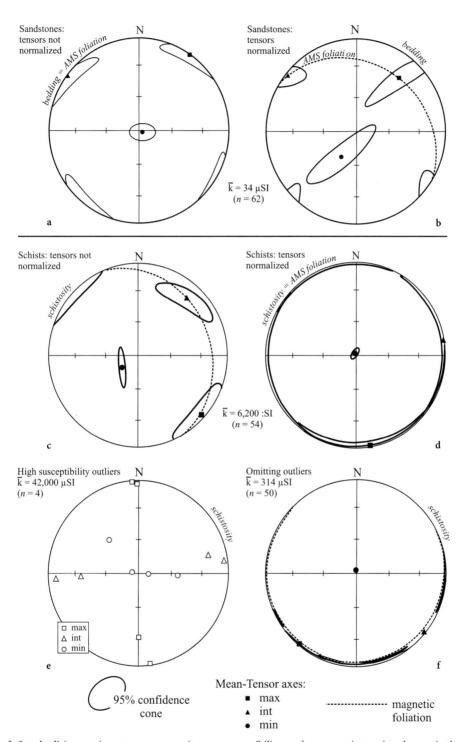

Fig. 11.17a–f. Standardising specimen tensors to a unit-mean susceptibility produces some interesting changes in the mean tensor for the orientation distribution of a suite of specimens. **a** Mean tensor for AMS of sandstones with a bedding-planar, flat-shaped (~oblate) AMS fabric. **b** Standardisation of the specimen tensors loses definition of the bedding-parallel fabric and produces non-orthorhombic confidence cones. **c** Mean tensor for AMS of some schists. **d** Standardising specimen tensors to common magnitude improves definition of their planar fabric. **e** Isolation of four anomalous high-susceptibility specimens in the schist sample. **f** Removing the anomalous specimens from the sample improves definition of the S-fabric and changes the location of the mean axes of maximum and intermediate susceptibility

actually confounds two sources of variability or uncertainty. In the interpretation of the orientation distribution, the primary source of variability is specimen-level orientation. The second source of variability is derived from weighting the contribution of individual specimens according to their mean susceptibility value. For example, we noted that the presence of a subfabric of differently oriented, high-susceptibility grains could deflect the matrix-anisotropy directions (Figs. 11.8, 11.16).

Jelinek's approach permits the interpretation of *specimen-level uncertainty* due to their variation in orientation. These larger confidence cones may be useful in identifying specimens which could be considered as outliers; their exclusion may then result in a more reasonable interpretation of the tensor mean for the remainder of the sample. For the purposes of sifting data into homogeneous subgroups for further statistical or interpretative treatment, it may be important to identify specimens whose axes lie within reasonable (95% confidence) limits of the mean tensor. For example, ob-

serve the density plots for maximum, intermediate and minimum susceptibility in granulite-facies gneisses (Fig. 11.19a). The axes of the mean tensor for the sample suite differ slightly from the peak densities, as expected, but the 95% confidence cones about those axes correspond quite well to the density peaks. For any of the three axes, the mean-tensor confidence limits are smaller than the confidence limit of the sample's axial distribution by a term involving the factor $(1/\sqrt{n})$. The *sample-confidence cone* may be used to isolate anomalous, incongruent specimens on the basis of unusual orientation. After removing such outlier specimens from the sample, it may be possible to recompute more meaningful tensor means and the confidence limits about the axes of the mean tensor. It may also assist in the isolation of a separate subfabric if the orientation distribution is multimodal, i.e. it would provide an objective criterion for deciding to which fabric we should assign a certain specimen.

We should not lose sight of the meaning of the 95% confidence cone: it merely signifies that there is a one-

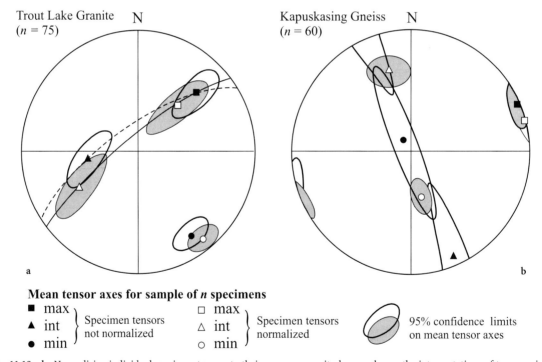

Fig. 11.18a, b. Normalising individual specimen tensors to their average magnitude may change the interpretations of tensor orientation distributions, as shown by these examples of anisotropy of magnetic susceptibility (AMS). **a** AMS of Trout Lake Granite, northern Ontario; standardising each specimen tensor produces a steeper foliation and slightly more elongate confidence ellipses for the mean orientations of the intermediate and maximum susceptibilities. **b** AMS of Kapuskasing Lower crustal gneisses, northern Ontario; standardisation of specimen tensors retains approximately the same orientation for the magnetic lineation (= maximum susceptibility). However, the original L-fabric had poorly constrained intermediate and minimum susceptibilities elongate confidence ellipses, perpendicular to the maximum susceptibility. Standardisation weights all samples equally, removing the contribution of spurious high-susceptibility outliers, and better constrains the intermediate and minimum orientations

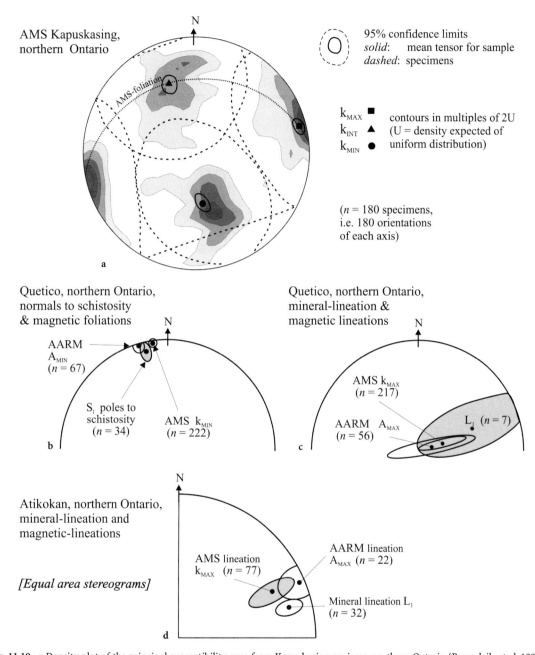

Fig. 11.19. a Density plot of the principal susceptibility axes from Kapuskasing gneisses, northern Ontario (Borradaile et al. 1999a). Each contoured cluster is associated with the appropriate, corresponding mean-tensor axis determined by Jelinek's statistics. The small cones are the 95% confidence limits for the axes of the mean tensor for the entire sample of 180 specimens. The larger 95% confidence cones show the uncertainty of the specimen-level distribution and may assist being useful to identify anomalous samples. Such outliers might be removed from the sample to refine calculations, or they may establish another subfabric. **b** Confidence cones for tensor-mean axes may be compared with one another, and with 95% confidence cones for other axes such as normals to schistosity in this example. There is some overlap of confidence limits for the minimum AMS susceptibility, minimum AARM susceptibility and the schistosity normals (from Borradaile et al. 1994; Werner and Borradaile 1996). **c** From the same suite of specimens, the mineral lineations and maximum orientations of AMS and AARM also show some overlap. **d** From another area in northern Ontario, AMS, AARM and quartz-feldspar mineral lineations have means that lie beyond each other's 95% confidence limits, suggesting that they are different at the 95% confidence level

in-twenty chance that the population mean (= true value) lies beyond that limit. This is important when we attempt to make fine distinctions between mean-orientations of tensor axes using the 95% confidence cone. We should remember that large *confidence ellipses* for the *mean tensor* raise suspicion on technical grounds because Jelinek's statistics showed they are strictly only valid for radii $\leq 25°$.

Apart from expressing confidence in the orientation of a mean tensor, the confidence cones may be used to compare different samples, e.g. representing suites of specimens from adjacent areas. We may wish to compare the confidence cones for k_{MAX} from two different samples. If the 95% confidence cones do not overlap, we may reject the hypothesis that the samples are drawn from the same population, with a 5% probability of type I error (Chap. 4). Informally expressed, their means are not the same, but there is a 1 in-20 chance that we have rejected a true hypothesis. Such mean-tensor confidence cones may also be compared with different anisotropies and with confidence cones for vectors or non-tensor axes, such as structural axes like mineral lineation or poles-to-schistosity. For example, consider the comparison of AMS fabrics, AARM fabrics and the structural features of mineral lineation and schistosity normals in some Archean schists (Fig. 11.19b, c). The AMS fabrics are mainly due to the orientation distribution of mica and chlorite whereas the AARM fabrics are due to accessory magnetite and pyrrhotite. If the schistosity-defined petrofabric alignment of minerals is due to a simple coaxial kinematic history, the AMS ellipsoid for mica-chlorite and the AARM ellipsoid for the magnetite-pyrrhotite fabrics should be coaxial. This may be the case since there is some overlap among the three fabric elements when we consider poles to the respective foliations which seem to imply parallelism with *schistosity normal* = $k_{MIN} = A_{MIN}$ (Fig. 11.19b). Since the three fabric elements are reasonably coaxial, one might assume they are synchronous or that they were formed in a coaxial strain history. The linear components of the fabric show less perfect parallelism with *mineral lineation* = $k_{MAX} = A_{MAX}$ (Fig. 11.19c). Although the number of field measurements is small for mineral lineations the concept of comparing confidence cones should be apparent. Another example from Archean schists of northern Ontario shows that there may be some non-coaxiality of magnetic subfabrics (Fig. 11.19d). Here, the mean maximum axes of the tensors for AMS (k_{MAX}) and for AARM (A_{MAX}) lie outside each other's 95% confidence cone. Moreover, their confidence cones do not even overlap with that for the visible mineral lineation in the field. In brief, the field mineral lineation is

due to feldspars or quartz, AMS is due to recrystallised metamorphic silicates, whereas the AARM signal is due to accessory magnetite and pyrrhotite. This is understood to be the order in which the minerals appeared in the rock; thus, the different mean orientations of their orientation distributions indicate non-coaxiality of stress and strain during their development.

We have seen that the confidence limits of mean-tensor principal axes is important in interpreting and comparing natural orientation distributions as well as the significance of the mean tensor. The larger specimen-level 95% confidence cone may be used to isolate outliers that are simply aberrant or erroneous measurements and perhaps classify orientations that belong to a separate subfabric. However, Jelinek's statistics have been primarily used with orientation uncertainty of the specimens and their influence on the mean tensor. We saw that approach lends itself to two clear choices in routine averaging a sample of tensors. First, all specimen tensors may be standardised to a uniform unit magnitude by dividing each specimen anisotropy by its mean susceptibility. Alternatively, each specimen is left in its original units of measurement. We saw that approach was useful in identifying subfabrics that were also characterised by anomalous mean susceptibility. However, it omitted any consideration of specimen-level uncertainty in the mean tensor, for example the measurement errors on individual specimen tensor magnitudes and specimen axial orientations. More complex weighting of individual specimen tensors is possible in a quite different manner, as discussed next.

11.12.2
Le Goff-Henry Bivariate-Fisher Statistics

An alternative approach to the incorporation of *specimen-level uncertainty* into the mean tensor was introduced by Henry and Le Goff (1995). This extended the bivariate-Fisher statistical model used previously by Le Goff (1990) to provide a simple means of determining elliptical confidence cones around paleomagnetic vectors that followed a Bingham-type distribution. The only limitation of their approach is that it requires each principal axis (k_{MAX} or k_{INT} or k_{MIN}) to be treated as though they were independent orientations, like vectors. Thus, we lose the ability to constrain orthogonality of the mean orientation of each set of principal axes. We saw this effect when, for the purposes of illustration, mean orientations were determined for each axis independently instead of using tensor statistics (Fig. 11.12d). However, this limitation should not be too great when we deal with highly symmetrical, well-constrained orientation distributions with well-defined mean tensor having small confidence cones (e.g. Fig. 11.11). At any rate, the best mean orientations of the principal axes may be determined by tensor statistics in order to retain orthogonality, and then bivariate-Fisher statistics may be applied to determine the confidence about each principal axis. The advantage of this is revealed next.

The following reasoning permits a wide selection of *tensor-uncertainty weighting*, not merely uncertainties in specimen-level AMS orientation and magnitude due to measurement error or intrinsic variation. In particular, it has unexplored potential beyond the consideration of conventional specimen-tensor measurement error. For example, the weighting could be any simple numerical value that represents mineralogy, specimen

location, metamorphic grade or a host of subject-specific factors that may be considered as secondary control variables influencing the orientation of the mean tensor for the sample of specimens.

Henry and Le Goff's (1995) example of anisotropy of magnetic susceptibility (AMS) for a sandstone showed quite different anisotropies for different specimens (Fig. 11.20a). As typical with planar (~oblate) depositional fabrics, k_{MIN} is perpendicular to bedding and usually well defined with equi-dimensional confidence cones. The maximum and intermediate susceptibilities are less precisely defined in the plane of bedding, simply as a function of the anisotropy. Therefore, they show strongly elliptical confidence cones parallel to the trace of bedding. In most studies, the mean tensor is calculated on the basis of the dispersion of the specimen's principal axes without regard for their indi-

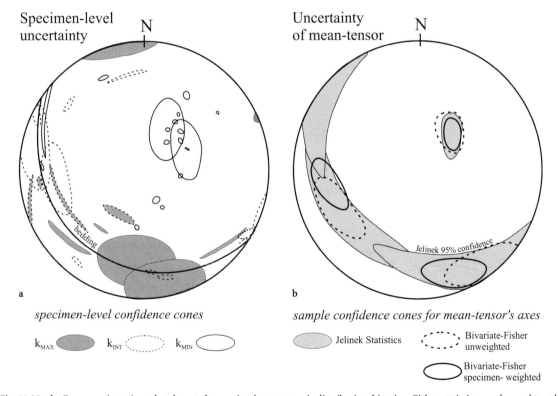

Fig. 11.20a, b. For any orientations that do not show a circular-symmetric distribution, bivariate-Fisher statistics may be used to calculate the mean orientation and elliptical confidence cones as explained in Chapter 10 (Le Goff 1990; Le Goff et al. 1992). The method may be applied to the maximum axes of a sample of specimen tensors, and then to their intermediate axes and finally to their minimum axes. This individual treatment of principal axes does not force the orientations of the mean axes to be orthogonal (e.g. see Fig. 11.12), but it has the advantage that the uncertainties of individual specimen tensors may be included in the orientational uncertainty of the mean tensor (Le Goff and Henry 1995). **a** AMS of some sandstone samples showing 95% confidence cones for the tensors of individual specimens. **b** For all specimens, the orientation of the mean tensor and its confidence cones calculated by different methods. This shows the conservative, large cones from the Jelinek method and Le Goff and Henry's smaller confidence cones that consider individual specimen uncertainties (weighted bivariate-Fisher statistics) or ignore them (unweighted)

vidual uncertainty. However, Henry and Le Goff applied the bivariate-Fisher approach (Chap. 10) to each group of principal axes (k_{MAX}, then k_{INT}, then k_{MIN}). Their comparison of Jelinek statistics and bivariate-Fisher statistics is revealing (Fig. 11.20b). Overall, there is little difference for the confidence cones around the tensor-mean k_{MIN}, largely because the specimen-level errors were generally smaller than for other axes, and almost circular. However, for k_{MAX} and k_{INT}, the Jelinek confidence cones are very large and very elliptical, to the extent that they overlap, implying an orientation-distribution tensor for AMS that is almost perfectly oblate. Considering all specimens to have equal precision (= equal weight), the unweighted bivariate-Fisher statistics yield k_{MAX} and k_{INT} confidence cones that isolate the tensor-mean axes more precisely, perhaps defining a sedimentary "current-alignment" lineation by tensor-mean k_{MAX}. However, the bivariate-Fisher model permits the inclusion of specimen-level uncertainty; when this is done, k_{MAX} and k_{INT} show slightly different orientations and smaller cones of confidence. As we discussed in connection with paleomagnetic vectors treated by bivariate-Fisher statistics (Chap. 10), the weighting may involve a wide range of uncertainty factors, not merely the errors of measurement of individual specimen tensors. For example, the mean value of the tensor, as well as attributes like location, rock type and metamorphic grade, could all be incorporated into an expression of confidence. In terms of fabric interpretation, especially concerning the role of competing subfabrics, Le Goff and Henry's approach may be useful in that each confidence cone is not constrained by the orthogonal symmetry of the mean axes.

CHAPTER 12

Appendix

12

Appendix I: Errors in Compound Quantities: Propagation of Error

If we measure *a* with an error of observation Δa, and *b* with an error of observation Δb, what will be the effect of these errors on *a quantity Q, where Q is some simple expression of the observed values, for example Q = a + b?* This topic is described as the confounding or propagation of errors and is discussed in many textbooks concerning applied statistics and experimental measurements (e. g. Topping 1965). First, let us consider how errors propagate through simple arithmetic operations that involve values with errors of observation or measurement. Note that the fractional error in *a* is given by $f = \Delta a/a$. Subsequently, more general situations will be mentioned, including the manner in which variances of samples of observations influence the variance of some derived quantity.

Multiplication of Quantities with Errors of Observation

If $Q = a.b$ and f_1 and f_2 are the *fractional* errors in a and b then $Q = a_o(1 + f_1) \times b_o(1 + f_2)$ or $Q = a_o b_o(1 + f_1 + f_2 + f_1 f_2)$ and $f_1 f_2$ may be neglected if both fractional errors, f_1 and f_2 are suitably small for the project in question, thus: $Q \cong a_o b_o(1 + f_1 + f_2)$

Note that the fractional error in the product is given approximately by the sum of the fractional errors in the observations.

Division of Quantities with Errors of Observation

$$Q = a / b$$

$$= \frac{a_o(1 + f_1)}{b(1 + f_2)} = \frac{a}{b}(1 + f_1)(1 - f_2 + f_2^2 - f_2^3 \ldots)$$

In the binomial expansion of the denominator we may neglect $f_2^2 >$ and higher powers, if f_2 is small enough. Then

$$Q \approx \frac{a}{b}(1 + f_1 - f_2)$$

Sum (or Difference) of Quantities with Errors of Observation

If *a* and *b* have fractional errors, Δa and Δb, what is the error *e* in a compound quantity $Q = a \, \Delta b$? It may be shown that the most likely confounded error is given by $e = \sqrt{(\Delta a^2 + \Delta b^2)}$. The fractional error in *Q* is then given by (e/Q).

Multiple Sources of Error, in General

In general, we may meet with a compound quantity *Q* that is any mathematical function *F* of observations *x, y, z,* etc., whose mean values are each subject to *standard* error $\Delta x, \Delta y, \Delta z$ etc. The *standard error ΔQ* of the compound quantity *Q* may be obtained in general from:

$$(\Delta Q^2) = \left(\frac{\partial F}{\partial x}\right)^2 \Delta x^2 + \left(\frac{\partial F}{\partial y}\right)^2 \Delta y^2 + \left(\frac{\partial F}{\partial z}\right)^2 \Delta z^2 + \ldots$$

For mean values, *x, y,* etc., with standard errors $\Delta x, \Delta y$, the standard error ΔQ of some common compound quantities (*Q*) is given as follows:

Compound quantity Q:	Standard error ΔQ given from:
$x \pm y \pm z \ldots$	$\sqrt{(\Delta x^2 + \Delta y^2 + \Delta z^2 \ldots)}$
$xyz \ldots$	$\left(\frac{\Delta Q}{Q}\right)^2 = \left(\frac{\Delta x}{x}\right)^2 + \left(\frac{\Delta y}{y}\right)^2 + \left(\frac{\Delta z}{z}\right)^2 + \ldots$
x^n	$nx^{n-1}\Delta x$

Confounding Contributions of Variances (Chapters 2 and 7)

Suppose that a variable quantity Q is a function of some other variables, x_1, x_2, whose variances (var x_1 etc.) are known. What then is the expected variance of Q? An approximation exists such that:

$$\text{var}(Q) = \sum_{i=1}^{n} \left(\frac{\partial Q}{\partial x_i} \right) \text{var}\, x_i + \sum_{i=1}^{n} \sum_{j=1}^{n} \left(\frac{\partial Q}{\partial x_i} \frac{\partial Q}{\partial x_j} \right) \text{cov}\, x_i y_i$$

If the variables are not co-dependent, the covariance is zero and the second term disappears, leaving:

$$\text{var}(Q) = \sum_{i=1}^{n} \left(\frac{\partial Q}{\partial x_i} \right) \text{var}\, x_i$$

which may be applied to common simple relationships as follows, using the conventional symbol σ^2 for variance, where σ is the standard deviation:

Function, U	Variance of function, (σ_U^2)
$U = cx$	$\sigma_U^2 = c^2 \sigma_x^2$
$U = x_1 \pm x_2 \pm \ldots$	$\sigma_U^2 = \sigma_1^2 + \sigma_2^2 + \ldots$
$U = x_1 x_2$ or $U = x_1/x_2$	$\left(\dfrac{\sigma_U}{U} \right)^2 = \left(\dfrac{\sigma_1}{x_1} \right)^2 + \left(\dfrac{\sigma_2}{x_2} \right)^2 + \ldots$

Appendix II: Notes on the Manual Use of Stereograms

Most students in the earth sciences have exposure to stereographic projection of orientation data, in mineralogy, structural geology and geophysics. This has been greatly facilitated by the use of computer programs that accelerate the rate of work, reduce errors and provide consistency in aspects that require judgement, notably the contouring of density distributions (Diggle and Fisher 1985; Robin and Jowett 1996; Starkey 1996a,b). One may even adapt the Excel spreadsheet software to plot stereonets (Tolson and Correa-Mora 1996).

Unfortunately, some universities have abandoned manual exercises completely. However, without some hands-on practice, it is quite difficult to appreciate the concepts and pitfalls involved, particularly regarding aspects of projection, distortion, rotation axes, and interpretation of density contours. Moreover, some computer programs may not be particularly easy to adapt to specific geometrical questions such as what rotation

axis could explain a distribution and what angle lies between two axes. It should also be noted that the accuracy of manually plotted stereograms suffices for all presentations and for almost all calculations or estimations in field geology. Many structural geology textbooks have short introductions, or appendices explaining the manipulation of stereonets (e.g. Ragan 1973; Hobbs et al. 1976; Nicolas 1987; Ramsay and Huber 1987; Lisle 1988) but some excellent old textbooks are available in libraries (e.g. Phillips 1960), including one that has a useful statistical component combined with an evaluation of many aspects of orientation diagrams (Vistelius 1966). To the best of the author's knowledge the most comprehensive and most modern account of the manual use of such projections is by Priest (1985), which is recommended.

Plotting Lines, Planes and Poles to Planes

Terminology for specifying the orientation of lines (axes or vectors) and planes is discussed fully in Chapter 10, but is simplified and repeated here in Fig. AII.1. A linear feature such as an axis (e.g. mineral lineation, fold axis, maximum magnetic susceptibility axis) requires only a trend and a plunge for its specification. It may always be plotted on one hemisphere and the lower hemisphere is most common in structural geology and geophysics. The upper hemisphere is used in crystallography and mineralogy. Vectors, such as paleomagnetic remanence, may occur on either hemisphere and thus their plunge may take a sign, negative for upwards-directed vectors and positive for downwards-directed vectors.

The orientation of the lineation (Fig. AII.1a) may be represented by the point at which the lineation touches the hemisphere. In plan projection this becomes a dot.

Planes may be represented by their intersection with the sphere along a great circle. (A great circle has the same diameter as the stereogram.) In projection, this curve is called the cyclographic trace (Fig. AII.2b). Such curves become cumbersome where many data are plotted, and their use is usually reserved for determinations of intersections of planes, etc. More commonly, planes are recorded in terms of the projections of their normals or poles. These plot as dots on the hemisphere on the opposite side of the net towards which the pane actually dips.

Consider the manual plotting of an axis (= lineation) of arbitrary orientation as shown in Fig. AII.2a. Suppose the axis has a trend and plunge of 065/60. Firstly, a tracing paper is fixed so that it is free to rotate about a thumbtack in the centre of the net. Secondly, a north arrow is marked on the tracing paper, to coincide

Plotting axes (lineations)

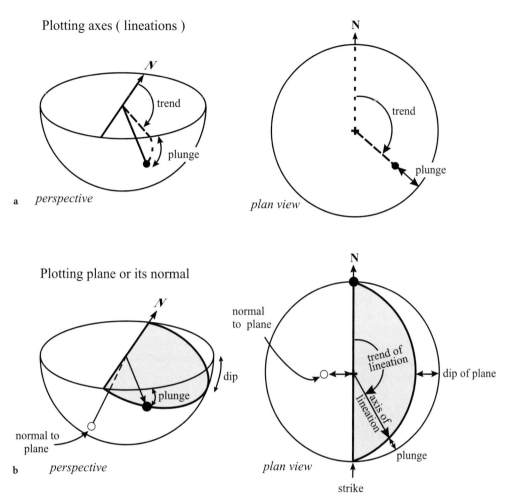

a *perspective*   *plan view*

Plotting plane or its normal

b *perspective*   *plan view*

Fig. AII.1. Perspective and plan views of the spherical projection (stereogram) **a** of an axis and of **b** a plane. Trend and plunge angles define the orientation of an axis, which is shown in both **a** and also in **b** where it lies in the plane. A plane requires careful specification of its strike or dip direction, as discussed in the text (Chap. 10, Fig. 10.5), plus the dip angle (maximum inclination). The plane is usually plotted in terms of its normal for statistical purposes; the cyclographic trace is usually preserved for calculations concerning the intersections of planes

with the top of the net. Thirdly, we mark the trend of 065° on the perimeter of the net. The tracing paper is then rotated until the 065° mark lies parallel to one of the straight lines with a linear scale on the stereonet template, either the north-south, or east-west line. In this example it was most convenient to use the east-west line. Now, we may take advantage of the linear scale of the template to count in 60° from the edge of the net to mark the point representing the plunge. Finally, the tracing paper is restored to its original orientation with North arrow at the top: the dot now shows the correct projection of the axis 065/60 (Fig. AII.1b).

Similarly, we may manually plot the orientation of a plane as shown in Fig. AII.2c. Suppose the strike is 315° and it dips 40° to the NE. Firstly, a north arrow is marked on the tracing paper as a reference. Secondly, a mark is made on the tracing paper corresponding to 315° on the underlying net. Thirdly, the tracing paper is now rotated until this mark is aligned with the north arrow of the underlying net. Fourthly, counting in from the east mark on the underlying net we locate the great circle which dips 40°. This great circle may now be traced as the cyclographic plot of the plane: it represents the plan-view projection of the intersection of the plane with the lower hemisphere (see Fig. AII.2d). The cyclographic trace is a useful visual presentation in the learning stages and it may be useful later when one needs to fix the orientation of the intersection of planes, or the orientation of linear features lying in the plane. However, the normal to the plane, summarises

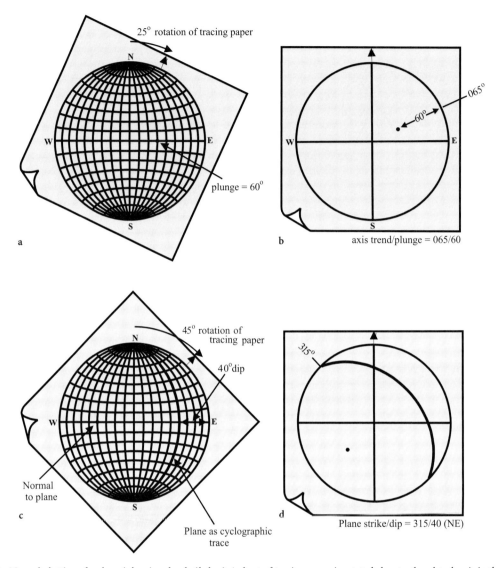

Fig. AII.2. Manual plotting of a plane (*above*) and pole (*below*). A sheet of tracing paper is rotated about a thumbtack axis in the centre of the net

all the necessary information about the plane's orientation in a single point. This is less cumbersome and it is the best presentation that allows us to easily estimate the mean orientation and dispersion of the planes from density contours. The pole is located on the opposite side of the net centre from the direction of (maximum) dip of the plane. It may be located by counting 90° along the EW line, from the maximum dip of the plane in Fig. AII.2c and marking the pole as a point. Alternatively, one may count around the perimeter of the net, 90° from the strike and then count in (90° – dip), in this case = 60° along a linear net scale.

Restoring the tracing paper to geographic coordinates, with "north" at the top, reveals the cyclographic trace and the pole of the plane, which is recorded with a strike and dip using the notation 315/40 NE (Fig. AII.2d). Finally, with regard to cyclographic plots of planes, the intersections of such planes are not all significant in a structural sense. For example, bedding planes measured around a fold should intersect to define the fold axis (e.g. consider Fig. 10.11). However, most intersections of cyclographic traces will be spurious: for n planes there will be $(1/2n(n-1))$ intersections, not just n.

Great Circles, Small Circles; Dip Plunge and Rake

Bearing in mind also that stereograms will normally be prepared by computer software, such as Spheristat (Pangaea Scientific), only a few further comments will be made concerning the use of spherical projections inasmuch as they affect interpretation and the appreciation of the underlying concepts. The first point, which may seem trivial, but can cause confusion when we consider rotations of axes or normals (below). This concerns the careful distinction between the use of small circles and great circles. As a readily comprehensible analogy, the earth's lines of longitude are great circles; they have the same diameter as the sphere on which they are drawn. In contrast, the earth's lines of latitude are small circles, with the exception of the equator, because they have diameters less than that of the sphere. To avoid confusion, the stereonet template only has one set of small circles and one set of great circles. In the most common examples, small circles are coaxial with the N-S line, and great circles all intersect at the north and south poles of the stereonet template. Other templates are available, notably the "polar" projection in which one looks down on the "north pole" of the net, with small circles appearing as circles about the centre and great circles all appearing as straight, ra-

dial lines (Fig. 10.4, inset). However, the version shown here is most popular and useful. The most important distinction that must be made between small and great circles is shown in Fig. AII.3a. The plunge of an axis, the angle of depression or inclination from the horizontal, must be measured along a radial great circle, following the shortest straight-line distance from the edge of the net to its centre. To accomplish this, it is always necessary to rotate the tracing paper so that the shortest straight-line distance lies above either an E-W or N-S net axis, because these are the only linear great-circle scales available. In contrast, the small-circle distance from the point to the edge of the net is not the plunge (Fig. AII.3a).

A further note on the manipulation of data on a stereonet concerns the use of cyclographic traces. A few computer programs fall short on this aspect but the cyclographic trace does have some practical use and more value in terms of instruction. The intersection of two planes defines a lineation and consequently, the intersection of cyclographic traces defines the trend and plunge of that feature (Fig. AII.3b). Moreover, the angle between a linear feature in a plane and the horizontal (= strike of the plane) is sometimes useful as it is a readily measured item in the field, e.g. striations on bedding, slickensides on a fault. This is referred to as the rake (Fig. AII.3b).

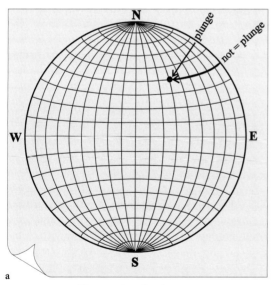

Plunge given by shortest distance to net-perimeter

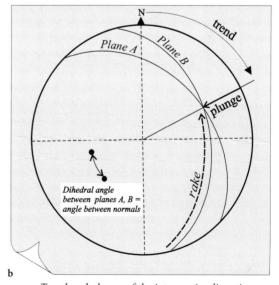

Trend and plunge of the intersection lineation formed by two planes

Fig. AII.3. Some points of attention in the initial manual use of stereograms. **a** Plunge of an axis (or normal) must be measured radially on the net. **b** Rotations of axes or poles must be accomplished using small circles. The tracing paper may be manipulated to bring the point onto the appropriate small circle. Multiple rotations may be accomplished in sequence, but their order is not arbitrary, and different results will be achieved due to the non-commutative nature of rotations. **c** The intersection of two planes defines an axis, whose trend and plunge are shown. The dihedral angle between the planes is defined by the inter-pole angular separation

Rotations: for Convenient Viewing or for Tectonic Restorations

Although the stereogram is a powerful visual presentation of orientation data, there may be three reasons for wishing to change the frame of reference of the orientations from the *geographical coordinates,* in which they are originally recorded and presented (i.e. N, E, S, W, up, down). Broadly these are referred to as "rotations" because they refer to spinning the data about some hypothetical axis to a new position on the hemisphere. The first reason for rotating the data may be to view concentrations from a less biased perspective point. For example, if the orientations of axes are subhorizontal, the maximum concentration will be divided on diametrically opposed sides of the stereogram. *Centering* data permits a clear appreciation of the orientation distribution, e.g. whether the maximum is circular or elongated. An example of this was shown in Fig. 10.16. A second reason to rotate orientations from their geographical frame of reference is that there may be a more logical frame of reference for the purposes of study. The geographical reference may be less important than internal structural markers. In petrofabrics and petrophysics, the actual geographic orientations of mineral-preferred orientations may be of less scientific interest than their association with other frames of reference, for example structural features such as schistosity or a fault orientation. Therefore, one may wish to turn the data set until it is viewed parallel to some important structural feature. An example is shown in Fig. 11.11: it is more important to examine the preferred mineral orientations with respect to cleavage and mineral lineation rather than to the original geographic coordinates. Thus, these data are plotted with cleavage horizontal and mineral lineation "north-south".

The third reason for replotting orientations in a new frame of reference is more fundamental. We may believe or suspect that the measured elements *were physically rotated* by some geological process and we need to undo this procedure in order to evaluate the original orientations. This is of major concern in paleomagnetism because the paleomagnetic vectors have commonly been disturbed tectonically. Without an accurate correction for later disturbances, we cannot determine the position of the ancient magnetic pole or make sensible comments on paleolatitude and plate movements. We must be careful to note that back-rotating data on the stereonet can only undo rigid-body tilts of orientations. If the feature in question was subject to shape change, a distortion known to structural geologists as finite strain, much more complicated and

unreliable corrections are required (Borradaile 1997). In general, the rigid body rotations that may be "untilted" by stereographic rotations are limited to tilted, or gently folded strata, fault blocks, micro-plates rotated about vertical axes and other non-penetratively strained geological units. The principle behind using the stereonet for rotations is that the tracing paper must be manipulated so that the data point to be restored lies on a small circle with the required axis of rotation lying above the N-S line of the underlying stereonet template (Fig. AII.4a). The rotation is then accomplished by replotting the point at some position ω from its original orientation, where ω is the angle of rotation. Clearly, this is trivial in the example illustrated because the rotation axis is already horizontal: in practice, the rotation axis will be inclined so that, together with the data, it must be rotated first to an intermediate stage where the rotation axis is horizontal. Then tracing paper with "horizontalised" data may be turned until the rotation axis lies above the north-south axis of the stereonet. Next, the required rotation may be applied about that axis (comparable to the one shown in Fig. AII.3a). Finally, the data must be restored to its original frame of reference by turning the rotation axis, *and all the rotated data,* back to the original frame of reference. *Manually,* the "bookkeeping" is quite tedious and subject to errors. Apart from that, there are two common pitfalls. *Firstly,* the rotations are sometimes not applied consistently "looking outwards" along the rotation axis from the centre of the net. (The convention for rotations or *axial vectors is* that rotations are considered looking outward from the centre of the net and positive rotations are clockwise.) If once is inconsistent, careless clockwise and anticlockwise corrections can be indiscriminately applied to different parts of the data set, in different parts of the diagram. *Secondly,* there is a fundamental issue that is neither trivial nor normally resolvable. Successive rotations are processes whose order cannot be interchanged indiscriminately. Consider the general case of two rotations about different axes and by different angles, let them be R_1 and R_2: it is easy to demonstrate that applying the rotations R_1 followed by R_2 to an orientation produces a different result from applying the sequence R_2 followed by R_1 (Fig. AII.4c, d). The different results are shown in the stereogram of Fig. AII.4b. That rotations are non-commutative is recognised in the structural and paleomagnetic literature, but there still occur cases where it is unfortunately forgotten or merely paid lip-service. In applications to real situations the problem becomes more complex (MacDonald 1980; Borradaile 1997, 2001). Although any general rotation can be decomposed into component rotations,

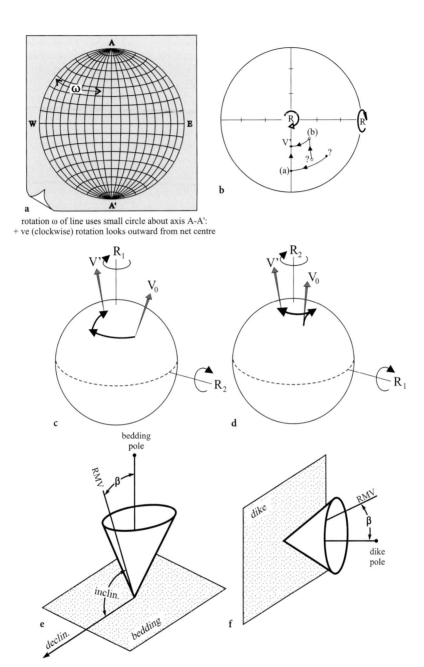

rotation ω of line uses small circle about axis A-A':
+ ve (clockwise) rotation looks outward from net centre

Fig. AII.4 Rotations may be part of geometrical procedures for viewing the data in different directions to perform calculations (**a, b**); or they may simulate natural tilting events (**e, f**) that exclude internal distortion (strain or "deformation). **a** Calculations involving manual rotations require the manipulation of the point (= line) so that it lies above a small circle with the axis of rotation horizontal and above the N-S line of the stereonet template. To achieve this, it may be necessary first to rotate the point *about some other axis* to bring it into a position so that the final desired rotation axis is as shown. *Unfortunately*, the sequence of rotations cannot be arbitrarily chosen. Thus, rotations are *non-commutative*; a rotation about R₁ followed by R₂ (**c**) produces a different result from the sequence R₂ followed by R₁ (**d**). Where real geological rotations have occurred, we commonly wish to restore the rocks to their original orientation. In paleomagnetism we may wish to untilt beds (**e**) or dykes (**d**) to determine the original orientation of their remanent-mean vector (RMV). However, even in this restricted set of possible circumstances, assuming the beds were originally horizontal (**e**) or the dykes vertical (**f**), it is clear that ambiguity arises. The RMV could lie anywhere along the cones shown: for restored beds (**e**), the declination is indeterminate. For dykes, neither the inclination nor the declination can be fixed (**f**)

we do not know the actual rotation or the rotation axis in many situations and it is that axis that we should be using to restore the data to a pre-tectonic configuration. It may be quite wrong to assume that beds were simply tilted about a horizontal axis; it may be better to use a full structural analysis to determine the real geological rotation axis, rather than make assumptions about component rotations, and still further assumptions about the sequence in which they occurred. Even when we can assume that tilted beds may be restored about a horizontal rotation axis, there will always exist ambiguity about the declination, also called azimuth or trend, of some simultaneously restored linear feature (consider Fig. AII.4e). Attempts to restore dykes to an assumed original vertical orientation about a horizontal axis result in an ambiguity of *both declination and inclination* of any integral linear feature, such as a paleomagnetic vector (Fig. AII.4f; Borradaile 2001). If this is not bad enough, we must resist confusing componental rotations chosen for convenience of geometrical reconstructions, with real rotation events in geological history. Accepting that a natural rotation need not occur about a unique, arbitrarily oriented axis, we may readily accept that rotations involved in heterogeneously deformed bodies may have occurred along multiple, ephemeral rotation axes, and any restoration may descend to the realms of guesswork.

Contouring

We understand topographic maps, mostly intuitively, as the representation of the elevation of a surface at different (x, y) coordinates. Lines joining points of equal elevation are thus topographic contours. Contours of equal values may be extended to other kinds of "map", for example contours of ore grade across a map, or of magnetic field strength, are fairly well grasped. However, some care is required in the construction of all contoured maps because the original data are normally made at discrete points. We must interpolate between points to produce a sufficient data density for contours and we must draw those contours at appropriate intervals, requiring us to interpolate certain integer values that will provide useful contour intervals. The simplest kind of contour is developed from a linear interpolation between values. Thus, if two adjacent discrete sample points have values z_1 and z_2, we may interpolate a value $z' = (z_1 + z_2)/2$, mid-way between the two samples. More sophisticated methods exist and are usually reserved for computer program methods; interpolated or extrapolated values may be "weighted" according to their distance from the actual sampled values. The otherwise linearly estimated value is multi-plied by a weighting factor. One logical weighting factor is to use a factor that is inversely proportional to distance from the site where a value is to be interpolated.

Below, we consider simplified manual techniques for contouring the density of orientations on stereograms. This may be useful for students who have just become familiar with the study of orientations, but is also useful for us to appreciate how published density contours were generated in the past. Contouring the density of orientations is a much more complicated problem than contouring scalar values on a map and, consequently, it is performed much better by computer than by hand. Computer programming of these procedures is difficult, but several good packages are available, especially Spheristat by Pangaea Scientific, and code and algorithms are available for most basic aspects of contouring (e.g. Diggle and Fisher 1985; Robin and Jowett 1986; Starkey 1983, 1993, 1996a). We shall focus on obtaining the values that we wish to contour. The orientations may be axes, vectors, or normals to planes, contoured per unit of orientation space (i.e. per solid angle).

The first item of concern is the choice of hemispherical projection. There are two common practical choices for the representation of an angular cone that would contain some concentration of orientations that we wish to plot. One may preserve the area of similar conical projections but sacrifice their shape; this is the so-called "equal-area" projection (Fig. AII.5a). Alternatively, we may preserve inter-angular relations between the orientations but grossly distort the areas of similar angular cones on the projection; this is the equal-angle projection (Fig. AII.5b). Clearly, in contouring densities' orientation distributions, we prefer the "equal-area" option because a concentration of ten axes per angular interval would appear equally concentrated wherever that interval fell on the projection. Of course, at the perimeter, the density of points in any interval would be "stretched" concentric with the perimeter, as shown by the shape of the elliptical shaded elements in (a). Had we chose projection (b), the density of a similar number of axes per shaded circle would appear much less at the perimeter of the projection. Vistelius (1966) recommended a different projection, the Kraviskii net, which provided a compromise. This was not widely adopted; it would require considerable encouragement to promote its use, and the diagrams would be difficult to compare with the many existing, published diagrams which are mostly equal-area diagrams in structural geology, tectonics and geophysics.

The essence of the contouring procedure is to count how many observations lie in a given angular interval

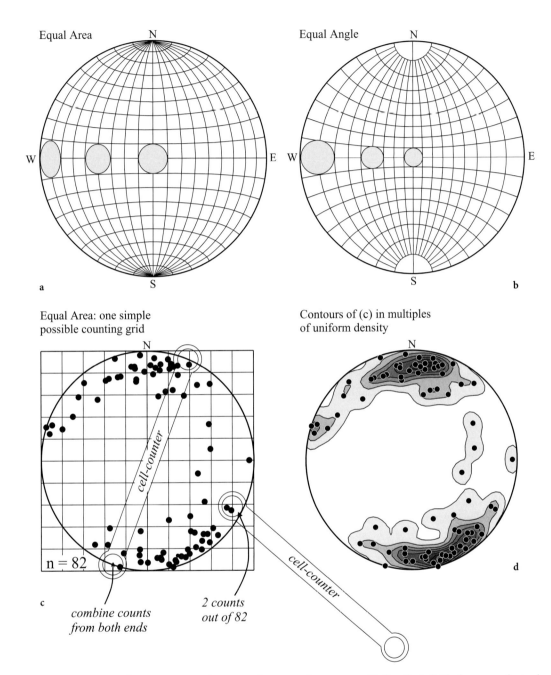

Fig. AII.5. Density contours of stereograms reveal patterns of concentrations more readily than the individual axes, or poles to planes.
a The equal-area projection is preferred in contouring because areas subtended by the same solid angle on the hemisphere have the same relative areas on the stereogram, despite progressive elliptical distortion toward the equator. Thus data are compressed radially, towards the perimeter, a point worth noting when inspecting contours. **b** The equal-angle net has unacceptable areal distortion for statistical work. **c** One simple, manual contouring procedure places a grid over the net, and the number of points per cell is determined as a fraction of the total counts. Most usefully, this density is normalised by comparing it to the density expected, if the same number of data were distributed uniformly over the sphere. The subject is quite complex and this diagram serves only to illustrate the basic principle: it does not do justice to the actual manual or computer procedures. Even manually, a finer grid, or "free-floating" counting cells, may be used. Manual techniques commonly keep the counting cell at 1% of the stereogram's area; computer methods decrease the size of the counting cell as the number of data increase. **d** the actual contours shown here were produced by the Spheristat program (Pangaea Scientific) using a cell size one fifth that shown in **c**; contours are in multiples of the expected uniform concentration

of the stereonet and record that number in the centre of that cell for subsequent contouring (Fig. AII.5c). As long as we use an "equal-area" projection, a square counting grid should suffice to record the data density. For simple manual contouring, the grid is normally set with cells that have ~1% of the stereonet area. If there were 200 observations in total and 20 fell in a single 1% square, its density would then be reported as 10% (= 20/200) per 1% unit area. This is the traditional nomenclature for approximate contouring by hand, and almost all data published before 1985 was produced in this way. After reading Chapter 10, the shortcomings of this rough-and-ready approach will be apparent. For serious work, of course, there is now no substitute for computer contouring.

In computer-generated plots, several important advantages arise. First, the counting grid may be adjusted to suit the number of observations (n). This optimises the smoothness and spacing of contours. Second, the concentrations are normally expressed as a multiple of the concentration expected (U) if there was a uniform distribution of the same number of orientations over the hemisphere. This normalises stereo plots so that plots differing in the number of observations and degree of preferred orientation may be readily compared. All the contoured plots in this book are presented this way, with contours at e.g. 2U, 4U, etc. (Fig. AII.5d).

In the older literature, most plots were generated manually in the following manner: The illustration (c) shows a coarse fixed grid. In practice, manual plots would not normally use a finer grid, but the *same grid* could be displaced, e.g. half-a-space left to produce another set of counts, then half-a-space down, etc. to produce finer spacing and smoother contours. It is completely valid to repeat the contouring procedure as many times as required, with a slightly displaced grid.

In this manner, a large number of counts would be generated, facilitating the construction of "eye-balled" contours. Another method of manual counting uses a mobile overlay circular cell of 1% of the stereonet area. This cell is moved over the net, counting the number of observations falling within its area. That count is then recorded at the centre of the cell. The circular counting cell may be moved to regular grid intersections, as in (a), or it may be moved freely over the net. Preferably both should be done; the grid-centered counts first, then free counting at the edges, and at places where interpolation needs improvement. Contrary to the impression given by some elementary manuals, there is no restriction on the choice or number of counting positions, as long as the cell size is constant. In fact, in manual contouring, it may be more efficient to use the "free-counter" method, moving the counting cell to any strategic position on the stereonet, in order to refine the contours.

One cautionary note is that any counting cell that straddles the net's perimeter must include counts from the diametrically opposed remainder of that cell projected onto the opposite side of the net. This is clear from the stationary grid method (c), but care must be taken with free-counter methods (usually one uses a pair of cells on either end of a transparent template that spins about the net centre "thumbtack"). As a final check of the validity of any set of contours, one should note that a contour touching the net perimeter must reappear on the opposite side of the net, approaching the net perimeter from the opposite direction. Furthermore, we should note that by whatever manual or computational methods the contours are produced with the equal-area net, the data concentration is always "stretched" around the edge of the net as one gets closer to the edge of the net (a).

CHAPTER 13

References

Recommended Textbooks for Earth-Science Statistics

(a) Recommended General Introductory Texts in Approximate Order of Increasing Difficulty

McGrew, JC , Monroe, CB (2000) An introduction to statistical problem solving in geography, 2nd edn. McGraw-Hill, Boston, 254 pp

Cheeney RF (1983) Statistical methods in geology. George Allen and Unwin, London, 169 pp

Till R (1974) Statistical methods for the earth scientist. MacMillan Press, London, 154 pp

Davis JC (1973) Statistics and data analysis in geology. Wiley, New York, 550 pp

Davis JC (2002) Statistics and data analysis in geology, 3rd edn. Wiley, New York, 638 pp

Marsal D (1987) Statistics for geoscientists. Pergamon Press, Oxford, 176 pp

(b) More Advanced Statistical Treatments Directed Toward Earth Science

Agterberg FP (1974) Geomathematics. Elsevier, Amsterdam, 596 pp

Chayes F (1971) Ratio correlation: a manual for students of petrology and geochemistry. University of Chicago Press, Chicago, 99 pp

Devillers J, Karcher W (1991) Applied multivariate analysis in structure-activity relationships (SAR) and environmental studies. Kluwer, Dordrecht, 530 pp

Gill R (1997) Modern analytical geochemistry. Longman Press, London, 329 pp

Koch GS, Link RF (1970) Statistical analysis of geological data, vol 1. Wiley, New York, 375 pp

Koch GS, Link RF (1971) Statistical analysis of geological data, vol 2. Wiley, New York, 438 pp

Krumbein WC, Graybill FA (1965) An introduction to statistical methods in geology. Wiley, New York

Le Maitre RW (1982) Numerical petrology: statistical interpretation of geochemical data. Elsevier, Amsterdam, 281 pp

Merriam DF (1970) Geostatics. Plenum Press, New York, 177 pp

Miller RL, Kahn JS (1962) Statistical analysis in the geological sciences. Wiley, New York, 483 pp

Panofsky HA, Brier GW (1965) Some applications of statistics to meteorology. University of Pennsylvania Press, Philadelphia, 224 pp

Reyment RA (1971) Introduction to quantitative palaeoecology. Elsevier, Amsterdam, 226 pp

Walden AT, Guttorp P (1992) Statistics in the environmental and Earth sciences. Edward Arnold, London, 306 pp

(c) Textbooks on Spatial Distributions, Especially "Geostatistics"

Armstrong M (1998) Basic linear geostatistics. Springer, Berlin Heidelberg New York, 153 pp

Bartlett MS (1975) The statistical analysis of spatial pattern. Chapman and Hall, London, 90 pp

Clark I (1979) Practical geostatistics. Applied Science Publishers, London, 129 pp

Clark I, Harper WV (2000) Practical geostatistics 2000. Ecosse North America, Columbus, OH, 342 pp

Cliff AD, Ord JK (1973) Spatial autocorrelation. Pion Ltd, London, 178 pp

Cressie NAC (1993) Statistical analysis for spatial data. Wiley, New York, 900 pp

Goovaerts P (1997) Geostatistics for natural resources evaluation. Oxford University Press, Oxford, 483 pp

Gregory S (1963) Statistical methods and the geographer. Longman, London, 271 pp

Isaaks EH, Srivastava RM (1989) Applied geostatistics. Oxford University Press, New York, 561 pp

Journel AG, Huijbregts CJ (1978) Mining geostatistics. Academic Press, London, 600 pp

King LJ (1969) Statistical analysis in geography. Prentice-Hall, Englewood Cliffs, 288 pp

Matheron G (1971) The theory of regionalised variables. Les cahiers de Centre Morphologie Mathématique. Fasc. 5. Paris ENSMP, Paris, 211 pp (Translation in English)

Royle A et al. (eds) (1980) Geostatistics. McGraw-Hill, New York, 168 pp

Wackernagel H (1998) Multivariate geostatistics. Springer, Berlin Heidelberg New York, 219 pp

Webster R, Oliver MA (2001) Geostatistics for environmental scientists. Wiley, Chichester, 271 pp

(d) Textbooks Concerning or Including Statistics of Orientations

Butler RF (1992) Paleomagnetism. Blackwell, Oxford, 319 pp

Fisher NI (1993) Statistical analysis of circular data. Cambridge University Press, Cambridge, 277 pp

Fisher NI, Lewis T, Embleton BJJ (1987) Statistical analysis of spherical data. Cambridge University Press, Cambridge, 329 pp

Mardia KV (1972) Statistics of directional data. Academic Press, London, 357 pp

Mardia KV, Jupp PE (2000) Directional statistics. Wiley, New York, 429 pp

Tauxe L (1998) Paleomagnetic principles and practice. Kluwer, Dordrecht, 299 pp

Vistelius AB (1966) Structural diagrams. Pergamon Press, Oxford, 178 pp

Textbooks for General Statistical Background

(a) Useful Introductions in Approximate Order of Increasing Difficulty

Topping J (1965) Errors of observations and their treatment. Chapman and Hall, London, 119 pp

Backhouse JK (1967) Statistics, an introduction to tests of significance. Longman, London, 197 pp

Barford NC (1967) Experimental measurements: precision, error and truth. Addison-Wesley, London, 143 pp

Sanders DH, Smidt RK, Adatia A, Larson G (2001) Statistics: a first course. McGraw-Hill Ryerson, Toronto, 664 pp plus tables

Chatfield C (1978) Statistics for technology. Chapman and Hall, London, 370 pp

Glantz SA (1997) A primer of biostatistics. McGraw-Hill, New York, 473 pp

Devore J, Peck R (2001) Statistics: the exploration and analysis of data. Duxberry Press, Pacific Grove, CA, 713 pp

Johnson RA (1994) Miller and Freund's probability and statistics for engineers, 5th edn. Prentice-Hall, Englewood Cliffs, 630 pp

(b) More Advanced Texts

Bevington PR (1969) Data reduction and error analysis for the physical sciences. McGraw-Hill, New York, pp 187–195

Collins C (1992) Introduction to multivariate analysis. Chapman and Hall, London, 246 pp

Hicks CR (1993) Fundamental concepts in the design of experiments. Saunders College Publishing, New York, 509 pp

Hoel PG (1971) Introduction to mathematical statistics, 4th edn. Wiley, New York

Kachigan SK (1986) Statistical analysis: an interdisciplinary approach to univariate and multivariate methods. Radius Press, New York, 589 pp

Manly BFJ (1986) Multivariate statistical methods: a primer. Chapman and Hall, London, 159 pp

McClave JT, Dietrich FH, Sincich T (1997) Statistics. Prentice-Hall, Englewood Cliffs, 823 pp

Miller R (1986) Beyond ANOVA: the basics of applied statistics. Wiley, New York

Spiegel MR (1961) Theory and problems of statistics. McGraw-Hill, New York, 359 pp

(c) Time Series

Gouriereux C, Monfort A (1997) Time series and dynamic models. Cambridge University Press, Cambridge

Kanasewich ER (1975) Time sequence analysis in geophysics. University of Alberta Press, Edmonton, Canada, 364 pp

Wiener N (1966) Time series. MIT Press, Cambridge, MA, 163 pp

(d) Bayesian Statistical Approaches

Lee PM (1997) Bayesian statistics: an introduction. Oxford University Press, New York, 2nd edn, 344 pp

Morgan BW (1968) An introduction to Bayesian statistical processes. Prentice Hall, Englewood Cliffs, 116 pp

Smith JQ (1988) Decision analysis: a Bayesian approach. Chapman and Hall, London, 138 pp

Spall JC (ed) (1988) Bayesian analysis of time series and dynamic models. Marcel Dekker, New York, 536 pp

Scientific Articles and Some Books Cited as References or Sources of Examples

Aitken MJ (1990) Science-based dating in archaeology. Longman, New York, 274 pp

Angelakis AN, Issar AS (eds) (1996) Diachronic impacts of water resources. Springer, Berlin Heidelberg New York

Barbetti M (1983) Archeomagnetic results from Australia. In: Creer KM, Tucholka P, Barton CE (eds) Geomagnetism of baked clays and recent sediments. Elsevier, Amsterdam, pp 173–175

Barrett TJ, Fralick PW (1989) Turbidites and iron formations, Beardmore-Geraldton, Ontario: application of a combined ramp fan model to archean clastic and chemical sedimentation. Sedimentology 36:221–234

Barton CE (1983) Analysis of paleomagnetic time series: techniques and applications. Geophys Surv 5:335–368

Batt CM (1997) The British archeometric calibration curve: an objective treatment. Archeometry 39:153–168

Bingham C (1964) Distributions on the sphere and on the projective plane. PhD Thesis, Yale University, New Haven, Connecticut

Blackwell GH, Johnston TG (1986) Short- and long-term open-pit grade control. In: David M, Sinclair AJ, Vallee M (eds) Ore reserve estimation: methods, models and reality. CIMM Montreal, Canada, pp 108–129

Blatt H, Tracy RJ (1996) Petrology: igneous, sedimentary and metamorphic. WH Freeman, New York, 529 pp

Borradaile GJ (1974) Bulk finite strain estimates from the deformation of neptunian dykes. Tectonophysics 22:127–139

Borradaile GJ (1976) A study of a granite/granite-gneiss transition and accompanying schistosity formation in SE Spain. J Geol Soc Lond 132:417–428

Borradaile GJ (1979a) Pretectonic reconstruction of the Islay Anticline: implications for the depositional history of Dalradian rocks in the SW Scottish Highlands. In: Harris AL, Holland CE, Leake BE (eds) The Caledonides reviewed. Special Publication no 8, Geol Soc of London, London, pp 229–238

Borradaile GJ (1979b) Strain study of the Caledonides in the Islay region, SW Scotland: implications for strain histories and deformation mechanisms in greenschists. J Geol Soc Lond 136:77–88

Borradaile GJ (1982) Tectonically deformed pillow lava as an indicator of bedding and way-up. J Struct Geol 4:469–479

Borradaile GJ (1987a) Analysis of strained sedimentary fabrics: review and tests. Can J Earth Sci 24:442–455

Borradaile GJ (1987b) Anisotropy of magnetic susceptibility: rock composition versus strain. Tectonophysics 138:327–329

Borradaile GJ (1988) Magnetic susceptibility, petrofabrics and strain. Tectonophysics 156:1–20

Borradaile GJ (1991a) Correlation of strain with anisotropy of magnetic susceptibility (AMS). Pure Appl Geophys 135:15–29

Borradaile GJ (1991b) Remanent magnetism and ductile deformation in an artificially deformed magnetite–bearing limestone. Phys Earth Planet Interiors 67:362–373

Borradaile GJ (1994) Paleomagnetism carried by crystal inclusions: the effect of preferred crystallographic orientations. Earth Planet Sci Lett 126:171–182

Borradaile GJ (1996a) Experimental stress remagnetization of magnetite. Tectonophysics 261:229–248

Borradaile GJ (1996b) An 1800–year Archeological experiment in remagnetization. Geophys Res Lett 23:1585–1588

Borradaile GJ (1997) Deformation and paleomagnetism. Surv Geophys 18:405–435.

Borradaile GJ (1999) Viscous remanent magnetization of high thermal stability in limestone. In: Tarling DH, Turner P (eds)

Palaeomagnetism and diagenesis in sediments. Geological Society London, Spec Publ 151, pp 27–42

Borradaile GJ (2001) Paleomagnetic vectors and tilted dikes. Tectonophysics 333:417–426

Borradaile GJ, Brann M (1997) Remagnetization – dating of Roman and mediaeval masonry. J Archeol Sci 24:813–824

Borradaile GJ, Dehls JF (1993) Regional kinematics inferred from magnetic subfabrics in Archean rocks of northern Ontario, Canada. J Struct Geol 15:887–894

Borradaile GJ, Henry B (1997) Tectonic applications of magnetic susceptibility and its anisotropy. Earth Sci Rev 42:49–93

Borradaile GJ, Kukkee KK (1996) Rock–magnetic study of gold mineralization near a weakly deformed Archean syenite, Thunder Bay, Canada. Explor Geophys 27:25–31

Borradaile GJ, Lagroix F (2000) Magnetic characterization of limestones using a new hysteresis projection. Geophys J Int 141:213–226

Borradaile GJ, Mothersill JS (1984) Coaxial deformed and magnetic fabrics without simply correlated magnitudes of principal values. Phys Earth Planet Interiors 35:294–300

Borradaile GJ, Poulsen KH (1981) Tectonic deformation of pillow lava. Tectonophysics 79:T17–T26

Borradaile GJ, Spark RN (1991) Deformation of the Archean Quetico–Shebandowan subprovince boundary in the Canadian Shield near Kashabowie, northern Ontario. Can J Earth Sci 28:116–125

Borradaile GJ, Stupavsky M (1995) Anisotropy of magnetic susceptibility: Measurement schemes. Geophys Res Lett 22:1957–1960

Borradaile GJ, Werner T (1994) Magnetic Anisotropy of some phyllosilicates. Tectonophysics 235:233–248

Borradaile GJ, MacKenzie A, Jensen E (1991) A study of colour changes in purple–green slate by petrological and rock–magnetic methods. Tectonophysics 200:157–172

Borradaile GJ, Puumala M, Stupavsky M (1992) Anisotropy of complex magnetic susceptibility (ACMS) as an indicator of strain and petrofabric in rocks bearing sulphides. Tectonophysics 202:309–318

Borradaile GJ, Stewart RA, Werner T (1994) Archean uplift of a subprovince boundary in the Canadian Shield, revealed by magnetic fabrics. Tectonophysics 227:1–15

Borradaile GJ, Lagroix F, King D (1998) Tilting and transpression of an Archean anorthosite in northern Ontario. Tectonophysics 293:239–254

Borradaile GJ, Werner T, Lagroix F (1999a) Magnetic fabrics and anisotropy-controlled thrusting in the Kapuskasing Structural Zone, Canada. Tectonophysics 302:241–256

Borradaile GJ, Lane T, Lagroix F, Maher L, Linford N, Linford P (1999b) Attempts to date Salt-making activity in Iron Age Britain using magnetic inclinations. J Archeol Sci 26:1377–1389

Borradaile GJ, Lagroix F, Trimble D (2001) Improved isolation of archeomagnetic signals by combined low temperature and alternating field demagnetization. Geophys J Int 147:176–182

Braitsch 0 (1956) Quantitative Auswertung einfacher Gefügediagramme. Heidelberger Beitr Mineral Petrogr 5:210–226

Briskin M, Harrell J (1980) Time series analysis of the Pleistocene deep-sea paleoclimatic record. Mar Geol 36:1–22

Broecker WS, van Donk J (1970) Insolation changes, ice volumes, and the O^{18} record in deep-sea cores. Rev Geophys Space Phys 8:169–198

Bukrinski VA (1965) Prakticheskiy kurs geometrii nedr. Nedra Press, Moscow, 244 pp

Bull WB (1996a) Dating San Andreas fault earthquakes with lichenometry. Geology 24:111–114

Bull WB (1996b) Prehistorical earthquakes on the Alpine Fault, New Zealand. J Geophys Res 101:6037–6050

Burger A (1988) Milankovitch theory and climate. Rev Geophys 26:624–657

Butler RF (1992) Paleomagnetism. Blackwell, Oxford, 319 pp

Campbell WH (2000) Earth magnetism: a guided tour through magnetic fields. Harcourt-Academic Press, San Diego, 151 pp

Clark AJ, Tarling DH, Noel M (1988) Developments in archeomagnetic dating in Britain. J Archeol Sci 15:645–667

Clifton E, Hunter R, Swanson FJ, Phillips RL (1969) Sampling size and meaningful gold analysis. USGS Professional Paper 625-C

Corfu C, Andrews AJ (1987) Geochronological constraints on the timing of magmatism, deformation and gold mineralization in the Red Lake Greenstone belt, northwestern Ontario. Can J Earth Sci 24:1302–1320

Cox A (1968) Lengths of polarity reversals. J Geophys Res 73:3247–3260

Creer KM, Tucolka P, Barton CE (1983) Geomagnetism of baked clays and sediments. Elsevier, Amsterdam, 324 pp

Daly L, Zinsser H (1973) Étude comparative des anisotropies de susceptibilité et d'aimantation rémanente isotherme: consequences pour l'analyse structurale et le paléomagnétisme. Ann Géophys 29:189–200

Davis DW (1982) Optimum linear regression and error estimation applied to U-Pb data. Can J Earth Sci 19:2141–2149

Davis DW, Edwards GR (1986) Crustal evolution of Archean rocks in the Kakagi Lake area, Wabigoon subprovince, Ontario, as interpreted from high-precision U-Pb geochronology. Can J Earth Sci 23:182–192

Davis DW, Blackburn CE, Krogh T (1982) Zircon U-Pb ages from the Wabigoon-Manitou Lakes region, Wabigoon subprovince, northwestern Ontario. Can J Earth Sci 19:254–266

De Young JH Jr (1981) The Lasky tonnage-grade relationship – a reexamination. Econ Geol 76:1067–1080

Debiche MG, Watson GS (1995) Confidence limits and their bias correction for estimating angles between directions with applications to paleomagnetism. J Geophys Res 100: 24405– 24429

Derder MEM, Henry B, Bayou B, Djelit H, Amena M (2001) New Moscovian paleomagnetic pole from the Edjeleh fold (Sahara craton, Algeria). Geophys J Int 147:343–355

Diggle PJ, Fisher NI (1985) Sphere: a contouring program for spherical data. Comput Geosci 1:725–766

Doornkamp JC, King CAM (1971) Numerical analysis in geomorphology: an introduction. Edward Arnold, London, 372 pp

Dragoni W (1996) Response of some hydrological systems in central Italy to climatic variations. In: Angelakis AN, Issar AS (eds) Diachronic impacts of water resources. Springer, Berlin Heidelberg New York, pp 193–229

Eicher DL (1976) Geologic time, 2nd edn. Prentice-Hall, Englewood Cliffs

Embleton BJJ (1972) The palaeomagnetism of some Palaeozoic sediments from central Australia. J Proc R Soc NSW 105:86–93

Embleton BJJ (1981) A review of the palaeomagnetism of Australia and Antarctica. In: McElhinny MW, Valencio DA (eds) Palaeoreconstruction of the continents, 2. American Geophysical Union and Geological Society of America, Washington, DC, pp 77–92

Epstein B (1947) The mathematical description of certain breakage mechanisms leading to the logarithmico-normal distribution. J Franklin Inst 244:471

Epstein B (1948) Statistical aspects of fracture problems. J Appl Physics 19:140–147

Ericson DB, Wollin G (1968) Pleistocene climates and chronology in deep-sea sediments. Science 162:1227–1234

Erikson RL, van Sickle GH, Nakagama HM, McCarthy JH, Leong KW (1966) Gold geochemical anomaly in the Cortez district, Nevada. US Geol Surv Bull 534, 1–9

Faure G (1986) Principles of isotope geology. Wiley, New York, 589 pp

Finnerty AA, Boyd FR (1984) Evaluation of thermobarometers for garnet peridotites. Geochim Cosmochim Acta 48:15–27

Fisher RA (1953) Dispersion on a sphere. Proc R Soc A217: 295–305

Fisher NI, Powell CMA (1989) Statistical analysis of two-dimensional paleocurrent data: methods and examples. Aust J Earth Sci 36:91–107

Fisher NI, Lewis T, Embleton BJJ (1987) Statistical analysis of spherical data. Cambridge University Press, Cambridge, 329 pp

Flanagan FJ (1960) The lead content of G-1: Second report on a cooperative investigation of the composition of two silicate rocks. US Geol Surv Bull 1113:113–121

Fletcher RA, Fletcher SW (1979) Clinical research in medical journals: a 30-year perspective. New Engl J Med 301:180–183

Flinn D (1962) On folding during three dimensional progressive deformation. Q J Geol Soc Lond 118:385–428

Flinn D (1965) On the symmetry principle and the deformation ellipsoid. Geol Mag 102:36–45

Folk (1975) A review of grain-size parameters. Sedimentology 6:73–93

Girdler RW (1961) The measurement and computation of anisotropy of magnetic susceptibility in rocks. Geophys J R Soc Astron 5:34–44

Gomez M, Hazen K (1970) Evaluation of sulphur and ash distribution in coal seams by statistical response surface regression analysis. Report of Investigation 7377. US Bureau of Mines, Washington, 120 pp

Gustafson LB, Hunt JP (1975) The porphyry-copper deposit at El Salvador, Chile. Econ Geol 70:859–912

Hamilton et al. (1979) USGS Open File report 79-681

Henry B (1989) Magnetic fabric and orientation tensor of minerals in rocks. Tectonophysics 165:21–27

Henry B (1997) The magnetic zone axis: a new element of magnetic fabric for the interpretation of the magnetic lineation. Tectonophysics 271:325–332

Henry B, Le Goff M (1995) Application de l'extension bivariate de la statistique Fisher aux donnés d'anisotropie de susceptibilité magnétique: intégration des incertitudes de mesure sur l'orientation des directions principales. C R Acad Sci Paris 320 Ser Iia:1037–1042

Henry B, Rouvier H, Le Goff M, Smati A, Hatira N, Laatar E, Mansouri A, Perthuisot V (2000) Paleomagnetism as a structural polarity criterion: application to Tunisian diapirs. J Struct Geol 22:323–334

Henry B, Rouvier H, Le Goff M, Leach D, Macquar J-C, Thibieroz J, Lewchuk M (2001) Paleomagnetic dating of widespread remagnetization on the southeastern border of the French Massif Central and implications for fluid flow and Mississippi-type mineralization. Geophys J Int 145:368–380

Hext G (1963) The estimation of second-order tensors, with related tests and designs. Biometrika 50:353–357

Hobbs BE, Means WD, Williams PF (1976) An outline of structural geology. Wiley, New York, 571 pp

Houghton RA, Woodwell GM (1989) Global climatic change. Sci Am 260:36–44

Hrouda F (1982) Magnetic anisotropy of rocks and its application in geology and geophysics. Geophys Surv 5:37–82

Hrouda F, Jelinek J (1999) Theoretical models for the relationship between magnetic anisotropy and strain: effect of triaxial magnetic grains. Tectonophysics 301:183–190

Hrouda F, Schulmann K (1990) Conversion of the magnetic susceptibility tensor into the orientation tensor in some rocks. Phys Earth Planet Interiors 63:71–77

Hurst HE (1952) The Nile. Constable Press, London, 326 pp

Hutchison CS (1975) Correlation of Indonesian active volcanic geochemistry with Benioff zone depth. Geol Mijnbouw 54:157–168

Imbrie J (1982) Astronomical theory of the ice ages: a brief historical review. Icarus 50:408–422

Imbrie J, Imbrie IK (1979) Ice ages. Hillside, New Jersey

Imbrie J, Purdy EG (1962) Classification of modern Bahamian carbonate sediments. Am Assoc Petrol Geol Mem No 1: 253–272

Irving E (1964) Paleomagnetism and its application to geological and geophysical problems. Wiley, New York, 399 pp

Ishihara S (1979) Lateral variation of magnetic susceptibility of the Japanese granitoids. J Geol Soc Jpn 85:509–523

Ishihara S, Sasaki A (1989) Sulfur-isotopic ratios of the magnetite-series and ilmenite-series granitoids of the Sierra Nevada Batholith – a reconnaissance study. Geology 17: 788– 791

Issar AS (1995) Climatic change and the history of the Middle East. Am Sci 350–355

Issar AS, Brown N (eds) (1998) Water, environment and society in times of climate change. Kluwer, Dordrecht, Netherlands

Jackson MJ (1990) Diagenetic sources of stable remanence in remagnetized Paleozoic cratonic carbonates: a rock magnetic study. J Geophys Res 95B:2753–2761

Jackson MJ (1991) Anisotropy of magnetic remanence: a brief review of mineralogical sources, physical origins, and geological applications, and comparison with susceptibility anisotropy. Pure Appl Geophys 136:1–28

Jackson M, Borradaile GJ (1991) On the origin of the magnetic fabric in Cambrian Purple Slates of North Wales. Tectonophysics 194:49–58

Jackson M, Tauxe L (1991) Anisotropy of magnetic susceptibility and remanence: developments in the characterization of tectonic, sedimentary and igneous fabric. Rev Geophysics Supplement, US National Report to International Union of Geodesy and Geophysics. pp 371–376

Jackson M, Worm H-U, Banerjee SJ (1990) Fourier analysis of digital hysteresis data: rock magnetic applications. Phys Earth Planet Interior 65:78–87

James DE (ed) (1989) the encyclopaedia of solid earth geophysics. Van Nostrand Reinhold, New York, 1328 pp

Jelinek V (1976) The statistical theory of measuring anisotropy of magnetic susceptibility of rocks and its application. AGICO, Brno, Geophysika 1–88

Jelinek V (1978) Statistical processing of anisotropy of magnetic susceptibility measured on groups of specimens. Stud Geoph-Geodetica 22:50–62

Jelínek V (1981) Characterization of the magnetic fabrics of rocks. Tectonophysics 79:T63–T67

Jelínek V (1996) Theory and measurement of the anisotropy of isothermal remanent magnetization of rocks. Travaux Géophys 37:124–134

Jin R-S (1992) Cross-correlation of the variations of the geomagnetic dipole moment and the fluctuations of Earth's rotation. J Geophys Res 97:17251–17260

Kaufman GM (1963) Statistical decision and related techniques in oil and gas exploration. Prentice-Hall, Englewood Cliffs, 307 pp

Kent JT (1982) The Fisher-Bingham distribution on the sphere. J R Astron Soc B44:71–80

Keys D (2000) Catastrophe: an investigation into the origins of the modern world. Arrow Press, London, 509 pp

Kistler RW, Peterman ZE (1973) Variation in Sr, Rb, K, Na and initial $^{87}Sr/^{86}Sr$ in Mesozoic granitic rocks and wall rocks in central California. Geol Soc Bull Am 84:3489–3512

Kolmorgoroff AN (1941) Über das logarithmisch normale Verteilungsgesetz der Teilchen bei Zerstückelung. C R Acad Sci URSS 31:99

Kottler F (1950) The distribution of particle sizes. J Franklin Inst 250:339–356; 419–441

Krige DG (1951) A statistical approach to some basic mine valuation problems on the Witwatersrand. J Chem Metall Min Soc S Afr 52:119–139

Krige DG (1978) Lognormal and de Wijsian geostatistics for ore evaluation. Geostatistics Series, South African IMM Monograph Series no. 1

Krige DG, Magri EJ (1982) Geostatistical case studies of the advantages of lognormal-Dewijsian kriging with mean for a base metal mine and a gold mine. J Math Geol 14:547–555

Krumbein WC, Graybill FA (1965) An introduction to statistical models in geology. McGraw-Hill, New York, 475 pp

Krumbein WC, Shreve RL (1970) Some statistical properties of dendritic channel networks. US Office of Naval Research, Technical Report 13, pp 1–117

Laj C, Nordemann D, Pomeau Y (1979) Correlation function analysis of geomagnetic field reversals. J Geophys Res 84: 4511–4515

Lagroix F, Borradaile GJ (2000a) Tectonics of the Circum-Troodos sedimentary cover of Cyprus, from rock magnetic and structural observations. J Struct Geol 22:453–469

Lagroix F, Borradaile GJ (2000b) Magnetic fabric interpretation complicated by inclusions in mafic silicates. Tectonophysics 325:207–225

Le Goff M (1990) Lissage et limites d'incertitude des courbes de migration polaire: pondération des données et extension bivariate de la statistique de Fisher. C R Acad Sci Paris 311: 1191–1198

Le Goff M, Henry B, Daly L (1992) Practical method for drawing a VGP path. Phys Earth Planet Interiors 70:201–204

Le Goff M, Henry B (1995)

Le Goff M et al. (1993)

Le Mouel J-L, Courtillot V, Jault D (1992) Changes in earth rotation. Nature 355:26

Leopold LB, Langbein WB (1966) River meanders. Sci Am 214:60–70

Lienert BR (1991) Monte–Carlo simulation of errors in the anisotropy of magnetic susceptibility: a second–rank symmetric tensor. J Geophys Res 96:19539–19544

Lisle RJ (1977) Estimation of the tectonic strain ratio from the mean shape of deformed elliptical markers. Geol Mijnbouw 56:140

Lisle RJ (1988) Geological structures and maps: a practical guide. Pergamon Press, Oxford, 150 pp

Lisle RJ (1989) The statistical analysis of orthogonal orientation data. J Geol 97:360–364

Lock J, McElhinny MW (1991) The global paleomagnetic database: design, installation, and use with ORACLE. Surv Geophys 12:317–506

Lovelock J (2000) Gaia: a new look at life on Earth. Oxford University Press, Oxford, 148 pp

Ludwig KR (1980) Calculation of uncertainties of U-Pb isotopic data. Earth Planet Sci Lett 46:212–220

Lund SP, Banerjee SK (1985) Late quaternary paleomagnetic field secular variation from two Minnesota lakes. J Geophys Res 90:803–825

MacDonald WD (1980) Net tectonic rotation, apparent tectonic rotation, and the structural tilt correction in paleomagnetic studies. J Geophys Res 85:3659–3669

Mardia KV, Zemroch PJ (1977) Table of maximum likelihood estimates for the Bingham distribution. J Statist Comput Simul 6:29–34

Maguire BA, Pearson ES, Wynn AHA (1952) The time intervals between industrial accidents. Biometrika 39:168–180

McElhinny MW (1973) Palaeomagnetism and plate tectonics. Cambridge University Press, Cambridge, 358 pp

McElhinny MW (1979) The earth, its structure, origin and evolution. Academic Press, London

McKelvey VE (1960) Relations of reserves of the elements to their crustal abundance. Am J Sci 258A:234–241

Means WD (1976) Stress and strain: basic concepts of continuum mechanics for geologists. Springer, Berlin Heidelberg New York, 338 pp

Meckel LD (1967) Origin of Pottsville conglomerates (Pennsylvanian) in the Central Appalachians. Geol Soc Am Bull 78:223–258

Merrill RT, McElhinny MW (1983) The Earth's magnetic field. Academic Press, London, 401 pp

Merriam DF, Robinson JE (1970) Trend analysis in geologic geophysical exploration. In: Mathematical models in geology and geophysics. Hornicka Pribram vi vide a tecnhic Pribram, Czechoslovakia

McElhinny MW, McFadden PL (2000) Paleomagnetism: continents and oceans. Academic Press, London, 382 pp

Miall AD (1973) Markov chain analysis applied to an ancient alluvial plain succession. Sedimentology 20:347–364

Milankovitch M (1938) Die Chronologie des Pleistocans. Bull Acad Sci Math Nat Belgrade 4:49

Milankovitch M (1938) Astronomische Mittel zur Erforschung der erdgeschichtlichen Klimate. Handb Geophys 9:593–698

Mitchell RH (1986) Kimberlites. Plenum Press, New York, NY

Miyashiro A (1972) Metamorphism and related magmatism in plate tectonics. Am J Sci 272:629–656

Mohler D, Dulberg CS, Wells GA (1994) Statistical power, sample size, and their reporting in clinical trials. J A M 272:122–124

Montgomery CW (1997) Environmental geology. McGraw-Hill, New York, 554 pp

Moore JG (1959) The quartz-diorite boundary line in the western United States. J Geol 67:198–210

Mothersill JS (1979) The paleomagnetic record of the Late Quaternary sediments of Thunder Bay. Can J Earth Sci 16:1016–1023

Mothersill JS (1983) Results form the Great Lakes. In: Creer KM, Tucholka P, Barton C (eds) Geomagnetism of baked clays and recent sediments. Elsevier, New York, pp 223–230

Naidu PS (1971) Statistical structure of geomagnetic field reversals. J Geophys Res 76:2649–2662

Nakamura N, Borradaile GJ (2001) Do reduction spheroids predate finite strain? A magnetic diagnosis of Cambrian slates in North Wales. Tectonophysics 304:133–139

Nakamura NG, Borradaile GJ (2002) Metamorphic control of magnetic susceptibility and its fabric anisotropies: a 3-D projection. Geophys J Int (in press)

Negi JG, Tiwari RK, Rao KNN (1996) Clean periodicity in secular variations of dolomite abundance in deep marine sediments. Mar Geol 133:113–121

Nickelson PN, Hough VJD (1967) Jointing in the Appalachian plateau of Pennsylvania. Geol Soc Am Bull 78:609

Nicolas A (1987) Principles of rock deformation. Riedel, Dordrecht, 208 pp

Nicolas A, Poirier JP (1976) Crystalline plasticity and solid state flow in metamorphic rocks. Wiley, New York, 444 pp

Nightingale F (1858) Notes on matters affecting the health, efficiency and hospital administration of the British Army. Harrison and Sons, London, 1092 pp

Nye JF (1957, 1985) Physical properties of crystals. Oxford Univ Press, New York, 329 pp

Onstott TC (1980) Application of the Bingham distribution function in paleomagnetic studies. J Geophys Res 85:1500–1510

Owens WH (2000a) Error estimates in the measurement of anisotropic magnetic susceptibility. Geophys J Int 142:516–526

Owens WH (2000b) Statistical applications to second-rank tensors in magnetic fabric analysis. Geophys J Int 142:527–538

Pangaea Scientific (1998) Spheristat: spherical projection software. Ontario, Canada

Parks JR (1966) Cluster analysis applied to multivariate geologic problems. J Geol 74:703–715

Pearce TH (1968) A contribution to the theory of variation diagrams. Contrib Min Petrol 19:142–157

Phillips FC (1960) The use of the stereographic projection in structural geology. Edward Arnold, London, 86 pp

Phillips J, Blakely RJ, Cox A (1975) Independence of geomagnetic polarity reversals. Geophys JR Astron Soc 43:747–754

Pierce C, Adams K, Stewart JD (1998) Determining the fuel constituents of ancient hearth ash via ICP-AES analysis. J Archeol Sci 25:493–505

Pilkington M, Percival JA (1999) Crystal magnetization and long-wavelength aeromagnetic anomalies of the Minto block, Quebec. J Geophys Res 104:7513–7526

Pitcher WS (1993) The nature and origin of granite. Blackie, Glasgow, 321 pp

Poirier JP (1985) Creep of crystals. Cambridge University Press, Cambridge, 260 pp

Potter PE, Pettijohn FJ (1977) Paleocurrents and basin analysis. Springer, Berlin Heidelberg New York, 423 pp

Priest SD (1985) Hemispherical projection methods in rock mechanics. George Allen and Unwin, London, 124 pp

Quinn TM, Mountain GS (2000) Shallow water science and ocean drilling face challenges. EOS 81:397; 404

Ragan DM (1973) Structural geology: an introduction to geometrical techniques, 2nd edn. Wiley, New York, 208 pp

Ramsay JG (1967) Folding and fracturing of rocks. McGraw-Hill, New York, 568 pp

Ramsay JG, Huber MI (1983) The techniques of modern structural geology, vol 1. Strain analysis. Academic Press, London, 307 pp

Ramsay JG, Huber MI (1987) The techniques of modern structural geology, vol 2. Folds and fractures. Academic Press, London, pp 309–700

Rapp RH (1989) Gravity anomalies: statistical analysis. In: James DE (ed) Encyclopedia of solid earth geophysics. pp 617–622

Rathore JS (1979) Magnetic susceptibility anisotropy in the Cambrian Slate Belt of North Wales and correlation with strain. Tectonophysics 53:83–97

Rathore JS (1980) The magnetic fabrics of some slates from the Borrowdale Volcanic Group in the English Lake District and their correlations with strain. Tectonophysics 67:207–220

Raymond GF, Armstrong WP (1986) Short and long-term open-pit grade control. In: David M, Sinclair AJ, Vallee M (eds) Ore reserve estimation: methods, models and reality. CIMM, Montreal, Canada, pp 108–129

Reiche P (1938) An analysis of cross lamination in the Coconino sandstone. J Geol 46:905–932

Robin P-YF (1977) Determination of geologic strain using randomly oriented strain markers of any shape. Tectonophysics 42:T7–T16

Robin P-YF, Jowett EC (1986) Computerized contouring and statistical evaluation of orientation data using counting circles and continuous weighting functions. Tectonophysics 121:207–223

Rochette P, Jackson MJ, Aubourg C (1992) Rock magnetism and the interpretation of anisotropy of magnetic susceptibility. Rev Geophys 30:209–226

Ruddiman WF (2001) Earth's climate: past and future. WH Freeman, New York, 465 pp

Runcorn SK (1959) On the theory of geomagnetic secular variation. Ann Geophys 15:87–92

Russell JK, Stanley CR (1990) Theory and application of Pearce Element Ratios to geochemical data analysis. Short course no 8. Geological Association of Canada, Ottawa, 315 pp

Sabourin R (1975) Geostatistical evaluation of sulphur content in Lingan coal mine, Cape Breton. Proc 13th Int APCOM Symposium, Clausthal; pp 1–16

Sandefur RL, Grant DC (1980) Applying geostatistics to Roll Front Uranium in Wyoming. In: Royle A et al. (eds) Geostatistics. McGraw-Hill, New York, pp 127–143

Sander B (1930) Gefügekunde der Gesteine. Springer, Berlin Heidelberg New York, 352 pp

Scheidegger AE (1965) On the statistics of the orientation of bedding planes, grain axes, and similar sedimentological data. US Geol Survey Prof Pap 525-C:164–167

Schenk PE (1969) Carbonate-sulfate-redbed facies and cyclic sedimentation of the Windsorian Stage (Middle Carboniferous), Maritime Provinces. Can J Earth Sci 6:1037–1066

Schmidt W (1917) Statistiche Methoden beim Gefügestudium Kristalliner Schiefer. Sitz Kaiserl Akad Wiss Wien Math 126:515–538

Scotese CR, Bambach RK, Barton C, van der Voo R, Ziegler AM (1979) Paleozoic base maps. J Geol 87:217–277

Sorby HC (z) On slaty cleavage as exhibited in the Devonian Limestones of Devonshire. Philos Mag 11:20–37

Soyer B (1984) In: DeVivo B, Ippolito F, Capaldi, G, Simpson, PR Uranium geochemistry, mineralogy, geology, exploration and resources. IMM Stephen Austin, Hertford, UK (201 pp), pp 109–116

Starkey J (1983) A crystallographic approach to the calculation of orientation diagrams. Can J Earth Sci 20:932–952

Starkey J (1993) The analysis of three-dimensional orientation data. Can J Earth Sci 30:1355–1362

Starkey J (1996a) A computer program to print inclined stereographic projections. In: dePaor DG (ed) Structural geology and computers. Pergamon Press, Oxford, pp 195–215

Starkey J (1996b) Microcomputers and the optical Universal stage. In: dePaor DG (ed) Structural geology and computers. Pergamon Press, Oxford, pp 217–232

Stephenson A, Sadikun S, Potter DK (1986) A theoretical and experimental comparison of the anisotropies of magnetic susceptibility and remanence in rocks and minerals. Geophys J R Astron Soc 84:185–200

Stesky RM (1985) Least-squares fitting of a non-circular cone. Comput Geosci 11:357–368

Stuiver M, Reimer P, Bard E, Beck JW, Burr GS, Hughen KA, Kromer B, McCormac G, van der Plicht J, Spurck M (1998) INTCAL98 radiocarbon age calibration, 24,000–0 cal BP. Radiocarbon 40:1041–1083

Stupavsky M, Symons DTA (1982) Isolation of Paleohelikian remanence in Grenville anorthosites of the French River area, Ontario. Can J Earth Sci 19:819–828

Symons DTA (1975a) Age and flow direction from magnetic measurements on the historic Aiyansh flow, British Columbia. J Geophys Res 80:2622–2626

Symons DTA (1975b) Polar paleolatitude and revised Huronian polar wander path from paleomagnetism of the Gowgonda formation, Ontario. Can J Earth Sci 12:940–948

Tarling DH (1983) Paleomagnetism. Chapman and Hall, London, 379 pp

Tarling DH, Hrouda F (1993) The magnetic anisotropy of rocks. Chapman and Hall, London, 217 pp

Taylor RE, Aitken MJ (1997) Chronometric dating in archeology. Plenum, New York, 395 pp

Thompson R, Oldfield F (1986) Environmental magnetism. Allen and Unwin, London, 227 pp

Thurston PC, Williams HR, Sutcliffe RH, Stott GM (1991) The geology of Ontario, part 1. Ministry of Northern Development and Mines, Ontario, Canada, 711 pp

Tolson G, Correa-Mora F (1996) Manipulation of orientation data using spreadsheet software. In: dePaor DG (ed) Structural geology and computers. Pergamon Press, Oxford, pp 237–243

Turner FJ, Weiss LE (1963) Structural analysis of metamorphic tectonites. McGraw-Hill, New York, 545 pp

Turner GM, Thompson R (1979) Behaviour of the earth's magnetic field as recorded in the sediment of Loch Lomond. Earth Planet Sci Lett 42:412–426

Turner GM, Thompson R (1981) Lake sediment record of the geomagnetic secular variations in Britain during Holocene times. Geophys J R Astron Soc 65:703–725 van der Voo R (1993) Paleomagnetism of the Atlantic Tethys, and Iapetus Oceans. Cambridge University Press, New York, 411 pp

Védes I (1970) Map plotting with weighted average on the surface of a circular disc. Pure Appl Geophys 78:5–17

Vernon R (1976) Metamorphic processes. George Allen and Unwin, London, 247 pp

Vitorello I, Pollack HN (1980) On the variation of continental heat-flow and the thermal evolution of continents. J Geophys Res 85:983–995

Visher G (1969) Grain-size distribution and depositional processes. J Sediment Petrol 39:1074–1106

Von Mises R (1918) Über die „Ganzzähligkeit" der Atomgewichte und verwandte Fragen. Phys Z 19:490–500

Watson GS (1956) Analysis of dispersion on a sphere. Mon Notes R Astron Soc Geophys Suppl 7:153–159

Watson GS (1972) Trend surface analysis and spatial correlation. Geol Soc Am Spec Paper 146:39–46

Watson GS, Irving E (1957) Statistical methods in rock magnetism. Mon Notes R Astron Soc Geophys Suppl 7:289–300

Weltje GJ (2002) Quantitative analysis of detrital modes: statistically rigorous confidence regions in ternary diagrams and their use in sedimentary petrology. Earth Sci Rev 57:211–253

Werner T (1997) Experimental designs for determination of the anisotropy of remanence-test of the efficiency of least-squares and bootstrap methods applied to metamorphic rocks from southern Poland. Phys Chem Earth 22:131–136

Werner T, Borradaile GJ (1996) Paleoremanence dispersal across a transpressed Archean terrain: deflection by anisotropy or by late compression? J Geophys Res 10:5531–5545

Whitten ETH (1963) Application of quantitative methods in the geochemical study of granite massifs. Royal Soc Canada, Spec Publ 6. Studies in analytical geochemistry. University of Toronto Press, Toronto, 139 pp

Wood A (1982) A bimodal distribution on the sphere. Appl Stat 31:52–58

Woodcock NH (1977) Specification of fabric shapes using an eigenvalue method. Geol Soc Am Bull 88:1231–1236

York D (1969) Least-squares fitting of a straight line with correlated errors. Earth Planet Sci Lett 5:320–324

Zijderveld JDA (1967) AC demagnetization of rocks: analysis of results. In: Collinson DW, Creer KM, Runcorn SK (eds) Methods in paleomagnetism. Elsevier, New York, 1967, pp 254–256

Zhu RX, Coe RS, Zhao XX (1998) Sedimentary record of two geomagnetic excursions within the last 15,000 years in Beijing, China. J Geophys Res 103:30323–30333

General Index

Index of Examples